# Medical Genetics

# Medical Genetics

## Fourth Edition

**Lynn B. Jorde, PhD**
Professor
H.A. and Edna Benning Presidential Chair
Department of Human Genetics
University of Utah Health Sciences Center
Salt Lake City, Utah

**John C. Carey, MD, MPH**
Professor
Division of Medical Genetics
Department of Pediatrics
University of Utah Health Sciences Center
Salt Lake City, Utah

**Michael J. Bamshad, MD**
Professor
Division of Genetic Medicine
Department of Pediatrics
University of Washington School of Medicine
Seattle Children's Hospital
Seattle, Washington

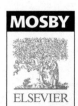

MOSBY

ELSEVIER

## MOSBY
### ELSEVIER

1600 John F. Kennedy Blvd
Ste 1800
Philadelphia, PA 19103-2899

MEDICAL GENETICS                                    ISBN: 978-0-323-05373-0

---

### Notice

Knowledge and best practice in this field are constantly changing. As new research and experience broaden our knowledge, changes in practice, treatment, and drug therapy may become necessary or appropriate. Readers are advised to check the most current information provided (i) on procedures featured or (ii) by the manufacturer of each product to be administered, to verify the recommended dose or formula, the method and duration of administration, and contraindications. It is the responsibility of the practitioner, relying on his or her experience and knowledge of the patient, to make diagnoses, to determine dosages and the best treatment for each individual patient, and to take all appropriate safety precautions. To the fullest extent of the law, neither the Publisher nor the Authors assume any liability for any injury and/or damage to persons or property arising out of or related to any use of the material contained in this book.

The Publisher

---

**Library of Congress Cataloging-in-Publication Data**
Jorde, Lynn B.
Medical genetics / Lynn B. Jorde, John C. Carey, Michael J. Bamshad. – 4th ed.
    p. ; cm.
  Rev. ed. of: Medical genetics / Lynn B. Jorde ... [et al.]. 3rd ed. 1996.
  Includes bibliographical references.
  ISBN 978-0-323-05373-0
  1. Medical genetics. I. Carey, John C., 1946- II. Bamshad, Michael J. III. Medical genetics. IV. Title.
  [DNLM: 1. Genetic Diseases, Inborn–genetics. 2. Genetics, Medical–methods. 3. Genetic Services.
  4. Genetic Techniques. QZ 50 J82m 2010]
  RB155.J67 2010
  616'.042–dc22

                                                    2009027335

*Acquisitions Editor:* Kate Dimock
*Developmental Editor:* Andrew Hall
*Publishing Services Manager:* Linda Van Pelt
*Project Manager:* Frank Morales
*Design Direction:* Steve Stave

Working together to grow
libraries in developing countries

www.elsevier.com | www.bookaid.org | www.sabre.org

ELSEVIER    BOOK AID International    Sabre Foundation

Printed in China

Last digit is the print number:   9  8  7  6  5  4  3  2  1

To Our Families

Debra, Eileen, and Alton Jorde
Leslie, Patrick, and Andrew Carey
Jerry and Joanne Bamshad

J.B.S. Haldane titled an anthology of some of his more dyspeptic writings "Everything Has a History," and this is clearly applicable to the field of medical genetics. More than 200 years ago scientists such as Buffon, Lamarck, Goethe, and Kielmeyer reflected on how the developmental history of each organism related to the history of life on Earth. Based on these ideas, the discipline of biology was born in 18th century Europe, enjoyed adolescence as morphology and comparative anatomy in the 19th century, and reached adulthood in the 20th century as the field of genetics. However, the late 19th century definition of genetics (heredity) as the science of variation (and its causes) is still valid. Thus, human genetics is the science of human variation, medical genetics the science of abnormal human variation, and clinical genetics that branch of medicine that cares for individuals and families with abnormal variation of structure and function.

In the late 19th and early 20th centuries, the unity of morphology-based science was gradually replaced by a pluralistic view of biology that splintered the field into many different, and often rivalrous, disciplines. However, thanks to the application of novel molecular biological methods to the analysis of development and to the understanding of the materials of heredity (i.e., genes), the various branches of biology are being reunited. This new discipline, termed molecular morphology, may be defined as the study of the form, formation, transformation, and malformation of living organisms. Indeed, ignorant as they may be of the traditional methods of historiography, geneticists have developed their own brilliant and highly effective methods. Consequently, they have achieved a perspective remarkably longer and much better documented than that of historians. This nearly 4-billion-year perspective unites living organisms into a single web of life related to one another in unbroken descent to a common ancestor. This makes the phylogenetic (i.e., the genetic relationships of different species to one another) and the ontogenetic (i.e., the genetic basis for the development of individual organisms) perspectives of development not only complementary but inseparable. Thus, it is now possible to effectively explore a key question of biology of the 19th and 20th centuries: What is the relationship between evolution and development?

In 1945 the University of Utah established the Laboratory for the Study of Hereditary and Metabolic Disorders (later called the Laboratory of Human Genetics). Here, an outstanding group of scientists performed pioneering studies on clefts of lips and palate, muscular dystrophy, albinism, deafness, hereditary polyposis of the colon (Gardner syndrome), and familial breast cancer. These predecessors would be enormously proud of their current peers at the University of Utah, whose successes have advanced knowledge in every aspect of the field of genetics.

In their attempts to synthesize the story of genetics and its applications to human variability, health and disease, development, and cancer, the authors of this text have succeeded admirably. This concise, well-written and -illustrated, carefully edited and indexed book is highly recommended to undergraduate students, new graduate students, medical students, genetic counseling students, nursing students, and students in the allied health sciences. Importantly, it is also a wonderful text for practicing physicians (primary care providers and specialists) who want an authoritative introduction to the basis and principles of modern genetics as applied to human health and development. This text, by distinguished and internationally respected colleagues and friends who love to teach, is a joy to read in its expression of enthusiasm and of wonder, which Aristotle said was the beginning of all knowledge.

Einstein once said, "The most incomprehensible thing about the world is that it is comprehensible." When I began to work in the field of medical genetics, the gene was widely viewed as incomprehensible. Indeed, some scientists, such as Goldschmidt, cast doubt on the very existence of the gene, although the great American biologist E.B. Wilson had predicted its chemical nature more than 100 years previously. In this text, genes and their function in health and disease are made comprehensible in a manner that should have wide appeal to all.

JOHN OPITZ, MD
Salt Lake City, Utah

# PREFACE

Medical genetics is a rapidly progressing field. No textbook can remain factually current for long, so we have attempted to emphasize the central principles of genetics and their clinical application. In particular, this textbook integrates recent developments in molecular genetics and genomics with clinical practice.

This new edition maintains the format and presentation that were well received in three previous editions. Basic principles of molecular biology are introduced early in the book so that they can be discussed and applied in subsequent chapters. The chapters on autosomal and X-linked disorders include updated discussions of topics such as genomic imprinting, anticipation, and expanded trinucleotide repeats. The chapter on cytogenetics highlights important advances in this area, including comparative genomic hybridization and newly described microdeletion syndromes. Gene mapping and identification, which constitute a central focus of modern medical genetics, are treated at length, and recent advances based on completion of the human genome project are discussed. Chapters are included on the rapidly developing fields of immunogenetics and cancer genetics. Considerable discussion is devoted to the genetics of common adult diseases, such as heart disease, diabetes, stroke, and hypertension. The book concludes with chapters on genetic diagnosis (again emphasizing current molecular approaches such as microarray analysis), gene therapy, personalized medicine, and clinical genetics and genetic counseling.

As in previous editions, a Web site is available to provide access to continually changing information in medical genetics (http://evolve.elsevier.com/Jorde/). The Web site includes downloadable versions of all of the figures in the textbook, many additional patient photographs, hyperlinks to other relevant sites, and a battery of test questions and answers.

Several pedagogical aids are incorporated in this book:

- Clinical Commentary boxes present detailed coverage of the most important genetic diseases and provide examples of modern clinical management.
- Mini-summaries, highlighted in red, are placed on nearly every page to help the reader understand and summarize important concepts.
- Study questions, provided at the end of each chapter, assist the reader in review and comprehension.

- A detailed glossary is included at the end of the book.
- Key terms are emphasized in boldface.
- Important references are listed at the end of each chapter.

Many major additions have been incorporated into this edition:

- All chapters have been thoroughly updated, with special attention given to rapidly changing topics such as genetic diagnosis, gene therapy, cancer genetics, and the genetics of other common diseases.
- A new chapter, entitled "Genetics and Personalized Medicine," has been added.
- More than 100 new clinical photographs and figures have been added or updated.
- To facilitate the creation of illustrations for teaching purposes, all images on the Web site (including line drawings from the textbook) can now be downloaded.
- An expanded comprehensive index includes all text citations of all diseases.

This textbook evolved from courses we teach for medical students, nursing students, genetic counseling students, and graduate and undergraduate students in human genetics. These students are the primary audience for this book, but it should also be useful for house staff, physicians, and other health care professionals who wish to become more familiar with medical genetics.

## ACKNOWLEDGMENTS

Many of our colleagues have generously donated their time and expertise in reading and commenting on portions of this book. We extend our sincere gratitude to Diane Bonner, PhD; Arthur Brothman, PhD; Peter Byers, MD; William Carroll, MD; Debbie Dubler, MS; Ruth Foltz, MS; Ron Gibson, MD, PhD; Sandra Hasstedt, PhD; Susan Hodge, PhD; Rajendra Kumar-Singh, PhD; James Kushner, MD; Jean-Marc Lalouel, MD, DSc; Claire Leonard, MD; Mark Leppert, PhD; William McMahon, MD; James Metherall, PhD; Dan Miller, MD, PhD; Sampath Prahalad, MD; Shige Sakonju, PhD; Gary Schoenwolf, PhD; Sarah South, PhD; Carl Thummel, PhD; Thérèse Tuohy, PhD; Scott Watkins, MS; John Weis, PhD; H. Joseph Yost, PhD; Maxine J. Sutcliffe, PhD; Leslie R. Schover, PhD; and Craig Smith, medical student. In addition, a number of colleagues provided photographs; they are acknowledged individually in the figure captions. We wish to thank Peeches Cedarholm, RN; Karin Dent, MS; Bridget Kramer, RN; and Ann Rutherford, BS, for their help in obtaining and organizing the photographs. The karyotypes in Chapter 6 were provided by Arthur Brothman, PhD, and Bonnie Issa, BS.

Our editors at Elsevier, Kate Dimock and Andrew Hall, offered ample encouragement and understanding.

Finally, we wish to acknowledge the thousands of students with whom we have interacted during the past three decades. Teaching involves communication in both directions, and we have undoubtedly learned as much from our students as they have learned from us.

LYNN B. JORDE
JOHN C. CAREY
MICHAEL J. BAMSHAD

# CONTENTS

## Chapter 1

# BACKGROUND AND HISTORY

Genetics is playing an increasingly important role in the practice of clinical medicine. Medical genetics, once largely confined to relatively rare conditions seen by only a few specialists, is now becoming a central component of our understanding of most major diseases. These include not only the pediatric diseases but also common adult diseases such as heart disease, diabetes, many cancers, and many psychiatric disorders. Because all components of the human body are influenced by genes, genetic disease is relevant to all medical specialties. Today's health care practitioners must understand the science of medical genetics.

## WHAT IS MEDICAL GENETICS?

Medical genetics involves any application of genetics to medical practice. It thus includes studies of the inheritance of diseases in families, mapping of disease genes to specific locations on chromosomes, analyses of the molecular mechanisms through which genes cause disease, and the diagnosis and treatment of genetic disease. As a result of rapid progress in molecular genetics, DNA-based diagnosis is available for hundreds of inherited conditions, and gene therapy—the insertion of normal genes into patients in order to correct genetic disease—is showing promise for some conditions. Medical genetics also includes genetic counseling, in which information regarding risks, prognoses, and treatments is communicated to patients and their families.

## WHY IS A KNOWLEDGE OF MEDICAL GENETICS IMPORTANT FOR TODAY'S HEALTH CARE PRACTITIONER?

There are several reasons health care practitioners must understand medical genetics. Genetic diseases make up a large percentage of the total disease burden in pediatric and adult populations (Table 1-1). This percentage will continue to grow as our understanding of the genetic basis of disease grows. In addition, modern medicine is placing increasing emphasis on prevention. Because genetics provides a basis for understanding the fundamental biological makeup of the organism, it naturally leads to a better understanding of the disease process. In some cases, this knowledge can lead to prevention of the disorder. It also leads to more effective disease treatment. Prevention and effective treatment are among the highest goals of medicine. The

chapters that follow provide many examples of the ways genetics contributes to these goals. But first, this chapter reviews the foundations upon which current practice is built.

## A BRIEF HISTORY

The inheritance of physical traits has been a subject of curiosity and interest for thousands of years. The ancient Hebrews and Greeks, as well as later medieval scholars, described many genetic phenomena and proposed theories to account for them. Many of these theories were incorrect. Gregor Mendel (Fig. 1-1), an Austrian monk who is usually considered the father of genetics, advanced the field significantly by performing a series of cleverly designed experiments on living organisms (garden peas). He then used this experimental information to formulate a series of fundamental principles of heredity.

Mendel published the results of his experiments in 1865 in a relatively obscure journal. It is one of the ironies of biological science that his discoveries, which still form the foundation of genetics, received little recognition for 35 years. At about the same time, Charles Darwin formulated his theories of evolution, and Darwin's cousin, Francis Galton, performed an extensive series of family studies (concentrating especially on twins) in an effort to understand the influence of heredity on various human traits. Neither scientist was aware of Mendel's work.

Genetics as it is known today is largely the result of research performed during the 20th century. Mendel's principles were independently rediscovered in 1900 by three different scientists working in three different countries. This was also the year in which Landsteiner discovered the ABO blood group system. In 1902, Archibald Garrod described alkaptonuria as the first "inborn error of metabolism." In 1909, Johannsen coined the term **gene** to denote the basic unit of heredity.

The next several decades were a period of considerable experimental and theoretical work. Several organisms, including *Drosophila melanogaster* (fruit flies) and *Neurospora crassa* (bread mold) served as useful experimental systems in which to study the actions and interactions of genes. For example, H. J. Muller demonstrated the genetic consequences of ionizing radiation in the fruit fly. During this period, much of the theoretical basis of population genetics

**TABLE 1-1**
## A Partial List of Some Important Genetic Diseases

| Disease | Approximate Prevalence | Disease | Approximate Prevalence |
|---|---|---|---|
| **Chromosome Abnormalities** | | **Multifactorial Disorders** | |
| Down syndrome | 1/700 to 1/1000 | **Congenital Malformations** | |
| Klinefelter syndrome | 1/1000 males | Cleft lip with or without cleft palate | 1/500 to 1/1000 |
| Trisomy 13 | 1/10,000 | Club foot (talipes equinovarus) | 1/1000 |
| Trisomy 18 | 1/6000 | Congenital heart defects | 1/200 to 1/500 |
| Turner syndrome | 1/2500 to 1/10,000 females | Neural tube defects (spina bifida, anencephaly) | 1/200 to 1/1000 |
| | | Pyloric stenosis | 1/300 |
| **Single-Gene Disorders** | | | |
| Adenomatous polyposis coli | 1/6000 | **Adult Diseases** | |
| Adult polycystic kidney disease | 1/1000 | Alcoholism | 1/10 to 1/20 |
| $\alpha_1$-Antitrypsin deficiency | 1/2500 to 1/10,000 (whites)* | Alzheimer disease | 1/10 (Americans older than 65 years) |
| Cystic fibrosis | 1/2000 to 1/4000 (whites) | Bipolar affective disorder | 1/100 to 1/200 |
| Duchenne muscular dystrophy | 1/3500 males | Cancer (all types) | 1/3 |
| Familial hypercholesterolemia | 1/500 | Diabetes (types 1 and 2) | 1/10 |
| Fragile X syndrome | 1/4000 males; 1/8000 females | Heart disease or stroke | 1/3 to 1/5 |
| Hemochromatosis (hereditary) | 1/300 whites are homozygotes; approximately 1/1000 to 1/2000 are affected | Schizophrenia | 1/100 |
| Hemophilia A | 1/5000 to 1/10,000 males | **Mitochondrial Diseases** | |
| Hereditary nonpolyposis colorectal cancer | Up to 1/200 | Kaerns–Sayre syndrome | Rare |
| | | Leber hereditary optic neuropathy (LHON) | Rare |
| Huntington disease | 1/20,000 (whites) | Mitochondrial encephalopathy, lactic acidosis, and stroke-like episodes (MELAS) | Rare |
| Marfan syndrome | 1/10,000 to 1/20,000 | | |
| Myotonic dystrophy | 1/7000 to 1/20,000 (whites) | Myoclonic epilepsy and ragged red fiber disease (MERRF) | Rare |
| Neurofibromatosis type 1 | 1/3000 to 1/5000 | | |
| Osteogenesis imperfecta | 1/5000 to 1/10,000 | | |
| Phenylketonuria | 1/10,000 to 1/15,000 (whites) | | |
| Retinoblastoma | 1/20,000 | | |
| Sickle cell disease | 1/400 to 1/600 blacks* in America; up to 1/50 in central Africa | | |
| Tay–Sachs disease | 1/3000 Ashkenazi Jews | | |
| Thalassemia | 1/50 to 1/100 (South Asian and circum-Mediterranean populations) | | |

*The term "white" refers to individuals of European descent living in Europe, America, Australia, or elsewhere. The term "black" refers to individuals of African descent living in Africa, America, or elsewhere. These terms are used for convenience; some of the challenges in accurately describing human populations are discussed in Chapter 14.

**FIGURE 1-1**
Gregor Johann Mendel.
*(From Raven PH, Johnson GB: Biology, 3rd ed. St Louis: Mosby, 1992.)*

was developed by three central figures: Ronald Fisher, J. B. S. Haldane, and Sewall Wright. In addition, the modes of inheritance of several important genetic diseases, including phenylketonuria, sickle cell disease, Huntington disease, and cystic fibrosis, were established. In 1944, Oswald Avery showed that genes are composed of **deoxyribonucleic acid (DNA)**.

Probably the most significant achievement of the 1950s was the specification of the physical structure of DNA by James Watson and Francis Crick in 1953. Their seminal paper, which was only one page long, formed the basis for what is now known as **molecular genetics** (the study of the structure and function of genes at the molecular level). Another significant accomplishment in that decade was the correct specification of the number of human chromosomes. Since the early 1920s, it had been thought that humans had 48 chromosomes in each cell. Only in 1956 was the correct number, 46, finally determined. The ability to count and identify chromosomes led to a flurry of new findings in cytogenetics, including the discovery in 1959 that Down syndrome is caused by an extra copy of chromosome 21.

Technological developments since 1960 have brought about significant achievements at an ever-increasing rate. The most spectacular advances have occurred in the field of molecular genetics. Thousands of genes have been mapped to specific chromosome locations. The Human Genome Project, a large collaborative venture begun in 1990, provided the complete human DNA sequence in 2003 (the term **genome** refers to all of the DNA in an organism). Important developments in computer technology have helped to decipher the barrage of data being generated by this and related projects. In addition to mapping genes, molecular geneticists have pinpointed the molecular defects underlying a number of important genetic diseases. This research has contributed greatly to our understanding of the ways gene defects can cause disease, opening paths to more effective treatment and potential cures. The next decade promises to be a time of great excitement and fulfillment.

## TYPES OF GENETIC DISEASES

Humans are estimated to have approximately 20,000 to 25,000 genes. Alterations in these genes, or in combinations of them, can produce genetic disorders. These disorders are classified into several major groups:

- **Chromosome disorders**, in which entire chromosomes (or large segments of them) are missing, duplicated, or otherwise altered. These disorders include diseases such as Down syndrome and Turner syndrome.
- Disorders in which single genes are altered; these are often termed *mendelian conditions*, or **single-gene disorders**. Well-known examples include cystic fibrosis, sickle cell disease, and hemophilia.
- **Multifactorial disorders**, which result from a combination of multiple genetic and environmental causes. Many birth defects, such as cleft lip and cleft palate, as well as many adult disorders, including heart disease and diabetes, belong in this category.
- **Mitochondrial disorders**, a relatively small number of diseases caused by alterations in the small cytoplasmic mitochondrial chromosome.

Table 1-1 provides some examples of each of these types of diseases.

Of these major classes of diseases, the single-gene disorders have probably received the greatest amount of attention. These disorders are classified according to the way they are inherited in families: autosomal dominant, autosomal recessive, or X-linked. These modes of inheritance are discussed extensively in Chapters 4 and 5. The first edition of McKusick's *Mendelian Inheritance in Man*, published in 1966,

listed only 1,368 autosomal traits and 119 X-linked traits. Today, the online version of McKusick's compendium lists more than 19,000 entries, of which more than 18,000 are autosomal, more than 1,000 are X-linked, 57 are Y-linked, and 63 are in the mitochondrial genome. DNA variants responsible for more than 2,500 of these traits, most of which are inherited diseases, have been identified. With continued advances, these numbers are certain to increase.

Although some genetic disorders, particularly the single-gene conditions, are strongly determined by genes, many others are the result of multiple genetic and nongenetic factors. One can therefore think of genetic diseases as lying along a continuum (Fig. 1-2), with disorders such as cystic fibrosis and Duchenne muscular dystrophy situated at one end (strongly determined by genes) and conditions such as measles situated at the other end (strongly determined by environment). Many of the most prevalent disorders, including many birth defects and many common diseases such as diabetes, hypertension, heart disease, and cancer, lie somewhere in the middle of the continuum. These diseases are the products of varying degrees of genetic and environmental influences.

## THE CLINICAL IMPACT OF GENETIC DISEASE

Genetic diseases are sometimes perceived as so rare that the average health care practitioner will seldom encounter them. That this is far from the truth is becoming increasingly evident as knowledge and technology progress. Less than a century ago, diseases of largely nongenetic causation (i.e., those caused by malnutrition, unsanitary conditions, and pathogens) accounted for the great majority of deaths in children. During the 20th century, however, public health vastly improved. As a result, genetic diseases have come to account for an ever-increasing percentage of deaths among children in developed countries. For example, the percentage of pediatric deaths due to genetic causes in various hospitals in the United Kingdom increased from 16.5% in 1914 to 50% in 1976 (Table 1-2).

In addition to contributing to a large fraction of childhood deaths, genetic diseases also account for a large share of admissions to pediatric hospitals. For example, a survey of Seattle hospitals showed that 27% of all pediatric inpatients presented with a genetic disorder, and a survey of admissions to a major pediatric hospital in Mexico showed that 37.8% involved a disease that was either genetic or "partly genetic."

Another way to assess the importance of genetic diseases is to ask, "What fraction of persons in the population will be found to have a genetic disorder?" This is not as simple a question as it might seem. A variety of factors can influence the answer. For example, some diseases are found more frequently in certain ethnic groups. Cystic fibrosis is especially

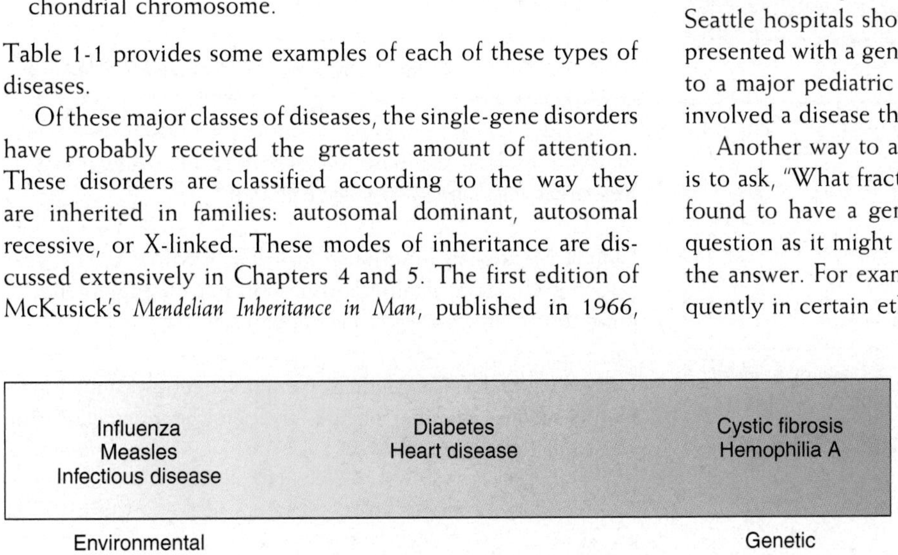

| Influenza<br>Measles<br>Infectious disease | Diabetes<br>Heart disease | Cystic fibrosis<br>Hemophilia A |
|---|---|---|
| Environmental | | Genetic |

**FIGURE 1-2**
Continuum of disease causation. Some diseases (e.g., cystic fibrosis) are strongly determined by genes, whereas others (e.g., infectious diseases) are strongly determined by environment.

**TABLE 1-2**
Percentages of Childhood Deaths in United Kingdom Hospitals Attributable to Nongenetic and Genetic Causes

| Cause | London 1914 | London 1954 | Newcastle 1966 | Edinburgh 1976 |
|---|---|---|---|---|
| **Nongenetic*** | | | | |
| All causes | 83.5 | 62.5 | 58.0 | 50.0 |
| **Genetic** | | | | |
| Single gene | 2.0 | 12.0 | 8.5 | 8.9 |
| Chromosomal | - | - | 2.5 | 2.9 |
| Multifactorial | 14.5 | 25.5 | 31.0 | 38.2 |

*Infections, for example.
Data from Rimoin DL, Connor JM, Pyeritz RE, Korf BR: Emery and Rimoin's Principles and Practice of Medical Genetics. London: Churchill Livingstone, 2007.

common among whites, whereas sickle cell disease is especially common among Africans. Some diseases are more common in older persons. For example, colon cancer, breast cancer, and Alzheimer disease are caused by dominant genes in a small fraction (5% to 10%) of cases but are not usually manifested until later in life. The prevalence estimates for these genetic diseases would be higher in an older population. Variations in diagnostic and recording practices can also cause prevalence estimates to vary. Accordingly, the prevalence figures shown in Table 1-3 are given as rather broad ranges. Keeping these sources of variation in mind, it is notable that a recognizable genetic disease will be diagnosed in 3% to 7% of the population at some point. This tabulation

**TABLE 1-3**
Approximate Prevalence of Genetic Disease in the General Population

| Type of Genetic Disease | Lifetime Prevalence per 1000 Persons |
|---|---|
| Autosomal dominant | 3-9.5 |
| Autosomal recessive | 2-2.5 |
| X-linked | 0.5-2 |
| Chromosome disorder | 6-9 |
| Congenital malformation* | 20-50 |
| Total | 31.5-73 |

*Congenital means "present at birth." Most congenital malformations are thought to be multifactorial and therefore probably have both genetic and environmental components.

does not include most cases of the more common adult diseases, such as heart disease, diabetes, and cancer, although it is known that these diseases also have genetic components. If such diseases are included, the clinical impact of genetic disease is considerable indeed.

## Suggested Readings

Baird PA, Anderson TW, Newcombe HB, Lowry RB. Genetic disorders in children and young adults: a population study. Am J Hum Genet 1988;42:677-93.

Dunn LC. A Short History of Genetics. New York: McGraw-Hill, 1965.

McKusick VA. History of medical genetics. In: Rimoin DL, Connor JM, Pyeritz RE, Korf BR (eds): Emery and Rimoin's Principles and Practice of Medical Genetics, 5th ed, vol. 1. London: Churchill Livingstone: 2007, pp. 3-32.

Passarge E. Color Atlas of Genetics, 3rd ed. Stuttgart: Georg Thieme Verlag, 2007.

Rimoin DL, Connor JM, Pyeritz RE, Korf BR. Emery and Rimoin's Principles and Practice of Medical Genetics, 5th ed. London: Churchill Livingstone, 2007.

Scriver CR, Sly WS, Childs G, et al. The Metabolic and Molecular Bases of Inherited Disease, 8th ed. New York: McGraw-Hill, 2001.

Seashore MS, Wappner RS. Genetics in Primary Care and Clinical Medicine. Stamford, Conn: Appleton & Lange, 1996.

Watson JD, Crick FHC. Molecular structure of nucleic acids: A structure for deoxyribose nucleic acid. Nature 1953;171:737.

## Internet Resources

Dolan DNA Learning Center, Cold Spring Harbor Laboratory (a useful online resource for learning and reviewing basic principles) http://www.dnalc.org/

Genetic Science Learning Center (another useful resource for learning and reviewing basic genetic principles) http://gslc.genetics.utah.edu/

Landmarks in the History of Genetics http://cogweb.ucla.edu/EP/DNA_history.html

National Human Genome Research Institute Educational Resources http://www.genome.gov/Education

Online Mendelian Inheritance in Man (OMIM) (a comprehensive catalog and description of single-gene conditions) http://www.ncbi.nlm.nih.gov/Omim/

University of Kansas Medical Center Genetics Education Center (a large number of links to useful genetics education sites) http://www.kumc.edu/gec/

## Chapter 2

# BASIC CELL BIOLOGY: STRUCTURE AND FUNCTION OF GENES AND CHROMOSOMES

All genetic diseases involve defects at the level of the cell. For this reason, one must understand basic cell biology to understand genetic disease. Errors can occur in the replication of genetic material or in the translation of genes into proteins. Such errors commonly produce single-gene disorders. In addition, errors that occur during cell division can lead to disorders involving entire chromosomes. To provide the basis for understanding these errors and their consequences, this chapter focuses on the processes through which genes are replicated and translated into proteins, as well as the process of cell division.

In the 19th century, microscopic studies of cells led scientists to suspect that the nucleus of the cell (Fig. 2-1) contains the important mechanisms of inheritance. They found that **chromatin**, the substance that gives the nucleus a granular appearance, is observable in the nuclei of nondividing cells. Just before a cell undergoes division, the chromatin condenses to form discrete, dark-staining bodies called **chromosomes** (from the Greek words for "colored bodies"). With the rediscovery of Mendel's breeding experiments at the beginning of the 20th century, it soon became apparent that chromosomes contain **genes**. Genes are transmitted from parent to offspring and are considered the basic unit of inheritance. It is through the transmission of genes that physical traits such as eye color are inherited in families. Diseases can also be transmitted through genetic inheritance.

Physically, genes are composed of **deoxyribonucleic acid (DNA)**. DNA provides the genetic blueprint for all proteins in the body. Thus, genes ultimately influence all aspects of body structure and function. Humans are estimated to have 20,000 to 25,000 genes (sequences of DNA that code for ribonucleic acid [RNA] or proteins). An error (or **mutation**) in one of these genes often leads to a recognizable genetic disease.

▶ Genes, the basic unit of inheritance, are contained in chromosomes and consist of DNA.

Each human **somatic cell** (cells other than the **gametes**, or sperm and egg cells) contains 23 pairs of different chromosomes, for a total of 46. One member of each pair is derived from the individual's father, and the other member is derived from the mother. One of the chromosome pairs consists of the **sex chromosomes**. In normal males, the sex chromosomes are a Y chromosome inherited from the father and an X chromosome inherited from the mother. Two X chromosomes are found in normal females, one inherited from each parent. The other 22 pairs of chromosomes are **autosomes**. The members of each pair of autosomes are said to be **homologs**, or **homologous**, because their DNA is very similar. The X and Y chromosomes are not homologs of each other.

Somatic cells, having two of each chromosome, are **diploid** cells. Human gametes have the **haploid** number of chromosomes, 23. The diploid number of chromosomes is maintained in successive generations of somatic cells by the process of **mitosis**, whereas the haploid number is obtained through the process of **meiosis**. Both of these processes are discussed in detail later in this chapter.

▶ Somatic cells are diploid, having 23 pairs of chromosomes (22 pairs of autosomes and one pair of sex chromosomes). Gametes are haploid and have a total of 23 chromosomes.

## DNA, RNA, AND PROTEINS: HEREDITY AT THE MOLECULAR LEVEL

### DNA

#### Composition and Structure of DNA

The DNA molecule has three basic components: the pentose sugar, deoxyribose; a phosphate group; and four types of nitrogenous **bases** (so named because they can combine with hydrogen ions in acidic solutions). Two of the bases, **cytosine** and **thymine**, are single carbon–nitrogen rings called **pyrimidines**. The other two bases, **adenine** and **guanine**, are double carbon–nitrogen rings called **purines** (Fig. 2-2). The four bases are commonly represented by their first letters: C, T, A, and G.

One of the contributions of Watson and Crick in the mid-20th century was to demonstrate how these three components are physically assembled to form DNA. They proposed the now-famous **double helix** model, in which DNA can be envisioned as a twisted ladder with chemical bonds as its rungs (Fig. 2-3). The two sides of the ladder are composed of the sugar and phosphate components, held together by strong phosphodiester bonds. Projecting from each side of the ladder, at regular intervals, are the nitrogenous bases. The

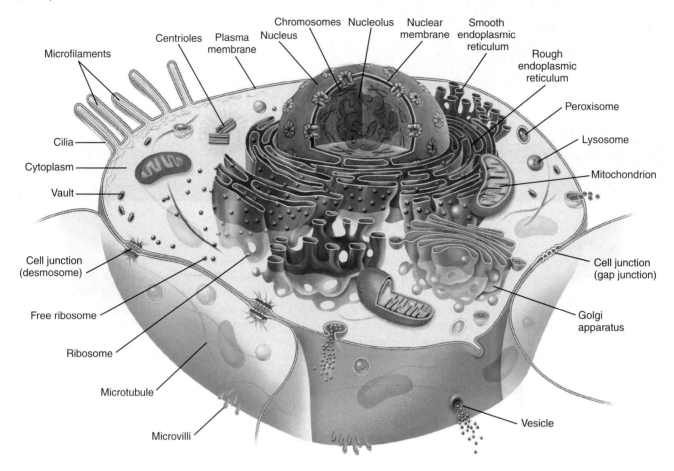

**FIGURE 2-1**
The anatomy of the cell.
(*From McCance KL, Huether SE: Pathophysiology: The Biologic Basis for Disease in Adults and Children, 5th ed. St. Louis: Mosby, 2006.*)

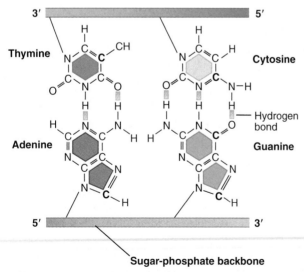

**FIGURE 2-2**
Chemical structure of the four bases, showing hydrogen bonds between base pairs. Three hydrogen bonds are formed between cytosine–guanine pairs, and two bonds are formed between adenine–thymine pairs.

base projecting from one side is bound to the base projecting from the other side by relatively weak hydrogen bonds. The paired nitrogenous bases therefore form the rungs of the ladder.

Figure 2-2 illustrates the chemical bonds between bases and shows that the ends of the ladder terminate in either 3' or 5'. These labels are derived from the order in which the five carbon atoms composing deoxyribose are numbered. Each DNA subunit, consisting of one deoxyribose, one phosphate group, and one base, is called a **nucleotide**.

Different sequences of nucleotide bases (e.g., ACCAAGTGC) specify different proteins. Specification of the body's many proteins must require a great deal of genetic information. Indeed, each haploid human cell contains approximately 3 billion nucleotide pairs, more than enough information to specify the composition of all human proteins.

▶ The most important constituents of DNA are the four nucleotide bases: adenine, thymine, cytosine, and guanine. DNA has a double helix structure.

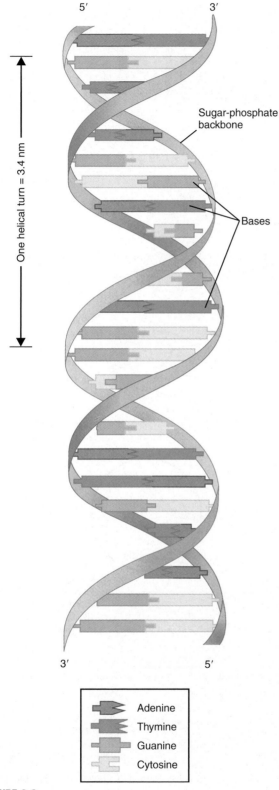

One helical turn = 3.4 nm

5′       3′

Sugar-phosphate backbone

Bases

3′       5′

Adenine
Thymine
Guanine
Cytosine

**FIGURE 2-3**
The DNA double helix, with sugar-phosphate backbone and nitrogenous bases.

## DNA Coiling

Textbook illustrations usually depict DNA as a double helix molecule that continues in a long, straight line. However, if the DNA in a cell were actually stretched out in this way, it would be about 2 meters long. To package all of this DNA into a tiny cell nucleus, it is coiled at several levels. First, the DNA is wound around a **histone** protein core to form a **nucleosome** (Fig. 2-4). About 140 to 150 DNA bases are wound around each histone core, and then 20 to 60 bases form a spacer element before the next nucleosome complex. The nucleosomes in turn form a helical **solenoid**; each turn of the solenoid includes about six nucleosomes. The solenoids themselves are organized into **chromatin loops**, which are attached to a protein scaffold. Each of these loops contains approximately 100,000 base pairs (bp), or 100 **kilobases (kb)**, of DNA. The end result of this coiling and looping is that the DNA, at its maximum stage of condensation, is only about 1/10,000 as long as it would be if it were fully stretched out.

> DNA is a tightly coiled structure. This coiling occurs at several levels: the nucleosome, the solenoid, and 100-kb loops.

## Replication of DNA

As cells divide to make copies of themselves, identical copies of DNA must be made and incorporated into the new cells. This is essential if DNA is to serve as the fundamental genetic material. DNA **replication** begins as the weak hydrogen bonds between bases break, producing single DNA strands with unpaired bases. The consistent pairing of adenine with thymine and guanine with cytosine, known as **complementary base pairing**, is the key to accurate replication. The principle of complementary base pairing dictates that the unpaired base will attract a free nucleotide only if that nucleotide has the proper complementary base. For example, a portion of a single strand with the base sequence ATTGCT will bond with a series of free nucleotides with the bases TAACGA. The single strand is said to be a **template** upon which the complementary strand is built. When replication is complete, a new double-stranded molecule identical to the original is formed (Fig. 2-5).

Several different enzymes are involved in DNA replication. One enzyme unwinds the double helix, and another holds the strands apart. Still another enzyme, **DNA polymerase**, travels along the single DNA strand, adding free nucleotides to the 3′ end of the new strand. Nucleotides can be added only to the 3′ end of the strand, so replication always proceeds from the 5′ to the 3′ end. When referring to the orientation of sequences along a gene, the 5′ direction is termed **upstream**, and the 3′ direction is termed **downstream**.

In addition to adding new nucleotides, DNA polymerase performs part of a **proofreading** procedure, in which a newly added nucleotide is checked to make certain that it is in fact complementary to the template base. If it is not, the nucleotide is excised and replaced with a correct complementary nucleotide base. This process substantially enhances the accuracy of DNA replication. When a DNA replication error is not successfully repaired, a mutation has occurred. As will be seen in Chapter 3, many such mutations cause genetic diseases.

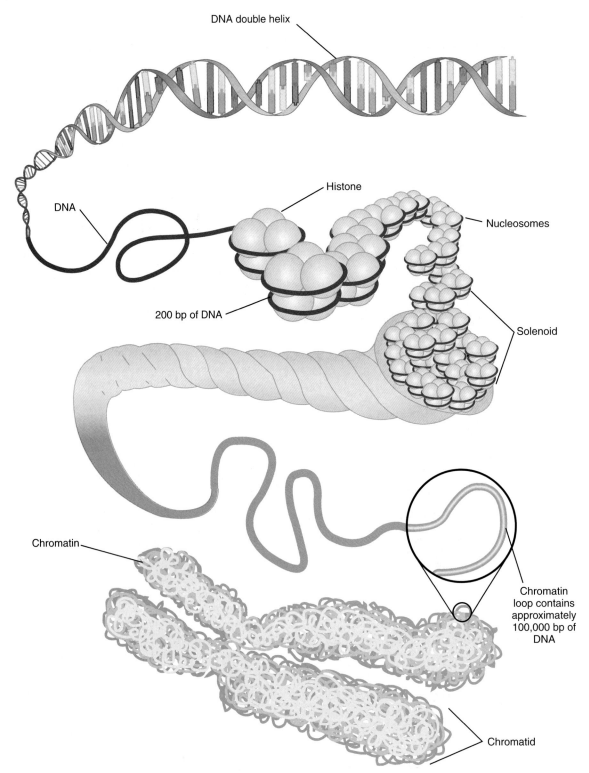

**FIGURE 2-4**
Patterns of DNA coiling. DNA is wound around histones to form nucleosomes. These are organized into solenoids, which in turn make up the chromatin loops.

DNA replication is critically dependent on the principle of complementary base pairing. This allows a single strand of the double-stranded DNA molecule to form a template for the synthesis of a new, complementary strand.

The rate of DNA replication in humans, about 40 to 50 nucleotides per second, is comparatively slow. In bacteria the rate is much higher, reaching 500 to 1000 nucleotides per second. Given that some human chromosomes have as many as 250 million nucleotides, replication would be an

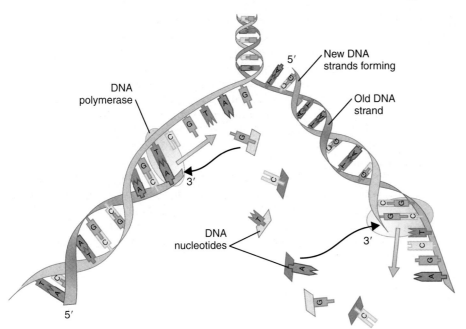

**FIGURE 2-5**
DNA replication. The hydrogen bonds between the two original strands are broken, allowing the bases in each strand to undergo complementary base pairing with free bases. This process, which proceeds in the 5' to 3' direction on each strand, forms two new double strands of DNA.

extraordinarily time-consuming process if it proceeded linearly from one end of the chromosome to the other: For a chromosome of this size, a single round of replication would take almost 2 months. Instead, replication begins at many different points along the chromosome, termed **replication origins**. The resulting multiple separations of the DNA strands are called **replication bubbles** (Fig. 2-6). By occurring simultaneously at many different sites along the chromosome, the replication process can proceed much more quickly.

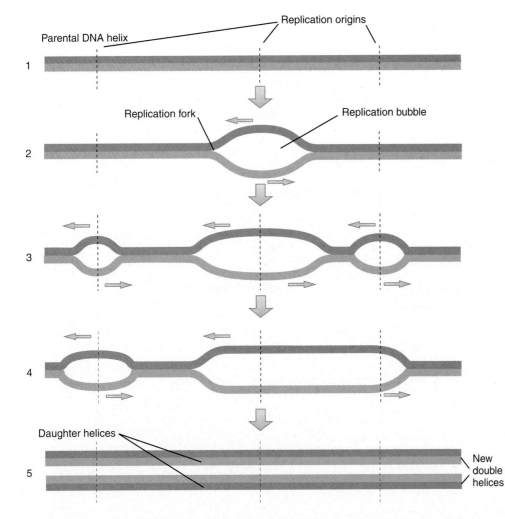

**FIGURE 2-6**
Replication bubbles form at multiple points along the DNA strand, allowing replication to proceed more rapidly.

▶ Replication bubbles allow DNA replication to take place at multiple locations on the chromosome, greatly speeding the replication process.

## From Genes to Proteins

While DNA is formed and replicated in the cell nucleus, protein synthesis takes place in the cytoplasm. The information contained in DNA must be transported to the cytoplasm and then used to dictate the composition of proteins. This involves two processes, transcription and translation. Briefly, the DNA code is **transcribed** into messenger RNA, which then leaves the nucleus to be **translated** into proteins. These processes, summarized in Figure 2-7, are discussed at length later in this chapter. Transcription and translation are both mediated by **ribonucleic acid (RNA)**, a type of nucleic acid that is chemically similar to DNA. Like DNA, RNA is composed of sugars, phosphate groups, and nitrogenous bases. It differs from DNA in that the sugar is ribose instead of deoxyribose, and uracil rather than thymine is one of the four bases. Uracil is structurally similar to thymine, so, like thymine, it can pair with adenine. Another difference between RNA and DNA is that whereas DNA usually occurs as a double strand, RNA usually occurs as a single strand.

▶ DNA sequences encode proteins through the processes of transcription and translation. These processes both involve RNA, a single-stranded molecule that is similar to DNA except that it has a ribose sugar rather than deoxyribose and a uracil base rather than thymine.

**FIGURE 2-7**
A summary of the steps leading from DNA to proteins. Replication and transcription occur in the cell nucleus. The mRNA is then transported to the cytoplasm, where translation of the mRNA into amino acid sequences composing a protein occurs.

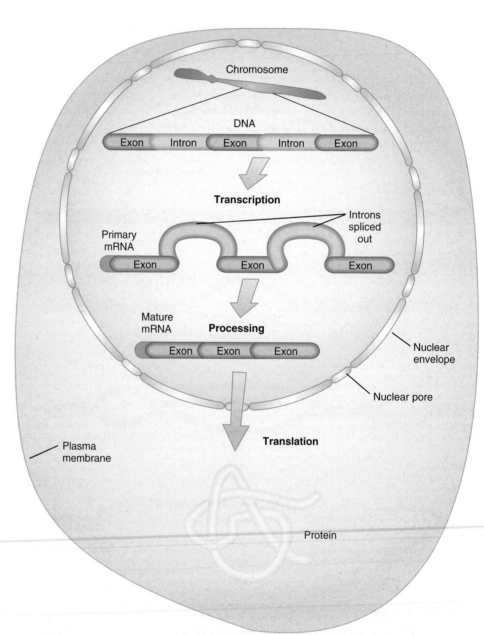

## Transcription

**Transcription** is the process by which an RNA sequence is formed from a DNA template (Fig. 2-8). The type of RNA produced by the transcription process is **messenger RNA (mRNA)**. To initiate mRNA transcription, one of the **RNA polymerase** enzymes (RNA polymerase II) binds to a **promoter** site on the DNA (a promoter is a nucleotide sequence that lies just upstream of a gene). The RNA polymerase then pulls a portion of the DNA strands apart from each other, exposing unattached DNA bases. One of the two DNA strands provides the template for the sequence of mRNA nucleotides. Although either DNA strand could in principle serve as the template for

mRNA synthesis, only one is chosen to do so in a given region of the chromosome. This choice is determined by the promoter sequence, which orients the RNA polymerase in a specific direction along the DNA sequence. Because the mRNA molecule can be synthesized only in the 5′ to 3′ direction, the promoter, by specifying directionality, determines which DNA strand serves as the template. This template DNA strand is also known as the **antisense** strand. RNA polymerase moves in the 3′ to 5′ direction along the DNA template strand, assembling the complementary mRNA strand from 5′ to 3′ (see Fig. 2-8). Because of complementary base pairing, the mRNA nucleotide sequence is identical to that of the DNA strand that does not serve as the template—the **sense strand**—except, of course, for the substitution of uracil for thymine.

Soon after RNA synthesis begins, the 5′ end of the growing RNA molecule is capped by the addition of a chemically modified guanine nucleotide. This **5′ cap** appears to help prevent the RNA molecule from being degraded during synthesis, and later it helps to indicate the starting position for translation of the mRNA molecule into protein. Transcription continues until a group of bases called a **termination sequence** is reached. Near this point, a series of 100 to 200 adenine bases are added to the 3′ end of the RNA molecule. This structure, known as the **poly-A tail**, may be involved in stabilizing the mRNA molecule so that it is not degraded when it reaches the cytoplasm. RNA polymerase usually continues to transcribe DNA for several thousand additional bases, but the mRNA bases that are attached after the poly-A tail are eventually lost. Finally, the DNA strands and the RNA polymerase separate from the RNA strand, leaving a transcribed single mRNA strand. This mRNA molecule is termed the **primary transcript**.

In some human genes, such as the one that can cause Duchenne muscular dystrophy, several different promoters exist and are located in different parts of the gene. Thus, transcription of the gene can start in different places, resulting in the production of somewhat different proteins. This allows the same gene sequence to code for variations of a protein in different tissues (e.g., muscle tissue versus brain tissue).

> In the process of transcription, RNA polymerase II binds to a promoter site near the 5′ end of a gene on the antisense strand and, through complementary base pairing, helps to produce an mRNA strand from the antisense DNA strand.

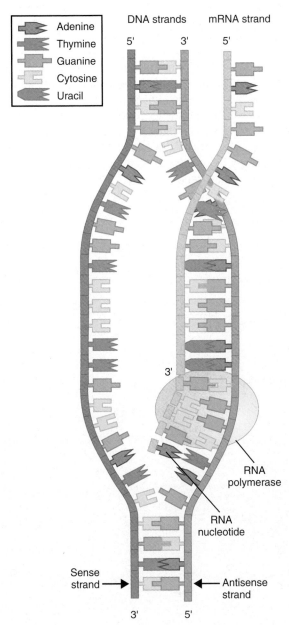

**FIGURE 2-8**
Transcription of DNA to mRNA. RNA polymerase II proceeds along the DNA strand in the 3′ to 5′ direction, assembling a strand of mRNA nucleotides that is complementary to the DNA template strand.

### Transcription and the Regulation of Gene Expression

Some genes are transcribed in all cells of the body. These **housekeeping genes** encode products that are required for a cell's maintenance and metabolism. Most genes, however, are transcribed only in specific tissues at specific points in time. Therefore, in most cells, only a small fraction of genes are actively transcribed. This specificity explains why there is a large variety of different cell types making different protein products, even though almost all cells have exactly the same DNA sequence. For example, the globin genes are transcribed in the progenitors of red blood cells (where they help to form hemoglobin), and the low-density lipoprotein receptor genes are transcribed in liver cells.

Many different proteins participate in the process of transcription. Some of these are required for the transcription of all genes, and these are termed **general transcription factors**. Others, labeled **specific transcription factors**, have more specialized roles, activating only certain genes at certain stages of development. A key transcriptional element is RNA polymerase II, which was described previously. Although this enzyme plays a vital role in initiating transcription by binding to the promoter region, it cannot locate the promoter region on its own. Furthermore, it is incapable of producing significant quantities of mRNA by itself. Effective transcription requires the interaction of a large complex of approximately 50 different proteins. These include general (basal) transcription factors, which bind to RNA polymerase and to specific DNA sequences in the promoter region (sequences such as TATA and others needed for initiating transcription). The general transcription factors allow RNA polymerase to bind to the promoter region so that it can function effectively in transcription (Fig. 2-9).

The transcriptional activity of specific genes can be greatly increased by interaction with sequences called **enhancers**, which may be located thousands of bases upstream or downstream of the gene. Enhancers do not interact directly with genes. Instead, they are bound by a class of specific transcription factors that are termed **activators**. Activators bind to a second class of specific transcription factors called **co-activators**, which in turn bind to the general transcription factor complex described previously (see Fig. 2-9). This chain of interactions, from enhancer to activator to co-activator to the general transcription complex and finally to the gene itself, increases the transcription of specific genes at specific points in time. Whereas enhancers help to increase the transcriptional activity of genes, other DNA sequences, known as **silencers**, help to repress the transcription of genes through a similar series of interactions.

Mutations in enhancer, silencer, or promoter sequences, as well as mutations in the genes that encode transcription factors, can lead to faulty expression of vital genes and consequently to genetic disease. Many examples of such diseases are discussed in the following chapters.

> Transcription factors are required for the transcription of DNA to mRNA. General transcription factors are used by all genes, and specific transcription factors help to initiate the transcription of genes in specific cell types at specific points in time. Transcription is also regulated by enhancer and silencer sequences, which may be located thousands of bases away from the transcribed gene.

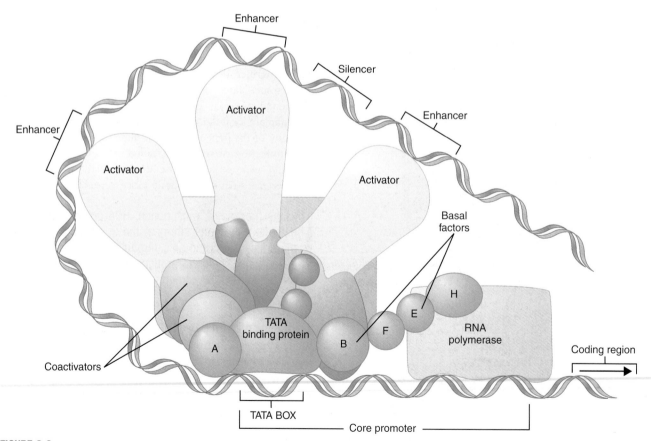

**FIGURE 2-9**
Key elements of transcription control include general (basal) transcription factors and specific enhancers and silencers. The activity of enhancers is mediated by activators and coactivators, which are specific transcription factors.
(*Data from Tjian R: Molecular machines that control genes. Sci Am 1995 Feb;272(2):54-61.*)

**TABLE 2-1**
## The Major Classes of DNA-Binding Motifs Found in Transcription Factors

| Motif | Description | Human Disease Examples |
|-------|-------------|------------------------|
| Helix–turn–helix | Two α helices are connected by a short chain of amino acids, which constitute the turn. The carboxyl-terminal helix is a recognition helix that binds to the DNA major groove. | Homeodomain proteins (HOX): mutations in human *HOXD13* and *HOXA13* cause synpolydactyly and hand–foot–genital syndrome, respectively. |
| Helix–loop–helix | Two α helices (one short and one long) are connected by a flexible loop. The loop allows the two helices to fold back and interact with one another. The helices can bind to DNA or to other helix–loop–helix structures. | Mutations in the *TWIST* gene cause Saethre–Chotzen syndrome (acrocephalosyndactyly type III) |
| Zinc finger | Zinc molecules are used to stabilize amino acid structures (e.g., α helices, β sheets), with binding of the α helix to the DNA major groove. | *BRCA1* (breast cancer gene); *WT1* (Wilms tumor gene); *GL13* (Greig syndrome gene); vitamin D receptor gene (mutations cause rickets) |
| Leucine zipper | Two leucine-rich α helices are held together by amino acid side chains. The α helices form a Y-shaped structure whose side chains bind to the DNA major groove. | *RB1* (retinoblastoma gene); *JUN* and *FOS* oncogenes |
| β Sheets | Side chains extend from the two-stranded β sheet to form contacts with the DNA helix. | TBX family of genes: *TBX5* (Holt–Oram syndrome); *TBX3* (ulnar–mammary syndrome) |

The large number and complexity of transcription factors allow fine-tuned regulation of gene expression. But how do the transcription factors locate specific DNA sequences? This is achieved by **DNA-binding motifs**: configurations in the transcription-factor protein that allow it to fit snugly and stably into a unique portion of the DNA double helix. Several examples of these binding motifs are listed in Table 2-1, and Figure 2-10 illustrates the binding of one such motif to DNA. Each major motif contains many variations that allow specificity in DNA binding.

An intriguing type of DNA-binding motif is contained in the high-mobility group (HMG) class of proteins. These proteins are capable of bending DNA and can facilitate interactions between distantly located enhancers and the appropriate basal factors and promoters (see Fig. 2-9).

> Transcription factors contain DNA-binding motifs that allow them to interact with specific DNA sequences. In some cases, they bend DNA so that distant enhancer sequences can interact with target genes.

Gene activity can also be related to patterns of chromatin coiling or condensation (a **chromatin** is the combination of DNA and the histone proteins around which the DNA is wound). Decondensed, or open, chromatin regions, termed **euchromatin**, are typically characterized by **histone acetylation**, the attachment of acetyl groups to lysine residues in the histones. Acetylation of histones reduces their binding to DNA, helping to decondense the chromatin so that it is more accessible to transcription factors. Euchromatin is thus transcriptionally active. In contrast, **heterochromatin** is usually less acetylated, more condensed, and transcriptionally inactive.

Gene expression can also be influenced by **microRNAs** (**miRNA**), which are small RNA molecules (17-27 nucleotides) that are not translated into proteins. Instead, because they are complementary to specific mRNA sequences, they can bind to and down-regulate these mRNAs, thus lowering their expression levels.

> Heterochromatin, which is highly condensed and hypoacetylated, tends to be transcriptionally inactive, whereas euchromatin, which is acetylated and less condensed, tends to be transcriptionally active.

### Gene Splicing

The primary mRNA transcript is exactly complementary to the base sequence of the DNA template. In **eukaryotes**,* an important step takes place before this RNA transcript leaves the nucleus. Sections of the RNA are removed by nuclear enzymes, and the remaining sections are spliced together to form the functional mRNA that will migrate to the cytoplasm. The excised sequences are called **introns**, and the sequences

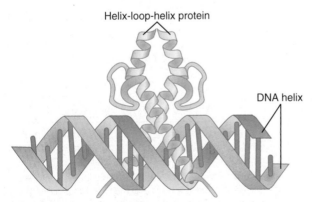

Helix-loop-helix protein

DNA helix

**FIGURE 2-10**
A helix–loop–helix motif binds tightly to a specific DNA sequence.

---

*Eukaryotes are organisms that have a defined cell nucleus, as opposed to **prokaryotes**, which lack a defined nucleus.

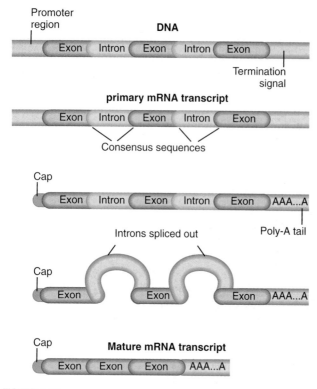

**FIGURE 2-11**
Gene splicing. Introns are precisely removed from the primary mRNA transcript to produce a mature mRNA transcript. Consensus sequences mark the sites at which splicing occurs.

that are left to code for proteins are called **exons** (Fig. 2-11). Only after gene splicing is completed does the **mature transcript** move out of the nucleus into the cytoplasm. Some genes contain **alternative splice sites**, which allow the same primary transcript to be spliced in different ways, ultimately producing different protein products from the same gene. Errors in gene splicing, like replication errors, are a form of mutation that can lead to genetic disease.

> Introns are spliced out of the primary mRNA transcript before the mature transcript leaves the nucleus. Exons contain the mRNA that specifies proteins.

## The Genetic Code

Proteins are composed of one or more **polypeptides**, which are in turn composed of sequences of **amino acids**. The body contains 20 different types of amino acids, and the amino acid sequences that make up polypeptides must in some way be designated by the DNA after transcription into mRNA.

Because there are 20 different amino acids and only four different RNA bases, a single base could not be specific for a single amino acid. Similarly, specific amino acids could not be defined by couplets of bases (e.g., adenine followed by guanine, or uracil followed by adenine) because only 16 (4 × 4) different couplets are possible. If triplet sets of bases are translated into amino acids, however, 64 (4 × 4 × 4)

combinations can be achieved—more than enough to specify each amino acid. Conclusive proof that amino acids are specified by these triplets of bases, or **codons**, was obtained by manufacturing synthetic nucleotide sequences and allowing them to direct the formation of polypeptides in the laboratory. The correspondence between specific codons and amino acids, known as the **genetic code**, is shown in Table 2-2.

Of the 64 possible codons, three signal the end of a gene and are known as **stop codons**. These are UAA, UGA, and UAG. The remaining 61 codons all specify amino acids. This means that most amino acids can be specified by more than one codon, as Table 2-2 shows. The genetic code is thus said to be **degenerate**. Although a given amino acid may be specified by more than one codon, each codon can designate only one amino acid.

**TABLE 2-2**
## The Genetic Code*

| First Position (5′ end) ↓ | Second Position | | | | Third Position (3′ end) ↓ |
|---|---|---|---|---|---|
| | U | C | A | G | |
| U | Phe | Ser | Tyr | Cys | U |
| U | Phe | Ser | Tyr | Cys | C |
| U | Leu | Ser | STOP | STOP | A |
| U | Leu | Ser | STOP | Trp | G |
| C | Leu | Pro | His | Arg | U |
| C | Leu | Pro | His | Arg | C |
| C | Leu | Pro | Gln | Arg | A |
| C | Leu | Pro | Gln | Arg | G |
| A | Ile | Thr | Asn | Ser | U |
| A | Ile | Thr | Asn | Ser | C |
| A | Ile | Thr | Lys | Arg | A |
| A | Met | Thr | Lys | Arg | G |
| G | Val | Ala | Asp | Gly | U |
| G | Val | Ala | Asp | Gly | C |
| G | Val | Ala | Glu | Gly | A |
| G | Val | Ala | Glu | Gly | G |

*Examples: UUG is translated into leucine; UAA is a stop codon; GGG is translated into glycine. Under some circumstances the UGA codon can specify an amino acid called selenocysteine, which is often called the 21st amino acid.

Ala, *Alanine;* Arg, *arginine;* Asn, *asparagine;* Asp, *aspartic acid;* Cys, *cysteine;* Gln, *glutamine;* Glu, *glutamic acid;* Gly, *glycine;* His, *histidine;* Ile, *isoleucine;* Leu, *leucine;* Lys, *lysine;* Met, *methionine;* Phe, *phenylalanine;* Pro, *proline;* Ser, *serine;* Thr, *threonine;* Trp, *tryptophan;* Tyr, *tyrosine;* Val, *valine.*

Individual amino acids, which compose proteins, are encoded by units of three mRNA bases, termed codons. There are 64 possible codons and only 20 amino acids, so the genetic code is degenerate.

A significant feature of the genetic code is that it is universal: virtually all living organisms use the same DNA codes to specify amino acids. One known exception to this rule occurs in **mitochondria**, cytoplasmic organelles that are the sites of cellular respiration (see Fig. 2-1). The mitochondria have their own extranuclear DNA molecules. Several codons of mitochondrial DNA encode different amino acids than do the same nuclear DNA codons.

## Translation

**Translation** is the process in which mRNA provides a template for the synthesis of a polypeptide. mRNA cannot, however, bind directly to amino acids. Instead, it interacts with molecules of **transfer RNA (tRNA)**, which are cloverleaf-shaped RNA strands of about 80 nucleotides. As Figure 2-12 illustrates, each tRNA molecule has a site at the 3' end for the attachment of a specific amino acid by a covalent bond. At the opposite end of the cloverleaf is a sequence of three nucleotides called the **anticodon**, which undergoes

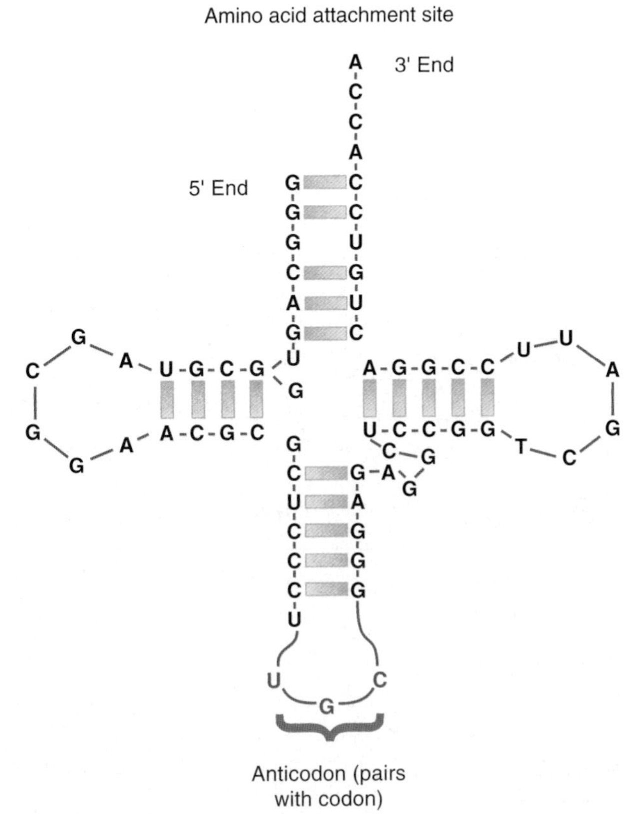

**FIGURE 2-12**
The structure of a tRNA molecule. In two dimensions, the tRNA has a cloverleaf shape. Note the 3' site of attachment for an amino acid. The anticodon pairs with a complementary mRNA codon.

complementary base pairing with an appropriate codon in the mRNA. The attached amino acid is then transferred to the polypeptide chain being synthesized. In this way, mRNA specifies the sequence of amino acids by acting through tRNA.

The cytoplasmic site of protein synthesis is the **ribosome**, which consists of roughly equal parts of enzymatic proteins and **ribosomal RNA (rRNA)**. The function of rRNA is to help bind mRNA and tRNA to the ribosome. During translation, depicted in Figure 2-13, the ribosome first binds to an initiation site on the mRNA sequence. This site consists of a specific codon, AUG, which specifies the amino acid methionine (this amino acid is usually removed from the polypeptide before the completion of polypeptide synthesis). The ribosome then binds the tRNA to its surface so that base pairing can occur between tRNA and mRNA. The ribosome moves along the mRNA sequence, codon by codon, in the 5' to 3' direction. As each codon is processed, an amino acid is translated by the interaction of mRNA and tRNA.

In this process, the ribosome provides an enzyme that catalyzes the formation of covalent peptide bonds between the adjacent amino acids, resulting in a growing polypeptide. When the ribosome arrives at a stop codon on the mRNA sequence, translation and polypeptide formation cease. The amino ($NH_2$) terminus of the polypeptide corresponds to the 5' end of the mRNA strand, and the carboxyl (COOH) terminus corresponds to the 3' end. After synthesis is completed, the mRNA, the ribosome, and the polypeptide separate from one another. The polypeptide is then released into the cytoplasm.

In the process of translation, the mRNA sequence serves as a template to specify sequences of amino acids. These sequences, which form polypeptides, are assembled by ribosomes. The tRNA and rRNA molecules interact with mRNA in the translation process.

Before a newly synthesized polypeptide can begin its existence as a functional protein, it often undergoes further processing, termed **posttranslational modification**. These modifications can take a variety of forms, including cleavage into smaller polypeptide units or combination with other polypeptides to form a larger protein. Other possible modifications include the addition of carbohydrate side chains to the polypeptide. Such modifications may be needed, for example, to produce proper folding of the mature protein or to stabilize its structure. An example of a clinically important protein that undergoes considerable posttranslational modification is type I collagen (Clinical Commentary 2-1).

Posttranslational modification consists of various chemical changes that occur in proteins shortly after they are translated.

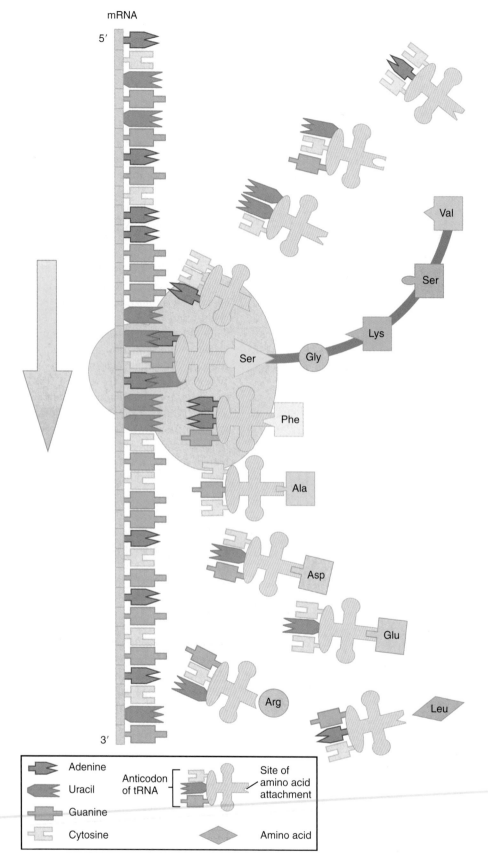

**FIGURE 2-13**

Translation of mRNA to amino acids. The ribosome moves along the mRNA strand in the 5′ to 3′ direction, assembling a growing polypeptide chain. In this example, the mRNA sequence GUG AGC AAG GGU UCA has assembled five amino acids (Val, Ser, Lys, Gly, and Ser, respectively) into a polypeptide.

## CLINICAL COMMENTARY 2-1
### *Osteogenesis Imperfecta, an Inherited Collagen Disorder*

As its name implies, osteogenesis imperfecta is a disease caused by defects in the formation of bone. This disorder, sometimes known as **brittle bone disease**, affects approximately 1 in 10,000 individuals in all ethnic groups.

Approximately 90% of osteogenesis imperfecta cases are caused by defects in type I collagen, a major component of bone that provides much of its structural stability. The function of collagen in bone is analogous to that of the steel bars incorporated in reinforced concrete. This is an especially apt analogy because the tensile strength of collagen fibrils is roughly equivalent to that of steel wires.

When type I collagen is improperly formed, the bone loses much of its strength and fractures easily. Patients with osteogenesis imperfecta can suffer hundreds of bone fractures, or they might experience only a few, making this disease highly variable in its expression (the reasons for this variability are discussed in Chapter 4). In addition to bone fractures, patients can have short stature, hearing loss, abnormal tooth development (dentinogenesis imperfecta), bluish sclerae, and various bone deformities. Osteogenesis imperfecta was traditionally classified into four major types; three additional types have recently been added. There is currently no cure for this disease, and management consists primarily of the repair of fractures and, in some cases, the use of external or internal bone support (e.g., surgically implanted rods). Additional therapies include the administration of bisphosphonates to decrease bone resorption and human growth hormone to facilitate growth. Physical rehabilitation also plays an important role in clinical management.

### Subtypes of Osteogenesis Imperfecta

| Type | Disease features |
|------|------------------|
| I | Mild bone fragility, blue sclerae, hearing loss in 50% of patients, normal or near-normal stature, few bone deformities, dentinogenesis imperfecta in some cases |
| II | Most severe form, with extreme bone fragility, long bone deformities, compressed femurs; lethal in the perinatal period (most die of respiratory failure) |
| III | Severe bone fragility, very short stature, variably blue sclerae, progressive bone deformities, dentinogenesis imperfecta is common |
| IV | Short stature, normal sclerae, mild to moderate bone deformity, hearing loss in some patients, dentinogenesis imperfecta is common; bone fragility is variable |
| V | Similar to type IV but also includes calcification of interosseous membrane of forearm, radial head dislocation, and hyperplastic callus formation |
| VI | More fractures than type IV, including vertebral compression fractures; no dentinogenesis imperfecta |
| VII | White sclerae, early lower limb deformities, congenital fractures, osteopenia |

*Types I-IV are caused by mutations in the two genes that encode type I collagen protein; types V-VII have been identified on the basis of distinct bone histology.*

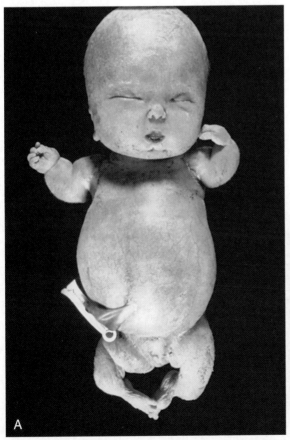

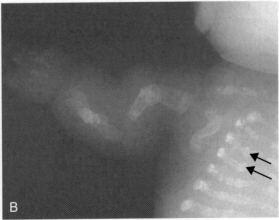

A, A stillborn infant with type II osteogenesis imperfecta (the perinatal lethal form). The infant had a type I procollagen mutation and short, slightly twisted limbs.
B, Radiograph of an infant with type II osteogenesis imperfecta. Note rib fractures, which are observable as "beads" on the ribs *(arrows)*.

*Continued*

**CLINICAL COMMENTARY 2-1**
*Osteogenesis Imperfecta, an Inherited Collagen Disorder—cont'd*

Type I collagen is a trimeric protein (i.e., having three subunits) with a triple helix structure. It is formed from a precursor protein, type 1 procollagen. Two of the three subunits of type 1 procollagen, labeled pro-α1(I) chains, are encoded by an 18-kb (kb = 1000 base pairs) gene on chromosome 17, and the third subunit, the pro-α2(I) chain, is encoded by a 38-kb gene on chromosome 7. Each of these genes contains more than 50 exons. After transcription and splicing, the mature mRNA formed from each gene is only 5 to 7 kb long. The mature mRNAs proceed to the cytoplasm, where they are translated into polypeptide chains by the ribosomal machinery of the cell.

At this point, the polypeptide chains undergo a series of posttranslational modifications. Many of the proline and lysine residues* are hydroxylated (i.e., hydroxyl groups are added) to form hydroxyproline and hydroxylysine, respectively. (Mutations in a gene that is required for this hydroxylation step were recently shown to cause osteogenesis imperfecta type VII.) The three polypeptides, two pro-α1(I) chains, and one pro-α2(I) chain, begin to associate with one another at their COOH termini. This association is stabilized by sulfide bonds that form between the chains near the COOH termini. The triple helix then forms, in zipper-like fashion, beginning at the COOH terminus and proceeding toward the NH₂ terminus. Some of the hydroxylysines are glycosylated (i.e., sugars are added), a modification that commonly occurs in the rough endoplasmic reticulum (see Fig. 2-1). The hydroxyl groups in the hydroxyprolines help to connect the three chains by forming hydrogen bonds, which stabilize the triple helix. Critical to proper folding of the helix is the presence of a glycine in every third position of each polypeptide.

Once the protein has folded into a triple helix, it moves from the endoplasmic reticulum to the Golgi apparatus (see Fig. 2-1) and is secreted from the cell. Yet another modification then takes place: The procollagen is cleaved by proteases near both the NH₂ and the COOH termini of the triple helix, removing some amino acids at each end. These amino acids performed essential functions earlier in the life of the protein (e.g., helping to form the triple helix structure, helping to thread the protein through the endoplasmic reticulum) but are no longer needed. This cleavage results in the mature protein, type I collagen. The collagen then assembles itself into fibrils, which react with adjacent molecules outside the cell to form the covalent cross-links that impart tensile strength to the fibrils.

The path from the DNA sequence to the mature collagen protein involves many steps. The complexity of this path provides many opportunities for mistakes (in replication, transcription, translation, or posttranslational modification) that can cause disease. One common mutation produces a replacement of glycine with another amino acid. Because only glycine is small enough to be accommodated in the center of the triple helix structure, substitution of a different amino acid causes instability of the structure and thus poorly formed fibrils. This type of mutation is often seen in severe forms of osteogenesis imperfecta. Other mutations can cause excess posttranslational modification of the polypeptide chains, again producing abnormal fibrils. Other examples of disease-causing mutations are provided in the suggested readings at the end of this chapter.

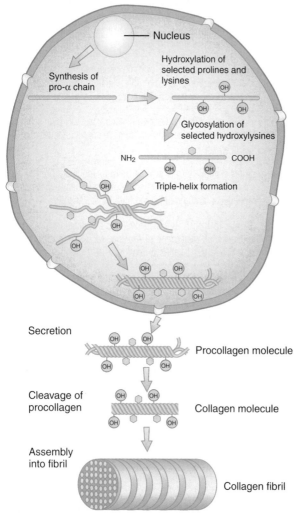

The process of collagen fibril formation. After the pro-α polypeptide chain is formed, a series of posttranslational modifications takes place, including hydroxylation and glycosylation. Three polypeptide chains assemble into a triple helix, which is secreted outside the cell. Portions of each end of the procollagen molecule are cleaved, resulting in the mature collagen molecule. These molecules then assemble into collagen fibrils.

_____

* A residue is an amino acid that has been incorporated into a polypeptide chain.

## THE STRUCTURE OF GENES AND THE GENOME

Some aspects of gene structure, such as the existence of introns and exons, have already been touched on. Alterations of different parts of genes can have quite distinct consequences in terms of genetic disease. It is therefore necessary to describe more fully the details of gene structure. A schematic diagram of gene structure is given in Figure 2-14.

### Introns and Exons

The intron–exon structure of genes, discovered in 1977, is one of the attributes that distinguishes eukaryotes from prokaryotes. Introns form the major portion of most eukaryotic genes. As noted previously, introns are spliced out of the mRNA before it leaves the nucleus, and this splicing must be under very precise control. Enzymes that carry out

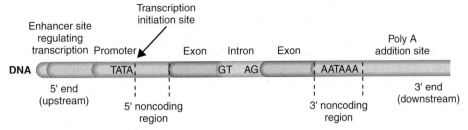

**FIGURE 2-14**
Details of gene structure, showing promoter and upstream regulation (enhancer) sequences and a poly-A addition site.

splicing are directed to the appropriate locations by DNA sequences known as **consensus sequences** (so named because they are common in all eukaryotic organisms), which are situated adjacent to each exon.

Because most eukaryotic genes are composed primarily of introns, it is natural to ask whether introns might have some function. At present, this is largely material for speculation. One interesting hypothesis is that introns, by lengthening genes, encourage the shuffling of genes when homologous chromosomes exchange material during meiosis (see later discussion). It has also been suggested that introns evolved to modify the amount of time required for DNA replication and transcription.

> The intron–exon structure is a key feature of most eukaryotic genes. The function of introns, if any, is currently unknown.

Surprisingly, some introns contain transcribed genes that are apparently unrelated to the gene in which the introns are contained. For example, introns of the human neurofibromatosis type 1 (NF1) gene contain three genes that are transcribed in the direction opposite that of the NF1 gene itself. These genes appear to have no functional relationship to the NF1 gene. Similar gene inserts have been found within the factor VIII gene (*F8*) on the human X chromosome.

## Types of DNA

Although most of the emphasis in genetics is given to the DNA that encodes proteins, only 34 million (1%) of the 3 billion nucleotide pairs in the human genome actually perform this role. Another 21 million nucleotides are transcribed into mRNA that is not translated into proteins. Most of our genetic material has no known function. To better understand the nature of all types of DNA, we briefly review the several categories into which it is classified (Fig. 2-15).

The first and most important class of DNA is termed **single-copy DNA**. As the name implies, single-copy DNA sequences are seen only once (or possibly a few times) in the genome. Single-copy DNA accounts for about 45% of the genome and includes the protein-coding genes. However, protein-coding DNA represents only a small fraction of all single-copy DNA, most of which is found in introns or in DNA sequences that lie between genes.

The remaining 55% of the genome consists of **repetitive DNA**, sequences that are repeated over and over again in the genome, often thousands of times. There are two major classes of repetitive DNA: **dispersed repetitive DNA** and **satellite DNA**. Satellite repeats are clustered together in certain chromosome locations, where they occur in tandem (i.e., the beginning of one repeat occurs immediately adjacent to the end of another). Dispersed repeats, as the name implies, tend to be scattered singly throughout the genome; they do not occur in tandem.

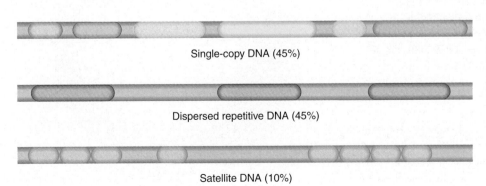

Single-copy DNA (45%)

Dispersed repetitive DNA (45%)

Satellite DNA (10%)

**FIGURE 2-15**
Single-copy DNA sequences are unique and are dispersed throughout the genome. Satellite DNA sequences are repetitive elements that occur together in clusters. Dispersed repeats are similar to one another but do not cluster together.

The term *satellite* is used because these sequences, owing to their composition, can easily be separated by centrifugation in a cesium chloride density gradient. The DNA appears as a satellite, separate from the other DNA in the gradient. This term is not to be confused with the satellites that can be observed microscopically on certain chromosomes (see Chapter 6). Satellite DNA accounts for approximately 10% of the genome and can be further subdivided into several categories. α-Satellite DNA occurs as tandem repeats of a 171-bp sequence that can extend to several million base pairs or longer. This type of satellite DNA is found near the centromeres of chromosomes. **Minisatellites** are blocks of tandem repeats (each 14 to 500 bp long) whose total length is much smaller, usually a few thousand base pairs. A final category, **microsatellites**, are smaller still: the repeat units are 1 to 13 bp long, and the total length of the array is usually less than a few hundred base pairs. Minisatellites and microsatellites are of special interest in human genetics because they vary in length among individuals, making them highly useful for gene mapping (see Chapter 8). A minisatellite or microsatellite is found at an average frequency of one per 2 kb in the human genome; altogether they account for about 3% of the genome.

Dispersed repetitive DNA makes up about 45% of the genome, and these repeats fall into several major categories. The two most common categories are short interspersed elements (**SINEs**) and long interspersed elements (**LINEs**). Individual SINEs range in size from 90 to 500 bp, and individual LINEs can be as large as 7000 bp. One of the most important types of SINEs is the *Alu* repeat. These *Alu* repeat units, which are about 300 bp long, contain a DNA sequence that can be cut by the *Alu* restriction enzyme (see Chapter 3 for further discussion). The *Alu* repeats are a **family** of genes, meaning that all of them have highly similar DNA sequences. About 1 million *Alu* repeats are scattered throughout the genome; they thus constitute approximately 10% of all human DNA. A remarkable feature of *Alu* sequences, as well as some LINEs, is that some of them can generate copies of themselves, which can then be inserted into other parts of the genome. This insertion can sometimes interrupt a protein-coding gene, causing genetic disease (examples are discussed in Chapter 4).

> There are several major types of DNA, including single-copy DNA, satellite DNA, and dispersed repetitive DNA. The latter two categories are both classes of repeated DNA sequences. Less than 5% of human DNA actually encodes proteins.

## THE CELL CYCLE

During the course of development, each human progresses from a single-cell **zygote** (an egg cell fertilized by a sperm cell) to a marvelously complex organism containing approximately 100 trillion ($10^{14}$) individual cells. Because few cells last for a person's entire lifetime, new ones must be generated to replace those that die. Both of these processes—development and replacement—require the manufacture of new cells. The cell division processes that are responsible for the creation of new diploid cells from existing ones are **mitosis** (nuclear division) and **cytokinesis** (cytoplasmic division). Before dividing, a cell must duplicate its contents, including its DNA; this occurs during **interphase**. The alternation of mitosis and interphase is referred to as the **cell cycle**.

As Figure 2-16 shows, a typical cell spends most of its life in interphase. This portion of the cell cycle is divided into three phases, **G1**, **S**, and **G2**. During G1 (gap 1, the interval between mitosis and the onset of DNA replication), synthesis of RNA and proteins takes place. DNA replication occurs during the S (synthesis) phase. During G2 (the interval between the S phase and the next mitosis), some DNA repair takes place, and the cell prepares for mitosis. By the time G2 has been reached, the cell contains two identical copies of each of the 46 chromosomes. These identical chromosomes are referred to as **sister chromatids**. Sister chromatids often exchange material during interphase, a process known as **sister chromatid exchange**.

> The cell cycle consists of the alternation of cell division (mitosis and cytokinesis) and interphase. DNA replication and protein synthesis take place during interphase.

The length of the cell cycle varies considerably from one cell type to another. In rapidly dividing cells such as those of epithelial tissue (found, for example, in the lining of the intestines and in the lungs), the cycle may be completed in less than 10 hours. Other cells, such as those of the liver, might divide only once each year or so. Some cell types,

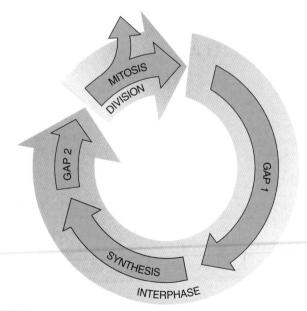

**FIGURE 2-16**
Major phases of the mitotic cell cycle, showing the alternation of interphase and mitosis (division).

such as skeletal muscle cells and neurons, largely lose their ability to divide and replicate in adults. Although all stages of the cell cycle have some variation in length, the great majority of variation is due to differences in the length of the G1 phase. When cells stop dividing for a long period, they are often said to be in the G0 stage.

Cells divide in response to important internal and external cues. Before the cell enters mitosis, for example, DNA replication must be accurate and complete and the cell must have achieved an appropriate size. The cell must respond to extracellular stimuli that require increased or decreased rates of division. Complex molecular interactions mediate this regulation. Among the most important of the molecules involved are the **cyclin-dependent kinases (CDKs)**, a family of kinases that phosphorylate other regulatory proteins at key stages of the cell cycle. To carry out this function, the CDKs must form complexes with various **cyclins**, proteins that are synthesized at specific cell-cycle stages and are then degraded when CDK action is no longer needed. The cyclins and CDKs, as well as the many proteins that interact with them, are subjects of intense study because of their vital role in the cell cycle and because their malfunction can lead to cancer (see Chapter 11).

> The length of the cell cycle varies in different cell types. Critical to regulation of the cell cycle are CDKs, which phosphorylate other proteins, and cyclins, which form complexes with CDKs. Faulty regulation of the cell cycle can lead to cancer.

### Mitosis

Although mitosis usually requires only 1 to 2 hours to complete, this portion of the cell cycle involves many critical and complex processes. Mitosis is divided into several phases (Fig. 2-17). During **prophase**, the first mitotic stage, the chromosomes become visible under a light microscope as they condense and coil (chromosomes are not clearly visible during interphase). The two sister chromatids of each chromosome lie together, attached at a point called the **centromere**. The nuclear membrane, which surrounds the nucleus, disappears during this stage. **Spindle fibers** begin to form, radiating from two **centrioles** located on opposite sides of the cell. The spindle fibers become attached to the centromeres of each chromosome and eventually pull the two sister chromatids in opposite directions.

The chromosomes reach their most highly condensed state during **metaphase**, the next stage of mitosis. Because they are highly condensed, they are easiest to visualize microscopically during this phase. For this reason, clinical diagnosis of chromosome disorders is usually based on metaphase chromosomes. During metaphase, the spindle fibers begin to contract and pull the centromeres of the chromosomes, which are now arranged along the middle of the spindle (the **equatorial plane** of the cell).

During **anaphase**, the next mitotic stage, the centromere of each chromosome splits, allowing the sister chromatids to separate. The chromatids are then pulled by the spindle fibers, centromere first, toward opposite sides of the cell. At the end of anaphase, the cell contains 92 separate chromosomes, half lying near one side of the cell and half near the other side. If all has proceeded correctly, the two sets of chromosomes are identical.

**Telophase**, the final stage of mitosis, is characterized by the formation of new nuclear membranes around each of the two sets of 46 chromosomes. Also, the spindle fibers disappear, and the chromosomes begin to decondense. Cytokinesis usually occurs after nuclear division and results in a roughly equal division of the cytoplasm into two parts. With the completion of telophase, two diploid **daughter cells**, both identical to the original cell, have been formed.

> Mitosis is the process through which two identical diploid daughter cells are formed from a single diploid cell.

### Meiosis

When an egg cell and a sperm cell unite to form a zygote, their chromosomes are combined into a single cell. Because humans are diploid organisms, there must be a mechanism to reduce the number of chromosomes in gametes to the haploid state. Otherwise the zygote would have 92, instead of the normal 46, chromosomes. The primary mechanism by which haploid gametes are formed from diploid precursor cells is **meiosis**.

Two cell divisions occur during meiosis. Each meiotic division has been divided into stages with the same names as those of mitosis, but the processes involved in some of the stages are quite different (Fig. 2-18). During meiosis I, often called the **reduction division stage**, two haploid cells are formed from a diploid cell. These diploid cells are the **oogonia** in females and the **spermatogonia** in males. After meiosis I, a second meiosis, the **equational division**, takes place, during which each haploid cell is replicated.

The first stage of the meiotic cell cycle is **interphase I**, during which important processes such as replication of chromosomal DNA take place. The second phase of meiosis I, **prophase I**, is quite complex and includes many of the key events that distinguish meiosis from mitosis. Prophase I begins as the chromatin strands coil and condense, causing them to become visible as chromosomes. During the process of **synapsis**, the homologous chromosomes pair up, side by side, lying together in perfect alignment (in males, the X and Y chromosomes, being mostly nonhomologous, line up end to end). This pairing of homologous chromosomes is an important part of the cell cycle that does not occur in mitosis. As prophase I continues, the chromatids of the two chromosomes intertwine. Each pair of intertwined homologous chromosomes is either **bivalent** (indicating two chromosomes in the unit) or **tetrad** (indicating four chromatids in the unit).

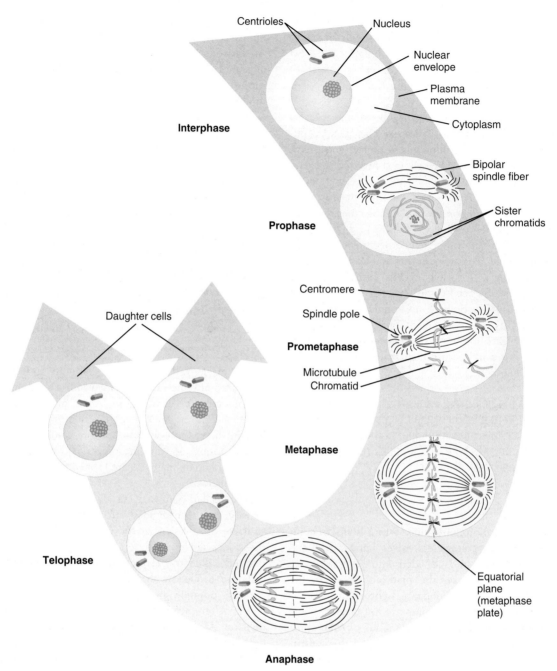

**FIGURE 2-17**
The stages of mitosis, during which two identical diploid cells are formed from one original diploid cell.

A second key feature of prophase I is the formation of **chiasmata** (plural of **chiasma**), cross-shaped structures that mark attachments between the homologous chromosomes (Fig. 2-19). Each chiasma indicates a point at which the homologous chromosomes exchange genetic material. This process, called **crossing over**, produces chromosomes that consist of combinations of parts of the original chromosomes. This chromosomal shuffling is important because it greatly increases the possible combinations of genes in each gamete and thereby increases the number of possible combinations of human traits. Also, as discussed in Chapter 8, this phenomenon is critically important in inferring the order of genes along chromosomes. At the end of prophase I, the bivalents begin to move toward the equatorial plane, a spindle apparatus begins to form in the cytoplasm, and the nuclear membrane dissipates.

**Metaphase I** is the next phase. As in mitotic metaphase, this stage is characterized by the completion of spindle formation and alignment of the bivalents, which are still attached at the chiasmata, in the equatorial plane. The two centromeres of each bivalent now lie on opposite sides of the equatorial plane.

**FIGURE 2-18**
The stages of meiosis, during which haploid gametes are formed from a diploid cell. For brevity, prophase II and telophase II are not shown. Note the relationship between meiosis and spermatogenesis and oogenesis.

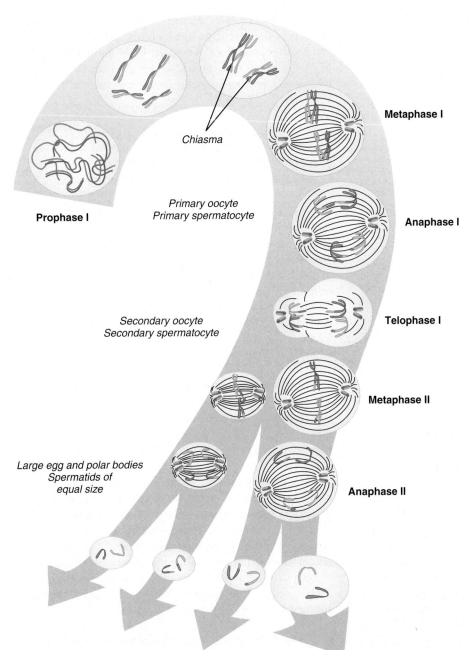

*Chiasma*

**Metaphase I**

*Primary oocyte*
*Primary spermatocyte*

**Anaphase I**

**Prophase I**

*Secondary oocyte*
*Secondary spermatocyte*

**Telophase I**

**Metaphase II**

*Large egg and polar bodies*
*Spermatids of*
*equal size*

**Anaphase II**

During **anaphase I**, the chiasmata disappear and the homologous chromosomes are pulled by the spindle fibers toward opposite poles of the cell. The key feature of this phase is that unlike the corresponding phase of mitosis, the centromeres do not duplicate and divide, so that only half of the original number of chromosomes migrates toward each pole. The chromosomes migrating toward each pole thus consist of one member of each pair of autosomes and one of the sex chromosomes.

The next stage, **telophase I**, begins when the chromosomes reach opposite sides of the cell. The chromosomes uncoil slightly, and a new nuclear membrane begins to form. The two daughter cells each contain the haploid number of chromosomes, and each chromosome has two sister chromatids. In humans, cytokinesis also occurs during this phase. The cytoplasm is divided approximately equally between the two daughter cells in the gametes formed in males. In those formed in females, nearly all of the cytoplasm goes into one daughter cell, which will later form the egg. The other daughter cell becomes a **polar body**, a small, nonfunctional cell that eventually degenerates.

**Homologous chromosomes**

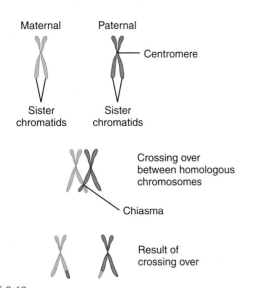

**FIGURE 2-19**
The process of chiasma formation and crossing over results in the exchange of genetic material between homologous chromosomes.

> Meiosis I (reduction division) includes a prophase I stage in which homologous chromosomes line up and exchange material (crossing over). During anaphase I the centromeres do not duplicate and divide. Consequently only one member of each pair of chromosomes migrates to each daughter cell.

The equational division, meiosis II, then begins with **interphase II**. This is a very brief phase. The important feature of interphase II is that, unlike interphase I, no replication of DNA occurs. **Prophase II**, the next stage, is quite similar to mitotic prophase, except that the cell nucleus contains only the haploid number of chromosomes. During prophase II, the chromosomes thicken as they coil, the nuclear membrane disappears, and new spindle fibers are formed. This is followed by **metaphase II**, during which the spindle fibers pull the chromosomes into alignment at the equatorial plane.

**Anaphase II** then follows. This stage resembles mitotic anaphase in that the centromeres split and each carries a single chromatid toward a pole of the cell. The chromatids have now separated, but, because of chiasma formation and crossing over, the newly separated sister chromatids might not be identical (see Fig. 2-18).

**Telophase II**, like telophase I, begins when the chromosomes reach opposite poles of the cell. There they begin to uncoil. New nuclear membranes are formed around each group of chromosomes, and cytokinesis occurs. In gametes formed in males, the cytoplasm is again divided equally between the two daughter cells. The end result of male meiosis is thus four functional daughter cells, each of which has

an equal amount of cytoplasm. In female gametes, unequal division of the cytoplasm again occurs, forming the egg cell and another polar body. The polar body formed during meiosis I sometimes undergoes a second division, so three polar bodies may be present when the second stage of meiosis is completed.

> Meiosis is a specialized cell division process in which a diploid cell gives rise to haploid gametes. This is accomplished by combining two rounds of division with only one round of DNA replication.

Most chromosome disorders are caused by errors that occur during meiosis. Gametes can be created that contain missing or additional chromosomes or chromosomes with altered structures. In addition, mitotic errors that occur early in the life of the embryo can affect enough of the body's cells to produce clinically significant disease. Mitotic errors occurring at any point in one's lifetime can, under some circumstances, cause cancer. Cancer genetics is discussed in Chapter 11, and chromosome disorders are the subject of Chapter 6.

### The Relationship between Meiosis and Gametogenesis

The stages of meiosis can be related directly to stages in **gametogenesis**, the formation of gametes (see Fig. 2-18). In mature males, the seminiferous tubules of the testes are populated by spermatogonia, which are diploid cells. After going through several mitotic divisions, the spermatogonia produce **primary spermatocytes**. Each primary spermatocyte, which is also diploid, undergoes meiosis I to produce a pair of **secondary spermatocytes**, each of which contains 23 double-stranded chromosomes. These undergo meiosis II, and each produces a pair of **spermatids** that contain 23 single-stranded chromosomes. The spermatids then lose most of their cytoplasm and develop tails for swimming as they become mature **sperm** cells. This process, known as **spermatogenesis**, continues throughout the life of the mature male.

> In spermatogenesis, each diploid spermatogonium produces four haploid sperm cells.

**Oogenesis**, the process by which female gametes are formed, differs in several important ways from spermatogenesis. Whereas the cycle of spermatogenesis is constantly recurring, much of female oogenesis is completed before birth. Diploid oogonia divide mitotically to produce **primary oocytes** by the third month of fetal development. More than 6 million primary oocytes are formed during gestation, and they are suspended in prophase I by the time of birth. Meiosis continues only when a mature primary oocyte is ovulated. In meiosis I, the primary oocyte produces one secondary oocyte (containing the cytoplasm) and one polar body. The **secondary oocyte** then emerges from the follicle and proceeds down the fallopian tube, with the polar body attached to it. Meiosis II begins only if the secondary oocyte is fertilized by a sperm cell. If this occurs, one haploid **mature ovum**, containing the cytoplasm, and another haploid polar body are produced. The polar bodies eventually disintegrate. About 1 hour after fertilization, the nuclei of the sperm cell and ovum fuse, forming a diploid zygote. The zygote then begins its development into an embryo through a series of mitotic divisions.

> In oogenesis, one haploid ovum and three haploid polar bodies are produced meiotically from a diploid oogonium. In contrast to spermatogenesis, which continues throughout the life of the mature male, the first phase of oogenesis is completed before the female is born; oogenesis is then halted until ovulation occurs.

## Study Questions

1. Consider the following double-stranded DNA sequence:

   5'-CAG AAG AAA ATT AAC ATG TAA-3'
   3'-GTC TTC TTT TAA TTG TAC ATT-5'

   If the bottom strand serves as the template, what is the mRNA sequence produced by transcription of this DNA sequence? What is the amino acid sequence produced by translation of the mRNA sequence?

2. Arrange the following terms according to their hierarchical relationship to one another: genes, chromosomes, exons, codons, nucleotides, genome.

3. Less than 5% of human DNA encodes proteins. Furthermore, in a given cell type only 10% of the coding DNA actively encodes proteins. Explain these statements.

4. What are the major differences between mitosis and meiosis?

5. The human body contains approximately $10^{14}$ cells. Starting with a single-cell zygote, how many mitotic cell divisions, on average, would be required to produce this number of cells?

6. How many mature sperm cells will be produced by 100 primary spermatocytes? How many mature egg cells will be produced by 100 primary oocytes?

## Suggested Readings

Alberts B, Johnson A, Lewis J, et al. Molecular Biology of the Cell, 4th ed. New York: Garland Science, 2002.

Berger SL. Histone modifications in transcriptional regulation. Curr Opin Genet Dev 2002;12:142-48.

Byers PH. Osteogenesis imperfecta: Perspectives and opportunities. Curr Opin Pediatr 2000;12:603-9.

Cho KS, Elizondo LI, Boerkoel CF. Advances in chromatin remodeling and human disease. Current Opin Genet Dev 2004;14:308-15.

Cook PR. The organization of replication and transcription. Science 1999;284:1790-5.

Johnson CA. Chromatin modification and disease. J Med Genet 2000;37:905-15.

Lander ES, Linton LM, Birren B, et al. Initial sequencing and analysis of the human genome. Nature 2001;409: 860-921.

Lemon B, Tjian R. Orchestrated response: A symphony of transcription factors for gene control. Genes Dev 2000;14: 2551-69.

Lewin B. Genes IX. Boston: Jones and Bartlett, 2008.

Mitchison TJ, Salmon ED. Mitosis: A history of division. Nat Cell Biol 2001;3:E17-21.

Page SL, Hawley RS. Chromosome choreography. The meiotic ballet. Science 2003;301:785-89.

Rauch F, Glorieux FH. Osteogenesis imperfecta. Lancet 2004;363:1377-85.

### Internet Resources

Mitosis and meiosis tutorials and animations *http://www.biology.arizona.edu/cell_bio/cell_bio.html*

Tutorial on DNA structure, replication, transcription, and translation *http://www.ncc.gmu.edu/dna/*

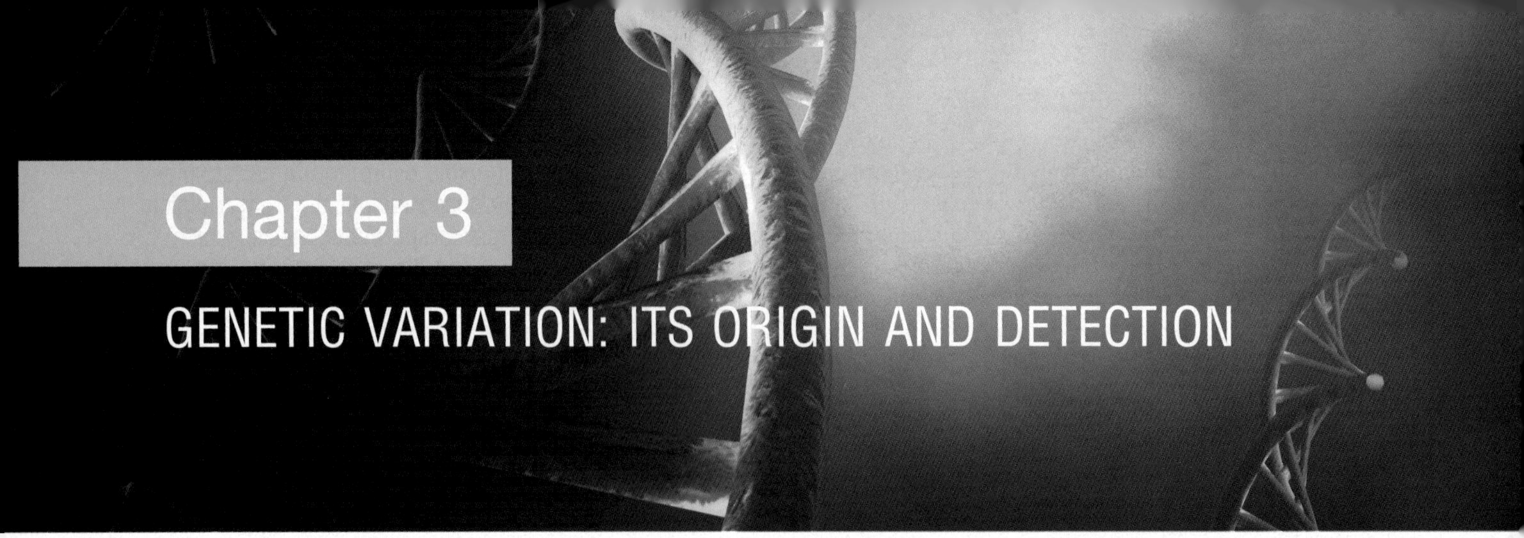

# Chapter 3

# GENETIC VARIATION: ITS ORIGIN AND DETECTION

Humans display a substantial amount of genetic variation. This is reflected in traits such as height, blood pressure, and skin color. Included in the spectrum of genetic variation are disease states, such as cystic fibrosis or type 1 neurofibromatosis (see Chapter 4). This aspect of genetic variation is the focus of medical genetics.

All genetic variation originates from the process known as **mutation**, which is defined as a change in DNA sequence. Mutations can affect either **germline** cells (cells that produce gametes) or **somatic** cells (all cells other than germline cells). Mutations in somatic cells can lead to cancer and are thus of significant concern. However, this chapter is directed primarily to germline mutations, because they can be transmitted from one generation to the next.

As a result of mutations, a gene can differ among individuals in terms of its DNA sequence. The differing sequences are referred to as **alleles**. A gene's location on a chromosome is termed a **locus** (from the Latin word for "place"). For example, it might be said that a person has a certain allele at the β-globin locus on chromosome 11. If a person has the same allele on both members of a chromosome pair, he or she is said to be a **homozygote**. If the alleles differ in DNA sequence, the person is a **heterozygote**. The alleles that are present at a given locus are the person's **genotype**.

In human genetics, the term *mutation* has often been reserved for DNA sequence changes that cause genetic diseases and are consequently relatively rare. DNA sequence variants that are more common in populations (i.e., in which two or more alleles at a locus each have frequencies exceeding 1%), are said to be **polymorphic** ("many forms"). Such **loci** (plural of locus) are termed **polymorphisms**, although nowadays alleles that have a frequency less than 1% are often called polymorphisms as well. Many polymorphisms are now known to influence the risks for complex, common diseases such as diabetes and heart disease (see Chapter 12), so the distinction between mutation and polymorphism has become increasingly blurred.

One of Gregor Mendel's important contributions to genetics was to show that the effects of one allele at a locus can mask those of another allele at the same locus. He performed crosses (matings) between pea plants homozygous for a "tall" allele (i.e., having two identical copies of an allele that we will label *H*) and plants homozygous for a "short" allele (having two copies of an allele labeled *b*). This cross, which can produce only heterozygous (*Hb*) offspring, is illustrated in the **Punnett square** shown in Figure 3-1. Mendel found that the offspring of these crosses, even though they were heterozygotes, were all tall. This is because the *H* allele is **dominant**, and the *b* allele is **recessive**. (It is conventional to label the dominant allele in upper case and the recessive allele in lower case.) The term *recessive* comes from a Latin root meaning "to hide." This describes the behavior of recessive alleles well: In heterozygotes, the consequences of a recessive allele are hidden. A dominant allele exerts its effect in both the homozygote (*HH*) and the heterozygote (*Hb*), whereas the presence of the recessive allele is detected only when it occurs in homozygous form (*bb*). Thus, short pea plants can be created only by crossing parent plants that each carry at least one *b* allele. An example is a heterozygote × heterozygote cross, shown in Figure 3-2.

In this chapter, we examine mutation as the source of genetic variation. We discuss the types of mutation, the causes and consequences of mutation, and the biochemical and molecular techniques that are now used to detect genetic variation in human populations.

**FIGURE 3-1**

Punnett square illustrating a cross between *HH* and *hh* homozygote parents.

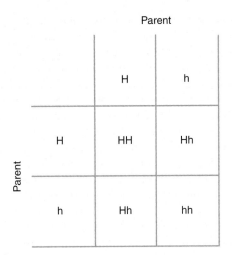

Parent

**FIGURE 3-2**
Punnett square illustrating a cross between two *Hh* heterozygotes.

## MUTATION: THE SOURCE OF GENETIC VARIATION
### Types of Mutation

Some mutations consist of an alteration of the number or structure of chromosomes in a cell. These major chromosome abnormalities can be observed microscopically and are the subject of Chapter 6. Here, the focus is on mutations that affect only single genes and are not microscopically observable. Most of our discussion centers on mutations that take place in coding DNA or in regulatory sequences, because mutations that occur in other parts of the genome usually have no clinical consequences.

One important type of **single-gene mutation** is the **base-pair substitution**, in which one base pair is replaced by another.* This can result in a change in the amino acid sequence. However, because of the redundancy of the genetic code, many of these mutations do not change the amino acid sequence and thus have no consequence. Such mutations are called **silent substitutions**. Base-pair substitutions that alter amino acids consist of two basic types: **missense** mutations, which produce a change in a single amino acid, and **nonsense** mutations, which produce one of the three stop codons (UAA, UAG, or UGA) in the messenger RNA (mRNA) (Fig. 3-3). Because stop codons terminate translation of the mRNA, nonsense mutations result in a premature termination of the polypeptide chain. Conversely, if a stop codon is altered so that it encodes an amino acid, an abnormally elongated polypeptide can be produced.

---

*In molecular genetics, base-pair substitutions are also termed **point mutations**. However, "point mutation" was used in classical genetics to denote any mutation small enough to be unobservable under a microscope.

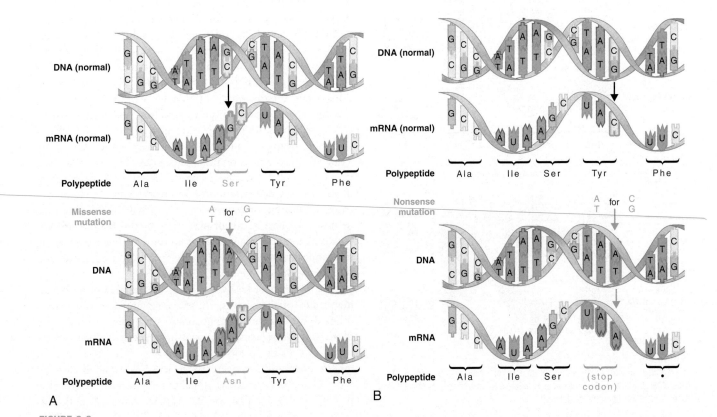

**A**     **B**

**FIGURE 3-3**
Base pair substitution. Missense mutations (**A**) produce a single amino acid change, whereas nonsense mutations (**B**) produce a stop codon in the mRNA. Stop codons terminate translation of the polypeptide.

Alterations of amino acid sequences can have profound consequences, and many of the serious genetic diseases discussed later are the result of such alterations.

A second major type of mutation consists of **deletions** or **insertions** of one or more base pairs. These mutations, which can result in extra or missing amino acids in a protein, are often detrimental. An example of such a mutation is the 3-bp deletion that is found in most persons with cystic fibrosis (see Chapter 4). Deletions and insertions tend to be especially harmful when the number of missing or extra base pairs is not a multiple of three. Because codons consist of groups of three base pairs, such insertions or deletions can alter all of the downstream codons. This is a **frameshift mutation** (Fig. 3-4). For example, the insertion of a single base (an A in the second codon) converts a DNA sequence read as 5'-ACT GAT TGC GTT-3' to 5'-ACT GAA TTG CGT-3'. This changes the amino acid sequence from Thr-Asp-Cys-Val to Thr-Glu-Leu-Arg. Often, a frameshift mutation produces a stop codon downstream of the insertion or deletion, resulting in a truncated polypeptide.

On a larger scale, **duplications** of whole genes can also lead to genetic disease. A good example is given by Charcot–Marie–Tooth disease. This disorder, named after the three physicians who described it more than a century ago, is a peripheral nervous system disease that leads to progressive atrophy of the distal limb muscles. It affects approximately 1 in 2500 persons and exists in several different forms. About 70% of patients who have the most common form (type 1A) display a 1.5 million-bp duplication on one copy of chromosome 17. As a result, they have three, rather than two, copies of the genes in this region. One of these genes, *PMP22*, encodes a component of peripheral myelin. The increased dosage of the gene product contributes to the demyelination that is characteristic of this form of the disorder. Interestingly, a deletion of this same region produces a distinct disease, hereditary neuropathy with liability to pressure palsies (paralysis). Because either a reduction (to 50%) or an increase (to 150%) in the gene product produces disease, this gene is said to display **dosage sensitivity**. Point mutations in *PMP22* itself can produce yet another disease: Dejerine–Sottas syndrome, which is characterized by distal muscle weakness, sensory alterations, muscular atrophy, and enlarged spinal nerve roots.

Other types of mutation can alter the regulation of transcription or translation. A **promoter mutation** can decrease the affinity of RNA polymerase for a promoter site, often resulting in reduced production of mRNA and thus decreased production of a protein. Mutations of transcription factor genes or enhancer sequences can have similar effects.

Mutations can also interfere with the splicing of introns as mature mRNA is formed from the primary mRNA transcript. **Splice-site mutations**, those that occur at intron–exon boundaries, alter the splicing signal that is necessary for proper excision of an intron. Splice-site mutations can occur at the GT sequence that defines the 5' splice site (the **donor site**) or at the AG sequence that defines the 3' splice site (the **acceptor site**). They can also take place in the sequences that lie near the donor and acceptor sites. When such mutations occur, the excision is often made within the next exon, at a splice site located in the exon. These splice sites, whose DNA sequences differ slightly from those of normal splice sites, are ordinarily unused and hidden within the exon. They are thus termed **cryptic splice sites**. The use of a cryptic site for splicing results in partial deletion of the exon or, in other cases, the deletion of an entire exon. As Figure 3-5 shows, splice-site mutations can also result in the abnormal inclusion of part or all of an intron in the mature mRNA. Finally, a mutation can occur at a cryptic splice site, causing it to appear as a normal splice site and thus to compete with the normal splice site.

Several types of DNA sequences are capable of propagating copies of themselves; these copies are then inserted in other locations on chromosomes (examples include the LINE and *Alu* repeats, discussed in Chapter 2). Such insertions can cause frameshift mutations. The insertion of **mobile elements** has been shown to cause isolated cases of type 1 neurofibromatosis, Duchenne muscular dystrophy, β-thalassemia, familial breast cancer, familial polyposis

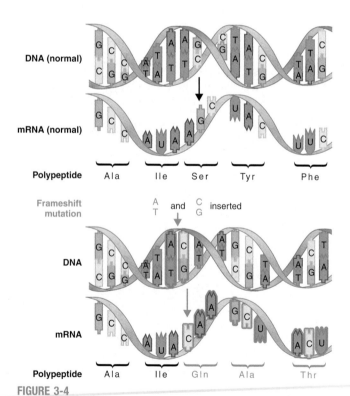

**FIGURE 3-4**

Frameshift mutations result from the addition or deletion of a number of bases that is not a multiple of three. This alters all of the codons downstream from the site of insertion or deletion.

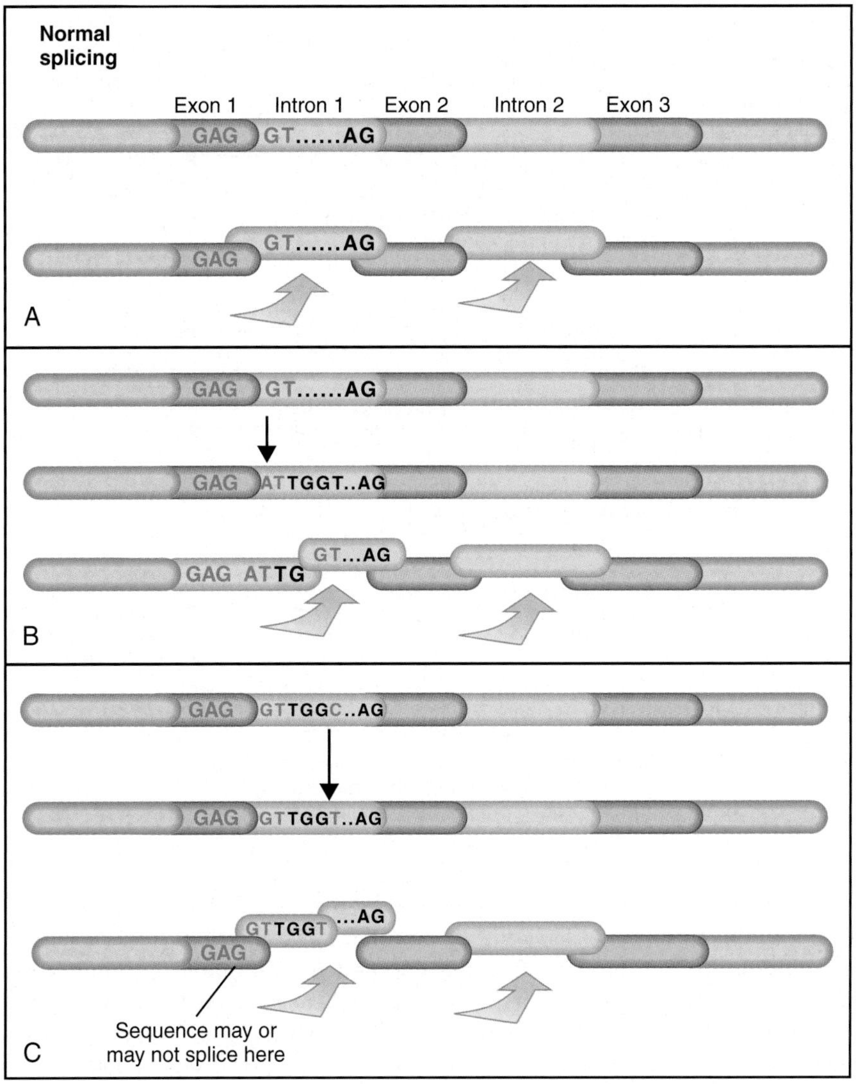

**FIGURE 3-5**
**A,** Normal splicing. **B,** Splice-site mutation. The donor sequence, GT, is replaced with AT. This results in an incorrect splice that leaves part of the intron in the mature mRNA transcript. In another example of splice-site mutation (**C**), a second GT donor site is created within the first intron, resulting in a combination of abnormally and normally spliced mRNA products.

(colon cancer), and hemophilia A and B (clotting disorders) in humans.

The final type of mutation to be considered here affects tandem repeated DNA sequences (see Chapter 2) that occur within or near certain disease-related genes. The repeat units are usually 3 bp long, so a typical example would be CAG-CAGCAG. A normal person has a relatively small number of these tandem repeats (e.g., 10 to 30 CAG consecutive elements) at a specific chromosome location. Occasionally, the number of repeats increases during meiosis or possibly during early fetal development, so that a newborn might have hundreds or even thousands of tandem repeats. When this occurs in certain regions of the genome, it causes genetic disease. Like other mutations, these **expanded repeats** can be transmitted to the patient's offspring. More than a dozen genetic diseases are now known to be caused by expanded repeats (see Chapter 4).

> Mutations are the ultimate source of genetic variation. Some mutations result in genetic disease, but most have no physical effects. The principal types of mutation are missense, nonsense, frameshift, promoter, and splice-site mutations. Mutations can also be caused by the random insertion of mobile elements, and some genetic diseases are known to be caused by expanded repeats.

## Molecular Consequences of Mutation

It is useful to think of mutations in terms of their effects on the protein product. Broadly speaking, mutations can produce either a **gain of function** or a **loss of function** of the protein product (Fig. 3-6). Gain-of-function mutations occasionally result in a completely novel protein product. More commonly, they result in overexpression of the product or

**FIGURE 3-6**
**A,** Gain-of-function mutations produce a novel protein product or an increased amount of protein product. **B,** Loss-of-function mutations decrease the amount of protein product. **C,** Dominant negative mutations produce an abnormal protein product that interferes with the otherwise normal protein product of the normal allele in a heterozygote.

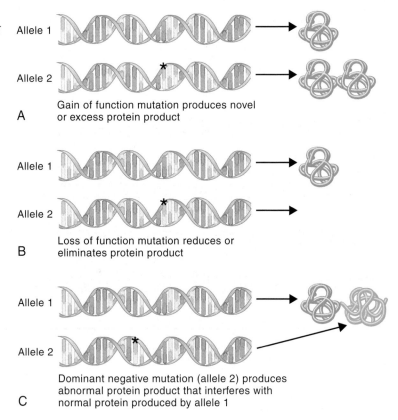

Allele 1
Allele 2
**A**    Gain of function mutation produces novel or excess protein product

Allele 1
Allele 2
**B**    Loss of function mutation reduces or eliminates protein product

Allele 1
Allele 2
**C**    Dominant negative mutation (allele 2) produces abnormal protein product that interferes with normal protein produced by allele 1

inappropriate expression (i.e., in the wrong tissue or in the wrong stage of development). Gain-of-function mutations produce dominant disorders. Charcot–Marie–Tooth disease can result from overexpression of the protein product and is considered a gain-of-function mutation. Huntington disease, discussed in Chapter 4, is another example.

Loss-of-function mutations are often seen in recessive diseases. In such diseases, the mutation results in the loss of 50% of the protein product (e.g., a metabolic enzyme), but the 50% that remains is sufficient for normal function. The heterozygote is thus unaffected, but the homozygote, having little or no protein product, is affected. In some cases, however, 50% of the gene's protein product is not sufficient for normal function (**haploinsufficiency**), and a dominant disorder can result. Haploinsufficiency is seen, for example, in the autosomal dominant disorder known as familial hypercholesterolemia (see Chapter 12). In this disease, a single copy of a mutation (heterozygosity) reduces the number of low-density lipoprotein (LDL) receptors by 50%. Cholesterol levels in heterozygotes are approximately double those of normal homozygotes, resulting in a substantial increase in the risk of heart disease. As with most disorders involving haploinsufficiency, the disease is more serious in affected homozygotes (who have few or no functional LDL receptors) than in heterozygotes.

A **dominant negative** mutation results in a protein product that not only is nonfunctional but also inhibits the function of the protein produced by the normal allele in the heterozygote. Typically, dominant negative mutations are seen in genes that encode multimeric proteins (i.e., proteins composed of two or more subunits). Type I collagen (see Chapter 2), which is composed of three helical subunits, is an example of such a protein. An abnormal helix created by a single mutation can combine with the other helices, distorting them and producing a seriously compromised triple-helix protein.

> Mutations can result in either a gain of function or a loss of function of the protein product. Gain-of-function mutations are sometimes seen in dominant diseases. Loss of function is seen in recessive diseases and in diseases involving haploinsufficiency, in which 50% of the gene product is insufficient for normal function. In dominant negative mutations, the abnormal protein product interferes with the normal protein product.

## Clinical Consequences of Mutation: The Hemoglobin Disorders

Genetic disorders of human hemoglobin are the most common group of single-gene diseases: An estimated 7% of the world's population carries one or more mutations of the genes involved in hemoglobin synthesis. Because almost all of the types of mutation described in this chapter have been observed in the hemoglobin disorders, these disorders serve as an important illustration of the clinical consequences of mutation.

The hemoglobin molecule is a tetramer composed of four polypeptide chains, two labeled α and two labeled β. The β chains are encoded by a gene on chromosome 11, and the α chains are encoded by two genes on chromosome 16 that are very similar to each other. A normal person has two normal β genes and four normal α genes (Fig. 3-7). Ordinarily, tight regulation of these genes ensures that roughly equal numbers of α and β chains are produced. Each of these **globin** chains is associated with a **heme** group, which contains an iron atom and binds with oxygen. This property allows hemoglobin to perform the vital function of transporting oxygen in erythrocytes (red blood cells).

The hemoglobin disorders can be classified into two broad categories: structural abnormalities, in which the hemoglobin molecule is altered, and thalassemias, a group of conditions in which either the α- or the β-globin chain is structurally normal but reduced in quantity. Another condition, hereditary persistence of fetal hemoglobin (HPFH), occurs when fetal hemoglobin, encoded by the α-globin genes and by two β-globin–like genes called $^A\gamma$ and $^G\gamma$ (see Fig. 3-7), continues to be produced after birth (normally, γ-chain production ceases and β-chain production begins at the time of birth). HPFH does not cause disease but instead can compensate for a lack of normal adult hemoglobin.

A large array of different hemoglobin disorders have been identified. The discussion that follows is a greatly simplified presentation of the major forms of these disorders. The hemoglobin disorders, the mutations that cause them, and their major features are summarized in Table 3-1.

## Sickle Cell Disease

Sickle cell disease, which results from an abnormality of hemoglobin structure, is seen in approximately 1 in 400 to 1 in 600 African American births. It is even more common

**TABLE 3-1**
## Summary of the Major Hemoglobin Disorders

| Disease | Mutation Type | Major Disease Features |
|---|---|---|
| Sickle cell disease | β-globin missense mutation | Anemia, tissue infarctions, infections |
| HbH disease | Deletion or abnormality of three of the four α-globin genes | Moderately severe anemia, splenomegaly |
| Hydrops fetalis (Hb Barts) | Deletion or abnormality of all four α-globin genes | Severe anemia or hypoxemia, congestive heart failure; stillbirth or neonatal death |
| β⁰-Thalassemia | Usually nonsense, frameshift, or splice-site donor or acceptor mutations; no β-globin produced | Severe anemia, splenomegaly, skeletal abnormalities, infections; often fatal during first decade if untreated |
| β⁺-Thalassemia | Usually missense, regulatory, or splice-site consensus sequence or cryptic splice-site mutations; small amount of β-globin produced | Features similar to those of β⁰-thalassemia but often somewhat milder |

in parts of Africa, where it can affect up to 1 in 50 births, and it is also seen occasionally in Mediterranean and Middle Eastern populations. Sickle cell disease is typically caused by a single missense mutation that effects a substitution of valine for glutamic acid at position 6 of the β-globin polypeptide chain. In homozygotes, this amino acid substitution alters the structure of hemoglobin molecules such that they form aggregates, causing erythrocytes to assume a characteristic sickle shape under conditions of low oxygen tension (Fig. 3-8A). These conditions are experienced in capillaries, the tiny vessels whose diameter is smaller than that of the erythrocyte. Normal erythrocytes (see Fig. 3-8B) can squeeze through capillaries, but sickled erythrocytes are less flexible and are unable to do so. In addition, the abnormal erythrocytes tend to stick to the vascular endothelium (the innermost lining of blood vessels).

The resultant vascular obstruction produces localized hypoxemia (lack of oxygen), painful sickling crises, and infarctions of various tissues, including bone, spleen, kidneys, and lungs (an infarction is tissue death due to hypoxemia). Premature destruction of the sickled erythrocytes decreases the number of circulating erythrocytes and the hemoglobin level, producing **anemia**. The spleen becomes enlarged (splenomegaly), but infarctions eventually destroy this organ, producing some loss of immune function. This contributes to the recurrent bacterial infections (especially pneumonia) that are commonly seen in persons with sickle cell disease and

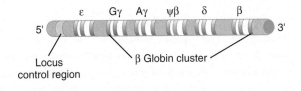

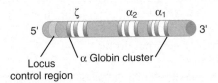

**FIGURE 3-7**
The α-globin gene cluster on chromosome 16 and the β-globin gene cluster on chromosome 11. The β-globin cluster includes the ε-globin gene, which encodes embryonic globin, and the γ-globin genes, which encode fetal globin. The ψβ gene is not expressed. The α-globin cluster includes the ζ-globin gene, which encodes embryonic α-globin.

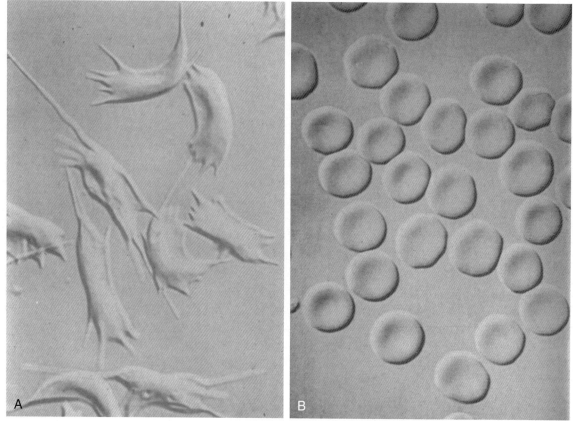

**FIGURE 3-8**
A, Erythrocytes from patients with sickle cell disease assume a characteristic shape under conditions of low oxygen tension. B, Compare with normal erythrocytes.

commonly cause death. In North America, it is estimated that the life expectancy of persons with sickle cell disease is reduced by about 30 years.

> Sickle cell disease, which causes anemia, tissue infarctions, and multiple infections, is the result of a single missense mutation that produces an amino acid substitution in the β-globin chain.

### Thalassemia

The term *thalassemia* is derived from the Greek word *thalassa* ("sea"); thalassemia was first described in populations living near the Mediterranean Sea, although it is also common in portions of Africa, the Mideast, India, and Southeast Asia. In contrast to sickle cell disease, in which a mutation alters the structure of the hemoglobin molecule, the mutations that cause thalassemia reduce the quantity of either α globin or β globin. Thalassemia can be divided into two major groups, α-thalassemia and β-thalassemia, depending on the globin chain that is reduced in quantity. When one type of chain is decreased in number, the other chain type, unable to participate in normal tetramer formation, tends to form molecules consisting of four chains of the excess type only. These are termed **homotetramers**, in contrast to the **heterotetramers** normally formed by α and β chains. In α-thalassemia, the α-globin chains are deficient, so the β chains (or γ chains in the fetus) are found in excess. They form homotetramers that have a greatly reduced oxygen-binding capacity, producing hypoxemia. In β-thalassemia, the excess α chains form homotetramers that precipitate and damage the cell membranes of red blood cell precursors (i.e., the cells that form erythrocytes). This leads to premature erythrocyte destruction and anemia.

Most cases of α-thalassemia are caused by deletions of the α-globin genes. The loss of one or two of these genes has no clinical effect. The loss or abnormality of three of the α genes produces moderately severe anemia and splenomegaly (HbH disease). Loss of all four α genes, a condition seen primarily among Southeast Asians, produces hypoxemia in the fetus and *hydrops fetalis* (a condition in which there is a massive buildup of fluid). Severe hydrops fetalis often causes stillbirth or neonatal death.

> The α-thalassemia conditions are usually caused by deletions of α-globin genes. The loss of three of these genes leads to moderately severe anemia, and the loss of all four is fatal.

Persons with a β-globin mutation in one copy of chromosome 11 (heterozygotes) are said to have β-thalassemia minor, a condition that involves little or no anemia and does not ordinarily require clinical management. Those in whom both copies of the chromosome carry a β-globin mutation develop either β-thalassemia major (also called Cooley's anemia) or a less-serious condition, β-thalassemia intermedia. β-Globin may be completely absent ($\beta^0$-thalassemia), or it may be reduced to about 10% to 30% of the normal amount ($\beta^+$-thalassemia). Typically, $\beta^0$-thalassemia produces a more severe disease phenotype, but because disease features are caused by an excess of α-globin chains, patients with $\beta^0$-thalassemia are less severely affected when they also have α-globin mutations that reduce the quantity of α-globin chains.

β-Globin is not produced until after birth, so the effects of β-thalassemia major are not seen clinically until the age of 2 to 6 months. These patients develop severe anemia. If the condition is left untreated, substantial growth retardation can occur. The anemia causes bone marrow expansion, which in turn produces skeletal changes, including a protuberant upper jaw and cheekbones and thinning of the long bones (making them susceptible to fracture). Splenomegaly (Fig. 3-9) and infections are common, and patients with untreated β-thalassemia major often die during the first decade of life. β-Thalassemia can vary considerably in severity, depending on the precise nature of the responsible mutation.

In contrast to α-thalassemia, gene deletions are relatively rare in β-thalassemia. Instead, most cases are caused by single-base mutations. Nonsense mutations, which result in premature termination of translation of the β-globin chain, usually produce $\beta^0$-thalassemia. Frameshift mutations also typically produce the $\beta^0$ form. In addition to mutations in the β-globin gene itself, alterations in regulatory sequences are often seen. β-Globin transcription is regulated by a promoter, two enhancers, and an upstream region known as the locus control region (LCR) (see Fig. 3-7). Mutations in these regulatory regions usually result in reduced synthesis of mRNA and a reduction, but not complete absence, of β-globin ($\beta^+$-thalassemia). Several types of splice-site mutations have also been observed. If a point mutation occurs at a donor or acceptor site, normal splicing is destroyed completely, producing $\beta^0$-thalassemia. Mutations in the surrounding consensus sequences usually produce $\beta^+$-thalassemia. Mutations also occur in the cryptic splice sites found in introns or exons of the β-globin gene, causing these sites to be available to the splicing mechanism. These additional splice sites then compete with the normal splice sites, producing some normal and some abnormal β-globin chains. The result is usually $\beta^+$-thalassemia.

> Many different types of mutations can produce β-thalassemia conditions. Nonsense, frameshift, and splice-site donor and acceptor mutations tend to produce more-severe disease. Regulatory mutations and those involving splice-site consensus sequences and cryptic splice sites tend to produce less-severe disease.

More than 300 different β-globin mutations have been reported. Consequently, most patients with β-thalassemia are not homozygotes in the strict sense: they usually have a *different* β-globin mutation on each copy of chromosome 11 and are termed **compound heterozygotes** (Fig. 3-10). Even though the mutations differ, each of the two β-globin genes is altered, producing a disease state. It is common to apply the term *homozygote* loosely to compound heterozygotes.

Patients with sickle cell disease or β-thalassemia major are sometimes treated with blood transfusions and with chelating agents that remove excess iron introduced by the transfusions. Prophylactic administration of antibiotics and antipneumococcal vaccine help to prevent bacterial infections in patients with sickle cell disease, and analgesics are administered for pain relief during sickling crises. Bone marrow transplantation, which provides donor stem cells that produce genetically normal erythrocytes, has been performed on patients with severe β-thalassemia or sickle cell disease. However, it is often impossible to find a suitably matched donor, and the mortality rate from this procedure is still fairly high (approximately 5% to 30%, depending on the severity of disease and the age of the patient). A lack of normal adult β-globin can be compensated for by reactivating the genes that encode fetal β-globin (the γ-globin genes, discussed previously). Agents such as hydroxyurea and butyrate can reactivate these genes and are being investigated. Also, the hemoglobin disorders are possible candidates for gene therapy (see Chapter 13).

### Causes of Mutation

A large number of agents are known to cause **induced mutations**. These mutations, which are attributed to known environmental causes, can be contrasted with **spontaneous mutations**, which arise naturally during the process of DNA replication. Agents that cause induced mutations are

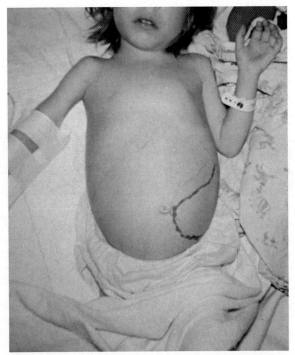

**FIGURE 3-9**
A child with β-thalassemia major who has severe splenomegaly.

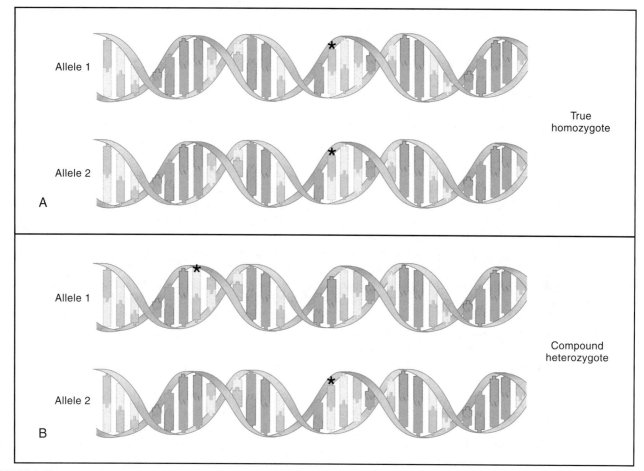

**FIGURE 3-10**
A, True homozygotes have two alleles that are identical in DNA sequence. Here, the homozygote has two copies of a single-base mutation, shown by the *asterisk* in the same position in the DNA sequence. Both mutations (alleles 1 and 2) have a loss-of-function effect, giving rise to a recessive disease.
B, The same effect is seen in a compound heterozygote, who has two different mutations *(asterisks)* in two different locations in the gene's DNA sequence. Each allele has a loss-of-function effect, again causing a recessive disease.

known collectively as **mutagens**. Animal studies have shown that **radiation** is an important class of mutagen (Clinical Commentary 3-1). **Ionizing radiation**, such as that produced by x-rays and nuclear fallout, can eject electrons from atoms, forming electrically charged ions. When these ions are situated within or near the DNA molecule, they can promote chemical reactions that change DNA bases. Ionizing radiation can also break the bonds of double-stranded DNA. This form of radiation can reach all cells of the body, including the germline cells.

## CLINICAL COMMENTARY   3-1
### *The Effects of Radiation on Mutation Rates*

Because mutation is a rare event, occurring less than once per 10,000 genes per generation, it is difficult to measure directly in humans. The relationship between radiation exposure and mutation is similarly difficult to assess. For a person living in a developed country, a typical lifetime exposure to ionizing radiation is about 6 to 7 rem.* About one third to one half of this amount is thought to originate from medical and dental x-ray procedures.

Unfortunate situations have arisen in which specific human populations have received much larger radiation doses. The most thoroughly studied such population consists of the survivors of the atomic bomb blasts that occurred in Hiroshima and Nagasaki, Japan, at the close of World War II. Many of those who were exposed to high doses of radiation died from radiation sickness. Others survived, and many of the survivors produced offspring.

*A rem is a standard unit for measuring radiation exposure. It is roughly equal to 0.01 joule of absorbed energy per kilogram of tissue.*

To study the effects of radiation exposure in this population, a large team of Japanese and American scientists conducted medical and genetic investigations of some of the survivors. A significant number developed cancers and chromosome abnormalities in their somatic cells, probably as a consequence of radiation exposure. To assess the effects of radiation exposure on the subjects' germlines, the scientists compared the offspring of those who suffered substantial radiation exposure with the offspring of those who did not. Although it is difficult to establish radiation doses with precision, there is no doubt that, in general, those who were situated closer to the blasts suffered much higher exposure levels. It is estimated that the exposed group received roughly 30 to 60 rem of radiation, many times the average lifetime radiation exposure.

In a series of more than 76,000 offspring of these survivors, researchers assessed a large number of factors, including stillbirths, chromosome abnormalities, birth defects, cancer before 20 years of age, death before 26 years of age, and various measures of growth and development (e.g., intelligence quotient). There were no statistically significant differences between the offspring of persons who were exposed to radiation and the offspring of those who were not exposed. In addition, direct genetic studies of mutations have been carried out using minisatellite polymorphisms and protein electrophoresis, a technique that detects mutations that lead to amino acid changes (discussed elsewhere in this chapter). Parents and offspring were compared to determine whether germline mutations had occurred at various loci. The numbers of mutations detected in the exposed and unexposed groups were statistically equivalent.

More recently, studies of those who were exposed to radiation from the Chernobyl nuclear power plant accident have demonstrated a significant increase in thyroid cancers among children exposed to radiation. This reflects the effects of somatic mutations. The evidence for increased frequencies of germline mutations in protein-coding DNA, however, remains unclear. A number of other studies of the effects of radiation on humans have been reported, including investigations of those who live near nuclear power plants. The radiation doses received by these persons are substantially smaller than those of the populations discussed previously, and the results of these studies are equivocal.

It is remarkable that even though there was substantial evidence for radiation effects on somatic cells in the Hiroshima and Nagasaki studies, no detectable effect could be seen for germline cells. What could account for this? Because large doses of radiation are lethal, many of those who would have been most strongly affected would not be included in these studies. Furthermore, because germline mutation rates are very small, even relatively large samples of radiation-exposed persons may be insufficient to detect increases in mutation rates. It is also possible that DNA repair compensated for some radiation-induced germline damage.

These results argue that radiation exposure, which is clearly associated with somatic mutations, should not be taken lightly. Above-ground nuclear testing in the American Southwest has produced increased rates of leukemia and thyroid cancer in a segment of the population. Radon, a radioactive gas that is produced by the decay of naturally occurring uranium, can be found at dangerously high levels in some homes and poses a risk for lung cancer. Any unnecessary exposure to radiation, particularly to the gonads or to developing fetuses, should be avoided.

---

**Nonionizing radiation** does not form charged ions but can move electrons from inner to outer orbits within an atom. The atom becomes chemically unstable. Ultraviolet (UV) radiation, which occurs naturally in sunlight, is an example of nonionizing radiation. UV radiation causes the formation of covalent bonds between adjacent pyrimidine bases (i.e., cytosine or thymine). These **pyrimidine dimers** (a dimer is a molecule having two subunits) are unable to pair properly with purines during DNA replication; this results in a base-pair substitution (Fig. 3-11). Because UV radiation is absorbed by the skin, it does not reach the germline but can cause skin cancer (Clinical Commentary 3-2).

## CLINICAL COMMENTARY 3-2
### Xeroderma Pigmentosum: A Disease of Faulty DNA Repair

An inevitable consequence of exposure to UV radiation is the formation of potentially dangerous pyrimidine dimers in the DNA of skin cells. Fortunately, the highly efficient nucleotide excision repair (NER) system removes these dimers in normal persons. Among those affected with the rare autosomal recessive disease xeroderma pigmentosum (XP), this system does not work properly, and the resulting DNA replication errors lead to base pair substitutions in skin cells. XP varies substantially in severity, but early symptoms are usually seen in the first 1 to 2 years of life. Patients develop dry, scaly skin (xeroderma) along with extensive freckling and abnormal skin pigmentation (pigmentosum). Skin tumors, which can be numerous, typically appear by 10 years of age. It is estimated that the risk of skin tumors in persons with XP is elevated approximately 1000-fold. These cancers are concentrated primarily in sun-exposed parts of the body. Patients are advised to avoid sources of UV light (e.g., sunlight), and cancerous growths are removed surgically. Neurological abnormalities are seen in about 30% of persons with XP. Severe, potentially lethal malignancies can occur before 20 years of age.

The NER system is encoded by at least 28 different genes, and inherited mutations in any of seven of these genes can give rise to XP. These genes encode helicases that unwind the double-stranded DNA helix, an endonuclease that cuts the DNA at the site of the dimer, an exonuclease that removes the dimer and nearby nucleotides, a polymerase that fills the gap with DNA bases (using the complementary DNA strand as a template), and a ligase that rejoins the corrected portion of DNA to the original strand.

It should be emphasized that the expression of XP requires inherited germline mutations of NER genes as well as subsequent uncorrected somatic mutations of genes in skin cells. Some of these somatic mutations can affect genes that promote cancer (see Chapter 11), resulting in tumor formation. The skin-cell mutations themselves are somatic and thus are not transmitted to future generations.

NER is but one type of DNA repair. The table below provides examples of a number of other diseases that result from defects in various types of DNA repair mechanisms.

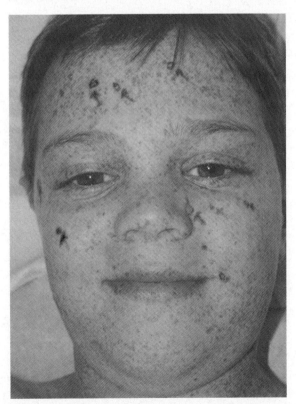

Xeroderma pigmentosum. This patient's skin has multiple hyperpigmented lesions, and skin tumors on the forehead have been marked for excision.

Continued

### Examples of Diseases that Are Caused by a Defect in DNA Repair

| Disease | Features | Type of Repair Defect |
|---|---|---|
| Xeroderma pigmentosum | Skin tumors, photosensitivity, cataracts, neurological abnormalities | Nucleotide excision repair defects, including mutations in helicase and endonuclease genes |
| Cockayne syndrome | Reduced stature, skeletal abnormalities, optic atrophy, deafness, photosensitivity, mental retardation | Defective repair of UV-induced damage in transcriptionally active DNA; considerable etiological and symptomatic overlap with xeroderma pigmentosum and trichothiodystrophy |
| Fanconi anemia | Anemia; leukemia susceptibility; limb, kidney, and heart malformations; chromosome instability | As many as eight different genes may be involved, but their exact role in DNA repair is not yet known |
| Bloom syndrome | Growth deficiency, immunodeficiency, chromosome instability, increased cancer incidence | Mutations in the reqQ helicase family |
| Werner syndrome | Cataracts, osteoporosis, atherosclerosis, loss of skin elasticity, short stature, diabetes, increased cancer incidence; sometimes described as "premature aging" | Mutations in the reqQ helicase family |
| Ataxia-telangiectasia | Cerebellar ataxia, telangiectases,* immune deficiency, increased cancer incidence, chromosome instability | Normal gene product is likely to be involved in halting the cell cycle after DNA damage occurs |
| Hereditary nonpolyposis colorectal cancer | Proximal bowel tumors, increased susceptibility to several other types of cancer | Mutations in any of six DNA mismatch-repair genes |

*Telangiectases are vascular lesions caused by the dilatation of small blood vessels. This typically produces discoloration of the skin.

**FIGURE 3-11**
**A**, Pyrimidine dimers originate when covalent bonds form between adjacent pyrimidine (cytosine or thymine) bases. This deforms the DNA, interfering with normal base pairing. **B**, The defect is repaired by removal and replacement of the dimer and bases on either side of it, with the complementary DNA strand used as a template.

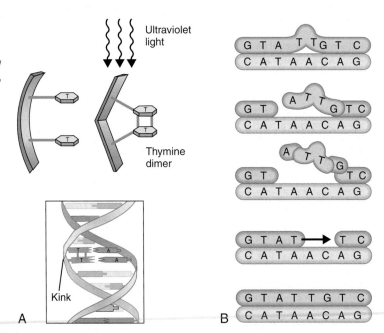

A variety of chemicals can also induce mutations, sometimes because of their chemical similarity to DNA bases. Because of this similarity, these **base analogs**, such as 5-bromouracil, can be substituted for a true DNA base during replication. The analog is not exactly the same as the base it replaces, so it can cause pairing errors during subsequent replications. Other chemical mutagens, such as acridine dyes, can physically insert themselves between existing bases, distorting the DNA helix and causing frameshift mutations. Still other mutagens can directly alter DNA bases, causing replication errors. An example of the latter is nitrous acid, which removes an amino group from cytosine, converting it to uracil. Although uracil is normally found in RNA, it mimics the pairing action of thymine in DNA. Thus, it pairs with adenine instead of guanine, as the original cytosine would have done. The end result is a base-pair substitution.

Hundreds of chemicals are now known to be mutagenic in laboratory animals. Among these are nitrogen mustard, vinyl chloride, alkylating agents, formaldehyde, sodium nitrite, and saccharin. Some of these chemicals are much more potent mutagens than others. Nitrogen mustard, for example, is a powerful mutagen, whereas saccharin is a relatively weak one. Although some mutagenic chemicals are produced by humans, many occur naturally in the environment (e.g., aflatoxin $B_1$, a common contaminant of foods).

> Many substances in the environment are known to be mutagenic, including ionizing and nonionizing radiation and hundreds of different chemicals. These mutagens are capable of causing base substitutions, deletions, and frameshifts. Ionizing radiation can induce double-stranded DNA breaks. Some mutagens occur naturally, and others are generated by humans.

## DNA Repair

Considering that 3 billion DNA base pairs must be replicated in each cell division, and considering the large number of mutagens to which we are exposed, DNA replication is surprisingly accurate. A primary reason for this accuracy is the process of **DNA repair**, which takes place in all normal cells of higher organisms. Several dozen enzymes are involved in the repair of damaged DNA. They collectively recognize an altered base, excise it by cutting the DNA strand, replace it with the correct base (determined from the complementary strand), and reseal the DNA. It is estimated that these repair mechanisms correct at least 99.9% of initial errors.

Because DNA repair is essential for the accurate replication of DNA, defects in DNA repair systems can lead to many types of disease. For example, inherited mutations in genes responsible for **DNA mismatch repair** result in the persistence of cells with replication errors (i.e., mismatches) and can lead

to some types of cancer (see Chapter 11). A diminished capacity to repair **double-stranded DNA breaks** can lead to ovarian and/or breast cancer. **Nucleotide excision repair** is necessary for the removal of larger changes in the DNA helix (e.g., pyrimidine dimers); defects in excision repair lead to a number of diseases, of which xeroderma pigmentosum is an important example (see Clinical Commentary 3-2).

> DNA repair helps to ensure the accuracy of the DNA sequence by correcting replication errors (mismatches), repairing double-stranded DNA breaks, and excising damaged nucleotides.

## Mutation Rates

How often do spontaneous mutations occur? At the nucleotide level, the mutation rate is estimated to be about $10^{-9}$ per base pair per cell division (this figure represents mutations that have escaped the process of DNA repair). At the level of the gene, the mutation rate is quite variable, ranging from $10^{-4}$ to $10^{-7}$ per locus per cell division. There are at least two reasons for this large range of variation: the size of the gene and the susceptibility of certain nucleotide sequences.

First, genes vary tremendously in size. The somatostatin gene, for example, is quite small, containing 1480 bp. In contrast, the gene responsible for Duchenne muscular dystrophy (*DMD*) spans more than 2 million bp. As might be expected, larger genes present larger targets for mutation and usually experience mutation more often than do smaller genes. The *DMD* gene, as well as the genes responsible for hemophilia A and type 1 neurofibromatosis, are all very large and have high mutation rates.

Second, it is well established that certain nucleotide sequences are especially susceptible to mutation. These are termed **mutation hot spots**. The best-known example is the two-base (**dinucleotide**) sequence CG. In mammals, about 80% of CG dinucleotides are **methylated**: A methyl group is attached to the cytosine base. A methylated cytosine, 5-methylcytosine, easily loses an amino group, converting it to thymine. The end result is a mutation from cytosine to thymine (Fig. 3-12). Surveys of mutations in human genetic diseases have shown that the mutation rate at CG dinucleotides is about 12 times higher than at other

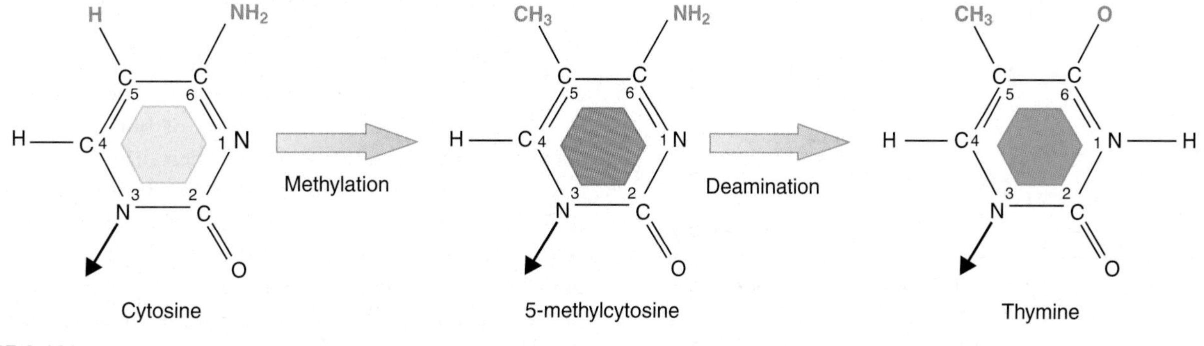

**FIGURE 3-12**
Cytosine methylation. The addition of a methyl group ($CH_3$) to a cytosine base forms 5-methylcytosine. The subsequent loss of an amino group (deamination) forms thymine. The result is a cytosine → thymine substitution.

dinucleotide sequences. Mutation hot spots, in the form of CG dinucleotides, have been identified in a number of important human disease genes, including the procollagen genes responsible for osteogenesis imperfecta (see Chapter 2). Other disease examples are discussed in Chapters 4 and 5.

Mutation rates also vary considerably with the age of the parent. Some chromosome abnormalities increase dramatically with maternal age (see Chapter 6). In addition, single-gene mutations can increase with paternal age. This increase is seen in several single-gene disorders, including Marfan syndrome and achondroplasia. As Figure 3-13 shows, the risk of producing a child with Marfan syndrome is approximately five times higher for a father older than 40 years than for a father in his 20s. This paternal age effect is usually attributed to the fact that the stem cells giving rise to sperm cells continue to divide throughout life, allowing a progressive buildup of DNA replication errors.

> Large genes, because of their size, are generally more likely to experience mutations than are small genes. Mutation hot spots, particularly methylated CG dinucleotides, experience elevated mutation rates. For some single-gene disorders, there is a substantial increase in mutation risk with advanced paternal age.

## DETECTION AND MEASUREMENT OF GENETIC VARIATION

For centuries, humans have been intrigued by the differences that can be seen among individuals. Attention was long focused on observable differences such as skin color or body shape and size. Only in the 20th century did it become possible to examine variation in genes, the consequence of mutations accumulated through time. The evaluation and measurement of this variation in populations and families are important for mapping genes to specific locations on chromosomes, a key step in

determining gene function (see Chapter 8). The evaluation of genetic variation also provides the basis for much of genetic diagnosis, and it is highly useful in forensics. In this section, several key approaches to detecting genetic variation in humans are discussed in historical sequence.

### Blood Groups

Several dozen **blood group** systems have been defined on the basis of antigens located on the surfaces of erythrocytes. Some are involved in determining whether a person can receive a blood transfusion from a specific donor. Because individuals differ extensively in terms of blood groups, these systems provided an important early means of assessing genetic variation.

Each of the blood group systems is determined by a different gene or set of genes. The various antigens that can be expressed within a system are the result of different DNA sequences in these genes. Two blood-group systems that have special medical significance—the ABO and Rh systems—are discussed here. The ABO and Rh systems are both of key importance in determining the compatibility of blood transfusions and tissue grafts. Some combinations of these systems can produce maternal-fetal incompatibility, sometimes with serious results for the fetus. These issues are discussed in detail in Chapter 9.

### The ABO Blood Group

Human blood transfusions were carried out as early as 1818, but they were often unsuccessful. After transfusion, some recipients suffered a massive, sometimes fatal, hemolytic reaction. In 1900 Karl Landsteiner discovered that this reaction was caused by the ABO antigens located on erythrocyte surfaces. The ABO system consists of two major antigens, labeled A and B. A person can have one of four major blood types: People with blood type A carry the A antigen on their erythrocytes, those with type B carry the B antigen, those with type AB carry both A and B, and those with type O carry neither antigen. Each individual has antibodies that react against any antigens that are not found on their own red blood cell surfaces. For example, a person with type A blood has anti-B antibodies, and transfusing type B blood into this person provokes a severe antibody reaction. It is straightforward to determine ABO blood type in the laboratory by mixing a small sample of a person's blood with solutions containing different antibodies and observing which combinations cause the clumping that is characteristic of an antibody–antigen interaction.

The ABO system, which is encoded by a single gene on chromosome 9, consists of three primary alleles, labeled $I^A$, $I^B$, and $I^O$. (There are also subtypes of both the $I^A$ and $I^B$ alleles, but they are not addressed here.) Persons with the $I^A$ allele have the A antigen on their erythrocyte surfaces (blood type A), and those with $I^B$ have the B antigen on their cell surfaces (blood type B). Those with both alleles express both antigens (blood type AB), and those with two copies of the $I^O$ allele have neither antigen (type O blood). Because the $I^O$ allele produces no antigen, persons who are $I^A I^O$ or $I^B I^O$ heterozygotes have blood types A and B, respectively (Table 3-2).

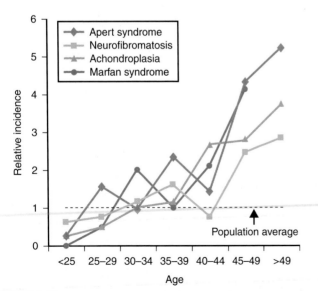

**FIGURE 3-13**
Paternal age effect. For some single-gene disorders, the risk of producing a child with the condition (*y-axis*) increases with the father's age (*x-axis*).

**TABLE 3-2**
## Relationship between ABO Genotype and Blood Type

| Genotype | Blood Type | Antibodies Present |
|---|---|---|
| $I^A I^A$ | A | Anti-B |
| $I^A I^O$ | A | Anti-B |
| $I^B I^B$ | B | Anti-A |
| $I^B I^O$ | B | Anti-A |
| $I^A I^B$ | AB | None |
| $I^O I^O$ | O | Anti-A and anti-B |

Because populations vary substantially in terms of the frequency with which the ABO alleles occur, the ABO locus was the first blood group system to be used extensively in studies of genetic variation among individuals and populations. For example, early studies showed that the A antigen is relatively common in western European populations, and the B antigen is especially common among Asians. Neither antigen is common among native South American populations, the great majority of whom have blood type O.

### The Rh System

Like the ABO system, the Rh system is defined on the basis of antigens that are present on erythrocyte surfaces. This system is named after the rhesus monkey, the experimental animal in which it was first isolated by Landsteiner in the late 1930s. It is typed in the laboratory by a procedure similar to the one described for the ABO system. The Rh system varies considerably among individuals and populations and thus has been another highly useful tool for assessing genetic variation. The molecular basis of variation in both the ABO and the Rh systems has been elucidated (for further details, see the suggested readings at the end of this chapter).

> The blood groups, of which the ABO and Rh systems are examples, have provided an important means of studying human genetic variation. Blood group variation is the result of antigens that occur on the surface of erythrocytes.

### Protein Electrophoresis

Although the blood group systems have been a useful means of measuring genetic variation, their number is quite limited. **Protein electrophoresis**, developed first in the 1930s and applied widely to humans in the 1950s and 1960s, increased the number of detectable polymorphic systems considerably. This technique makes use of the fact that a single amino acid difference in a protein (the result of a mutation in the corresponding DNA sequence) can cause a slight difference in the electrical charge of the protein.

An example is the common sickle cell disease mutation discussed earlier. The replacement of glutamic acid with valine in the β-globin chain produces a difference in electrical charge because glutamic acid has two carboxyl groups, whereas valine

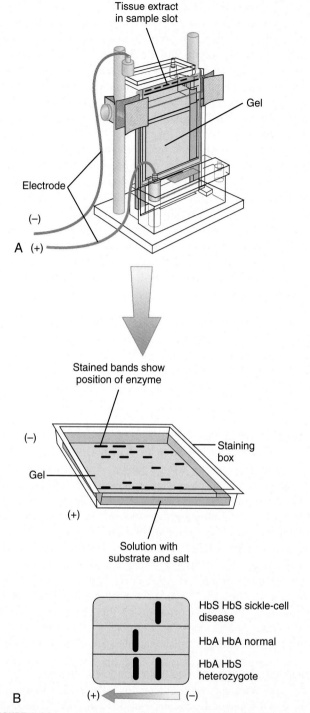

**FIGURE 3-14**
The process of protein electrophoresis. **A,** A tissue sample is loaded in the slot at the top of the gel, and an electrical current is run though the gel. After staining, distinct bands, representing molecules with different electrical charges and therefore different amino acid sequences, are visible. **B,** HbA homozygotes show a single band closer to the positive pole, whereas HbS homozygotes show a single band closer to the negative pole. Heterozygotes, having both alleles, show two bands.

has only one carboxyl group. Electrophoresis can be used to determine whether a person has normal hemoglobin (HbA) or the mutation that causes sickle cell disease (HbS). The hemoglobin is placed in an electrically charged gel composed of starch, agarose, or polyacrylamide (Fig. 3-14A). The slight difference in charge resulting from the amino acid difference causes the HbA and HbS forms to migrate at different rates through the gel. The protein molecules are allowed to migrate for several hours and are then stained with chemical solutions so that their positions can be seen (Fig. 3-14B). From the resulting pattern it can be determined whether the person is an HbA homozygote, an HbS homozygote, or a heterozygote having HbA on one chromosome copy and HbS on the other.

Protein electrophoresis has been used to detect amino acid variation in hundreds of human proteins. However, silent substitutions, which do not alter amino acids, cannot be detected by this approach. In addition, some amino acid substitutions do not alter the electrical charge of the protein molecule. For these reasons, it is estimated that protein electrophoresis detects only about one third of the mutations that occur in coding DNA. In addition, single-base substitutions in noncoding DNA are not usually detected by protein electrophoresis.

> Protein electrophoresis detects variations in genes that encode certain serum proteins. These variations are observable because proteins with slight differences in their amino acid sequence migrate at different rates through electrically charged gels.

### Detecting Variation at the DNA Level

It is estimated that variation in human DNA occurs at an average of 1 in every 300 to 500 bp. Thus, approximately 10 million polymorphisms exist among the 3 billion base pairs that compose the human genome. Because there are only 100 or so blood group and protein electrophoretic polymorphisms, these approaches have detected only a tiny fraction of human DNA variation, yet the assessment of this variation is critical to gene mapping and genetic diagnosis (see Chapters 8 and 13). Fortunately, molecular techniques developed since the 1980s have enabled the detection of millions of new polymorphisms at the DNA level. These techniques, which have revolutionized both the practice and the potential of medical genetics, are discussed next.

### Southern Blotting and Restriction Fragment Analysis

An early approach to the detection of genetic variation at the DNA level took advantage of the existence of bacterial enzymes known as **restriction endonucleases**, or **restriction**

**enzymes**. These enzymes cleave human DNA at specific sequences, termed **restriction sites**. For example, the intestinal bacterium *Escherichia coli* produces a restriction enzyme, called *Eco*RI, that recognizes the DNA sequence GAATTC. Each time this sequence is encountered, the enzyme cleaves the sequence between the G and the A (Fig. 3-15). A **restriction digest** of human DNA using this enzyme will produce more than 1 million DNA fragments (**restriction fragments**). These fragments are then subjected to gel electrophoresis, in which the smaller ones migrate more quickly through the gel than do the larger ones (Fig. 3-16). The DNA is denatured (i.e., converted from a double-stranded to a single-stranded form) by exposing it to alkaline chemical solutions. To fix their positions permanently, the DNA fragments are transferred from the gel to a solid membrane, such as nitrocellulose (this is a **Southern transfer**, after the man who invented the process in the mid-1970s). At this point, the solid membrane, often called a **Southern blot**, contains many thousands of fragments arrayed according to their size. Because of their large number, the fragments are indistinguishable from one another.

To visualize only the fragments corresponding to a specific region of DNA, a **probe**, consisting of a small piece of single-stranded human DNA (a few kilobases [kb] in length), is constructed using recombinant DNA techniques (Box 3-1). The probe is labeled, often with a radioactive isotope, and then exposed to the Southern blot. The probe undergoes complementary base pairing only with the corresponding complementary single-stranded DNA fragments on the blot identifying one or a few fragments from a specific portion of the DNA. To visualize the position on the blot at which the probe hybridizes, the blot is exposed to x-ray film, which darkens at the probe's position due to the emission of radioactive particles from the labeled probe. These darkened positions are usually referred to as **bands**, and the film is termed an **autoradiogram** (Fig. 3-17).

Southern blotting can be used in several ways. For example, it can detect insertions or deletions in DNA sequences, which cause specific fragments to become larger or smaller. If a disease-causing mutation alters a specific restriction site, as in the case of sickle cell disease (Fig. 3-18), this technique can be used as a cheap and efficient diagnostic tool. Because most disease-causing mutations do not affect restriction sites, this approach is somewhat limited, and other, newer techniques can be used. Finally, Southern blotting was instrumental in analyzing **restriction fragment length polymorphisms** (**RFLPs**), which are found throughout the human genome

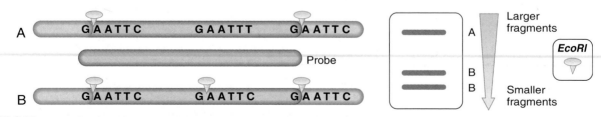

**FIGURE 3-15**
Cleavage of DNA by the *Eco*RI restriction enzyme. In **B**, the enzyme cleaves the three GAATTC recognition sequences, producing two smaller fragments. In **A**, the middle sequence is GAATTT instead of GAATTC, so it cannot be cleaved by the enzyme. The result is a single, longer fragment.

Restriction enzymes can cut DNA into fragments, which are sorted according to their length by electrophoresis, transferred to a solid membrane (Southern blotting), and visualized through the use of labeled probes. This process can detect deletions or duplications of DNA, as well as polymorphisms in restriction sites (RFLPs).

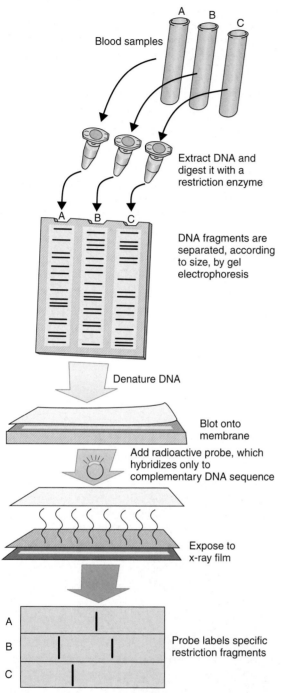

## Tandem Repeat Polymorphisms

The approach just described can detect polymorphisms that reflect the presence or absence of a restriction site. These polymorphisms have only two possible alleles, placing a limit on the amount of genetic diversity that can be seen. More diversity could be observed if a polymorphic system had many alleles, rather than just two. One such system exploits the microsatellites and minisatellites that exist throughout the genome. As discussed in Chapter 2, these are regions in which the same DNA sequence is repeated over and over, in tandem (Fig. 3-19). Microsatellites are typically composed of units that are only 2 to 5 bp long, whereas minisatellites contain longer repeat units. The genetic variation measured is the number of repeats in a given region, which varies substantially from individual to individual: a specific region could have as few as two or three repeats or as many as 20 or more. These polymorphisms can therefore reveal a high degree of genetic variation. Minisatellite polymorphisms are termed **variable number of tandem repeats** (VNTR), and microsatellite polymorphisms are termed **short tandem repeat polymorphisms** (STRPs). The latter are especially easy to assay, and thousands of them are distributed throughout the human genome. These properties make them useful for mapping genes by the process of linkage analysis, discussed in Chapter 8. Both types of polymorphisms are useful in forensic applications, such as paternity testing and the identification of criminal suspects (Box 3-2).

VNTRs are a type of polymorphism that results from varying numbers of minisatellite repeats in a specific DNA region. STRPs are a similar type of polymorphism that results from varying numbers of smaller, microsatellite repeats. Because VNTRs and STRPs can have many different alleles, they are especially useful in medical genetics and forensics.

## Single Nucleotide Polymorphisms

The most numerous type of polymorphism in the human genome consists of variants at single nucleotide positions on a chromosome, or **single nucleotide polymorphisms** (SNPs). It is estimated that SNPs account for approximately 3 million differences, on average, between individual pairs of humans. Because the human genome consists of 3 billion base pairs of DNA, this means that individual humans differ at roughly 1 in 1000 single bases. RFLPs, which are usually caused by single-base differences that occur only at restriction sites, are a subset of the more general set of SNPs. These polymorphisms, when they occur in functional DNA sequences, can cause inherited diseases, although most are

**FIGURE 3-16**
Restriction enzyme digestion and Southern blotting. DNA is extracted from blood samples from subjects A, B, and C. The DNA is digested by a restriction enzyme and then loaded on a gel. Electrophoresis separates the DNA fragments according to their size. The DNA is denatured and transferred to a solid membrane (Southern blot), where it is hybridized with a radioactive probe. Exposure to x-ray film (autoradiography) reveals specific DNA fragments (bands) of different sizes in individuals A, B, and C.

as a result of normal DNA sequence variation. These sequence variants were used to localize many important disease-causing genes, including those responsible for cystic fibrosis, Huntington disease, and type 1 neurofibromatosis (see Chapter 8).

BOX 3-1
# Genetic Engineering, Recombinant DNA, and Cloning

In the last 2 decades, most of the lay public has acquired at least a passing familiarity with the terms "recombinant DNA," "cloning," and "genetic engineering." Indeed, these techniques lie at the heart of what is often called the "new genetics."

**Genetic engineering** refers to the laboratory alteration of genes. An alteration that is of special importance in medical genetics is the creation of **clones**. Briefly, a clone is an identical copy of a DNA sequence. The following description outlines one approach to the cloning of human genes.

Our goal is to insert a human DNA sequence into a rapidly reproducing organism so that copies (clones) of the DNA can be made quickly. One system commonly used for this purpose is the **plasmid**, which is a small, circular, self-replicating piece of DNA that resides in many bacteria. Plasmids can be removed from bacteria or inserted into them without seriously disrupting bacterial growth or reproduction.

To insert human DNA into the plasmid, we need a way to cut DNA into pieces so that it can be manipulated. Restriction enzymes, discussed earlier in the text, perform this function efficiently. The DNA sequence recognized by the restriction enzyme *Eco*RI, GAATTC, has the convenient property that its complementary sequence, CTTAAG, is the same sequence, except backwards. Such sequences are called **palindromes**. When plasmid or human DNA is cleaved with *Eco*RI, the resulting fragments have sticky ends. If human DNA and plasmid DNA are both cut with this enzyme, both types of DNA fragments contain exposed ends that can undergo complementary base pairing with each other. Then, when the human and plasmid DNA are mixed together, they recombine (hence the term **recombinant DNA**). The resulting plasmids contain human DNA **inserts**. The plasmids are inserted back into bacteria, where they reproduce rapidly through natural cell division. The human DNA sequence, which is reproduced along with the other plasmid DNA, is thus cloned.

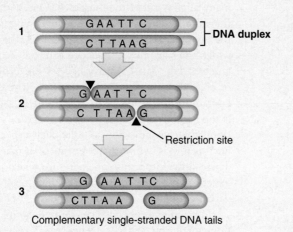

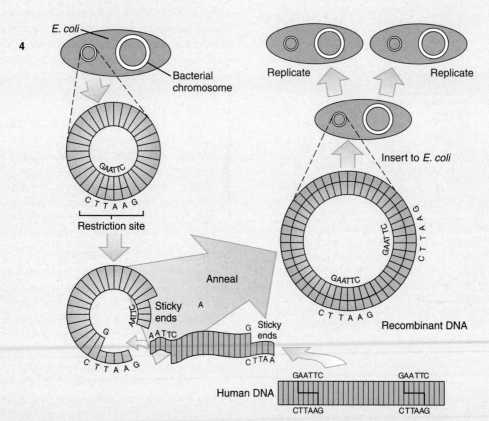

Recombinant DNA technology. Human and circular plasmid DNA are both cleaved by a restriction enzyme, producing sticky ends (*1-3*). This allows the human DNA to anneal and recombine with the plasmid DNA. Inserted into the plasmid DNA, the human DNA is now replicated when the plasmid is inserted into the *Escherichia coli* bacterium (*4*).

The plasmid is referred to as a **vector**. Several other types of vectors may also be used as cloning vehicles, including **bacteriophages** (viruses that infect bacteria), **cosmids** (phage–plasmid hybrids capable of carrying relatively large DNA inserts), **yeast artificial chromosomes** (YACs; vectors that are inserted into yeast cells and that behave much like ordinary yeast chromosomes), **bacterial artificial chromosomes** (BACs), and **human artificial chromosomes** (see Chapters 8 and 13). Although plasmids and bacteriophages can accommodate only relatively small inserts (about 10 and 20 kb, respectively), cosmids can carry inserts of approximately 50 kb, and YACs can carry inserts up to 1000 kb in length.

Cloning can be used to create the thousands of copies of human DNA needed for Southern blotting and other experimental applica-tions. In addition, this approach is now used to produce genetically engineered therapeutic products, such as insulin, interferon, human growth hormone, clotting factor VIII (used in the treatment of hemophilia A, a coagulation disorder), and tissue plasminogen activator (a blood clot–dissolving protein that helps to prevent heart attacks and strokes). When these genes are cloned into bacteria or other organisms, the organism produces the human gene product along with its own gene products. In the past, these products were obtained from donor blood or from other animals. The processes of obtaining and purifying them were slow and costly, and the resulting products sometimes contained contaminants. Genetically engineered gene products are rapidly becoming a cheaper, purer, and more efficient alternative.

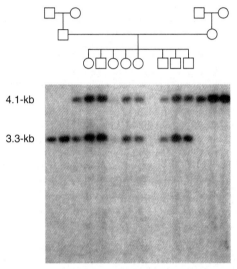

**FIGURE 3-17**
An autoradiogram, showing the positions of a 4.1-kb band and a 3.3-kb band. Each lane represents DNA from a subject in the family whose pedigree is shown above the autoradiogram.

harmless. Increasingly, they are being detected by microarray methods, which are discussed later in this chapter.

## Copy Number Variants

Another type of DNA variation has been shown to occur with substantial frequency in the human genome. **Copy number variants** (CNVs), which consist of DNA segments longer than 1000 base pairs, are present in some persons but absent in others. They can also be present in more than one copy in a person. CNVs are estimated to account for at least 4 million base pairs of difference, on average, between individual humans. Thus, they account for slightly more total variation in human DNA sequences than do SNPs. Some CNVs have been shown to be associated with inherited diseases. Figure 3-20 highlights the differences among RFLPs, tandem repeats, SNPs, and CNVs.

> SNPs are the most common type of variation in the human genome. CNVs consist of differences in the number of repeated DNA sequences longer than 1000 bp.

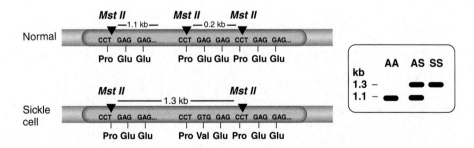

**FIGURE 3-18**
Cleavage of β-globin DNA by the *Mst*II restriction enzyme. Normal individuals have glutamic acid at position 6 of the β-globin polypeptide. Glutamic acid is encoded by the DNA sequence GAG. The sickle cell mutation results in the sequence GTG at this site instead of GAG, causing valine to be substituted for glutamic acid. The restriction enzyme *Mst*II recognizes the DNA sequence CCTNAGG (the *N* signifies that the enzyme will recognize any DNA base, including G, in this position). Thus, *Mst*II recognizes and cleaves the DNA sequence of the normal chromosome at this site as well as at the restriction sites on either side of it. The sickle cell mutation removes an *Mst*II recognition site, producing a longer, 1.3-kb fragment. The normal DNA sequence includes the restriction site (i.e., the sequence CCTGAG instead of CCTGTG), so a shorter, 1.1-kb fragment is produced. Therefore, on the autoradiogram, sickle cell homozygotes have a single 1.3-kb band, normal homozygotes have a single 1.1-kb band, and heterozygotes have both the 1.1-kb and the 1.3-kb bands. Because shorter fragments migrate farther on a gel, the two fragment sizes can easily be distinguished after hybridization of the blot with a probe containing DNA from the β-globin gene. Note that the banding pattern here, based on DNA sequence differences, resembles the banding pattern shown in Figure 3-14, which is based on hemoglobin amino acid sequences detected by protein electrophoresis.

**FIGURE 3-19**
Tandem repeat polymorphisms. Bands of differing length (**A** and **B**) are created by different numbers of tandem repeats in the DNA on the two copies of a chromosome. Following amplification and labeling of the region that contains the polymorphism, different fragment lengths are separated by electrophoresis and visualized on an autoradiogram.

---

BOX 3-2
## DNA Profiles in the Forensic Setting

Because of the large number of polymorphisms observed in the human genome, it is virtually certain that each of us is genetically unique (with the exception of identical twins, whose DNA sequences are nearly always identical). It follows that genetic variation could be used to identify individuals, much as a conventional fingerprint does. Because DNA can be found in any tissue sample, including blood, semen, and hair,* genetic variation has substantial potential in forensic applications (e.g., criminal cases, paternity suits, identification of accident victims). VNTRs and STRPs, with their many alleles, are very useful in establishing a highly specific **DNA profile**.

The principle underlying a DNA profile is quite simple. If we examine enough polymorphisms in a given individual, the probability that any other individual in the population has the same

allele at each examined locus becomes extremely small. DNA left at the scene of a crime in the form of blood or semen, for example, can be typed for a series of VNTRs and/or STRPs. Because of the extreme sensitivity of the PCR approach, even a tiny sample several years old can yield enough DNA for laboratory analysis (although extreme care must be taken to avoid contamination when using PCR with such samples). The detected alleles are then compared with the alleles of a suspect. If the alleles in the two samples match, the suspect is implicated.

A key question is whether another person in the general population might have the same alleles as the suspect. Could the DNA profile then falsely implicate the wrong person? In criminal cases, the probability of obtaining an allele match with a random member of the population is calculated. Because of the high

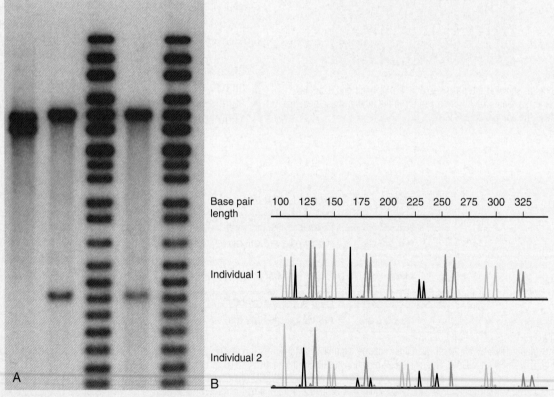

DNA profiles. **A,** An autoradiogram shows that the band pattern of DNA from suspect *A* does not match the DNA taken from the crime scene (*C*), whereas the band pattern from suspect *B* does match. In practice, several such variable number of tandem repeats (VNTRs) or short tandem repeat polymorphisms (STRPs) are assayed to reduce the possibility of a false match. (Courtesy Jay Henry, Criminalistics Laboratory, Department of Public Safety, State of Utah.). **B,** STRPs are now commonly assayed using a capillary gel apparatus. The resulting STRP profile is displayed as an electropherogram, in which the locations of peaks indicate the lengths of each STRP allele.

degree of allelic variation in VNTRs and STRPs, this probability is usually very small. A set of just four VNTR loci typically provides random match probabilities of about 1 in 1 million. The use of 13 STRPs, which is now common practice, yields random match or more probabilities in the neighborhood of 1 in 1 trillion. Provided that a large enough number of loci are used under well-controlled laboratory conditions, and provided that the data are collected and evaluated carefully, DNA profiles can furnish highly useful forensic evidence. DNA profiles are now used in many thousands of criminal court cases each year.

Although we tend to think of such evidence in terms of identifying the guilty party, it should be kept in mind that when a match is not obtained, a suspect may be exonerated. In addition, postconviction DNA testing has resulted in the release of hundreds of persons who were wrongly imprisoned. Thus, DNA profiles can also benefit the innocent.

*Even fingerprints left at a crime scene sometimes contain enough DNA for PCR amplification and DNA profiling.*

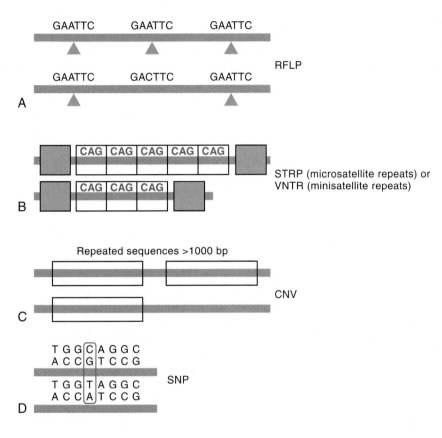

**FIGURE 3-20**
A, Restriction fragment length polymorphisms (RFLPs) result from DNA sequence differences that occur at restriction sites in the human genome. The locations of these sites are identified by hybridizing restriction fragments with cloned probes. B, Tandem repeats consist of short segments of DNA (microsatellites) or somewhat longer segments (minisatellites whose lengths can be 14 to 500 bp) that are repeated over and over, in tandem. C, Copy number variants (CNVs) represent differences in the numbers of larger repeated segments of DNA (>1000 bp to 2 million bp). D, Single nucleotide polymorphisms (SNPs) are single-base variations in the genome.

## DNA Amplification Using the Polymerase Chain Reaction

Because the DNA molecule is tiny, it is not possible to visualize DNA variation (variation in base pairs) directly. All methods of assessing DNA variation involve indirect assessment, as in the use of labeled probes to bind to specific DNA regions in Southern blotting.

Nearly all methods of visualizing DNA variation require indirect labeling of DNA. To observe the labels, multiple copies must be made. For example, bacteria can be used to make thousands of cloned copies of the labeled probes used in Southern blotting. However, this process (see Box 3-1) is time-consuming, often requiring several days or more, and it typically requires a relatively large amount of DNA from the subject (several micrograms). An alternative process, the **polymerase chain reaction** (PCR), was developed in the mid-1980s and has made the detection of genetic variation

at the DNA level much more efficient. Essentially, PCR is an artificial means of replicating a short, specific **DNA sequence** (several kb or less) very quickly, so that millions of copies of the sequence are made.

The PCR process, summarized in Figure 3-21, requires four components:

- Two **primers**, each consisting of 15 to 20 bases of DNA. These small DNA sequences are termed **oligonucleotides** (*oligo* means "a few"). The primers correspond to the DNA sequences immediately adjacent to the sequence of interest (such as a sequence that contains a tandem repeat polymorphism or a mutation that causes disease). The oligonucleotide primers are synthesized using a laboratory instrument.
- DNA polymerase. A thermally stable form of this enzyme, initially derived from the bacterium *Thermus aquaticus*, performs the vital process of DNA replication (here termed **primer extension**).

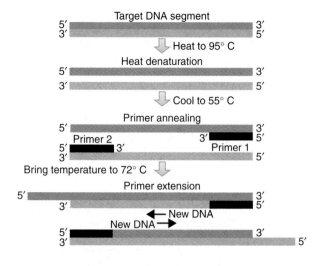

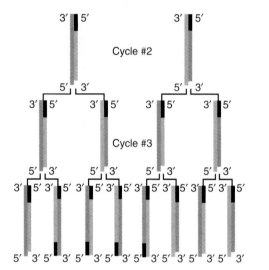

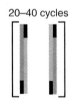

**FIGURE 3-21**

The polymerase chain reaction (PCR) process. Genomic DNA is first heated and denatured to form single strands. In the annealing phase, the DNA is cooled, allowing hybridization with primer sequences that flank the region of interest. Then the reaction is heated to an intermediate primer extension temperature, in which DNA polymerase adds free bases in the 3′ direction along each single strand, starting at the primer. Blunt-ended DNA fragments are formed, and these provide a template for the next cycle of heating and cooling. Repeated cycling produces a large number of DNA fragments bounded on each end by the primer sequence.

- A large number of free DNA nucleotides.
- Genomic DNA from an individual. Because of the extreme sensitivity of PCR, the quantity of this DNA can be very small.

The genomic DNA is first heated to a relatively high temperature (approximately 95° C) so that it denatures and

becomes single-stranded. As the single-stranded DNA is then cooled to a temperature of approximately 35° C to 65° C, it is exposed to large quantities of single-stranded primers, which hybridize, or anneal, to a specific location in the genomic DNA that contains the appropriate complementary bases. The DNA is then heated to an intermediate temperature (70° C to 75° C). In the presence of a large number of free DNA bases, a new DNA strand is synthesized by the DNA polymerase at this temperature, extending from the primer sequence. The newly synthesized DNA consists of a double strand that has the 5′ end of the primer at one end, followed by the bases added through primer extension by DNA polymerase. This double-stranded DNA is heated to a high temperature again, causing it to denature. The heating-and-cooling cycle is then repeated. Now, the newly synthesized DNA serves as the template for further synthesis. As the cooling-and-heating cycles are repeated, the primer-bounded DNA products are amplified geometrically: The number of copies doubles in each cycle (i.e., 2, 4, 8, 16, etc.). This is why the process is termed a *chain reaction*. Typically, the cycles are repeated 20 to 30 times, producing millions of copies of the original DNA. In summary, the PCR process consists of three basic steps: DNA denaturing at high temperature, primer hybridization at a low temperature, and primer extension at an intermediate temperature. The result is a product that consists almost entirely of a specific DNA sequence.

Because each heating-and-cooling cycle requires only a few minutes or less, a single molecule of DNA can be amplified to make millions of copies in only a few hours. Because the procedure is simple and entirely self-contained, inexpensive machines have been developed to automate it completely. Once the DNA is amplified, it can be analyzed in a variety of ways.

PCR has several advantages over older techniques. First, it can be used with extremely small quantities of DNA (usually nanogram amounts, as opposed to the micrograms required for cloning). The amount of DNA in a several-year-old blood stain, a single hair, or even the back of a licked postage stamp is often sufficient for analysis. Second, because it does not require cloning, the procedure is much faster than older techniques. Genetic testing for sickle cell disease, for example, can be done in a single day with PCR. Finally, because PCR can make large quantities of very pure DNA, it is less often necessary to use radioactive probes to detect specific DNA sequences or mutations. Instead, safer, nonradioactive substances, such as biotin, can be used.

PCR does have some disadvantages. First, primer synthesis obviously requires knowledge of the DNA sequence flanking the DNA of interest. If no sequence information is available, other techniques must be used. Second, the extreme sensitivity of PCR makes it susceptible to contamination in the laboratory. A number of precautions are commonly taken to guard against contamination. Finally, because it can be difficult to apply PCR to sequences longer than a few kilobases, it is not typically useful for detecting larger deletions (i.e., it is difficult or impossible to amplify the longer, normal sequence). Southern blotting or other techniques are used instead.

Because PCR is such a powerful and versatile technique, it is now used extensively in genetic disease diagnosis, forensic medicine, and evolutionary genetics. It has supplanted the Southern blotting technique in many applications and is now used to assay RFLPs and VNTRs. PCR is so sensitive that it has been used to analyze DNA from ancient mummies and even from more than a dozen Neanderthal specimens more than 30,000 years old. Analysis of these specimens showed that modern humans are genetically quite distinct from Neanderthals and are thus unlikely to have descended directly from them.

> PCR provides a convenient and efficient means of making millions of copies of a short DNA sequence. Heating-and-cooling cycles are used to denature DNA and then build new copies of a specific, primer-bounded sequence. Because of its speed and ease of use, this technique is now widely used for assessing genetic variation, for diagnosing genetic diseases, and for performing forensic investigations.

## DNA Sequencing

In many genetic studies, a primary goal is to determine the actual array of DNA base pairs that makes up a gene or part of a gene. Such a **DNA sequence** can indicate a great deal about the nature of a specific mutation, the function of a gene, and the gene's degree of similarity to other known genes. We first discuss a technique that has been widely used to determine DNA sequences.

The **dideoxy method** of DNA sequencing, invented by Frederick Sanger, makes use of chain-terminating dideoxynucleotides. These are chemically quite similar to ordinary deoxynucleotides, except that they are missing one hydroxyl group. This prevents the subsequent formation of phosphodiester bonds with free DNA bases. Thus, although dideoxynucleotides can be incorporated into a growing DNA helix, no additional nucleotides can be added once they are included.

Four different dideoxynucleotides are used, each corresponding to one of the four nucleotides (A, C, G, and T). The single-stranded DNA whose sequence we wish to determine is mixed with labeled primers, DNA polymerase, ordinary nucleotides, and one type of dideoxynucleotide (Fig. 3-22). The primer hybridizes to the appropriate complementary position in the single-stranded DNA, and DNA polymerase adds free bases to the growing DNA molecule, as in the PCR process. At any given position, either an ordinary nucleotide or the corresponding dideoxynucleotide may be incorporated into the chain; this is a random process. However, once a dideoxynucleotide is incorporated, the chain is terminated. The procedure thus yields DNA fragments of varying length, each ending with the same dideoxynucleotide.

The DNA fragments can be separated according to length by electrophoresis, as discussed previously. Four different sequencing reactions are run, one for each base. The fragments obtained from each reaction are electrophoresed side by side on the same gel, so that the position of each

fragment can be compared. Because each band corresponds to a DNA chain that terminates with a unique base, the DNA sequence can be read by observing the order of the bands on the gel after autoradiography or other detection methods (on an autoradiogram, a radioactive label attached to the primer indicates the position of the fragment on the film). Several hundred base pairs can usually be sequenced in one reaction series.

> DNA sequencing can be accomplished using the dideoxy method. This method depends on the fact that dideoxynucleotides behave in a fashion similar to ordinary deoxynucleotides, except that once they are incorporated into the DNA chain, they terminate the chain. They thus mark the positions of specific bases.

It should be apparent that this method of sequencing DNA is a relatively slow, laborious, and error-prone process. More recently, strategies for **automated DNA sequencing** using fluorescent, chemiluminescent, or colorimetric detection systems have been developed. The use of fluorochrome-labeled primers or dideoxynucleotides has become the most popular method, partly because it can easily be adapted for rapid automation.

Typically, a DNA template is sequenced using a method similar to the primer extension step in PCR. Each of the four different nucleotides can be labeled with a fluorochrome that emits a distinct spectrum of light. The fluorochrome-labeled reaction products are electrophoresed through a very thin polyacrylamide gel; as they migrate past a window, they are excited by a beam of light from a laser. The emitted light is captured by a digital camera for translation into an electronic signal, and a composite gel image is generated. This gel image is analyzed to produce a graph in which each of the four different nucleotides is depicted by a different-colored peak (Fig. 3-23). Automated sequencers can also be adapted to assay STRPs, single-nucleotide polymorphisms, and other types of polymorphisms.

In another approach to automated sequencing, DNA samples are electrophoresed in thin glass tubes (capillaries) rather than polyacrylamide gels. Because these tubes are very thin, relatively little heat is generated during electrophoresis. As a consequence, **capillary sequencing** is very rapid.

By using computers and advanced automated technology, approaches such as these have greatly increased the potential speed of DNA sequencing. These techniques have permitted the completion of the entire 3-billion-bp human DNA sequence.

> Automated DNA sequencing, using fluorescent labels and laser detection, greatly increases the speed and efficiency of the sequencing process.

## Detection of Mutations at the DNA Level

The detection of mutations or polymorphisms in DNA sequences is often a critical step in understanding how a

**FIGURE 3-22**

DNA sequencing by the dideoxy (Sanger) method. Labeled primer is added to the single-stranded DNA whose sequence is unknown. DNA polymerase adds free bases to the single strand, using complementary base pairing. Four different reactions are carried out, corresponding to the four dideoxynucleotides (ddATP, ddCTP, ddGTP, and ddTTP). Each of these terminates the DNA sequence whenever it is incorporated in place of the normal deoxynucleotide (dATP, dCTP, dGTP, and dTTP, corresponding to the bases A, C, G, and T, respectively). The process results in fragments of varying length, which can be separated by electrophoresis. The position of each fragment is indicated by the emission of radioactive particles from the label, which allows the DNA sequence to be read directly.

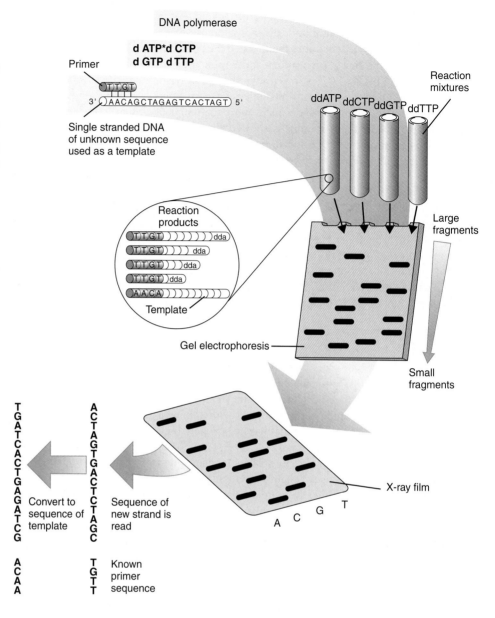

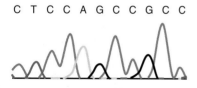

**FIGURE 3-23**

Analyzed data from a single DNA template sequenced on an automated DNA sequencer. Peaks of different colors represent the identity and relative location of different nucleotides in the DNA sequence. For example, the peak on the upper left is blue and identifies the position of a cytosine. The next peak is red, indicating the presence of a thymine. This base-calling continues until the end of the DNA template is reached (typically a few hundred base pairs).

gene causes a specific disease. New molecular methods have spawned a number of techniques for detecting DNA sequence variation. Many of the techniques summarized in Table 3-3 can provide rapid and efficient screening for the presence of mutations. These methods can indirectly indicate the existence and location of a mutation, after which the DNA in the indicated region can be sequenced to identify the specific mutation. Direct sequencing of DNA is a useful and accurate means of detecting mutations and is regarded as the definitive method of identifying and verifying mutations. As it becomes less expensive, direct DNA sequencing is being used with increasing frequency.

A great deal of progress has been made in fabricating **DNA microarrays** (also known as **DNA chips**) and using them for detection of mutations (Fig. 3-24). To make a DNA microarray, robots place single-stranded oligonucleotides on a small glass slide. A single slide (1 cm$^2$) can contain millions of different oligonucleotides. These oligonucleotides consist of normal DNA sequences as well as DNA sequences that contain known disease-causing mutations. Fluorescently labeled single-stranded DNA from a subject is hybridized with the oligonucleotides on the slide to determine whether the DNA hybridizes with the normal or with

**TABLE 3-3**
## Methods of Mutation Detection

| Technique | Brief Description | Application |
|---|---|---|
| Southern blotting | Digestion of test DNA with restriction enzyme; resolution of fragments with agarose gel electrophoresis; transfer of DNA to nylon membrane and hybridization of labeled probe to DNA fragments | Detection of insertions, deletions, rearrangements; ordering of DNA fragments into physical map |
| Analysis of PCR product size | PCR products are sorted by size using electrophoresis on an agarose or polyacrylamide gel | Detection of small insertions and deletions and triplet repeat expansions |
| Direct DNA sequencing | Determination of linear order of nucleotides of test DNA; specific nucleotide detected by chemical cleavage, dideoxy-chain termination, or fluorochrome dye | Detection of insertions, deletions, point mutations, rearrangements |
| DNA mismatch cleavage | Hybridization of a labeled probe to test DNA; cleavage of DNA at site of base-pairing mismatch | Detection of small insertions or deletions, point mutations |
| Allele-specific oligonucleotide (ASO) hybridization | Preferential hybridization of labeled probe to test DNA with uniquely complementary base composition | Detection of alleles of known composition |
| Multiplex ligation-dependent probe amplification (MLPA) | Ligation of DNA fragments after hybridization of probes specific to a region | Detection of deletions and duplications of exons or whole genes |
| Mass spectrometry | Detection of physical mass of sense and antisense strands of test DNA | Detection of small insertions or deletions, point mutations |
| DNA microarray hybridization | Hybridization of test DNA to arrays of oligonucleotides ordered on silicone chip or glass slide | Detection of SNPs, CNVs, expression differences |
| Protein truncation | Test DNA used to make complementary DNA (cDNA) by RT-PCR with 5′ primer containing T7 promoter; cDNA translated and product resolved by SDS-PAGE | Detection of frameshift, splice site, or nonsense mutations that truncate the protein product |

*RT-PCR, reverse transcriptase–polymerase chain reaction; SDS-PAGE, sodium dodecyl sulfate polyacrylamide gel electrophoresis.*

the mutation-containing oligonucleotides, and the pattern of hybridization signals is analyzed by a computer. With current technology, enough probes can be placed on a single microarray to analyze variation in one million SNPs in an individual. Microarrays are also used to examine copy number variants, methylation patterns in a person's genome, and genetic variation in various pathogenic organisms. A key difference between microarrays and the methods summarized in the preceding paragraph is that microarrays typically test for known mutations that are incorporated in oligonucleotide probes. A rare, previously unidentified mutation cannot be detected by conventional microarrays.

Still another application of DNA microarrays is to determine which genes are being expressed (i.e., transcribed) in a given tissue sample (e.g., from a tumor). mRNA from the tissue is extracted and used as a template to form a complementary DNA sequence, which is then hybridized on the slide with oligonucleotides representing many different genes. The pattern of positive hybridization signals indicates which genes are expressed in the tissue sample. The DNA microarray approach offers the extraordinary speed, miniaturization, and accuracy of computer-based mutation analysis. Tests for specific mutations, an important aspect of genetic diagnosis, are discussed further in Chapter 13.

> Many techniques can be used to detect mutations at the DNA sequence level. These include Southern blotting, direct DNA sequencing, and microarray analysis. Microarrays are used in mutation detection, gene expression analysis, and a wide variety of other applications.

## GENETIC VARIATION IN POPULATIONS

Although mutation is the ultimate source of genetic variation, it cannot alone account for the substantial differences in the incidence of many genetic diseases among different ethnic groups. Why, for example, is sickle cell disease seen in approximately 1 of every 600 African Americans, but seldom in northern Europeans? Why is cystic fibrosis 40 times more common in Europeans than in Asians? In this section, concepts are introduced that explain these differences. The study of genetic variation in population is an important focus of **population genetics**.

### Basic Concepts of Probability

Probability plays a central role in genetics, because it helps us to understand the transmission of genes through generations, and it helps to explain and analyze genetic variation

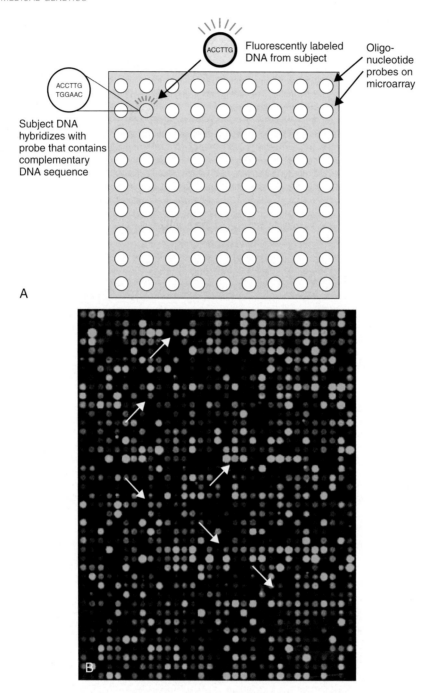

**FIGURE 3-24**

**A,** Schematic diagram of a microarray. Oligonucleotides are placed or synthesized on a chip. They are then exposed to labeled DNA from a subject. Hybridization occurs only if the oligonucleotide contains a DNA sequence that is complementary to that of the subject's DNA. The fluorescent label marks the location of the complementary oligonucleotide sequence on the chip. **B,** A microarray containing 36,000 oligonucleotides. This microarray was exposed to DNA from normal fibroblasts (*red,* see *arrows*) and fibroblasts from a patient with Niemann–Pick disease, type C (*green*). Arrows point to regions in which there was a strong hybridization signal with either normal or disease DNA. This microarray was used to search for genes that are highly expressed in the fibroblasts of patients.

in populations. It also aids in risk assessment, an important part of medical genetics. For example, the physician or genetic counselor commonly informs couples about their risk of producing a child with a genetic disorder. A **probability** is defined as the proportion of times that a specific outcome occurs in a series of events. Thus, we may speak of the probability of obtaining a 4 when a die is tossed, or the probability that a couple will produce a son rather than a daughter.

Because probabilities are proportions, they lie between 0 and 1, inclusive.

During meiosis, one member of a chromosome pair is transmitted to each sperm or egg cell. The probability that a given member of the pair will be transmitted is 1/2, and the probability that the other member of the pair will be transmitted is also 1/2. (Note that the probabilities of all possible events must add to 1 for any given experiment.)

Because this situation is directly analogous to coin tossing, in which the probabilities of obtaining heads or tails are each 1/2, we will use coin tossing as our illustrative example.

When a coin is tossed repeatedly, the outcome of each toss has no effect on subsequent outcomes. Each event (toss) is said to be **independent**. Even if we have obtained 10 heads in a row, the probability of obtaining heads or tails on the next toss remains 1/2. Similarly, the probability that a parent will transmit one of the two alleles at a locus is independent from one reproductive event to the next.

The independence principle allows us to deduce two fundamental concepts of probability, the **multiplication rule** and the **addition rule**. The multiplication rule states that if two trials are independent, then the probability of obtaining a given outcome in both trials is the product of the probabilities of each outcome. For example, we may wish to know the probability of obtaining heads on both tosses of a fair coin. Because the tosses are independent events, this probability is given by the product of the probabilities of obtaining heads in each individual toss: $1/2 \times 1/2 = 1/4$. Similarly, the probability of obtaining two tails in a row is $1/2 \times 1/2 = 1/4$.

The multiplication rule can be extended for any number of trials. Suppose a couple wants to know the probability that all three of their planned children will be girls. Because the probability of producing a girl is approximately 1/2, and because reproductive events are independent of one another, the probability of producing three girls is $1/2 \times 1/2 \times 1/2 = 1/8$. However, if the couple has already produced two girls and then wants to know the probability of producing a third girl, it is simply 1/2. This is because the previous two events are no longer probabilities; they have actually occurred. Because of independence, these past events have no effect on the outcome of the third event.

The addition rule states that if we want to know the probability of either one outcome or another, we can simply add the respective probabilities together. For example, the probability of getting two heads in a row is $1/2 \times 1/2$, or 1/4, and the probability of getting two tails in a row is the same. The probability of getting either two heads or two tails in a total of two tosses is the sum of the probabilities: $1/4 + 1/4 = 1/2$. As another example, imagine that a couple plans to have three children, and they have a strong aversion to having three children all of the same sex. They can be reassured somewhat by knowing that the probability of producing three girls or three boys is only $1/8 + 1/8$, or 1/4. The probability that they will have some combination of boys and girls is 3/4, because the sum of the probabilities of all possible outcomes must add to 1.

> Basic probability enables us to understand and estimate genetic risks and to understand genetic variation among populations. The multiplication rule is used to estimate the probability that two events will occur together. The addition rule is used to estimate the probability that one event or another will occur.

## Gene and Genotype Frequencies

The prevalence of many genetic diseases varies considerably from one population to another. The concepts of **genotype frequency** and **gene frequency** help us to measure and understand population variation in the incidence of genetic disease.

Imagine that we have typed 200 persons in a population for the MN blood group. This blood group, which is encoded by a locus on chromosome 2, has two major alleles, labeled *M* and *N*. In the *MN* system, the effects of both alleles can be observed in the heterozygote. *M* and *N* are therefore said to be **codominant**: The heterozygote can be distinguished from both homozygotes. Any individual in the population can have one of three possible genotypes (recall that the genotype is one's genetic makeup at a locus): He or she could be homozygous for *M* (genotype *MM*), heterozygous (*MN*), or homozygous for *N* (*NN*). After typing each person in our sample, we find the following distribution of genotypes: *MM*, 64; *MN*, 120; *NN*, 16. The genotype frequency is obtained simply by dividing each genotype count by the total number of subjects. The frequency of *MM* is 64/200, or 0.32; the frequency of *MN* is 120/200, or 0.60; and the frequency of *NN* is 16/200, or 0.08. The sum of these frequencies must equal 1.

The gene frequency for each allele, *M* and *N*, can be obtained here by the process of gene counting. Each *MM* homozygote has two *M* alleles, and each heterozygote has one *M* allele. Similarly, *NN* homozygotes have two *N* alleles, and heterozygotes have one *N* allele. In the example described, there are

$$(64 \times 2) + 120 = 248\ M \text{ alleles}$$

$$(16 \times 2) + 120 = 152\ N \text{ alleles}$$

In total, there are 400 alleles at the *MN* locus (i.e., twice the number of subjects, because each subject has two alleles). To obtain the frequency of *M*, we divide the number of *M* alleles by the total number of alleles at that locus: 248/400 = 0.62. Likewise, the frequency of *N* is 152/400, or 0.38. The sum of the two frequencies must equal 1.

> Gene and genotype frequencies specify the proportions of each allele and each genotype, respectively, in a population. Under simple conditions these frequencies can be estimated by direct counting.

## The Hardy–Weinberg Principle

The example given for the *MN* locus presents an ideal situation for gene frequency estimation, because, owing to codominance, the three genotypes can easily be distinguished and counted. What happens when one of the homozygotes is indistinguishable from the heterozygote (i.e., when there is dominance)? Here the basic concepts of probability can be used to specify a predictable relationship between gene frequencies and genotype frequencies.

Imagine a locus that has two alleles, labeled $A$ and $a$. Suppose that, in a population, we know the frequency of allele $A$, which we will call $p$, and the frequency of allele $a$, which we will call $q$. From these data, we wish to determine the expected population frequencies of each genotype, $AA$, $Aa$, and $aa$. We will assume that individuals in the population mate at random with regard to their genotype at this locus (**random mating** is also referred to as **panmixia**). Thus, the genotype has no effect on mate selection. If men and women mate at random, then the assumption of independence is fulfilled. This allows us to apply the addition and multiplication rules to estimate genotype frequencies.

Suppose that the frequency, $p$, of allele $A$ in our population is 0.7. This means that 70% of the sperm cells in the population must have allele $A$, as must 70% of the egg cells. Because the sum of the frequencies $p$ and $q$ must be 1, 30% of the egg and sperm cells must carry allele $a$ (i.e., $q = 0.30$). Under panmixia, the probability that a sperm cell carrying $A$ will unite with an egg cell carrying $A$ is given by the product of the gene frequencies: $p \times p = p^2 = 0.49$ (multiplication rule). This is the probability of producing an offspring with the $AA$ genotype. Using the same reasoning, the probability of producing an offspring with the $aa$ genotype is given by $q \times q = q^2 = 0.09$.

What about the frequency of heterozygotes in the population? There are two ways a heterozygote can be formed. Either a sperm cell carrying $A$ can unite with an egg carrying $a$, or a sperm cell carrying $a$ can unite with an egg carrying $A$. The probability of each of these two outcomes is given by the product of the gene frequencies, $pq$. Because we want to know the overall probability of obtaining a heterozygote (i.e., the first event or the second), we can apply the addition rule, adding the probabilities to obtain a heterozygote frequency of $2pq$. These operations are summarized in Figure 3-25. The relationship between gene frequencies and genotype frequencies was established independently by Godfrey Hardy and Wilhelm Weinberg and is termed the **Hardy–Weinberg principle**.

As already mentioned, this principle can be used to estimate gene and genotype frequencies when dominant homozygotes and heterozygotes are indistinguishable. This is often the case for recessive diseases such as cystic fibrosis. Only the affected homozygotes, with genotype $aa$, are distinguishable. The Hardy–Weinberg principle tells us that the frequency of $aa$ should be $q^2$. For cystic fibrosis in the European population, $q^2 = \sqrt{1/2500}$ (i.e., the prevalence of the disease among newborns). To estimate $q$, we take the square root of both sides of this equation: $q = \sqrt{1/2500} = 1/50 = 0.02$. Because $p + q = 1$, $p = 0.98$. We can then estimate the genotype frequencies of $AA$ and $Aa$. The latter genotype, which represents heterozygous carriers of the disease allele, is of particular interest. Because $p$ is almost 1.0, we can simplify the calculation by rounding $p$ up to 1.0 without a significant loss of accuracy. We then find that the frequency of heterozygotes is $2pq = 2q = 2/50 = 1/25$. This tells us something rather remarkable about cystic fibrosis and about recessive diseases in general. Whereas the incidence of affected homozygotes is only 1 in 2500, heterozygous carriers of the disease gene are much more common (1 in 25 individuals). The vast majority of recessive disease alleles, then, are effectively "hidden" in the genomes of heterozygotes.

> Under panmixia, the Hardy–Weinberg principle specifies the relationship between gene frequencies and genotype frequencies. It is useful in estimating gene frequencies from disease prevalence data and in estimating the incidence of heterozygous carriers of recessive disease genes.

## Causes of Genetic Variation

Mutation is the source of all genetic variation, and new genetic variants can be harmful (evolutionary mistakes), can be beneficial, or can have no effect whatsoever. **Natural selection** is often described as the "editor" of genetic variation. It increases the population frequency of favorable mutations (i.e., those who carry the mutation will produce more surviving offspring), and it decreases the frequency of variants that are unfavorable in a given environment (i.e., gene carriers produce fewer surviving offspring). Typically, disease-causing mutations are continually introduced into a population through the error processes described earlier. At the same time, natural selection removes these mutations.

Certain environments, however, can confer a selective advantage for a disease mutation. Sickle cell disease again provides an example. As discussed previously, persons who are homozygous for the sickle cell mutation are much more likely to die early. Heterozygotes ordinarily have no particular advantage or disadvantage. However, it has been shown that sickle cell heterozygotes have a distinct survival advantage in environments in which *Plasmodium falciparum* malaria is common (e.g., west-central Africa) (Fig. 3-26). Because the malaria parasite does not survive well in the erythrocytes of sickle cell heterozygotes, these persons are less likely to

Male population

|  |  | A (p) | a (q) |
|---|---|---|---|
| Female population | A (p) | AA (p²) | Aa (pq) |
|  | a (q) | Aa (pq) | aa (q²) |

**FIGURE 3-25**

The Hardy–Weinberg principle. The population frequencies of genotypes $AA$, $Aa$, and $aa$ are predicted on the basis of gene frequencies ($p$ and $q$). It is assumed that the gene frequencies are the same in males and females.

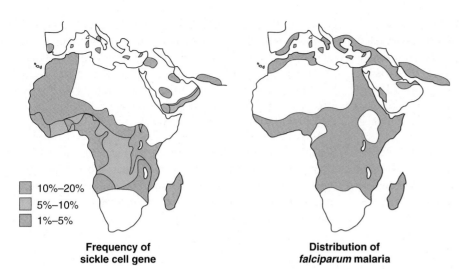

**FIGURE 3-26**
Correspondence between the frequency of the sickle cell allele and the distribution of *Plasmodium falciparum* malaria.

10%–20%
5%–10%
1%–5%

**Frequency of
sickle cell gene**

**Distribution of
*falciparum* malaria**

succumb to malaria than are normal homozygotes, conferring a selective advantage on the sickle cell mutation in this environment. Although there is selection against the mutation in sickle cell homozygotes, there is selection for the mutation in heterozygotes. The result is that the disease-causing mutation persists at a relatively high frequency in many African and Mediterranean populations. In nonmalarial environments (e.g., northern Europe), the sickle-cell mutation has no advantage, so natural selection acts strongly against it by eliminating homozygotes. This example illustrates the concept that variation in genetic disease incidence among populations can be caused by natural selection operating differentially in different environments.

> Natural selection is the evolutionary process in which alleles that confer survival or reproductive advantages in a specific environment are selected positively to increase in frequency, and alleles that confer lower survival or reproductive disadvantages are selected negatively so that they decrease in frequency.

**Genetic drift** is another force that can cause disease genes to vary in frequency among populations. To understand the process of genetic drift, consider a coin-tossing exercise in which 10 coins are tossed. Because heads and tails are equally likely, the expected number of heads and tails in this exercise would be 5 each. However, it is intuitively clear that, by chance, a substantial departure from this expectation could be observed. It would not be surprising to see 7 heads and 3 tails in 10 throws, for example. However, if 1000 coins are tossed, the degree of departure from the expected ratio of 50% heads and 50% tails is much smaller. A reasonable outcome of 1000 throws might be 470 heads and 530 tails, but it would be quite unlikely to obtain 700 heads and 300 tails. Therefore, there is less random fluctuation in larger samples.

The same principle applies to gene frequencies in populations. In a very small population, a gene frequency can deviate substantially from one generation to the next, but this is unlikely in a large population. Thus, genetic drift is greater in smaller populations. As a result, genetic diseases that are

otherwise uncommon may be seen fairly frequently in a small population. For example, Ellis–van Creveld syndrome, a rare disorder that involves reduced stature, polydactyly (extra digits), and congenital heart defects, is seen with greatly elevated frequency among the Old Order Amish population of Pennsylvania. The Amish population was founded in the United States by about 50 couples. Because of this small population size, there was great potential for genetic drift, resulting in increased frequencies of certain disease-causing alleles.

It is common to observe the effect of genetic drift in small, isolated populations throughout the world. Even relatively large populations might have experienced drift effects in the recent past if they underwent severe population bottlenecks or were established by a small number of founders (**founder effect**). For example, more than 30 otherwise rare genetic diseases are found with elevated frequency in Finland's population, which is thought to have been founded primarily by a small number of individuals some 100 generations ago. Phenylketonuria and cystic fibrosis, which are common in other Western European populations, are relatively rare in Finland, illustrating the fact that genetic drift can both increase and decrease the frequency of disease genes. Several genetic diseases (e.g., torsion dystonia, Tay–Sachs disease, Gaucher disease) occur with increased frequency in the Ashkenazi Jewish population (see Chapter 7); this may be the result of population bottlenecks that have occurred in the history of this population.

> Genetic drift is a random evolutionary process that produces larger changes in gene frequencies in smaller populations. Founder effect, in which small founder populations can experience large changes in gene frequency because of their small size, is a special case of genetic drift.

**Gene flow** occurs when populations exchange migrants who mate with one another. Through time, gene flow between populations tends to make them genetically more similar to each other. One reason sickle cell disease is less common in African Americans than in many African populations is because of gene flow between African Americans and European Americans (this same process is likely to have increased the frequency of cystic fibrosis in the African American population). In addition, because *P. falciparum* malaria is not found in North America, natural selection does not favor the sickle cell mutation.

The forces of mutation, natural selection, genetic drift, and gene flow interact in complex and sometimes unexpected ways to influence the distribution and prevalence of genetic diseases in populations. The interplay of mutation, which constantly introduces new variants, and natural selection, which often eliminates them, is an important and medically relevant example of such an interaction. A simple analysis of the relationship between mutation and selection helps us to understand variation in gene frequencies. Consider, for example, a dominant disease that results in death before the person can reproduce. This is termed a **genetic lethal mutation** because, even though the individual might survive for some time, he or she contributes no genes to the next generation. Each time mutation introduces a new copy of the lethal dominant disease allele into a population, natural selection eliminates it. In this case, $p$, the gene frequency of the lethal allele in the population, is equal to $\mu$, the mutation rate ($p = \mu$). Now, suppose that those who inherit the allele can survive into their reproductive years, but, on average, they produce 30% fewer children than those who do not inherit the allele. This reduction in offspring represents the **selection coefficient**, $s$, of the allele. In this case, $s = 0.30$. When the allele is completely lethal, $s = 1$ (i.e., no children are produced). We can now estimate the

gene frequency for this allele as $p = \mu/s$. As we would expect, the predicted frequency of an allele that merely reduces the number of offspring is higher (given the mutation rate) than the frequency of an allele that is completely lethal, where $p = \mu/s = \mu$. This predictable relationship between the effects of mutation and selection on gene frequencies is termed **mutation–selection balance.**

We can use the same principles to predict the relationship between mutation and selection against recessive alleles. The Hardy–Weinberg principle showed that most copies of harmful recessive alleles are found in heterozygotes and are thus protected from the effects of natural selection. We would therefore expect their gene frequencies to be higher than those of harmful dominant alleles that have the same mutation rate. Indeed, under mutation–selection balance, the predicted frequency of a recessive allele, $q$, that is lethal in homozygotes is $q = \sqrt{\mu}$ (because $\mu < 1$, $\sqrt{\mu} > \mu$, resulting in a relatively higher allele frequency for lethal recessive alleles). If the allele is not lethal in homozygotes, then $q = \sqrt{\mu/s}$, where $s$ is again the selection coefficient for those who have a homozygous affected genotype. Thus, understanding the principle of mutation–selection balance helps to explain why, in general, the gene frequencies for recessive disease-causing alleles are higher than are the frequencies of dominant disease-causing alleles.

Mutation–selection balance predicts a relatively constant gene frequency when new mutations introduce harmful alleles, whereas natural selection removes them. This process predicts that the gene frequencies should be lower for dominant diseases, in which most alleles are exposed to natural selection, than in recessive diseases, where most alleles are found in heterozygotes and are thus protected from natural selection.

## Study Questions

1. In the following list, the normal amino acid sequence is given first, followed by sequences that are produced by different types of mutations. Identify the type of mutation most likely to cause each altered amino acid sequence.
   Normal: Phe-Asn-Pro-Thr-Arg
   Mutation 1: Phe-Asn-Pro
   Mutation 2: Phe-Asn-Ala-His-Thr
   Mutation 3: Phe-His-Pro-Thr-Arg

2. Missense and transcription (promoter, enhancer, transcription factor) mutations often produce milder disease conditions than do frameshift, donor/acceptor site, and nonsense mutations. Using the globin genes as examples, explain why this is so.

3. Persons who have mutations that lower their production of both α- and β-globin often present with milder disease

symptoms than do those who have mutations lowering the production of only one type of chain. Why?

4. Outline the major differences between SNPs, VNTRs, and STRPs. Which of these three types of polymorphism is represented in the autoradiogram in Figure 3-27?

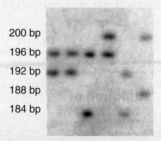

200 bp
196 bp
192 bp
188 bp
184 bp

**FIGURE 3-27**
Autoradiogram for study question 4.

**5.** $\alpha_1$-Antitrypsin deficiency is a disease that arises when both copies of the $\alpha_1$-antitrypsin gene are altered by mutations. Liver disease, chronic emphysema, and pulmonary failure can result. One of the mutations that causes $\alpha_1$-antitrypsin deficiency occurs in exon 3 of the gene and destroys a recognition site for the restriction enzyme *Bst*EII. RFLP analysis was performed on three members of a family, producing the autoradiogram in Figure 3-28. Determine the disease status of each individual.

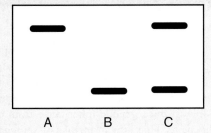

**FIGURE 3-28**
Autoradiogram for study question 5.

**6.** Using protein electrophoresis, 100 members of a population were studied to determine whether they carry genes for normal hemoglobin (HbA) or sickle hemoglobin (HbS). The following genotypes were observed:
HbA/HbA: 88
HbA/HbS: 10
Hbs/Hbs: 2
What are the gene frequencies of HbA and HbS? What are the observed genotype frequencies? Assuming Hardy–Weinberg proportions, what are the expected genotype frequencies?

**7.** Approximately 1 in 10,000 Europeans is born with PKU. What is the frequency of the disease-causing allele? What is the frequency of heterozygous carriers in the population?

## Suggested Readings

Crow JF. The origins, patterns and implications of human spontaneous mutation. Nat Rev Genet 2000; 1(1):40–7.

Driscoll MC. Sickle cell disease. Pediatr Rev 2007; 28(7):259–68.

Ellegren H, Smith NG, Webster MT. Mutation rate variation in the mammalian genome. Curr Opin Genet Develop 2003; 13(6):562–68.

Gill P. DNA as evidence—the technology of identification. N Engl J Med 2005;352:2669–71.

Graham CA, Hill AJ. Introduction to DNA sequencing. Methods Mol Biol 2001;167:1–12.

Hanawalt PC. Paradigms for the three rs: DNA replication, recombination, and repair. Mol Cell 2007;28(5): 702–7.

Heller C. Principles of DNA separation with capillary electrophoresis. Electrophoresis 2001;22(4):629–43.

Jorde LB. Human genetic variation and disease. In: Meyers RA (ed): Encyclopedia of Molecular Biology and Molecular Medicine, 2nd ed. Weinheim, Germany: Wiley-VCH, 2005, pp. 323–37.

Kraemer KH, Patronas NJ, Schiffmann R, et al. Xeroderma pigmentosum, trichothiodystrophy and Cockayne syndrome. A complex genotype–phenotype relationship. Neuroscience 2007;145(4):1388–96.

Mouro I, Colin Y, Cherif-Zahar B. Molecular genetic basis of the human Rhesus blood group system. Nature Genet 1993;5:62–5.

Neel JV. New approaches to evaluating the genetic effects of the atomic bombs. Am J Hum Genet 1995;57:1263–6.

Parman Y. Hereditary neuropathies. Curr Opin Neurol 2007;20(5):542–7.

Rund D, Rachmilewitz E. R-Thalassemia. N Engl J Med 2005;353(11):1135–46.

Shendure J, Ji H. Next-generation DNA sequencing. Nat Biotechnol 2008;26(10):1135–45.

Strachan T, Read AP. Human Molecular Genetics 3. New York: Garland Science, 2004.

Stuart MJ, Nagel RL. Sickle-cell disease. Lancet 2004;364 (9442):1343–60.

Syvanen AC. Accessing genetic variation. Genotyping single nucleotide polymorphisms. Nature Genet Rev 2001;2: 930–42.

Trevino V, Falciani F, Barrera-Saldana HA. DNA microarrays: a powerful genomic tool for biomedical and clinical research. Mol Med 2007;13(9–10):527–41.

Yamamoto F, Clausen H, White T, et al. Molecular genetic basis of the histo-blood group ABO system. Nature 1990; 345:229–33.

## Internet Resources

Science Primer (basic tutorials on microarrays, molecular genetics, and genetic variation) *http://www.ncbi.nih.gov/About/primer*

Sickle Cell Information Center *http://www.scinfo.org/*

Thalassemia (information on thalassemias and their management) *http://sickle.bwh.harvard.edu/menu_thal.html*

## Chapter 4

# AUTOSOMAL DOMINANT AND RECESSIVE INHERITANCE

Many important and well-understood genetic diseases are the result of a mutation in a single gene. The 2009 on-line edition of McKusick's *Mendelian Inheritance in Man* (http://www.ncbi.nlm.nih.gov/Omim/) lists more than 19,000 known **single-gene,** or **monogenic,** traits defined thus far in humans. Of these, more than 18,000 are located on autosomes, no more than 1000 are located on the X chromosome, and 57 are located on the Y chromosome. Monogenic traits have been the focus of the research done thus far in medical genetics. In many cases, these genes have been mapped to specific chromosome locations, cloned, and sequenced. This research has led to new and exciting insights not only in genetics but also in the basic pathophysiology of disease.

In this chapter we focus on single-gene disorders caused by mutations on the autosomes. (Single-gene disorders caused by mutations on the sex chromosomes are the subject of Chapter 5.) We discuss the patterns of inheritance of these diseases in families, as well as factors that complicate these patterns. When appropriate, the molecular mechanism that causes genetic disease is addressed. We also discuss the risks of transmitting single-gene diseases to one's offspring, because this is usually an important concern for at-risk couples.

## BASIC CONCEPTS OF FORMAL GENETICS

### Gregor Mendel's Contributions

Monogenic traits are also known as **mendelian** traits, after Gregor Mendel, the 19th-century Austrian monk who deduced several important genetic principles from his well-designed experiments with garden peas. Mendel studied seven traits in the pea, each of which is determined by a single gene. These traits included attributes such as height (tall versus short plants) and seed shape (rounded versus wrinkled). The variation in each of these traits is caused by the presence of different alleles at individual loci.

Two central principles emerged from Mendel's work. The first is the **principle of segregation,** which states that sexually reproducing organisms possess genes that occur in pairs and that only one member of this pair is transmitted to the offspring (i.e., it segregates). The prevalent thinking during Mendel's time was that hereditary factors from the two parents are blended in the offspring. In contrast, the principle of segregation states that genes remain intact and distinct.

An allele for "rounded" seed shape can be transmitted to an offspring in the next generation, which can, in turn, transmit the same allele to its own offspring. If, instead of remaining distinct, genes were somehow blended in offspring, it would be impossible to trace genetic inheritance from one generation to the next. Thus, the principle of segregation was a key development in modern genetics.

Mendel's **principle of independent assortment** was his second great contribution to genetics. This principle states that genes at different loci are transmitted independently. Consider the two loci mentioned previously. One locus can have either the "rounded" or the "wrinkled" allele, and the other can have either the "tall" or the "short" allele. In a reproductive event, a parent transmits one allele from each locus to its offspring. The principle of independent assortment dictates that the transmission of a specific allele at one locus ("rounded" or "wrinkled") has no effect on which allele is transmitted at the other locus ("tall" or "short").

The principle of segregation describes the behavior of chromosomes in meiosis. The genes on chromosomes segregate during meiosis, and they are transmitted as distinct entities from one generation to the next. When Mendel performed his critical experiments, he had no direct knowledge of chromosomes, meiosis, or genes (indeed, the last term was not coined until 1909, long after Mendel's death). Although his work was published in 1865 and cited occasionally, its fundamental significance was unrecognized for several decades. Yet Mendel's research, which was eventually replicated by other researchers at the turn of the 20th century, forms the foundation of much of modern genetics.

▶ Mendel's key contributions were the principles of segregation and independent assortment.

### The Concept of Phenotype

The term **genotype** has been defined as an individual's genetic constitution at a locus. The **phenotype** is what is actually observed physically or clinically. Genotypes do not uniquely correspond to phenotypes. Individuals with two different genotypes, a dominant homozygote and a heterozygote, can have the same phenotype. An example is cystic fibrosis (Clinical Commentary 4-1), an autosomal recessive condition in which only the recessive homozygote is affected. Conversely, the same genotype can produce different phenotypes in different environments. An example is the recessive disease phenylketonuria (PKU), which is seen in approximately 1 of every 10,000 European births. Mutations at the locus encoding the metabolic enzyme phenylalanine hydroxylase render the homozygote unable to metabolize the amino acid phenylalanine. Although babies with PKU are normal at birth, their metabolic deficiency produces a buildup of phenylalanine and its toxic metabolites. This process is highly destructive to the central nervous system, and it eventually produces severe mental retardation. It has been estimated that babies with untreated PKU lose, on average, 1 to 2 IQ points per week during the first year of life. Thus, the PKU genotype can produce a severe disease phenotype. However, it is straightforward to screen for PKU at birth (see Chapter 13), and mental retardation can be avoided by initiating a low-phenylalanine diet within 1 month after birth. The child still has the PKU genotype, but the phenotype has been profoundly altered by environmental modification.

## CLINICAL COMMENTARY 4-1
### *Cystic Fibrosis*

Cystic fibrosis (CF) is one of the most common single-gene disorders in North America, affecting approximately 1 in 2000 to 1 in 4000 European American newborns. It is less common in other populations. The prevalence among African Americans is about 1 in 15,000 births, and it is less than 1 in 30,000 among Asian Americans. Approximately 30,000 Americans suffer from this disease.

CF was first identified as a distinct disease entity in 1938 and was termed "cystic fibrosis of the pancreas." This refers to the fibrotic lesions that develop in the pancreas, one of the principal organs affected by this disorder. Approximately 85% of CF patients have pancreatic insufficiency (i.e., the pancreas is unable to secrete digestive enzymes, contributing to chronic malabsorption of nutrients). The intestinal tract is also affected, and approximately 15% to 20% of newborns with CF have meconium ileus (thickened, obstructive intestinal matter). The sweat glands of CF patients are abnormal, resulting in high levels of chloride in the sweat. This is the basis for the sweat chloride test, commonly used in the diagnosis of this disease. More than 95% of males with CF are sterile due to absence or obstruction of the vas deferens.

The major cause of morbidity and mortality in CF patients is pulmonary disease. Patients with CF have intense lower airway inflammation and chronic bronchial infection, progressing to end-stage lung disease characterized by extensive airway damage and fibrosis of lung tissue. Airway obstruction and lung injury are thought to be caused by a dehydrated airway surface and reduced clearance, resulting in thick airway mucus. This is associated with infection by bacteria such as *Staphylococcus aureus* and *Pseudomonas aeruginosa*. The combination of airway obstruction, inflammation, and infection leads to destruction of the airways and lung tissue, resulting eventually in death from pulmonary disease in more than 90% of CF patients.

As a result of improved nutrition, airway clearance techniques, and antibiotic therapies, the survival rate of CF patients has improved substantially during the past 3 decades. Median survival time is now nearly 40 years. This disease has highly variable expression, with some patients experiencing only mild respiratory difficulty and nearly normal survival. Others have much more severe respiratory problems and may survive less than 2 decades.

CF is caused by mutations in a gene, *CFTR*,* that encodes the cystic fibrosis transmembrane conductance regulator. *CFTR* encodes cyclic AMP-regulated chloride ion channels that span the membranes of specialized epithelial cells, such as those that line the bowel and lung. In addition, CFTR is involved in regulating the transport of sodium ions across epithelial cell membranes. The role of CFTR in sodium and chloride transport helps us to understand the multiple effects of mutations at the CF locus. Defective ion transport results in salt imbalances, depleting the airway of water and producing the thick, obstructive secretions seen in the lungs. The pancreas is also obstructed by thick secretions, leading to fibrosis and pancreatic insufficiency. The chloride ion transport defect explains the abnormally high concentration of chloride in the sweat secretions of CF patients: Chloride cannot be reabsorbed from the lumen of the sweat duct.

DNA sequence analysis has revealed more than 1500 different mutations at the *CFTR* locus. The most common of these is a three-base deletion that results in the loss of a phenylalanine residue at position 508 of the CFTR protein. This mutation is labeled ΔF508 (i.e., deletion of phenylalanine at position 508). ΔF508 accounts for nearly 70% of all CF mutations. This mutation, along with several dozen other relatively common ones, is assayed in the genetic diagnosis of CF (see Chapter 13).

Identification of the specific mutation or mutations that are responsible for CF in a patient can help to predict the severity of the disease. For example, the most severe classes of mutations (of which ΔF508 is an example; see the figure below) result in a complete lack of chloride ion channel production or in channels that cannot migrate to the cell membrane. Patients homozygous for these mutations nearly always have pancreatic insufficiency. In contrast, other mutations (e.g., R117H, a missense mutation) result in ion channels that do proceed to the cell membrane but respond poorly to cyclic AMP and consequently do not remain open as long as they should. The phenotype is thus milder: Patients who have this mutation are less likely to have pancreatic insufficiency. Patients with other mild *CFTR*

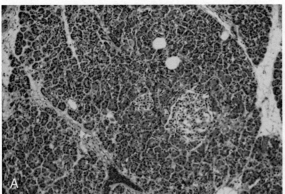
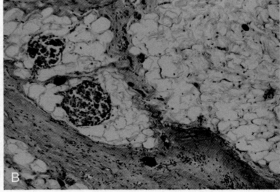

A, Normal pancreas. B, Pancreas from a cystic fibrosis patient, showing infiltration of fat and fibrotic lesions.

*Continued*

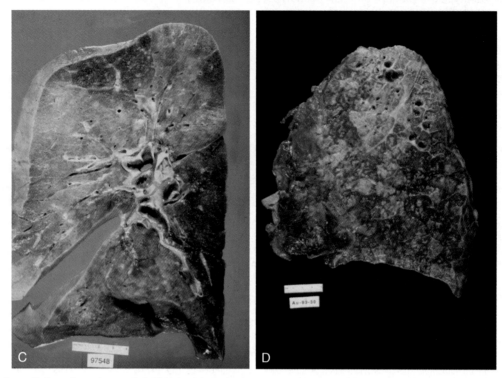

C, Normal lung tissue. D, Lung tissue from a cystic fibrosis patient, showing extensive destruction as a result of obstruction and infection. *(Courtesy of Dr. Edward Klatt, Florida State University School of Medicine.)*

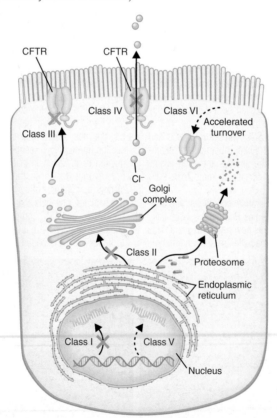

Classes of mutations in the *CFTR* gene and their effects in cells. Class I mutations result in no synthesis of the gene product. Class II mutations produce a defective protein product that is destroyed in proteasomes. Class III mutations produce a protein that gets to the cell surface but is abnormally regulated. Class IV mutations result in defective chloride ion conductance. Class V mutations are typically promoter or intron–exon splicing mutations that reduce the number of mRNA transcripts, allowing some normal protein products. Class VI mutations result in increased rates of turnover of the chloride channel at the cell surface.

mutations (generally, classes IV and V) tend to have less-severe pulmonary disease and lower mortality rates. Some males with mild *CFTR* mutations have only congenital bilateral absence of the vas deferens (CBAVD) but little, if any, lung or gastrointestinal disease. The correlation between genotype and phenotype is far from perfect, however, indicating that modifier loci and environmental factors must also influence expression of the disease (see text). In general, there is a reasonably good correlation between genotype and pancreatic function and a more variable relationship between genotype and pulmonary function.

The ability to identify *CFTR* mutations has led to surveys of persons who have one (heterozygous) or two (homozygous) *CFTR* mutations but who do not have cystic fibrosis. They have increased risks for a number of disease conditions, including CBAVD, bronchiectasis (chronic dilatation of the bronchi and abnormal mucus production), and pancreatitis (pancreatic inflammation).

By enhancing our understanding of the pathophysiology of CF, identification of *CFTR* has opened the possibility of new treatments for this disease. Examples include administration of drugs, such as gentamicin, that cause ribosomes to read through the premature stop codons that account for approximately 7% of *CFTR* mutations. Other drugs can increase the activity of chloride channels in patients with class III or IV mutations. Gene therapy, in which the normal *CFTR* gene is placed in viral or other vectors that are then introduced to the patient's airway (see Chapter 13), is also being actively investigated. This strategy, however, has encountered difficulties because viral vectors often induce an inflammatory immune response.

---

*Conventionally, the symbol for a gene, such as *CFTR*, is shown in italics, and the symbol for the protein product is not.

---

This example shows that the phenotype is the result of the interaction of genotype and environmental factors. It should be emphasized that "environment" can include the genetic environment (i.e., genes at other loci whose products can interact with a specific gene or its product).

▶ The phenotype, which is physically observable, results from the interaction of genotype and environment.

## Basic Pedigree Structure

The **pedigree** is one of the most commonly used tools in medical genetics. It illustrates the relationships among family members, and it shows which family members are affected or unaffected by a genetic disease. Typically, an arrow denotes the **proband**, the first person in whom the disease is diagnosed in the pedigree. The proband is sometimes also referred to as the **index case** or **propositus** (proposita for a female). Figure 4-1 describes the features of pedigree notation.

When discussing relatives in families, one often refers to degrees of relationship. First-degree relatives are those who are related at the parent–offspring or sibling (brother and sister) level. Second-degree relatives are those who are removed by one additional generational step (e.g., grandparents and their grandchildren, uncles or aunts and their nieces or nephews). Continuing this logic, third-degree relatives would include, for example, one's first cousins, great-grandchildren, and so on.

## AUTOSOMAL DOMINANT INHERITANCE

### Characteristics of Autosomal Dominant Inheritance

Autosomal dominant diseases are seen in roughly 1 of every 200 individuals (see Table 1-3 in Chapter 1). Individually, each autosomal dominant disease is rather rare in populations, however, with the most common ones having gene frequencies of about 0.001. For this reason, matings between two individuals who are both affected by the same autosomal dominant disease are uncommon. Most often, affected offspring are produced by the union of an unaffected parent with an affected heterozygote. The **Punnett square** in Figure 4-2 illustrates

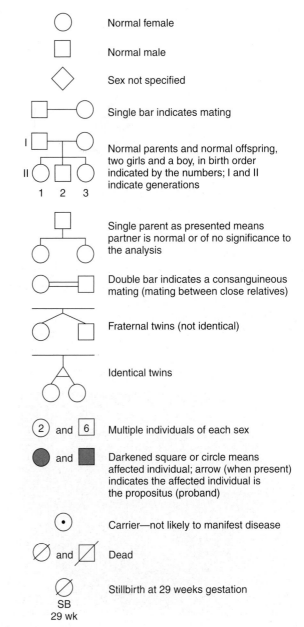

**FIGURE 4-1**
Basic pedigree notation. For further details, see Bennett et al: J Genet Counsel 2008; 17:424-433.

Unaffected parent

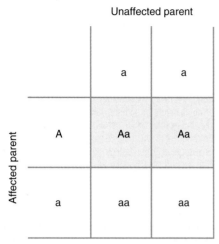

**FIGURE 4-2**
Punnett square illustrating the mating of an unaffected individual (*aa*) with an individual who is heterozygous for an autosomal dominant disease gene (*Aa*). The genotypes of affected offspring are shaded.

such a mating. The affected parent can pass either a disease gene or a normal gene to his or her children. Each event has a probability of 0.5. Thus, on the average, half of the children will be heterozygotes and will express the disease, and half will be unaffected homozygotes.

Postaxial polydactyly, the presence of an extra digit next to the fifth digit (Fig. 4-3), can be inherited as an autosomal dominant trait. Let *A* symbolize the allele for polydactyly, and let *a* symbolize the normal allele. An idealized pedigree for this disease is shown in Figure 4-4. This pedigree illustrates several important characteristics of autosomal dominant inheritance. First, the two sexes exhibit the trait in approximately equal ratios, and males and females are equally likely to transmit the trait to their offspring. This is because postaxial polydactyly is an *autosomal* disease (as opposed to a disease caused by an X chromosome mutation, in which these ratios typically differ). Second, there is no skipping of generations:

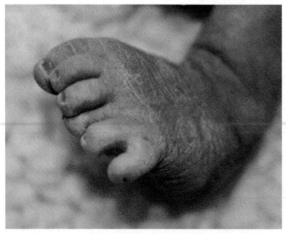

**FIGURE 4-3**
Postaxial polydactyly. An extra digit is located next to the fifth digit.

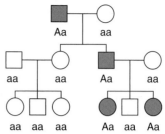

**FIGURE 4-4**
Pedigree illustrating the inheritance pattern of postaxial polydactyly, an autosomal dominant disorder. Affected individuals are represented by shading.

If an individual has polydactyly, one parent must also have it. This leads to a vertical transmission pattern, in which the disease phenotype is usually seen in one generation after another. Also, if neither parent has the trait, none of the children will have it. Third, father-to-son transmission of the disease gene is observed. Although father-to-son transmission is not *required* to establish autosomal dominant inheritance, its presence in a pedigree rules out some other modes of inheritance (particularly X-linked inheritance; see Chapter 5). Finally, as we have already seen, an affected heterozygote transmits the trait to approximately half of his or her children. However, because gamete transmission, like coin tossing, is subject to chance fluctuations, it is possible that all or none of the children of an affected parent will have the trait. When large numbers of matings of this type are studied, the ratio of affected children closely approaches 1/2.

> Autosomal dominant inheritance is characterized by vertical transmission of the disease phenotype, a lack of skipped generations, and roughly equal numbers of affected males and females. Father-to-son transmission may be observed.

### Recurrence Risks

Parents at risk for producing children with a genetic disease are often concerned with the question: What is the chance that our future children will have this disease? The probability that an individual offspring will be affected by the disease in question is termed the **recurrence risk**. If one parent is affected by an autosomal dominant disease (heterozygote) and the other is normal, the recurrence risk for each child is 1/2. It is important to keep in mind that each birth is an independent event, as in the coin-tossing examples. Thus, even if the parents have already had a child with the disease, their recurrence risk remains 1/2. Even if they have had several children, all affected (or all unaffected) by the disease, the law of independence dictates that the probability that their next child will have the disease is still 1/2. Although this concept seems intuitively obvious, it is commonly misunderstood by the lay population. Further aspects of communicating risks to families are discussed in Chapter 15.

> The recurrence risk for an autosomal dominant disorder is 50%. Because of independence, this risk remains constant no matter how many affected or unaffected children are born.

## AUTOSOMAL RECESSIVE INHERITANCE

Like autosomal dominant diseases, autosomal recessive diseases are fairly rare in populations. As shown previously, heterozygous carriers for recessive disease alleles are much more common than affected homozygotes. Consequently, the parents of individuals affected with autosomal recessive diseases are usually both heterozygous carriers. As the Punnett square in Figure 4-5 demonstrates, one fourth of the offspring of two heterozygotes will be unaffected homozygotes, half will be phenotypically unaffected heterozygous carriers, and one fourth will be homozygotes affected with the disease (on average).

### Characteristics of Autosomal Recessive Inheritance

Figure 4-6 is a pedigree showing the inheritance pattern of an autosomal recessive form of albinism that results from mutations in the gene that encodes tyrosinase, a tyrosine-metabolizing enzyme.* The resulting tyrosinase deficiency creates a block in the metabolic pathway that normally leads to the synthesis of melanin pigment. Consequently, the affected person has very little pigment in the skin, hair, and eyes (Fig. 4-7). Because melanin is also required for the normal development of the optic fibers, albinos can also display nystagmus (rapid uncontrolled eye movement), strabismus (deviation of the eye from its normal axis), and reduced visual acuity. The pedigree demonstrates most of the important criteria for distinguishing autosomal recessive

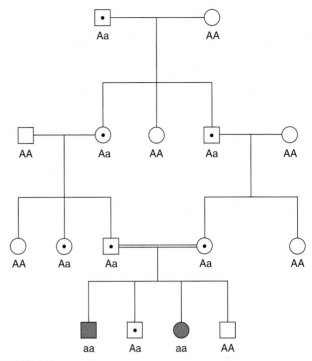

**FIGURE 4-6**
Pedigree showing the inheritance pattern of tyrosinase-negative albinism, an autosomal recessive disease. Consanguinity in this pedigree is denoted by a double bar connecting the parents of the affected individuals.

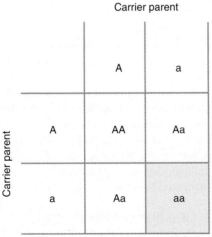

Carrier parent

|  | A | a |
|---|---|---|
| A | AA | Aa |
| a | Aa | aa |

Carrier parent

**FIGURE 4-5**
Punnett square illustrating the mating of two heterozygous carriers of an autosomal recessive gene. The genotype of the affected offspring is shaded.

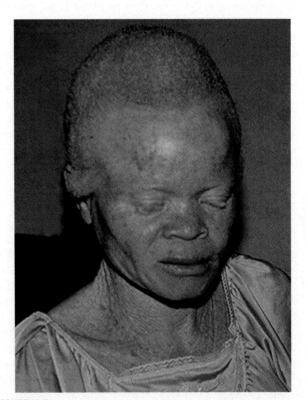

**FIGURE 4-7**
An African woman with oculocutaneous albinism, illustrating a lack of pigmentation in the hair and skin. She is looking away from the camera because her eyes are more sensitive to light than are those of persons with normally pigmented retinas.

*(Courtesy of Dr. Phil Fischer, Mayo Clinic.)*

---

*This form of albinism, termed *tyrosinase-negative* oculocutaneous albinism (OCA1), is distinguished from a second, milder form termed *tyrosinase-positive* oculocutaneous albinism (OCA2). OCA2 is typically caused by mutations in a gene on chromosome 15 (the "P" gene) whose protein product is thought to be involved in the transport and processing of tyrosinase.

**TABLE 4-1**

## A Comparison of the Major Attributes of Autosomal Dominant and Autosomal Recessive Inheritance Patterns

| Attribute | Autosomal Dominant | Autosomal Recessive |
|---|---|---|
| Usual recurrence risk | 50% | 25% |
| Transmission pattern | Vertical; disease phenotype seen in generation after generation | Disease phenotype may be seen in multiple siblings, but usually not in earlier generations |
| Sex ratio | Equal number of affected males and females (usually) | Equal number of affected males and females (usually) |
| Other | Father-to-son transmission of disease gene is possible | Consanguinity is sometimes seen, especially for rare recessive diseases |

inheritance (Table 4-1). First, unlike autosomal dominant diseases, in which the disease phenotype is seen in one generation after another, autosomal recessive diseases are usually observed in one or more siblings but not in earlier generations. Second, as in autosomal dominant inheritance, males and females are affected in equal proportions. Third, on average, one fourth of the offspring of two heterozygous carriers will be affected with the disorder. Finally, **consanguinity** is present more often in pedigrees involving autosomal recessive diseases than in those involving other types of inheritance (see Fig. 4-6). The term consanguinity (Latin, "with blood") refers to the mating of related persons. It is sometimes a factor in recessive disease because related persons are more likely to share the same disease-causing mutations. Consanguinity is discussed in greater detail later in this chapter.

> Autosomal recessive inheritance is characterized by clustering of the disease phenotype among siblings, but the disease is not usually seen among parents or other ancestors. Equal numbers of affected males and females are usually seen, and consanguinity may be present.

### Recurrence Risks

As already discussed, the most common mating type seen in recessive disease involves two heterozygous carrier parents. This reflects the relative commonness of heterozygous carriers and the fact that many autosomal recessive diseases are severe enough that affected individuals are less likely to become parents.

The Punnett square in Figure 4-5 demonstrates that one fourth of the offspring from this mating will be homozygous for the disease gene and therefore affected. The recurrence risk for the offspring of carrier parents is then 25%. As

before, these are *average* figures. In any given family chance fluctuations are likely, but a study of a large number of families would yield a figure quite close to this fraction.

Occasionally, a carrier of a recessive disease-causing allele mates with a person who is homozygous for this allele. In this case, roughly half of their children will be affected, and half will be heterozygous carriers. The recurrence risk is 50%. Because this pattern of inheritance mimics that of an autosomal dominant trait, it is sometimes referred to as **quasidominant** inheritance. With studies of extended pedigrees in which carrier matings are observed, quasidominant inheritance can be distinguished from true dominant inheritance.

When two persons affected by a recessive disease mate, all of their children must also be affected. This observation helps to distinguish recessive from dominant inheritance because two parents who are both affected by a dominant disease are almost always both heterozygotes. Thus one fourth of their children, on average, will be unaffected.

> The recurrence risk for autosomal recessive diseases is usually 25%. Quasidominant inheritance, with a recurrence risk of 50%, is seen when an affected homozygote mates with a heterozygote.

### "Dominant" Versus "Recessive": Some Cautions

The preceding discussion has treated dominant and recessive disorders as though they belong in rigid categories. However, these distinctions are becoming less strict as our understanding of these diseases increases. Many (probably most) of the so-called dominant diseases are actually more severe in affected homozygotes than in heterozygotes. An example is achondroplasia, an autosomal dominant disorder in which heterozygotes have reduced stature (Fig. 4-8). Heterozygotes enjoy a nearly normal life span, estimated to be only 10 years less than average. Affected homozygotes are much more severely affected and usually die in infancy from respiratory failure (see Chapter 10 for further discussion of achondroplasia).

Although heterozygous carriers of recessive disease genes are clinically normal, the effects of recessive genes can often be detected in heterozygotes because they result, for example, in reduced levels of enzyme activity. This is usually the basis for biochemical carrier detection tests (see Chapter 13). A useful and valid way to distinguish dominant and recessive disorders is that heterozygotes are clinically affected in most cases of dominant disorders, whereas they are almost always clinically unaffected in recessive disorders.

> Although the distinction between dominant and recessive diseases is not rigid, a dominant disease allele will produce disease in a heterozygote, whereas a recessive disease allele will not.

Another caution is that a disease may be inherited in autosomal dominant fashion in some cases and in autosomal recessive fashion in others. Familial isolated growth hormone deficiency (IGHD), another disorder that causes reduced

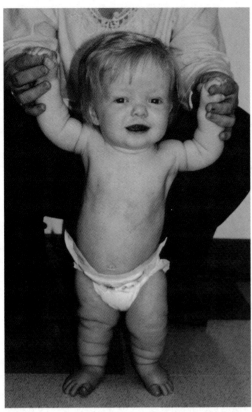

**FIGURE 4-8**
Achondroplasia. This girl has short limbs relative to trunk length. She also has a prominent forehead, low nasal root, and redundant skin folds in the arms and legs.

stature, is one such disease. DNA sequencing of a pituitary growth hormone gene on chromosome 17 (*GH1*) has revealed a number of different mutations that can produce IGHD. *Recessive* IGHD can be caused by nonsense, frameshift, or splice-site mutations that have a loss-of-function effect (a mature protein product is not synthesized). Because they have one normal copy of *GH1*, heterozygotes still produce half of the normal amount of growth hormone. This is sufficient for normal stature. Homozygotes for these mutations produce no GH1 product and have reduced stature.

How can a mutation at this locus produce *dominant* inheritance? In one form of dominantly inherited IGHD, a splice site mutation deletes the third exon of the *GH1* gene, producing a protein that proceeds to the secretory granules. Here, the abnormal GH1 product encoded by the mutated chromosome interacts with the normal product encoded by the normal chromosome. Acting as a dominant negative (see Chapter 3), the abnormal molecules disable the normal growth hormone molecules, resulting in greatly reduced production of GH1 product and thus reduced stature.

Another example is given by β-thalassemia, a condition discussed in Chapter 3. Although the great majority of β-thalassemia cases occur as a result of autosomal recessive mutations, a small fraction of cases are inherited in autosomal dominant fashion. Some of these are caused by nonsense or frameshift mutations that terminate translation in exon

3 or in downstream exons. The resulting messenger RNA (mRNA) proceeds to the cytoplasm and produces unstable β-globin chains. In heterozygotes, these abnormal chains exert a dominant negative effect on the normal β-globin chains produced by the normal allele (see Chapter 3). In contrast, frameshift or nonsense mutations that result in termination of translation in exons 1 or 2 of the gene result in very little abnormal mRNA in the cytoplasm, leaving the product of the normal allele intact. Hence, the heterozygote is unaffected.

These examples illustrate some of the complexities involved in applying the terms "dominant" and "recessive." They also show how molecular analysis of a gene can help to explain important disease features.

> In some cases, a disease may be inherited in either autosomal dominant or autosomal recessive fashion, depending on the nature of the mutation that alters the gene product.

A final caution is that the terms dominant and recessive, strictly speaking, apply to traits, not genes. To see why, consider the sickle cell mutation, discussed in Chapter 3. Homozygotes for this mutation develop sickle cell disease. Heterozygotes, who are said to have sickle cell *trait*, are usually clinically normal. However, a heterozygote has an increased risk for splenic infarctions at very high altitude. Is the mutant gene then dominant or recessive? Clearly, it makes more sense to refer to sickle cell *disease* as recessive and sickle cell *trait* as dominant. Nonetheless, it is common (and often convenient) to apply the terms dominant and recessive to genes.

## FACTORS THAT AFFECT EXPRESSION OF DISEASE-CAUSING GENES

The inheritance patterns described previously for conditions like postaxial polydactyly, cystic fibrosis, and albinism are quite straightforward. However, most genetic diseases vary in their degree of expression, and sometimes a person has a disease-causing genotype but never manifests the phenotype. Genetic diseases are sometimes seen in the absence of any previous family history. These phenomena, and the factors responsible for them, are discussed next.

### New Mutation

If a child has been born with a genetic disease that has not occurred previously in the family, it is possible that the disease is the product of a **new** (or **de novo**) **mutation**. That is, the gene transmitted by one of the parents underwent a change in DNA sequence, resulting in a mutation from a normal to a disease-causing allele. The alleles at this locus in the parent's other germ cells would still be normal. In this case the recurrence risk for the parents' subsequent offspring would not be elevated above that of the general population. However, the offspring of the affected child might have a substantially elevated risk (e.g., it would be 50% for an autosomal dominant disease). A large fraction of the observed cases of many autosomal dominant diseases are the result of new mutations. For example, it is estimated that 7/8 of all cases of achondroplasia are caused by new mutations, and only 1/8 are inherited from an affected parent. This is primarily because the disease tends to limit the potential for reproduction. To provide accurate risk estimates, it is essential to know whether a patient's disease is due to an inherited mutation or a new mutation. This can be done only if an adequate family history has been taken.

> New mutations are a common cause of the appearance of a genetic disease in a person with no previous family history of the disorder. The recurrence risk for the person's siblings is very low, but the recurrence risk for the person's offspring may be substantially increased.

## Germline Mosaicism

Occasionally, two or more offspring present with an autosomal dominant or X-linked disease when there is no family history of the disease. Because mutation is a rare event, it is unlikely that this would be due to multiple new mutations in the same family. The mechanism most likely to be responsible is termed **germline mosaicism** (mosaicism describes the presence of more than one genetically distinct cell line in the body). During the embryonic development of one of the parents, a mutation occurred that affected all or part of the germline but few or none of the somatic cells of the embryo (Fig. 4-9). Thus, the parent carries the mutation in his or her germline but does not actually express the disease because the mutation is absent in other cells of the body. As a result, the parent can transmit the mutation to numerous offspring. Although this phenomenon is relatively rare, it can have significant effects on recurrence risks when it does occur.

Germline mosaicism has been studied extensively in the lethal perinatal form of osteogenesis imperfecta (OI type II; see Chapter 2), which is caused by mutations in the type 1 procollagen genes. The fact that unaffected parents sometimes produced multiple offspring affected with this disease led to the conclusion that type II OI was an autosomal recessive trait. This was disputed by studies in which the polymerase chain reaction (PCR) technique was used to amplify DNA from the sperm of a father of two children with type II OI. This DNA was compared with DNA extracted from his somatic cells (skin fibroblasts). Although procollagen mutations were not detected in the fibroblast DNA, they were found in approximately one of every eight sperm cells. This was a direct demonstration of germline mosaicism in this man. Although germline mosaicism has been demonstrated for type II OI, most noninherited cases (approximately 95%) are thought to be caused by new mutations. Some cases of true autosomal recessive inheritance have also been documented, and two different genes that can each cause autosomal recessive OI have been documented.

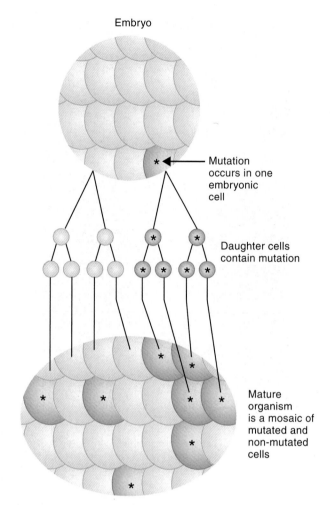

**FIGURE 4-9**
A mutation occurs in one cell of the developing embryo. All descendants of that cell have the same mutation, resulting in mosaicism. If the first mutated cell is part of the germline lineage, then germline mosaicism results.

Other diseases in which germline mosaicism has been observed include achondroplasia, neurofibromatosis type 1, Duchenne muscular dystrophy, and hemophilia A (the latter two diseases are discussed in Chapter 5). It has been estimated that germline mosaicism accounts for up to 15% of Duchenne muscular dystrophy cases and 20% of hemophilia A cases in which there is no previous family history.

> Germline mosaicism occurs when all or part of a parent's germline is affected by a disease mutation but the somatic cells are not. It elevates the recurrence risk for offspring of the mosaic parent.

## Reduced Penetrance

Another important characteristic of many genetic diseases is reduced (or incomplete) **penetrance**: A person who has a disease-causing genotype might not exhibit the disease phenotype at all, even though he or she can transmit the disease-causing mutation to the next generation. Retinoblastoma, a malignant eye tumor (Clinical Commentary 4-2), is a good example of an autosomal dominant disorder in which reduced penetrance is seen. The transmission pattern of this disorder is illustrated in Figure 4-10. Family studies have shown that about 10% of the **obligate carriers** of a retinoblastoma-causing mutation (i.e., those who have an affected parent and affected children and therefore must themselves carry the mutation) do not have the disease. The penetrance of the disease-causing genotype is then said to be 90%. Penetrance rates are usually estimated by examining a large number of families and determining what percentage of the obligate carriers (or obligate

homozygotes, in the case of recessive disorders) develop the disease phenotype.

Reduced penetrance describes the situation in which persons who have a disease-causing genotype do not develop the disease phenotype.

### Age-Dependent Penetrance

Although some genetic diseases are expressed at birth or shortly afterward, many others do not become apparent until well into adulthood. A delay in the age of onset of a genetic disease is known as **age-dependent penetrance**.

## CLINICAL COMMENTARY 4-2
### Retinoblastoma

Retinoblastoma is the most common childhood eye tumor, affecting approximately 1 in 20,000 children. The tumor typically initiates between 3 months after conception and 4 years of age, when retinal cells are actively dividing and proliferating. It nearly always presents clinically by the age of 5 years.

Approximately 60% of retinoblastoma cases are caused by somatic mutations that occur in early development and are therefore not transmitted to the affected individual's offspring. The remaining 40% are caused by inherited mutations: About 3/4 of these (30% of total cases) are the result of new mutations, most often transmitted by the father. The other 1/4 of inherited cases (10% of the total) are inherited from a parent who carries a retinoblastoma-causing mutation in all of his or her cells. About 10% of those who have inherited a disease-causing mutation never develop a tumor (reduced penetrance).

The analysis of changes in DNA in and near the disease-causing gene, *RB1*, finally explained the mechanism responsible for reduced penetrance. Briefly, an individual who has inherited a disease-causing *RB1* mutation carries the mutation in every cell of his or her body. However, this is not sufficient to cause tumor formation (if it were, every cell in the body would give rise to a tumor). In any cell, the presence of one normal *RB1* allele is sufficient to prevent tumor formation. To initiate a tumor in a developing retinal cell, a second somatic event must occur that disables the other, normal *RB1* allele (this two-hit process is discussed further in Chapter 11). The second event, which can be considered a somatic mutation, has a relatively low probability of occurring in any given cell. However, there are at least 1 million retinal cells in the developing fetus, each representing a potential target for the event. Usually, an individual who has inherited one disease-causing mutation will experience a second somatic mutation in several different retinal cells, giving rise to several tumors. Inherited retinoblastoma is thus usually multifocal (consisting of several tumor foci) and bilateral (affecting both eyes). Because the second hits are random events, a small fraction of persons who inherit the disease allele never experience a second hit in any retinal cell

and they do not develop a retinoblastoma. The requirement for a second hit thus explains the reduced penetrance seen in this disorder.

The retinoblastoma gene, *RB1*, encodes a protein product, pRb, that has been studied extensively. A major function of pRb, when hypophosphorylated, is to bind and inactivate members of the E2F family of nuclear transcription factors. The cell requires active E2F to proceed from the G1 to the S phase of mitosis. By inactivating E2F, pRb applies a brake to the cell cycle. When cell division is required, pRb is phosphorylated by cyclin-dependent kinase complexes (see Chapter 2). Consequently, E2F is released by pRb and activated. A loss-of-function mutation in pRb can cause a permanent loss of E2F-binding capacity. The cell, having lost its brake, will undergo repeated, uncontrolled mitosis, potentially leading to a tumor. Because of its controlling effect on the cell cycle, the *Rb* gene belongs to a class of genes known as tumor suppressors (see Chapter 11).

If untreated, retinoblastomas can grow to considerable size and can metastasize to the central nervous system or other organ systems. Fortunately, these tumors are now usually detected and treated before they become large. If found early enough through ophthalmological examination, the tumor may be treated successfully with cryotherapy (freezing) or laser photocoagulation. In more advanced cases, radiation, chemotherapy, or enucleation (removal) of the eye may be necessary. Currently, the 5-year survival rate for retinoblastoma patients in the United States is nearly 95%. Because persons with familial retinoblastoma have inherited an *RB1* mutation in all cells of their body, they are also susceptible to other types of cancer later in life. In particular, about 15% of those who inherit the mutation later develop osteosarcomas (malignant bone tumors). Other common second cancers include soft tissue sarcomas and cutaneous melanomas. Careful monitoring for subsequent tumors and avoidance of agents that could produce a second mutation (e.g., x-rays) are thus important aspects of management for the patient with inherited retinoblastoma.

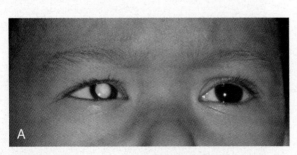
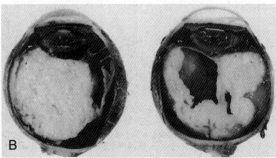

A, A white reflex (leukocoria) can be seen in the right eye of this individual on ophthalmoscopic examination. B, Bilateral retinoblastoma, showing presence of neoplastic tissue.
*(From Rosai J: Ackerman's Surgical Pathology, 8th ed. St Louis: Mosby, 1996.)*

**FIGURE 4-10**

Pedigree illustrating the inheritance pattern of retinoblastoma, a disorder with reduced penetrance. The unaffected obligate carrier, denoted by a dot, has the same genotype as the affected pedigree members.

One of the best-known examples is Huntington disease, a neurological disorder whose main features are progressive dementia and increasingly uncontrollable movements of the limbs (Clinical Commentary 4-3). The latter feature is known as *chorea* (from the Greek word for "dance," *khoreia*), and the disease is sometimes called Huntington chorea. This autosomal dominant disorder is named after Dr. George Huntington, who first described the disease in 1872. Symptoms are not usually seen until age 30 years or later (Fig. 4-11). Thus, those who develop the disease often have children before they are aware that they carry the disease-causing allele. If the disease were present at birth, nearly all affected persons would die before reaching reproductive age, and the frequency of the disease in the population would be much lower. Delaying the age of onset of the disease thus reduces natural selection against a disease-causing allele, increasing its frequency in a population. Age-dependent penetrance can cause difficulties in deducing the mode of inheritance of a disease because it is not possible to determine until later in life whether a person carries a disease-causing mutation.

## CLINICAL COMMENTARY 4-3
### Huntington Disease

Huntington disease (HD) affects approximately 1 in 20,000 persons of European descent. It is substantially less common among Japanese and Africans. The disorder usually manifests between the ages of 30 and 50 years, although it has been observed as early as 1 year of age and as late as 80 years of age.

HD is characterized by a progressive loss of motor control, dementia, and psychiatric disorders. There is a substantial loss of neurons in the brain, which is detectable by imaging techniques such as magnetic resonance imaging (MRI). Decreased glucose uptake in the brain, an early sign of the disorder, can be detected by positron-emission tomography (PET). Although many parts of the brain are affected, the area most noticeably damaged is the corpus striatum. In some patients, the disease leads to a loss of 25% or more of total brain weight.

The clinical course of HD is protracted. Typically, the interval from initial diagnosis to death is 15 to 20 years. As in many neurological disorders, patients with HD experience difficulties in swallowing; aspiration pneumonia is the most common cause of death. Cardiorespiratory failure and subdural hematoma (due to head trauma) are other frequent causes of death. The suicide rate among HD patients is 5 to 10 times higher than in the general population. Treatment includes drugs such as benzodiazepines to help control the choreic movements. Affective disturbances, which are seen in nearly half of the patients, are sometimes controlled with antipsychotic drugs and tricyclic antidepressants. Although these drugs help to control some of the symptoms of HD, there is currently no way to alter the outcome of the disease.

HD has the distinction of being the first genetic disease mapped to a specific chromosome using an RFLP marker, in 1983. Subsequent cloning and sequencing of the disease-causing gene showed that the mutation is a CAG expanded repeat (see Chapter 3) located in exon 1. In 90% to 95% of cases, the mutation is inherited from an affected parent. The normal repeat number ranges from 10 to 26. Persons with 27 to 35 repeats are unaffected but are more likely to transmit a still larger number of repeats to their offspring. The inheritance of 36 or more copies of the repeat can produce disease, although incomplete penetrance of the disease phenotype is seen in those who have 36 to 40 repeats. As in many disorders caused by trinucleotide repeat expansion, a larger number of repeats is correlated with earlier

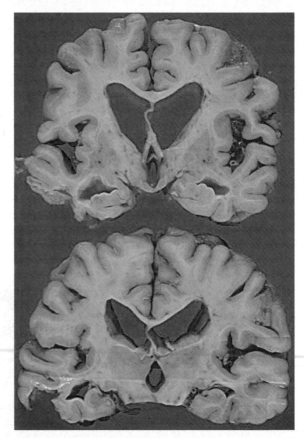

Two cross sections of the brain of an adult with Huntington disease, illustrating severe caudate atrophy and enlarged lateral ventricles.

*(Courtesy of Dr. Thomas Bird, University of Washington.)*

age of onset of the disorder. About 60% to 70% of the variation in age of onset of HD can be predicted by repeat number. There is a tendency for greater repeat expansion when the father, rather than the mother, transmits the disease-causing mutation, which helps to explain the difference in ages of onset for maternally and paternally transmitted disease seen in Figure 4-11. In particular, 80% of cases with onset before 20 years of age (juvenile Huntington disease) are due to paternal transmission, and these cases are characterized by especially large repeat expansions. It remains to be determined why the degree of repeat instability in the HD gene is greater in paternal transmission than in maternal transmission.

Cloning of the HD gene led quickly to the identification of the gene product, huntingtin. This protein is involved in the transport of vesicles in cellular secretory pathways. In addition, there is evidence that huntingtin

is necessary for the normal production of brain-derived neurotrophic factor. The CAG repeat expansion produces a lengthened series of glutamine residues near huntingtin's amino terminal. Although the precise role of the expanded glutamine tract in disease causation is unclear, it is correlated with a buildup of toxic protein aggregates within and near neuronal nuclei. These aggregates are thought to be toxic and are associated with early neuronal death. HD is notable in that affected homozygotes appear to display a clinical course very similar to that of heterozygotes (in contrast to most dominant disorders, in which homozygotes are more severely affected). This attribute, and the fact that mouse models in which one copy of the gene is inactivated are perfectly normal, supports the hypothesis that the mutation causes a harmful gain of function (see Chapter 3).

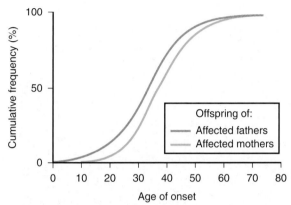

**FIGURE 4-11**
Distribution of the age of onset for Huntington disease. The age of onset tends to be somewhat earlier when the affected parent is the father.
*(Data from Conneally PM: Huntington disease: Genetics and epidemiology. Am J Hum Genet 1984;36:520.)*

A person whose parent has Huntington disease has a 50% chance of inheriting the disease allele. Until recently, this person would be confronted with a torturous question: Should I have children, knowing that there is a 50% chance that I might have this mutation and pass it to half of my children? With the identification of the mutation responsible for Huntington disease, it is now possible for at-risk persons to know with a high degree of certainty whether they carry a disease-causing allele.

As mentioned previously, a number of important genetic diseases exhibit age-dependent penetrance. These include

hemochromatosis, a recessive disorder of iron storage (see Chapter 7); familial Alzheimer disease (see Chapter 12); and many inherited cancers, including autosomal dominant breast cancer.*

▶ Age-dependent penetrance is observed in many genetic diseases. It complicates the interpretation of inheritance patterns in families.

### Variable Expression

Penetrance and expression are distinct entities. Penetrance is an all-or-none phenomenon: One either has the disease phenotype or does not. **Variable expression** refers to the degree of severity of the disease phenotype.

The severity of expression of many genetic diseases can vary greatly. A well-studied example of variable expression in an autosomal dominant disease is neurofibromatosis type 1, or von Recklinghausen disease (after the German physician who described the disorder in 1882). Clinical Commentary 4-4 provides further discussion of this disorder. A parent with mild expression of the disease—so mild that he or she is not aware of it—can transmit the disease-causing allele to a child, who may have severe expression. As with reduced penetrance, variable expression provides a mechanism for disease alleles to survive at higher frequencies in populations.

*Epidemiological studies indicate that about 5% of breast cancer cases in the United States are caused by genes inherited in autosomal dominant fashion. See chapters 11 and 12 for further discussion.

## CLINICAL COMMENTARY 4-4
### *Neurofibromatosis: A Disease with Highly Variable Expression*

Neurofibromatosis type 1 (NF1) is one of the most common autosomal dominant disorders, affecting approximately 1 in 3000 individuals in all populations. It provides a good example of variable expression in a genetic disease. Some patients have only *café-au-lait* spots (French for "coffee with milk," describing the color of the hyperpigmented skin patches), Lisch nodules (benign growths on the iris), and a few neurofibromas (nonmalignant peripheral nerve tumors). These persons are often unaware that they have the condition. Other patients have a much more severe expression

of the disorder, including hundreds to thousands of neurofibromas, plexiform neurofibromas, optic pathway gliomas (benign tumors of the optic nerve), learning disabilities, hypertension, scoliosis (lateral curvature of the spine), and malignancies. Fortunately, about two thirds of patients have only a mild cutaneous involvement. Approximately 10% develop malignant peripheral nerve sheath tumors (MPNSTs), which typically arise from plexiform neurofibromas. Expression can vary significantly within the same family: A mildly affected parent can produce a severely affected offspring.

*Continued*

A standard set of diagnostic criteria for NF1 has been developed. Two or more of the following must be present:

1. Six or more *café-au-lait* spots greater than 5 mm in diameter in prepubertal patients and greater than 15 mm in postpubertal patients
2. Freckling in the armpits or groin area
3. Two or more neurofibromas of any type or one plexiform neurofibroma (i.e., an extensive growth that occurs along a large nerve sheath)
4. Two or more Lisch nodules
5. Optic glioma
6. Distinctive bone lesions, particularly an abnormally formed sphenoid bone or tibial pseudarthrosis*
7. A first-degree relative with neurofibromatosis diagnosed using the previous six criteria

Although NF1 has highly variable expression, the penetrance of disease-causing mutations is virtually 100%. The *NF1* gene has one of the highest known mutation rates, about 1 in 10,000 per generation. Approximately 50% of patients with NF1 have the condition because of new mutations. *NF1* is a large gene, spanning approximately 350 kb of DNA. Its large size, which presents a sizable target for mutation, might help to account for the high mutation rate. The gene product, neurofibromin, acts as a tumor suppressor (see Chapter 11 for further details). *NF1* mutations can be detected in approximately 90% of cases, using a combination of detection methods including DNA sequencing, cytogenetic analysis, and analysis of abnormal (truncated) products. Persons in whom the entire *NF1* gene is deleted tend to be severely affected, with large numbers of neurofibromas and an increased risk of developing MPNSTs.

A mutation in the *NF1* gene that occurs during embryonic development will affect only some cells of the individual, resulting in somatic mosaicism. In this case, the disease features may be confined to only one part of the body (segmental neurofibromatosis).

Neurofibromatosis type 2 (NF2) is much rarer than NF1 and is characterized by vestibular schwannomas (tumors that arise in Schwann cells and affect the eighth cranial nerve) and, occasionally, *café-au-lait* spots. Patients who have NF2 do not, however, have true neurofibromas, so the term "neurofibromatosis type 2" is a misnomer. The *NF2* gene, which was mapped to chromosome 22, encodes a tumor suppressor protein called merlin or schwannomin.

Mild cases of neurofibromatosis can require very little clinical management. However, surgery may be required if malignancies develop or if benign tumors interfere with normal function. Scoliosis, tibial pseudarthrosis, and/or tibial bowing, seen in less than 5% of cases, can require orthopedic management. Hypertension can develop and is often secondary to a pheochromocytoma or a stenosis (narrowing) of the renal artery. The most common clinical problems in children are learning disabilities (seen in about 50% of persons with NF1), short stature, and optic gliomas (which can lead to vision loss). Close follow-up can help to detect these problems and minimize their effects. Recent clinical trials designed to reduce or eliminate the tumors seen in patients with NF1 have provided hope for better treatment options.

---

*Pseudarthrosis can occur when a long bone, such as the tibia, undergoes a loss of bone cortex, leading to weakening and fracture. Abnormal callus formation causes a false joint in the bone, leading to the term (*arthron* = "joint").

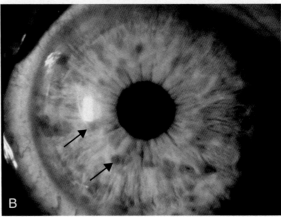

Neurofibromatosis type 1 (NF1). **A,** Multiple neurofibromas in an adult with type 1 neurofibromatosis. **B,** Lisch nodules (benign hamartomas of the iris) visible on a slit-lamp examination of an individual with type 1 neurofibromatosis.

*(A from Habif T, Campbell J, Chapman M, et al.: Skin Disease: Diagnosis and Treatment, 2nd ed. St. Louis, Mosby, 2005; B from Jones KL: Smith's Recognizable Patterns of Human Malformation, 6th ed. Philadelphia: Mosby, 2006.)*

---

Many factors can affect the expression of a genetic disease. They include environmental (i.e., nongenetic) influences such as diet, exercise, or exposure to harmful agents such as tobacco smoke. In the absence of a certain environmental factor, the disease-causing gene is expressed with diminished severity or not at all (e.g., the diminished expression of PKU under a low-phenylalanine diet). Another possible factor is the interaction of other genes, called **modifier loci**, with the disease-causing gene. Finally, variable expression can result from different types of mutations (i.e., different alleles) at the same disease locus. This is termed **allelic heterogeneity**. Efforts are often made to establish **genotype–phenotype correlations** to better predict the severity of a genetic disease, given the patient's genotype. In some cases, clinically distinct diseases may be the result of allelic heterogeneity, as in the β-globin mutations that can cause either sickle cell disease or various forms of β-thalassemia.

Cystic fibrosis, discussed in Clinical Commentary 4-1, illustrates ways these factors can influence disease severity. *CFTR* mutations that result in a complete absence of chloride channels on cell surfaces tend to produce more-severe disease than do mutations that result in partially active chloride ion channels (allelic heterogeneity). Some of the variation in severity of lung disease among CF patients with identical *CFTR* genotypes can be explained by variation in the *TGFB1* gene (transforming growth factor β), a modifier locus. Patients with CF who suffer from more-frequent and severe bacterial infections, a nongenetic (environmental) factor, experience accelerated lung damage. This disease thus provides examples of all three major causes of variable expression: allelic heterogeneity, modifier loci, and environmental factors.

Because of the many factors that can influence the expression of a genetic disease, it should be apparent that the commonly used term "single-gene disease" is an oversimplification. Although a mutation in a single gene may be sufficient to cause such a disease, its severity—always an important concern for clinicians—is typically influenced by many genetic and nongenetic factors.

> ▶ Variable expression of a genetic disease may be caused by environmental effects, modifier loci, or allelic heterogeneity.

### Locus Heterogeneity

Quite commonly, a single disease phenotype is caused by mutations at different loci in different families, which is termed **locus heterogeneity** (compare with allelic heterogeneity, discussed in the previous section, in which different mutations are seen within the same disease locus). A good example is adult polycystic kidney disease (APKD), an autosomal dominant disorder in which a progressive accumulation of renal cysts is seen. Patients can also develop liver cysts, hypertension, cerebral aneurysms, and cardiac valvular defects. Occurring in about 1 of every 1000 persons of European descent, this disorder is responsible for 8% to 10% of end-stage renal disease in North America. APKD can be caused by mutations in genes on either chromosome 16 (*PKD1*) or chromosome 4 (*PKD2*). Both of these genes encode membrane-spanning glycoproteins that interact with one another and may be involved in

**FIGURE 4-12**
Structure of the triple helix type 1 collagen protein. The two $\alpha_1$ chains are encoded by a gene on chromosome 17, and the $\alpha_2$ chain is encoded by a gene on chromosome 7.

cellular signaling. (When this signaling goes awry, it is thought that cellular growth regulation is compromised, resulting in cyst formation.) In one family, the disease may be caused by a *PKD1* mutation, whereas in another family it may be caused by a *PKD2* mutation. The disease states produced by mutations in these two genes may be clinically indistinguishable.

Osteogenesis imperfecta provides a second example of locus heterogeneity. Recall from Chapter 2 that the subunits of the procollagen triple helix are encoded by two genes, one on chromosome 17 and the other on chromosome 7 (Fig. 4-12). A mutation occurring in either of these genes can alter the normal structure of the triple helix, resulting ultimately in osteogenesis imperfecta. Table 4-2 lists some additional examples of diseases in which there is locus heterogeneity.

> ▶ A disease that can be caused by mutations at different loci in different families is said to exhibit locus heterogeneity.

### Pleiotropy

Genes that have more than one discernible effect on the body are said to be **pleiotropic**. A good example of a gene with pleiotropic effects is given by Marfan syndrome. First described in 1896 by Antoine Marfan, a French pediatrician, this autosomal dominant disorder affects the eye, the skeleton, and the cardiovascular system (Clinical Commentary 4-5). Most of the observed features of Marfan syndrome are caused by unusually stretchable connective tissue. The great majority of Marfan syndrome cases are caused by mutations in the gene that encodes fibrillin, a component of connective tissue that is expressed in most of the tissues and organs affected by Marfan syndrome (see Clinical Commentary 4-5). We have already discussed several other single-gene diseases in which pleiotropy is seen, including cystic fibrosis, in which sweat glands, lungs, and pancreas can be affected; osteogenesis imperfecta, in which bones, teeth, and sclerae can be affected; and albinism, in which pigmentation and optic fiber development are affected.

**TABLE 4-2**
## Some Examples of Diseases in Which There Is Locus Heterogeneity

| Disease | Description | Chromosomes on Which Known Loci Are Located |
|---|---|---|
| Retinitis pigmentosa* | Progressive retinopathy and loss of vision (see Chapter 8) | More than 20 chromosome regions identified |
| Osteogenesis imperfecta | Brittle bone disease | 7, 17 |
| Charcot–Marie–Tooth disease | Peripheral neuropathy | 1, 5, 8, 10, 11, 17, 19, X |
| Familial Alzheimer disease | Progressive dementia | 1, 10, 12, 14, 19, 21 |
| Familial melanoma | Autosomal dominant melanoma (skin cancer) | 1, 9 |
| Hereditary nonpolyposis colorectal cancer | Autosomal dominant colorectal cancer | 2p, 2q, 3, 7 |
| Autosomal dominant breast cancer | Predisposition to early-onset breast and ovarian cancer | 13, 17 |
| Tuberous sclerosis | Seizures, facial angiofibromas, hypopigmented macules, mental retardation, multiple hamartomas | 9, 16 |
| Adult polycystic kidney disease | Accumulation of renal cysts leading to kidney failure | 4, 16 |

## CLINICAL COMMENTARY   4-5
### *Marfan Syndrome: An Example of Pleiotropy*

Marfan syndrome is an autosomal dominant condition seen in approximately 1 of every 10,000 North Americans. It is characterized by defects in three major systems: ocular, skeletal, and cardiovascular. The ocular defects include myopia, which is present in most patients with Marfan syndrome, and displaced lens (ectopia lentis), which is observed in about half of Marfan syndrome patients. The skeletal defects include dolichostenomelia (unusually long and slender limbs), pectus excavatum ("hollow chest"), pectus carinatum ("pigeon chest"), scoliosis, and arachnodactyly (literally "spider fingers," denoting the characteristically long, slender fingers). Marfan patients also typically exhibit joint hypermobility.

The most life-threatening defects are those of the cardiovascular system. Most patients with Marfan syndrome develop prolapse of the mitral valve, a condition in which the cusps of the mitral valve protrude upward into the left atrium during systole. This can result in mitral regurgitation (leakage of blood back into the left atrium from the left ventricle). Mitral valve prolapse, however, is seen in 1% to 3% of the general population and is often of little consequence. A more serious complication is dilatation (widening) of the ascending aorta, which is seen in 90% of Marfan patients. As dilatation increases, the aorta becomes susceptible to dissection or rupture, particularly when cardiac output is high (as in heavy exercise or pregnancy). As the aorta widens, the left ventricle enlarges, and cardiomyopathy (damage to the heart muscle) ensues. The end result is congestive heart failure, a common cause of death among Marfan syndrome patients.

Most cases of Marfan syndrome are caused by mutations in a gene, *FBN1*, that is expressed in the aorta, the periosteum, and the suspensory ligament of the lens. Because *FBN1* encodes a connective tissue protein, fibrillin, mutations of this gene alter the structure of connective tissue. This helps to explain some of the cardiovascular and ocular features of this disorder. Hundreds of different *FBN1* mutations have been identified in Marfan syndrome patients. Most of these are missense mutations, but frameshifts and nonsense mutations producing a truncated fibrillin protein are also seen. In many cases, the missense mutations produce a more severe disease

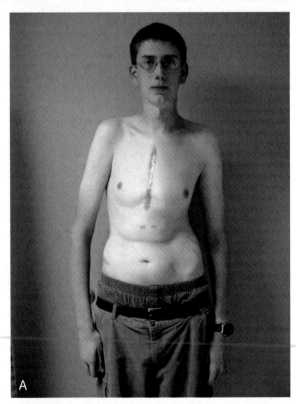

A, A young man with Marfan syndrome, showing characteristically long limbs and narrow face.

*(From Jones KL: Smith's Recognizable Patterns of Human Malformation, 6th ed, pp 549. Philadelphia: Saunders, 2006.)*

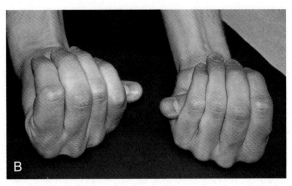

**B,** Arachnodactyly in an 8-year-old girl with Marfan syndrome. Note projection of the thumb well beyond the edge of the palm (Steinberg thumb sign).

*(From Jones KL: Smith's Recognizable Patterns of Human Malformation, 6th ed. Philadelphia: Mosby, 2006.)*

phenotype because of a dominant negative effect (i.e., the abnormal fibrillin proteins bind to and disable many of the normal fibrillin proteins produced by the normal allele in a heterozygote). A severe neonatal form of the disease is produced by mutations in exons 24 to 32. At least one Marfan syndrome compound heterozygote has been reported. This infant, who inherited a disease-causing allele from each of its affected heterozygous parents, had severe congestive heart failure, and died from cardiac arrest at 4 months of age.

Specific mutations in *FBN1* can cause familial arachnodactyly (with no other symptoms of Marfan syndrome), whereas others can cause familial ectopia lentis. A disease called *congenital contractural arachnodactyly* exhibits many of the skeletal features of Marfan syndrome but does not involve cardiac or ocular defects. This disease is caused by mutations in a second gene, *FBN2*, that encodes another form of fibrillin.

A small percentage of persons with Marfan syndrome do not have mutations in *FBN1* or *FBN2* but instead have mutations in the gene that encodes transforming growth factor β receptor 2 (*TGFBR2*). These mutations increase the signaling activity of transforming growth factor β (TGF-β), contributing to aortic dilatation and abnormal bone growth. It is interesting that the fibrillin protein is also thought to interact with TGF-β, such that mutations that disrupt fibrillin might also increase TGF-β signaling. Thus, *FBN1* mutations can produce structural connective tissue abnormalities as well as abnormal TGF-β activity, accounting for the pleiotropic features of this disorder.

Treatment for Marfan syndrome includes regular ophthalmological examinations and, for individuals with aortic dilatation, the avoidance of heavy exercise and contact sports. In addition, β-adrenergic blockers (e.g., atenolol) can be administered to decrease the strength and abruptness of heart contractions. This reduces stress on the aorta, although it is unclear whether these drugs reduce aortic dilatation. In some cases, the aorta and aortic valve are surgically replaced with a synthetic tube and artificial valve. With such treatment, persons with Marfan syndrome can achieve nearly normal life spans.

Another possible avenue for treatment has been opened by the discovery of elevated TGF-β signaling in Marfan syndrome. In mouse models of this disorder, administration of TGF-β antagonists has been shown to prevent aortic dilatation. One of these agents, losartan, is an angiotensin II type 1 receptor antagonist and is commonly used to treat high blood pressure. This drug is now being evaluated in clinical trials for the treatment of Marfan syndrome.

A number of historical figures might have had Marfan syndrome, including Niccolo Paganini, the violinist, and Sergei Rachmaninoff, the composer and pianist. Most controversial is the proposal that Abraham Lincoln might have had Marfan syndrome. He had skeletal features consistent with the disorder, and examination of his medical records has shown that he may well have had aortic dilatation. Some have suggested that he was in congestive heart failure at the time of his death and that, had he not been assassinated, he still would not have survived his second term of office.

> Genes that exert effects on multiple aspects of physiology or anatomy are pleiotropic. Pleiotropy is a common feature of human genes.

## CONSANGUINITY IN HUMAN POPULATIONS

Although consanguinity is relatively rare in Western populations, it is common in many populations of the world. For example, first-cousin unions are seen in 20% to 50% of marriages in many countries of the Middle East, and uncle–niece and first-cousin marriages are common in some parts of India. Because relatives more often share disease genes inherited from a common ancestor, consanguineous unions are more likely to produce offspring affected by autosomal recessive disorders. It is possible to quantify the percentage of genes shared by a pair of relatives by estimating the **coefficient of relationship** (Box 4-1). Estimation of this quantity shows, for example, that siblings share 1/2 of their DNA sequences, on average, because they share two parents. Uncles and nieces share 1/4 of their DNA sequences because of common ancestry, first cousins share 1/8, first cousins once removed share 1/16, second cousins share 1/32, and so on.*

---

*First cousins are the offspring of two siblings and thus share a set of grandparents. A first cousin once removed is the offspring of one's first cousin. Second cousins are the offspring of two different first cousins and thus share a set of great-grandparents.

## Consanguinity and the Frequency of Recessive Diseases

Recall that about 1 in 25 whites is a heterozygous carrier of a mutation that causes cystic fibrosis. A man who carries this allele thus has a 1 in 25 chance of meeting another carrier if he mates with somebody in the general population. He only triples his chance of meeting another carrier if he mates with a first cousin, who has a 1/8 chance of carrying the same gene. In contrast, a carrier of a relatively rare recessive disease, such as classic galactosemia (a metabolic disorder discussed in Chapter 7), has only a 1/170 chance of meeting another carrier in the general population. Because he shares 1/8 of his DNA with his first cousin, the chance that his first cousin also has a galactosemia mutation is still 1/8. With this rarer disease, a carrier is 21 times more likely to mate with another carrier in a first-cousin marriage than in a marriage with an unrelated individual. This illustrates an important principle: The rarer the recessive disease, the more likely that the parents of an affected individual are consanguineous.

This principle has been substantiated empirically. A French study showed that the frequency of first-cousin marriages in that country was less than 0.2%. Among patients with cystic fibrosis, a relatively common recessive disorder, 1.4% were the offspring of first-cousin matings. This percentage rose

## BOX 4-1
## Measurement of Consanguinity: The Coefficient of Relationship

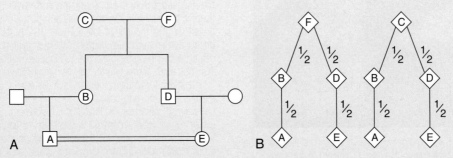

**A,** Pedigree for a first-cousin mating. **B,** The Pedigree is condensed to show only those individuals who are related to both of the first cousins.

To determine the possible consequences of a consanguineous mating, it is useful to know what percentage of genes are shared by two related individuals. The coefficient of relationship is a measure of this percentage. Clearly, individuals who are more closely related must share a greater percentage of their genes. To begin with a simple example, an individual receives half of his or her genes from each parent. Thus, the coefficient of relationship between a parent and offspring is 1/2. This also means that the probability that the parent and offspring share a given gene (e.g., a disease allele) is 1/2.

To continue with a more complex example, suppose that a man is known to be a heterozygous carrier for galactosemia, a relatively rare autosomal recessive metabolic disorder. If he mates with his first cousin, what is the probability that she also carries this disease gene? We know that this probability must be higher than that of the general population, because first cousins share one set of grandparents. There is thus a possibility that the grandparent who transmitted the galactosemia gene to the known carrier also transmitted it to the carrier's cousin. The coefficient of relationship specifies this probability. A pedigree for a first-cousin mating is shown below in Figure A on the left. The male carrier is labeled A, and his female cousin is labeled E. Because we are interested only in the family members who are related to both the man and his cousin, the pedigree is condensed, in Figure B on the right, to include only those individuals who form a path between the man and his cousin.

To estimate the coefficient of relationship, we begin with the carrier and ascend the pedigree. We know that there is a probability of 1/2 that the known carrier inherited the gene from the parent in the path (labeled B). There is also a probability of 1/2 that he inherited the gene from his other parent, who is not related to his cousin and is thus not included in the diagram. By similar reasoning, the probability that individual B inherited the disease gene from his parent, individual C, is also 1/2. The probability that individual C in turn passed on the disease gene to his offspring, D, is 1/2, and the probability that D passed the disease gene to E is also 1/2. Thus, for E to share a disease gene with A, each of these four events must have taken place. The multiplication

rule dictates that, to find the probability that all four events have taken place, we take the product of all four probabilities. Because each of these probabilities is 1/2, the result is $(1/2)^4 = 1/16$.

If individuals A and E shared only one grandparent, the coefficient of relationship would be 1/16. But, as with most first cousins, they share a common grandfather and grandmother. Thus, there are two paths through which the disease gene could have passed. To obtain the probability that the gene passed through the second path, we use the same procedure as in the previous paragraph and obtain a probability of 1/16. Now we need to estimate the probability that the gene went through either the first path or the second (i.e., through one grandparent or the other). The addition rule states that we can add these two probabilities together to get the overall probability that A and E share a disease gene: $1/16 + 1/16 = 1/8$. The probability that the carrier's cousin shares his disease allele, as a result of their descent from a common set of grandparents, is thus 1/8. This is the coefficient of relationship for first cousins.*

It should be recognized that individual E could also inherit a disease allele from an ancestor not included in either of these paths. However, for disease alleles that are relatively rare in populations, this probability is small and can usually be disregarded.

The rules for calculating the coefficient of relationship can be summarized as follows:

1. Each individual can appear in a route only once.
2. Always begin with one individual, proceed up the pedigree to the common ancestor, then down the pedigree to the other individual.
3. The coefficient of relationship for one route is given by $(1/2)^{n-1}$, where n is the number of individuals in the route.
4. If there are multiple routes (i.e., multiple common ancestors), the probabilities estimated for each route are added together.

---

*A related quantity, often used in population genetics, is the **inbreeding coefficient**. This coefficient is the probability that an individual is homozygous at a locus as a result of consanguinity in his or her parents. For a given type of mating, the inbreeding coefficient of an individual always equals the parents' coefficient of relationship multiplied by 1/2 (e.g., the inbreeding coefficient for the offspring of a first-cousin mating is 1/16).

---

to 7.1% for cystinosis and 12.5% for achromatopsia, both of which are less-common recessive disorders.

> Consanguinity increases the chance that a mating couple will both carry the same disease-causing mutation. It is seen more often in pedigrees involving rare recessive diseases than in those involving common recessive diseases.

### Health Consequences of Consanguinity

It has been estimated that each person carries the equivalent of one to five recessive mutations that would be lethal to offspring if matched with another copy of the mutation (i.e., homozygosity). It would therefore be expected that matings between relatives would more often produce offspring with genetic diseases. In fact, most empirical studies do show that mortality rates among the offspring of first-cousin marriages

**TABLE 4-3**

## Mortality Levels among Cousin and Unrelated Control Marriages in Selected Human Populations

| Population | Mortality Type | 1.0 Cousin | | 1.5 Cousin* | | 2.0 Cousin | | Unrelated | |
|---|---|---|---|---|---|---|---|---|---|
| | | % | N | % | N | % | N | % | N |
| Amish (Old Order) | Prereproductive | 14.4 | 1218[†] | - | - | 13.3 | 6064 | 8.2 | 17,200 |
| Bombay, India | Perinatal | 4.8 | 3309 | 2.8 | 176 | 0 | 30 | 2.8 | 35,620 |
| France (Loir-et-Cher) | Prereproductive | 17.7 | 282 | 6.7 | 105 | 11.7 | 240 | 8.6 | 1117 |
| Fukuoka, Japan | 0 to 6 yr | 10.0 | 3442 | 8.3 | 1048 | 9.2 | 1066 | 6.4 | 5224 |
| Hirado, Japan | Prereproductive | 18.9 | 2301 | 15.3 | 764 | 14.7 | 1209 | 14.3 | 28,569 |
| Kerala, India | Prereproductive | 18.6 | 391 | - | - | 11.8 | 34 | 8.7 | 770 |
| Punjab, Pakistan | Prereproductive | 22.1 | 3532 | 22.9 | 1114 | 20.1 | 57 | 16.4 | 4731 |
| Sweden | Prereproductive | 14.1 | 185 | 13.7 | 227 | 11.4 | 79 | 8.6 | 625 |
| Utah Mormons | Prereproductive | 22.4 | 1048 | 15.3 | 517 | 12.2 | 1129 | 13.2 | 302,454 |

*First cousins once removed.

[†]Includes 1.5 cousins.

Modified from Jorde LB: Inbreeding in human populations. In: Dulbecco R (ed): Encyclopedia of Human Biology, vol 5. New York: Academic Press, 1997, pp 1-13.

are substantially greater than those of the general population (Table 4-3). Similarly, the prevalence of genetic disease is roughly twice as high among the offspring of first-cousin marriages as among the offspring of unrelated persons. First-cousin marriages are illegal in most states of the United States. Marriages between closer relatives (except double first cousins, who share both sets of grandparents) are prohibited throughout the United States.

Very few data exist for matings between siblings or parents and offspring (defined as **incest**). The limited data indicate that the fraction of abnormal offspring produced by incestuous matings is very high: between 1/4 and 1/2. Mental retardation is particularly common among these offspring. Because of small sample sizes in these studies, it is difficult to separate the effects of genetics from those of a substandard environment. It is likely that the problems experienced by the offspring of incestuous matings are caused by both genetic and environmental influences.

▶ At the population level, consanguinity increases the frequency of genetic disease and mortality. The closer the degree of consanguinity, the greater the increase.

## Study Questions

**1.** A man who has achondroplasia marries a phenotypically normal woman. If they have four children, what is the probability that none of their children will be affected with this disorder? What is the probability that *all* of them will be affected?

**2.** The estimated penetrance for familial retinoblastoma is approximately 90%. If a man has had familial retinoblastoma and mates with a woman who does not have a retinoblastoma mutation, what is the risk that their offspring will develop retinoblastoma?

**3.** A 30-year-old woman had a sister who died from infantile Tay–Sachs disease, an autosomal recessive disorder that is fatal by age 6 years. What is the probability that this woman is a heterozygous carrier of the Tay–Sachs mutation?

**4.** A man has neurofibromatosis type 1. His mother also has this condition. What is the probability that his sister also has this disease? In the absence of knowledge of his sister's phenotype, what is the probability that his sister's daughter has neurofibromatosis type 1?

**5.** Consider a woman who is a known heterozygous carrier of a mutation that causes PKU (autosomal recessive). What is the probability that her two grandchildren, who are first cousins, are both heterozygous carriers of this PKU-causing allele? Suppose instead that the woman is affected with PKU. Now what is the probability that both of her grandchildren carry the disease-causing allele?

*Continued*

## Study Questions—cont'd

**6.** Two mating individuals, labeled A and B in Figure 4-13, share a single great-grandparent. What is their coefficient of relationship? Suppose that one member of this couple is a heterozygous carrier for PKU. What is the probability that this couple will produce a child affected with PKU?

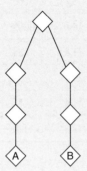

**FIGURE 4-13**
Diagram for study question 6.

**7.** A suspect in a rape case has been typed for three STR (short tandem repeat) loci. His alleles match those of the evidentiary sample (semen taken from the rape victim) for each locus. He is a heterozygote for the first two loci and a homozygote for the third. The allele frequencies for locus 1 in the general population are 0.05 and 0.10. For locus 2, they are 0.07 and 0.02. For locus 3, the allele frequency in the general population is 0.08. What is the probability that a random individual in the general population would match the evidentiary sample?

**8.** A man implicated in a paternity suit has had his DNA tested to establish whether or not he is the father of the baby. Four STR loci were tested for him, the mother, and the baby. The baby's alleles and the man's alleles match for all four loci. The frequencies of these alleles in the general population are 0.05, 0.01, 0.01, and 0.02. What is the probability that someone else in the general population could be the father of the baby?

## Suggested Readings

Balmer A, Zografos L, Munier F. Diagnosis and current management of retinoblastoma. Oncogene 2006;25:5341–49.

Bittles A. Consanguinity and its relevance to clinical genetics. Clin Genet 2001;60:89–98.

Borrell-Pages M, Zala D, Humbert S, Saudou F. Huntington's disease: From huntingtin function and dysfunction to therapeutic strategies. Cell Mol Life Sci 2006; 63:2642–60.

Ferner RE, Huson SM, Thomas N, et al. Guidelines for the diagnosis and management of individuals with neurofibromatosis 1. J Med Genet 2007;44:81–8.

Grantham JJ. Autosomal dominant polycystic kidney disease. N Engl J Med 2008;359(14):1477–85.

Gusella JF, Macdonald ME. Huntington's disease: Seeing the pathogenic process through a genetic lens. Trends Biochem Sci 2006;31:533–40.

Jorde LB. Inbreeding in human populations. In: Dulbecco R, (ed): Encyclopedia of Human Biology, vol. 5. New York: Academic Press, 1997, pp. 1–13.

Judge DP, Dietz HC. Marfan's syndrome. Lancet 2005;366:1965–76.

Keane MG, Pyeritz RE. Medical management of Marfan syndrome. Circulation 2008;117(21):2802–13.

Knowles MR. Gene modifiers of lung disease. Curr Opin Pulm Med 2006;12:416–21.

Li SH, Li XJ. Huntingtin-protein interactions and the pathogenesis of Huntington's disease. Trends Genet 2004;20:146–54.

Melamud A, Palekar R, Singh A. Retinoblastoma. Am Fam Physician 2006;73:1039–44.

Modell B, Darr A. Science and society: Genetic counselling and customary consanguineous marriage. Nat Rev Genet 2002;3:225–29.

Nadeau JH. Modifier genes in mice and humans. Nature Rev Genet 2001;2:165–74.

Potter A, Phillips JA, Rimoin DL. Genetic disorders of the pituitary gland. In: Rimoin DL, Connor JM, Pyeritz RE, Korf BR, (eds): Emery and Rimoin's Principles and Practice of Medical Genetics, vol. 2. Philadelphia: Churchill Livingstone, 2007, pp. 1889–931.

Ramirez F, Dietz HC. Marfan syndrome: From molecular pathogenesis to clinical treatment. Curr Opin Genet Dev 2007;17:252–58.

Reynolds RM, Browning G, Nawroz I, Campbell IW. Von Recklinghausen's neurofibromatosis: Neurofibromatosis type 1. Lancet 2003;361:1552–54.

Rowe SM, Clancy JP. Advances in cystic fibrosis therapies. Curr Opin Pediatr 2006;18:604–13.

Rowe SM, Miller S, Sorscher EJ. Cystic fibrosis. N Engl J Med 2005;352:1992–2001.

Scriver CR, Waters PJ. Monogenic traits are not simple. Lessons from phenylketonuria. Trends Genet 1999; 15: 267–72.

Sturm RA, Teasdale RD, Box NF. Human pigmentation genes: Identification, structure and consequences of polymorphic variation. Gene 2001;277:49–62.

Theos A, Korf BR. Pathophysiology of neurofibromatosis type 1. Ann Intern Med 2006;144:842–9.

Torres VE, Harris PC, Pirson Y. Autosomal dominant polycystic kidney disease. Lancet 2007;369:1287–1301.

Walker FO. Huntington's disease. Lancet 2007; 369: 218–28.

Zlotogora J. Germ line mosaicism. Hum Genet 1998; 102:381–86.

## Internet Resources

Cystic Fibrosis Mutation Database (also contains links to other useful cystic fibrosis websites) *http://www.genet.sickkids.on.ca/cftr/*

Eye Cancer Network: Retinoblastoma (descriptions, photographs, and useful links) *http://www.eyecancer.com/Patient/Condition.aspx?nID=53&Category=Retinal+Tumors&Condition=Retinoblastoma*

National Center for Biotechnology Information: Genes and Disease (brief summaries of many of the genetic diseases discussed in this text) *http://www.ncbi.nlm.nih.gov/books/bv.fcgi?call=bv.View..ShowTOC&rid=gnd.TOC&depth=2*

National Institute of Neurological Diseases and Stroke: Huntington Disease Information Page *http://www.ninds.nih.gov/disorders/huntington/huntington.htm*

National Marfan Foundation (basic information about Marfan syndrome, with links to other sites) *http://www.marfan.org/*

National Neurofibromatosis Foundation (many useful links to online resources) *http://www.ctf.org/*

## Chapter 5

# SEX-LINKED AND NONTRADITIONAL MODES OF INHERITANCE

The previous chapter dealt with genes located on the 22 autosomes; their mode of inheritance was elucidated by Gregor Mendel. In this chapter we discuss disease-causing mutations that are inherited in ways that were unknown to Mendel and are thus sometimes termed **nonmendelian**.

The first mutations to be discussed are DNA variants of the sex chromosomes (X and Y), known as **sex-linked mutations**. The human X chromosome is a large chromosome, containing about 5% of the nuclear genome's DNA (approximately 155 million base pairs [155 megabases, 155 Mb]). Almost 1100 genes have been localized to the X chromosome. Diseases caused by genes on this chromosome are said to be **X-linked**. In contrast to the X chromosome, the Y chromosome is quite small (60 Mb) and contains only a few dozen genes.

The next group of disease-causing mutations is located in the mitochondrial genome, which is inherited only from one's mother. Mitochondrial diseases thus display a unique pattern of inheritance in families. Extensive analyses have revealed a growing number of disease-causing mutations in the mitochondrial genome.

Finally, we discuss two processes that have been elucidated only in the past 2 to 3 decades: anticipation and imprinting. **Anticipation** refers to earlier age-of-onset of some genetic diseases in more recent generations of families. **Imprinting** refers to the fact that some genes are expressed only on paternally transmitted chromosomes and others are expressed only on maternally transmitted chromosomes. Our understanding of both of these recently discovered processes has been greatly enhanced by detailed molecular analyses of humans and model organisms.

## X INACTIVATION

The X chromosome contains many important protein-coding genes, and it has long been known that human females have two X chromosomes and males have only one. Thus, females have two copies of each X-linked gene, and males have only one copy. Yet males and females do not differ in terms of the amounts of protein products (e.g., enzyme levels) encoded by most of these genes. What could account for this?

In the early 1960s Mary Lyon hypothesized that one X chromosome in each somatic cell of the female is inactivated.

This would result in **dosage compensation**, an equalization of the amount of X-linked gene products in males and females. The **Lyon hypothesis** stated that X inactivation occurs early in female embryonic development and that the X chromosome contributed by the father is inactivated in some cells, whereas in other cells the X chromosome contributed by the mother is inactivated. In each cell, one of the two X chromosomes is chosen at random for inactivation, so the maternally and paternally derived X chromosomes are each inactivated in about half of the embryo's cells. Thus, inactivation, like gamete transmission, is analogous to a coin-tossing experiment. Once an X chromosome is inactivated in a cell, it will remain inactive in all descendants of that cell. X inactivation is therefore a randomly determined, but *fixed* (or permanent), process. As a result of X inactivation, all normal females have two distinct populations of cells: One population has an active paternally derived X chromosome, and the other has an active maternally derived X chromosome. (Fig. 5-1 provides a summary of this process.) Because they have two populations of cells, females are mosaics (see Chapter 4) for X chromosome activity. Males, having only one copy of the X chromosome, are not mosaics but are **hemizygous** for the X chromosome (*hemi* means "half").

> The Lyon hypothesis states that one X chromosome in each cell is randomly inactivated early in the embryonic development of females. This ensures that females, who have two copies of the X chromosome, will produce X-linked gene products in quantities roughly similar to those produced in males (dosage compensation).

The Lyon hypothesis relied on several pieces of evidence, most of which were derived from animal studies. First, it was known that females are typically mosaics for some X-linked traits and males are not. For example, female mice that are heterozygous for certain X-linked coat-color genes exhibit a dappled coloring of their fur, whereas male mice do not. A similar example is given by the calico cat. These female cats have alternating black and orange patches of fur that correspond to two populations of cells: one that contains

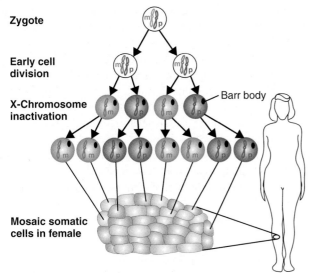

**FIGURE 5-1**

The X inactivation process. The maternal (*m*) and paternal (*p*)
X chromosomes are both active in the zygote and in early embryonic cells.
X inactivation then takes place, resulting in cells having either an active
paternal X or an active maternal X chromosome. Females are thus
X chromosome mosaics, as shown in the tissue sample at the bottom of
the figure.

X chromosomes in which an "orange" allele is active and one
that contains X chromosomes in which a "black" allele is
active. Male cats of this breed do not exhibit alternating colors. A final example, seen in humans, is X-linked ocular albinism. This is an X-linked recessive condition characterized
by a lack of melanin production in the retina and by ocular
problems such as nystagmus (rapid involuntary eye movements) and decreased visual acuity. Males who inherit the

mutation show a relatively uniform lack of melanin in their
retinas, and female heterozygotes exhibit alternating patches
of pigmented and nonpigmented tissue (Fig. 5-2).

The Lyon hypothesis was also supported by biochemical
evidence. The enzyme glucose-6-phosphate dehydrogenase
(G6PD) is encoded by a gene on the X chromosome and
is present in equal quantities in males and females (dosage
compensation). In females who are heterozygous for two
common G6PD alleles (labeled *A* and *B*), some skin cells
produce only the A variant of the enzyme and others produce only the B variant. This is further proof of X chromosome mosaicism in females.

Finally, cytogenetic studies in the 1940s showed that
interphase cells of female cats often contained a densely
staining chromatin mass in their nuclei. These masses were
not seen in males. They were termed **Barr bodies**, after
Murray Barr, one of the scientists who described them. Barr
and his colleague Ewart Bertram hypothesized that the Barr
body represented a highly condensed X chromosome. It is
now known that Barr and Bertram were correct, and that
the inactive X chromosome is observable as a Barr body
in the somatic cells of normal females. Its condensed state
is correlated with transcriptional activity, and its DNA
is replicated later in the S phase than that of other
chromosomes.

> The Lyon hypothesis is supported by cytogenetic
> evidence: Barr bodies, which are inactive
> X chromosomes, are seen only in cells with two
> or more X chromosomes. It is also supported
> by biochemical and animal studies that reveal
> mosaicism of X-linked traits in female
> heterozygotes.

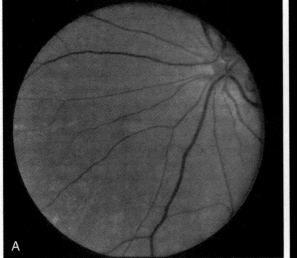
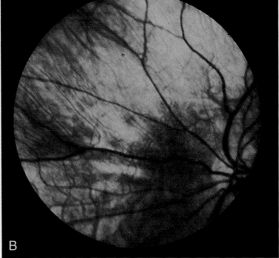

**FIGURE 5-2**

Fundus photos of X-linked ocular albinism. **A,** Fundus photograph of a female heterozygous carrier for X-linked ocular albinism. The pigmented and
nonpigmented patches demonstrate mosaicism of the X chromosome as a result of random X inactivation. **B,** Fundus photograph of the heterozygous carrier's
son, showing a much greater lack of melanin pigment.

*(Courtesy of Dr. Donnell J. Creel, University of Utah Health Sciences Center.)*

Further study has largely verified the Lyon hypothesis. Messenger RNA (mRNA) is transcribed from only one X chromosome in each somatic cell of a normal female. The inactivation process takes place within approximately 7 to 10 days after fertilization, when the embryonic inner-cell mass contains no more than a few dozen cells. Inactivation is initiated in a single 1-Mb region on the X chromosome long arm, the **X inactivation center**, and then spreads along the chromosome. Although inactivation is random among cells that make up the embryo itself, only the paternally derived X chromosome is inactivated in cells that will become extraembryonic tissue (e.g., the placenta). X inactivation is permanent for all somatic cells in the female, but the inactive X chromosome must later become reactivated in the female's germline so that each of her egg cells will receive one active copy of the X chromosome.

An important implication of the Lyon hypothesis is that the number of Barr bodies in somatic cells is always one less than the number of X chromosomes. Normal females have one Barr body in each somatic cell, and normal males have none. Females with Turner syndrome (see Chapter 6), having only one X chromosome, have no Barr bodies. Males with Klinefelter syndrome (two X chromosomes and a Y chromosome) have one Barr body in their somatic cells, and females who have three X chromosomes per cell have two Barr bodies in each somatic cell. This pattern leads to another question: if the extra X chromosomes are inactivated, why aren't people with extra (or missing) X chromosomes phenotypically normal?

The answer to this question is that X inactivation is *incomplete*. Some regions of the X chromosome remain active in all copies. For example, the tips of the short and long arms of the X chromosome do not undergo inactivation. The tip of the short arm of the X chromosome is homologous to the distal short arm of the Y chromosome (see Chapter 6). In total, about 15% of the genes on the X chromosome escape inactivation, and relatively more genes on the short arm escape inactivation than on the long arm. Some of the X-linked genes that remain active on both copies of the X chromosome have homologs on the Y chromosome, preserving equal gene dosage in males and females. Thus, having extra (or missing) copies of active portions of the X chromosome contributes to phenotypic abnormality.

> X inactivation is random, fixed, and incomplete. The last fact helps to explain why, despite X inactivation, most persons with abnormal numbers of sex chromosomes have a disease phenotype.

The X inactivation center contains a gene, *XIST*, that is transcribed only on the inactive X chromosome; its 17-kb mRNA transcripts are detected in normal females but not in normal males. The RNA transcript, however, is not translated into a protein. Instead, it remains in the nucleus and coats the inactive X chromosome. This coating process could act as a signal that leads to other aspects of inactivation, including late replication and condensation of the inactive X chromosome.

Methylation and histone deacetylation are additional features of the inactive X chromosome. Many CG dinucleotides in the 5′ regions of genes on the inactive X are heavily methylated, and the administration of demethylating agents, such as 5-azacytidine, can partially reactivate an inactive X chromosome in vitro. However, methylation does not appear to be involved in spreading the inactivation signal from the inactivation center to the remainder of the X chromosome. It is more likely to be responsible for maintaining the inactivation of a specific X chromosome in a cell and all of its descendants.

> The *XIST* gene is located in the X inactivation center and is required for X inactivation. It encodes an RNA product that coats the inactive X chromosome. X inactivation is also associated with methylation of the inactive X chromosome, a process that might help to ensure the long-term stability of inactivation.

## SEX-LINKED INHERITANCE

Sex-linked genes are those that are located on either the X or the Y chromosome. Because only a few dozen genes are known to be located on the human Y chromosome, our attention will be focused mostly on X-linked diseases. These have traditionally been grouped into X-linked recessive and X-linked dominant categories, and these categories are used here for consistency with other literature. However, because of variable expression, incomplete penetrance, and the effects of X inactivation, the distinction between X-linked dominant and X-linked recessive inheritance is sometimes ambiguous.

### X-Linked Recessive Inheritance

A number of well-known diseases and traits are caused by X-linked recessive genes. These include hemophilia A (Clinical Commentary 5-1), Duchenne muscular dystrophy (Clinical Commentary 5-2), and red–green colorblindness (Box 5-1). Additional X-linked diseases are listed in Table 5-1. The inheritance patterns and recurrence risks for X-linked recessive diseases differ substantially from those for diseases caused by autosomal genes.

Because females inherit two copies of the X chromosome, they can be homozygous for a disease-causing allele at a given locus, heterozygous at the locus, or homozygous for the normal allele at the locus. In this way, X-linked loci in females are much like autosomal loci. However, for most X-linked loci, there is only one copy of the allele in an individual somatic cell (because of X inactivation). This means that about half of the cells in a heterozygous female will express the disease allele and half will express the normal allele.

## CLINICAL COMMENTARY 5-1
### *Hemophilia A*

Hemophilia A is caused by mutations in the gene that encodes clotting factor VIII and affects approximately 1 in 5000 to 1 in 10,000 males worldwide. It is the most common of the severe bleeding disorders and has been recognized as a familial disorder for centuries. The Talmud states that boys whose brothers or cousins bled to death during circumcision are exempt from the procedure (this may well be the first recorded example of genetic counseling).

Queen Victoria carried a factor VIII mutation, transmitting it to one son and two carrier daughters. They in turn transmitted the allele to many members of the royal families of Germany, Spain, and Russia. One of the males affected by the disease was the Tsarevitch Alexei of Russia, the son of Tsar Nicholas II and Alexandra. Grigori Rasputin, called the "mad monk," had an unusual ability to calm the Tsarevitch during bleeding episodes, probably through hypnosis. As a result, he came to have considerable influence in the royal court, and some historians believe that his destabilizing effect helped to bring about the 1917 Bolshevik revolution. (Recently, the Russian royal family was again touched by genetics. Using the polymerase chain reaction, autosomal DNA microsatellites and mitochondrial DNA sequences were assayed in the remains of several bodies exhumed near Yekaterinburg, the reputed murder site of the royal family. Analysis of this genetic variation and comparison with living maternal relatives showed that the bodies were indeed those of the Russian royal family.)

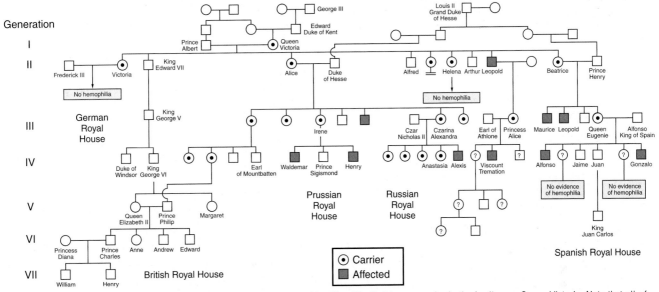

A pedigree showing the inheritance of hemophilia A in the European royal families. The first known carrier in the family was Queen Victoria. Note that all of the affected individuals are male.

*(Modified from McCance K, Huether S: Pathophysiology: The Biologic Basis for Disease in Adults and Children, 5th ed. St. Louis, Mosby, 2005.)*

Hemophilia A is caused by deficient or defective factor VIII, a key component of the clotting cascade. Fibrin formation is affected, resulting in prolonged and often severe bleeding from wounds and hemorrhages in the joints and muscles. Bruising is often seen. Hemarthroses (bleeding into the joints) are common in the ankles, knees, hips, and elbows. These events are often painful, and repeated episodes can lead to destruction of the synovium and diminished joint function. Intracranial hemorrhages can occur and are a leading cause of death. Platelet activity is normal in hemophiliacs, so minor lacerations and abrasions do not usually lead to excessive bleeding.

Hemophilia A varies considerably in its severity, and this variation is correlated directly with the level of factor VIII. About half of hemophilia A patients fall into the severe category, with factor VIII levels that are less than 1% of normal. These persons experience relatively frequent bleeding episodes, often several per month. Patients with moderate hemophilia (1%-5% of normal factor VIII) generally have bleeding episodes only after mild trauma and typically experience one to several episodes per year. Persons with mild hemophilia have 5% to 30% of the normal factor VIII level and usually experience bleeding episodes only after surgery or relatively severe trauma.

Historically, hemophilia A was often fatal before 20 years of age, but a major advance in treatment came in the early 1960s with the ability to purify factor VIII from donor plasma. Factor VIII is usually administered at the first

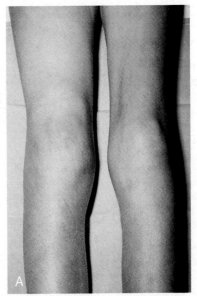

A, The enlarged right knee joint of a patient with hemophilia A, demonstrating the effects of hemarthrosis.

*Continued*

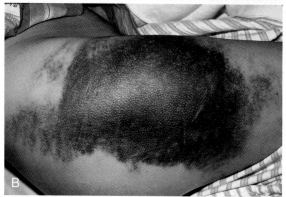

B, Extensive bruising of the right outer thigh.

*(From Hoffbrand VA: Color Atlas of Clinical Hematology, 3rd ed, Philadelphia: Mosby, 2000, pp 281-283.)*

sign of a bleeding episode and is a highly effective treatment. Prophylactic factor VIII administration in severe hemophiliacs is effective in preventing loss of joint function. By the 1970s, the median age at death of persons with hemophilia had increased to 68 years.

The major drawback of donor-derived factor VIII was the fact that, because a typical infusion contained plasma products from hundreds or thousands of different donors, it was often contaminated by viruses. Consequently, patients often suffered from hepatitis B and C infections. More seriously, human immunodeficiency virus (HIV) can be transmitted in this manner, and it is estimated that half of American hemophilia patients treated with donor-derived factor VIII between 1978 and 1985 became infected with HIV. From 1979 to 1998, acquired immune deficiency syndrome (AIDS) accounted for nearly half of deaths among Americans with hemophilia A, resulting in a decrease in the median age at death to 49 years in the 1980s. Donor blood has been screened for HIV since 1985, and heat treatment of donor-derived factor VIII kills HIV and hepatitis B virus, nearly eliminating the threat of infection. Consequently, AIDS mortality among those with hemophilia A has decreased markedly since 1995.

Cloning and sequencing of the factor VIII gene has led to a number of insights. Patients with nonsense or frameshift mutations usually develop severe hemophilia, and those with missense mutations usually have mild to moderate disease. This is expected because nonsense and frameshift mutations typically produce a truncated protein that is degraded and lost. Missense mutations produce a single amino acid substitution without a dominant negative effect, usually resulting in an altered but partially functional protein product. Many of the point mutations take place at methylated CG sequences, which are hot spots for mutation (see Chapter 3). About 45% of severe cases of hemophilia A are caused by a chromosome inversion (see Chapter 6) that disrupts the factor VIII gene. An additional 5% of patients have deletions, which usually lead to relatively severe disease. About 10% of female heterozygotes have factor VIII levels less than 35%, and some of these are manifesting heterozygotes, with mild symptoms of hemophilia A.

Cloning of the factor VIII gene has enabled the production of human factor VIII using recombinant DNA techniques. Extensive clinical testing showed that recombinant factor VIII works as effectively as the donor-derived form, and it was approved for commercial use in 1994. Recombinant factor VIII has, of course, the advantage that there is no possibility of viral contamination. However, as with other forms of factor VIII, recombinant factor VIII generates antifactor VIII antibody production in approximately 10% to 15% of patients. (This response is most common in patients who have no native factor VIII production.)

Two other major bleeding disorders are hemophilia B and von Willebrand disease. Hemophilia B, sometimes called Christmas disease,* is also an X-linked recessive disorder and is caused by a deficiency of clotting factor IX. This condition is about one fifth as common as hemophilia A and can be treated with donor-derived or recombinant factor IX. von Willebrand disease is an autosomal dominant disorder that is highly variable in expression. Although it can affect as many as 1% of individuals of European descent, it reaches severe expression in fewer than 1 in 10,000. The von Willebrand factor, which is encoded by a gene on chromosome 12, acts as a carrier protein for factor VIII. In addition, it binds to platelets and to damaged blood vessel endothelium, thus promoting the adhesion of platelets to damaged vessel walls.

*Christmas was the name of the first reported patient.

Muscular dystrophy, defined as a progressive weakness and loss of muscle, exists in dozens of different forms. Of these, Duchenne muscular dystrophy (DMD), named after the French neurologist who provided the first comprehensive description in 1868, is one of the most severe and common forms. It affects approximately 1 of every 3500 males, a prevalence figure that is similar among all ethnic groups studied thus far.

The symptoms of DMD are usually seen before the age of 5 years, with parents often noticing clumsiness and muscle weakness. Pseudohypertrophy of the calves, the result of infiltration of muscle by fat and connective tissue, is often seen early in the course of the disease. All skeletal muscle degenerates eventually, and most patients with DMD are confined to a wheelchair by 11 years of age. The heart and respiratory musculature become impaired, and death usually results from respiratory or cardiac failure. Survival beyond age 25 years is uncommon; little can be done to alter the ultimate course of this disease.

As muscle cells die, the enzyme creatine kinase (CK) leaks into the blood stream. In DMD patients, serum CK is elevated at least 20 times above the upper limit of the normal range. This elevation can be observed presymptomatically, before clinical symptoms such as muscle wasting are seen. Other traditional diagnostic tools include electromyography, which reveals reduced action potentials, and muscle biopsy.

Female heterozygous carriers of the DMD-causing mutations are usually free of disease, although 8% to 10% have some degree of muscle weakness. In addition, serum CK exceeds the 95th percentile in approximately two thirds of heterozygotes.

Until the gene responsible for DMD was isolated and cloned in 1986, little was known about the mechanism responsible for muscle deterioration. Cloning of the gene and identification of its protein product have advanced our knowledge tremendously. The *DMD* gene covers approximately 2.5 Mb of DNA, making it by far the largest gene known in the human. It contains

79 exons that produce a 14-kb mRNA transcript. Because of this gene's huge size, transcription of an mRNA molecule can take as long as 24 hours. The mRNA is translated into a mature protein of 3685 amino acids. The protein product, named **dystrophin**, was unknown before the cloning of *DMD*. Dystrophin accounts for only about 0.002% of a striated muscle cell's protein mass and is localized on the cytoplasmic side of the cell membrane. Although its function is still being explored, it is likely to be involved in maintaining the structural integrity of the cell's cytoskeleton. The amino terminus of the protein binds F-actin, a key cytoskeletal protein. The carboxyl terminus of dystrophin binds a complex of glycoproteins, known as the **dystroglycan–sarcoglycan complex**, that spans the cell membrane and binds to extracellular proteins. Dystrophin thus links these two cellular components and plays a key role in maintaining the structural integrity of the muscle cell. Lacking dystrophin, the muscle cells of the DMD patient gradually die as they are stressed by muscle contractions.

The large size of *DMD* helps to explain its high mutation rate, about $10^{-4}$ per locus per generation. As with the gene responsible for neurofibromatosis type 1, the *DMD* gene presents a large target for mutation. A slightly altered form of the *DMD* gene product is normally found in brain cells. Its absence in DMD patients helps to explain why approximately 25% have an intelligence quotient (IQ) less than 75. In brain cells, the transcription initiation site is farther downstream in the gene, and a different promoter is used. Thus, the mRNA transcript and the resulting gene product differ from the gene product found in muscle cells. Several additional promoters have also been found, providing a good example of a single gene that can produce different gene products as a result of modified transcription.

Becker muscular dystrophy (BMD), another X-linked recessive dystrophic condition, is less severe than the Duchenne form. The progression is also much slower, with onset at 11 years of age, on average. One study showed that whereas 95% of DMD patients are confined to a wheelchair before 12 years of age, 95% of those with BMD become wheelchair bound after 12 years of age. Some never lose their ability to walk. BMD is less common than DMD, affecting about 1 in 18,000 male births.

For some time it was unclear whether BMD and DMD are caused by distinct X-linked loci or by different mutations at the same locus. Cloning of *DMD* showed the latter to be the case. Both diseases usually result from deletions (65% of DMD cases and 85% of BMD cases) or duplications (6% to 7% of DMD and BMD cases) in *DMD*. But, whereas the great majority of DMD-causing deletions and duplications produce frameshifts, the majority of BMD-causing mutations are in-frame alterations (i.e., a multiple of three bases is deleted or duplicated). One would expect that a frameshift, which is likely to produce a premature stop codon (see Chapter 3) and no protein product, would produce more-severe disease than would an in-frame alteration.

The consequences of these different mutations can be observed in the gene product. Although dystrophin is absent in almost all DMD patients, it is usually present in reduced quantity (or as a shortened form of the protein) in BMD patients. Thus, a dystrophin assay can help to distinguish between the two diseases. This assay also helps to distinguish both diseases from other forms of muscular dystrophy, because several of these forms (e.g., various limb-girdle muscular dystrophies) result from mutations in genes that encode proteins of the dystroglycan–sarcoglycan complex, whereas dystrophin appears to be affected only in BMD and DMD.

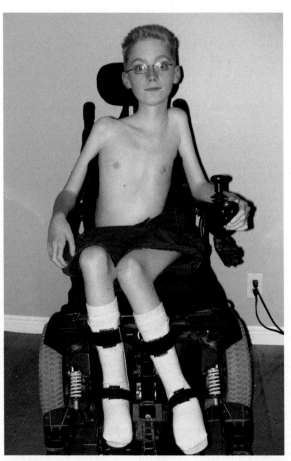

A patient with late-stage Duchenne muscular dystrophy, showing severe muscle loss.

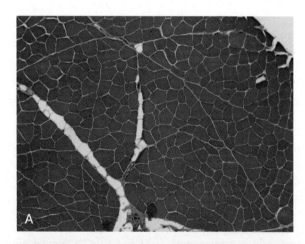

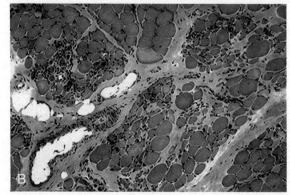

Transverse section of gastrocnemius muscle from (**A**) a normal boy and (**B**) a boy with Duchenne muscular dystrophy. Normal muscle fiber is replaced with fat and connective tissue.

*Continued*

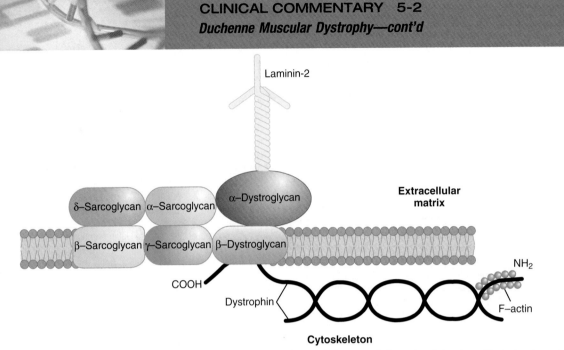

The amino terminus of the dystrophin protein binds to F-actin in the cell's cytoskeleton, and its carboxyl terminus binds to elements of the dystroglycan–sarcoglycan complex. The latter complex of glycoproteins spans the cell membrane and binds to proteins in the extracellular matrix, such as laminin.

The identification of *DMD* has led to mouse and dog models for the disease, and these are contributing considerably to our understanding of the human form of the disease. For example, a small-molecule drug, PTC124, causes ribosomes to read through premature stop codons and has shown promise in mouse models. It is now undergoing clinical trials in humans. In addition, work is progressing on gene therapy for DMD (see Chapter 13). However, because all muscles of the body, including the heart, are affected, this type of therapy faces formidable challenges.

Thus, as with autosomal recessive traits, the heterozygote will produce about 50% of the normal level of the gene product. Ordinarily this is sufficient for a normal phenotype. The situation is different for males, who are hemizygous for the X chromosome. If a male inherits a recessive disease gene on the X chromosome, he will be affected by the disease because the Y chromosome does not carry a normal allele to compensate for the effects of the disease allele.

An X-linked recessive disease with gene frequency $q$ will be seen in a fraction $q$ of males. This is because a male, having only one X chromosome, will manifest the disease if his X chromosome contains the disease-causing mutation. Females, needing two copies of the mutant allele to express the disease, will have a disease frequency of only $q^2$, as in autosomal recessive diseases. For example, hemophilia A (see Clinical Commentary 5-1) is seen in about 1 of every 10,000 males in some populations. Thus, in a collection of 10,000 male X chromosomes, one chromosome would contain the disease-causing mutation ($q = 0.0001$). Affected female homozygotes are almost never seen, because $q^2 = 0.00000001$, or 1 in 100,000,000. This example shows that, in general, males are more frequently affected with X-linked recessive diseases than are females, with this difference becoming more pronounced as the disease becomes rarer.

> Because females have two copies of the X chromosome and males have only one (hemizygosity), X-linked recessive diseases are much more common among males than among females.

Pedigrees for X-linked recessive diseases display several characteristics that distinguish them from pedigrees for autosomal dominant and recessive diseases (Fig. 5-3). As just mentioned, the trait is seen much more frequently in males than in females. Because a father can transmit only a Y chromosome to his son, X-linked genes are not passed from father to son. (In contrast, father-to-son transmission can be observed for autosomal disease alleles.) An X-linked disease allele can be transmitted through a series of phenotypically normal heterozygous females, causing the appearance of skipped generations. The gene is passed from an affected father to all of his daughters, who, as carriers, transmit it to approximately half of their sons, who are affected.

> X-linked recessive inheritance is characterized by an absence of father-to-son transmission, skipped generations when genes are passed through female carriers, and a preponderance of affected males.

The most common mating type involving X-linked recessive genes is the combination of a carrier female and a

## BOX 5-1
## Color Vision: Molecular Biology and Evolution

Human vision depends on a system of retinal photoreceptor cells, about 95% of which are **rod cells**. They contain the light-absorbing protein rhodopsin and allow us to see in conditions of dim light. In addition, the retina contains three classes of **cone cells**, which contain light-absorbing proteins (opsins) that react to light wavelengths corresponding to the three primary colors—red, green, and blue. Color vision depends on the presence of all four of these cell types. Because three major colors are involved, normal color vision is said to be **trichromatic**.

There are many recognized defects of human color vision. The most common of these involve red and green color perception and have been known since 1911 to be inherited in

X-linked recessive fashion. Thus, they are much more common in males than in females. Various forms of red–green colorblindness are seen in about 8% of European males, 4% to 5% of Asian males, and 1% to 4% of African and Native American males. Among European males, 2% are **dichromatic**: they are unable to perceive one of the primary colors, usually red or green. The inability to perceive green is termed **deuteranopia**, and the inability to perceive red is termed **protanopia**. About 6% of European males can detect green and red, but with altered perception of the relative shades of these colors. These are respectively termed **deuteranomalous** and **protanomalous** conditions.

**A,** Image perceived by a person with normal color vision. **B,** The predicted perception by a person with protanopia, a form of red–green colorblindness. Copyright George V. Kelvin.

It should be apparent that dichromats are not really color blind, because they can still perceive a fairly large array of different colors. True colorblindness (**monochromacy**, the ability to perceive only one color) is much less common, affecting approximately 1 in 100,000 persons. There are two major forms of monochromatic vision. **Rod monochromacy** is an autosomal recessive condition in which all visual function is carried out by rod cells. **Blue cone monochromacy** is an X-linked recessive condition in which both the red and green cone cells are absent.

Cloning of the genes responsible for color perception has revealed a number of interesting facts about both the biology and the evolution of color vision in humans. In the 1980s, Jeremy Nathans and colleagues reasoned that the opsins in all four types of photoreceptor cells might have similar amino acid sequences because they carry out similar functions. Thus, the DNA sequences of the genes encoding these proteins should also be similar. But none of these genes had been located, and the precise nature of the protein products was unknown. How could they locate these genes?

Fortunately, the gene encoding rhodopsin in cattle had been cloned. Even though humans and cattle are separated by millions of years of evolution, their rhodopsin proteins still share about 40% of the same amino acid sequence. Thus, the cattle (bovine) rhodopsin gene could be used as a probe to search for a similar DNA sequence in the human genome. A portion of the bovine rhodopsin gene was converted to single-strand form, radioactively labeled, and hybridized with human DNA (much in the same way that a probe is used in Southern blotting [see Chapter 3]). Low-stringency hybridization conditions were used: Temperature and other conditions were manipulated so that complementary base pairing would occur despite some sequence differences between the two species. In this way, the human rhodopsin gene was identified and mapped to chromosome 3.

The next step was to use the human rhodopsin gene as a probe to identify the cone-cell opsin genes. Each of the cone-cell opsin amino acid sequences shares 40% to 45% similarity

with the human rhodopsin amino acid sequence. By probing with the rhodopsin gene, the gene for blue-sensitive opsin was identified and mapped to chromosome 7. This gene was expected to map to an autosome because variants in blue sensitivity are inherited in autosomal recessive fashion. The genes for the red- and green-sensitive opsins were also identified in this way and, as expected, were found to be on the X chromosome. The red and green genes are highly similar, sharing 98% of their DNA sequence.

Initially, many investigators expected that people with color vision defects would display the usual array of deletions and missense and nonsense mutations seen in other disorders. But further study revealed some surprises. It was found that the red and green opsin genes are located directly adjacent to each other on the distal long arm of the X chromosome and that normal persons have one copy of the red gene but can have one to several copies of the green gene. The multiple green genes are 99.9% identical in DNA sequence, and the presence of multiple copies of these genes has no effect on color perception because only the red gene and the first green gene are expressed in the retina. However, when there are no green genes, deuteranopia is produced. Persons who lack the single red gene have protanopia.

The unique aspect of these deletions is that they are the result of **unequal crossover** during meiosis. Unlike ordinary crossover, in which equal segments of chromosomes are exchanged (see Chapter 2), unequal crossover results in a loss of chromosome material on one chromosome homolog and a gain of material on the other. Unequal crossover seems to be facilitated by the high similarity in DNA sequence among the red and green genes: It is relatively easy for the cellular machinery to make a mistake in deciding where the crossover should occur. Thus, a female with one red gene and two green genes could produce one gamete containing a red gene with one green gene and another gamete containing a red gene with three green genes. Unequal crossover could also result in gametes with no copies of a gene, producing protanopia or deuteranopia.

*Continued*

**BOX 5-1**
**Color Vision: Molecular Biology and Evolution—cont'd**

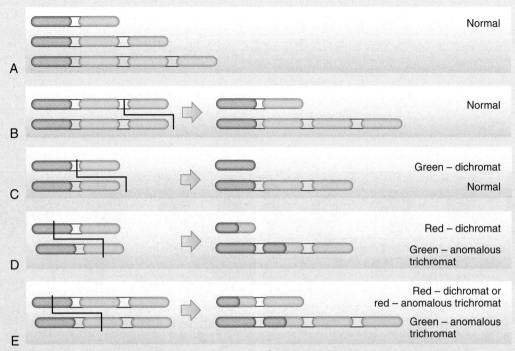

A, Normal individuals have one red gene and one to several green genes. B, Unequal crossover causes normal variation in the number of green genes. C, Unequal crossover can produce a green dichromat with no green genes (deuteranopia). D, Unequal crossover that occurs within the red and green genes can produce a red dichromat (protanopia) or a green anomalous trichromat (deuteranomaly). E, Crossovers within the red and green genes can also produce red anomalous trichromats (protanomaly). The degree of red and green color perception depends on where the crossover occurs within the genes.
*(Modified from Nathans J, Merbs SL, Sung C, et al: The genes for color vision. Sci Am 1989;260:42–49.)*

Unequal crossover also explains protanomalous and deuteranomalous color vision. Here, crossover takes place within the red or green genes, resulting in new chromosomes with hybrid genes (e.g., a portion of the red gene fused with a portion of the green gene). The relative ratio of red and green components of these **fusion genes** determines the extent and nature of the red–green anomaly.

Because the opsin genes have DNA sequence similarity and perform similar functions, they are members of a gene family, much like the globin genes (see Chapter 3). This suggests that they evolved from a single ancestral gene that, through time, duplicated and diverged to encode different but related proteins. Evidence for this process is provided by comparing these genes in humans and other species. Because the X-linked red and green

opsin genes share the greatest degree of DNA sequence similarity, we would expect that these two genes would be the result of the most recent duplication. Indeed, humans share all four of their opsin genes with apes and Old World monkeys, but the less closely related New World monkeys have only a single opsin gene on their X chromosomes. It is therefore likely that the red–green duplication occurred sometime after the split of the New and Old World monkeys, which took place about 30 to 40 million years ago. Similar comparisons date the split of the X-linked and autosomal cone opsin genes to approximately 500 million years ago. And finally, comparisons with the fruit fly, *Drosophila melanogaster*, indicate that the duplication that produced the rod and cone visual pigment genes may have occurred as much as 1 billion years ago.

normal male. The carrier mother will transmit the disease gene to half of her sons and half of her daughters, on average. As Figure 5-4 shows, approximately half of the daughters of such a mating will be carriers, and half will be normal. Half of the sons will be normal, and half, on average, will have the disease.

The other common mating type is an affected father and a homozygous unaffected mother (Fig. 5-5). Here, all of the sons must be normal, because the father can transmit only his Y chromosome to them. Because all of the daughters must receive the father's X chromosome, they will all be heterozygous carriers. None of the children will manifest

the disease, however. Because the father must transmit his X chromosome to his daughters and cannot transmit it to his sons, these risks, unlike those in the previous paragraph, are exact figures rather than probability estimates.

A much less common mating type is that of an affected father and a carrier mother (Fig. 5-6). Half of the daughters will be heterozygous carriers, and half, on average, will be homozygous for the disease gene and thus affected. Half of the sons will be normal, and half will be affected. It may appear that father-to-son transmission of the disease has occurred, but the affected son has actually received the disease allele from his mother.

## TABLE 5-1
### Additional Examples of X-Linked Recessive Disorders

| Name | Gene | Clinical Characteristics |
|------|------|--------------------------|
| Juvenile retinoschisis | RS1 | Progressive visual impairment caused by splitting of the nerve fiber layer of the retina; begins in the first or second decade of life; impairment typically 20/60 to 20/120 |
| Leri–Weill dyschondrosis | SHOX | Madelung deformity of the radius and ulna; mesomelia (shortening of the forearms and lower legs); short stature |
| ATR-X | ATRX | Mental retardation; genital anomalies; and α-thalassemia without abnormalities of the α-globin gene complex |
| Hypohydrotic ectodermal hypoplasia | EDA | Diminished sweating and heat intolerance; sparse and light-colored hair, eyelashes, and eyebrows; abnormal and/or missing teeth; recurrent upper airway infections |
| Vitamin D–resistant rickets | PHEX | Hypophosphatemia due to reduced renal phosphate reabsorption; short stature; bowed legs; poor teeth formation |
| Aarskog–Scott syndrome (faciogenital dysplasia) | FGD1 | Short stature; hypertelorism; genital anomalies |
| Cleft palate with ankyloglossia | TBX22 | Cleft palate with or without ankyloglossia (tongue tie) |
| Pelizaeus–Merzbacher disease | PLP1 | Defect of myelination; typically manifests in infancy or early childhood; characterized by nystagmus, hypotonia, spasticity, early death |
| Nephrogenic diabetes insipidus | AVPR2 | Impaired response to antidiuretic hormone that leads to inability to concentrate urine, polydipsia (excessive thirst), polyuria (excessive urine production) |
| Otopalatoldigital spectrum disorders | FLNA | Skeletal dysplasia ranging from mild to lethal; males more severely affected than females |

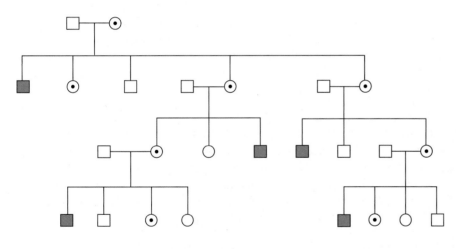

### FIGURE 5-3
A pedigree showing the inheritance of an X-linked recessive trait. *Solid symbols* represent affected individuals, and *dotted symbols* represent heterozygous carriers.

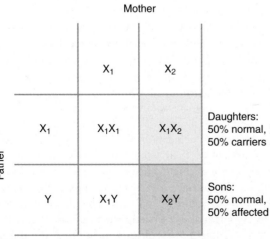

### FIGURE 5-4
Punnett square representation of the mating of a heterozygous female who carries an X-linked recessive disease gene with a normal male. $X_1$, chromosome with normal allele; $X_2$, chromosome with disease allele.

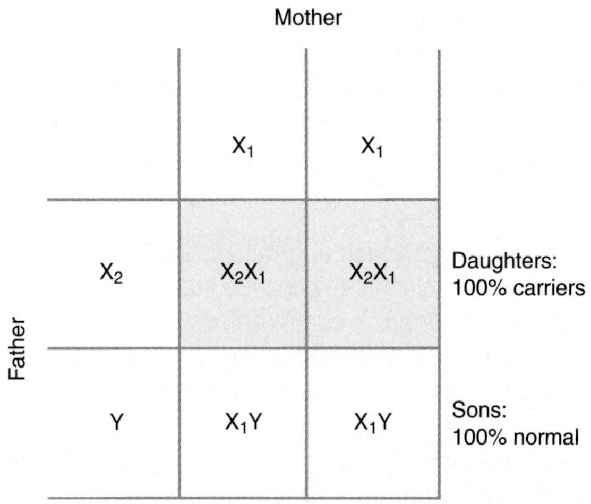

### FIGURE 5-5
Punnett square representation of the mating of a normal female with a male who is affected by an X-linked recessive disease. $X_1$, chromosome with normal allele; $X_2$, chromosome with disease allele.

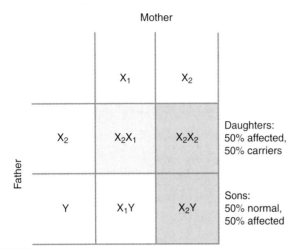

**FIGURE 5-6**
Punnett square representation of the mating of a carrier female with a male affected with an X-linked recessive disease. $X_1$, chromosome with normal allele; $X_2$, chromosome with disease allele.

> Recurrence risks for X-linked recessive disorders are more complex than for autosomal disorders. The risk depends on the genotype of each parent and the sex of the offspring.

Occasionally, females who inherit only a single copy of an X-linked recessive disease allele can be affected with the disease. Imagine a female embryo that has received a normal clotting factor VIII allele from one parent and a mutated clotting factor VIII allele from the other. Ordinarily, X inactivation will result in approximately equal numbers of cells having active paternal and maternal X chromosomes. In this case, the female carrier would produce about 50% of the normal level of factor VIII and would be phenotypically normal. However, because X inactivation is a random process, it sometimes results in a heterozygous female in whom nearly all of the active X chromosomes happen to be the ones carrying the disease-causing mutation. These females exhibit hemophilia A and are termed **manifesting heterozygotes**. Because such females usually maintain at least a small fraction of active normal X chromosomes, they tend to be relatively mildly affected. For example, approximately 5% of females who are heterozygous for a factor VIII mutation experience mild hemophilia because of factor VIII deficiency.

> Because X inactivation is a random process, some female heterozygotes experience inactivation of most of the normal X chromosomes in their cells. These manifesting heterozygotes are usually mildly affected.

Less commonly, females having only a single X chromosome (Turner syndrome) have been seen with X-linked recessive diseases such as hemophilia A. Females can also be affected with X-linked recessive diseases as a result of

translocations or deletions of X chromosome material (see Chapter 6). These events are rare.

### X-Linked Dominant Inheritance

X-linked dominant diseases are fewer and less prevalent than are X-linked recessive diseases. An example of an X-linked dominant disease is hypophosphatemic rickets, a disease in which the kidneys are impaired in their ability to reabsorb phosphate. This results in abnormal ossification, with bending and distortion of the bones (rickets). Another example is incontinentia pigmenti type 1, a disorder characterized by abnormal skin pigmentation, conical or missing teeth, and ocular and, in some cases, neurological abnormalities. This disorder is seen only in females. It is thought that hemizygous males are so severely affected that they do not survive to term. Heterozygous females, having one normal X chromosome, tend generally to have milder expression of X-linked dominant traits (just as heterozygotes for most autosomal dominant disease genes are less severely affected than are homozygotes).

X-linked dominant inheritance is also observed in Rett syndrome, a neurodevelopmental disorder seen in 1 in 10,000 females and in a much smaller fraction of males, most of whom do not survive to term. Rett syndrome is characterized by autistic behavior, mental retardation, seizures, and gait ataxia. The severity of this condition varies substantially among affected females, reflecting the effects of random X inactivation: in mildly affected females, a large percentage of the X chromosomes that bear the disease-causing mutation have been randomly inactivated. About 95% of classic Rett syndrome cases are caused by mutations in the *MECP2* gene, and most of these mutations are de novo events that occur in the paternal germline (consistent with a higher mutation rate in male gamete formation, discussed in Chapter 3). The protein product encoded by *MECP2* binds to methylated CG sequences found in the 5′ regions of other genes. After binding to these sequences, the protein helps to recruit other proteins that repress transcription by causing chromatin condensation. Loss-of-function mutations in *MECP2* result in the inappropriate expression of genes thought to be involved in brain development.

Figure 5-7 illustrates a pedigree for X-linked dominant inheritance. As with autosomal dominant diseases, a person need inherit only a single copy of an X-linked dominant disease gene to manifest the disorder. Because females have two X chromosomes, either of which can potentially carry the disease gene, they are about twice as commonly affected as males (unless the disorder is lethal in males, as in incontinentia pigmenti). Affected fathers cannot transmit the trait to their sons. All of their daughters must inherit the disease gene, so all are affected. Affected females are usually heterozygotes and thus have a 50% chance of passing the disease allele to their daughters and sons. The characteristics of X-linked dominant and X-linked recessive inheritance are summarized in Table 5-2. As already mentioned, the distinction between these categories can be blurred by incomplete penetrance in heterozygotes for X-linked dominant mutations

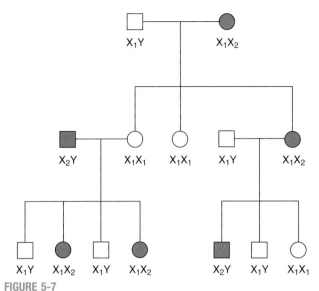

FIGURE 5-7
Pedigree demonstrating the inheritance of an X-linked dominant trait. $X_1$, chromosome with normal allele; $X_2$, chromosome with disease allele.

and by the presence of disease in heterozygotes for X-linked recessive mutations (manifesting heterozygotes).

> X-linked dominant diseases display characteristic patterns of inheritance. They are about twice as common in females as in males, skipped generations are uncommon, and father-to-son transmission is not seen.

## Y-Linked Inheritance

Although it consists of approximately 60 Mb of DNA, the Y chromosome contains relatively few genes. Only a few dozen Y-linked, or **holandric**, genes have been identified. These include the gene that initiates differentiation of the embryo into a male (see Chapter 6), several genes that encode testis-specific spermatogenesis factors, and a minor histocompatibility antigen (termed HY). Several housekeeping genes are located on the Y chromosome, and they all have inactivation-escaping homologs on the X chromosome. Transmission of Y-linked traits is strictly from father to son (Fig. 5-8).

**TABLE 5-2**
## Comparison of the Major Attributes of X-Linked Dominant and X-Linked Recessive Inheritance Patterns*

| Attribute | X-Linked Dominant | X-Linked Recessive |
|---|---|---|
| Recurrence risk for heterozygous female × normal male mating | 50% of sons affected; 50% of daughters affected | 50% of sons affected; 50% of daughters heterozygous carriers |
| Recurrence risk for affected male × normal female mating | 0% of sons affected; 100% of daughters affected | 0% of sons affected; 100% of daughters heterozygous carriers |
| Transmission pattern | Vertical; disease phenotype seen in generation after generation | Skipped generations may be seen, representing transmission through carrier females |
| Sex ratio | Twice as many affected females as affected males (unless disease is lethal in males) | Much greater prevalence of affected males; affected homozygous females are rare |
| Other | Male-to-male transmission is not seen; expression is less severe in female heterozygotes than in affected males | Male-to-male transmission not seen; manifesting heterozygotes may be seen in females |

*Compare with the inheritance patterns for autosomal diseases shown in Table 4-1.

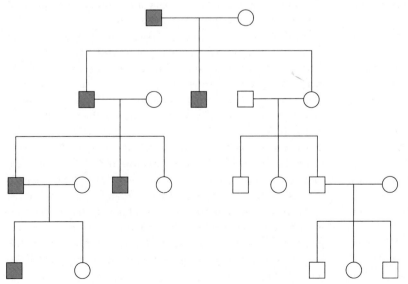

FIGURE 5-8
Pedigree demonstrating the inheritance of a Y-linked trait. Transmission is exclusively male to male.

## SEX-LIMITED AND SEX-INFLUENCED TRAITS

Confusion sometimes exists regarding traits that are sex-linked and those that are **sex-limited** or **sex-influenced**. A sex-limited trait occurs in only one of the sexes—due, for instance, to anatomical differences. Inherited uterine or testicular defects would be examples. A good example of a sex-influenced trait is male-pattern baldness, which occurs in both males and females but much more commonly in males. This is related in part to sex differences in hormone levels. Contrary to oft-stated belief, male-pattern baldness is not strictly X-linked, although variation in the X-linked androgen receptor gene is associated with baldness. Autosomal genes are also thought to influence male-pattern baldness, helping to explain apparent father-to-son transmission of this trait.

## MITOCHONDRIAL INHERITANCE

The great majority of genetic diseases are caused by defects in the nuclear genome. However, a small but significant number of diseases can be caused by mutations in mitochondrial DNA. Because of the unique properties of the mitochondria, these diseases display characteristic modes of inheritance and a large degree of phenotypic variability.

Each human cell contains several hundred or more mitochondria in its cytoplasm. Through the complex process of oxidative phosphorylation, these organelles produce adenosine triphosphate (ATP), the energy source essential for cellular metabolism. Mitochondria are thus critically important for cell survival.

The mitochondria have their own DNA molecules, which occur in several copies per organelle and consist of 16,569 base pairs arranged on a double-stranded circular molecule (Fig. 5-9). The mitochondrial genome encodes two ribosomal RNAs (rRNAs), 22 transfer RNAs (tRNAs), and 13 polypeptides involved in oxidative phosphorylation. (Another 90 or so nuclear DNA genes also encode polypeptides that are transported into the mitochondria to participate in oxidative phosphorylation.) Transcription of mitochondrial DNA (mtDNA) takes place in the mitochondrion, independently of the nucleus. Unlike nuclear genes, mtDNA genes contain no introns. Because it is located in the cytoplasm, mtDNA is inherited exclusively through the maternal line (Fig. 5-10). Males do not transmit mtDNA to their offspring because sperm cells contain only a small number of mtDNA molecules, which are not incorporated into the developing embryo. (One isolated case of paternal transmission of a mitochondrial DNA mutation has been reported, but such events appear to be extremely rare.)

The mutation rate of mtDNA is about 10 times higher than that of nuclear DNA. This is caused by a relative lack of DNA repair mechanisms in the mtDNA and also by damage from free oxygen radicals released during the oxidative phosphorylation process.

Because each cell contains a *population* of mtDNA molecules, a single cell can harbor some molecules that have an mtDNA mutation and other molecules that do not. This

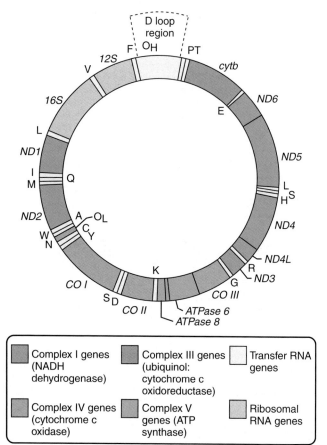

**FIGURE 5-9**

The circular mitochondrial DNA genome. Locations of protein-encoding genes (for reduced nicotinamide adenine dinucleotide [NADH] dehydrogenase, cytochrome c oxidase, cytochrome c oxidoreductase, and adenosine triphosphate [ATP] synthase) are shown, as are the locations of the two ribosomal RNA genes and 22 transfer RNA genes (designated by single letters). The replication origins of the heavy (OH) and light (OL) chains and the noncoding D loop (also known as the control region) are shown.
*(Modified from MITOMAP: A Human Mitochondrial Genome Database. http://www.mitomap.org, 2008.)*

heterogeneity in DNA composition, termed **heteroplasmy**, is an important cause of variable expression in mitochondrial diseases. The larger the percentage of mutant mtDNA molecules, the more severe the expression of the disease. As cells divide, changes in the percentage of mutant alleles can occur through chance variation (identical in concept to genetic drift, discussed in Chapter 3) or because of a selective advantage (e.g., deletions produce a shorter mitochondrial DNA molecule that can replicate more quickly than a full-length molecule).

Each tissue type requires a certain amount of ATP for normal function. Although some variation in ATP levels may be tolerated, there is typically a threshold level below which cells begin to degenerate and die. Organ systems with large ATP requirements and high thresholds tend to be the ones most seriously affected by mitochondrial diseases. For example, the central nervous system consumes about 20% of the body's ATP production and therefore is often affected by mtDNA mutations.

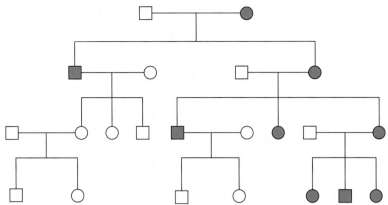

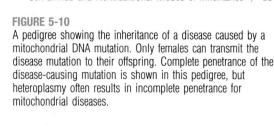

**FIGURE 5-10**
A pedigree showing the inheritance of a disease caused by a mitochondrial DNA mutation. Only females can transmit the disease mutation to their offspring. Complete penetrance of the disease-causing mutation is shown in this pedigree, but heteroplasmy often results in incomplete penetrance for mitochondrial diseases.

Like the globin disorders, mitochondrial disorders can be classified according to the type of mutation that causes them. Missense mutations in protein-coding mtDNA genes cause one of the best-known mtDNA diseases, Leber hereditary optic neuropathy (LHON). This disease, which affects about one in 10,000 persons, is characterized by rapid loss of vision in the central visual field as a result of optic nerve death. Vision loss typically begins in the third decade of life and is usually irreversible. Heteroplasmy is minimal in LHON, so expression tends to be relatively uniform and pedigrees for this disorder usually display a clear pattern of mitochondrial inheritance.

Single-base mutations in a tRNA gene can result in myoclonic epilepsy with ragged-red fiber syndrome (MERRF), a disorder characterized by epilepsy, dementia, ataxia (uncoordinated muscle movement), and myopathy (muscle disease). MERRF is characterized by heteroplasmic mtDNA and is thus highly variable in its expression. Another example of a mitochondrial disease caused by a single-base tRNA mutation is mitochondrial encephalomyopathy and stroke-like episodes (MELAS). Like MERRF, MELAS is heteroplasmic and highly variable in expression.

The final class of mtDNA mutations consists of duplications and deletions. These can produce Kearns–Sayre disease (muscle weakness, cerebellar damage, and heart failure), Pearson syndrome (infantile pancreatic insufficiency, pancytopenia, and lactic acidosis), and chronic progressive external ophthalmoplegia (CPEO). To date, the disease-causing mutations seen in mtDNA include more than 100 point mutations and more than 100 deletions or duplications.

Mitochondrial mutations are also associated with some common human diseases. A mitochondrial mutation causes a form of late-onset deafness, and the MELAS mutation is seen in 1% to 2% of persons with type 2 diabetes mellitus. Mitochondrial defects may also be associated with some cases of Alzheimer disease, although it remains unclear whether mitochondrial mutations are a primary cause or a secondary event. It has also been suggested that mtDNA mutations, which accumulate through the life of an individual as a result of free radical formation, could contribute to the aging process.

> The mitochondria, which produce ATP, have their own unique DNA. Mitochondrial DNA is maternally inherited and has a high mutation rate. A number of diseases are known to be caused by mutations in mitochondrial DNA.

## GENOMIC IMPRINTING

Mendel's experimental work with garden peas established that the phenotype is the same whether a given allele is inherited from the mother or the father. Indeed, this principle has long been part of the central dogma of genetics. Recently, however, it has become increasingly apparent that this principle does not always hold. For some human genes, one of the alleles is transcriptionally inactive (no mRNA is produced), depending upon the parent from whom the allele was received. For example, an allele transmitted by the mother would be inactive, and the same allele transmitted by the father would be active. The normal individual would have only one transcriptionally active copy of the gene. This process of gene silencing is known as **imprinting**, and the transcriptionally silenced genes are said to be imprinted. At least several dozen human genes, and perhaps as many as 200 or so, are known to be imprinted.

Imprinted alleles tend to be heavily methylated (in contrast to the nonimprinted copy of the allele, which typically is not methylated). The attachment of methyl groups to 5′ regions of genes, along with histone hypoacetylation and condensation of chromatin, inhibit the binding of proteins that promote transcription. It should be apparent that this process is similar in many ways to X-inactivation, discussed earlier in this chapter.

### Prader–Willi and Angelman Syndromes

A striking disease example of imprinting is provided by a deletion of about 4 Mb of the long arm of chromosome 15. When this deletion is inherited from the father, the child manifests a disease known as Prader–Willi syndrome (PWS). The features of PWS include short stature, hypotonia (poor muscle tone), small hands and feet, obesity, mild to moderate mental retardation, and hypogonadism (Fig. 5-11,*A*). When the same deletion is inherited from the

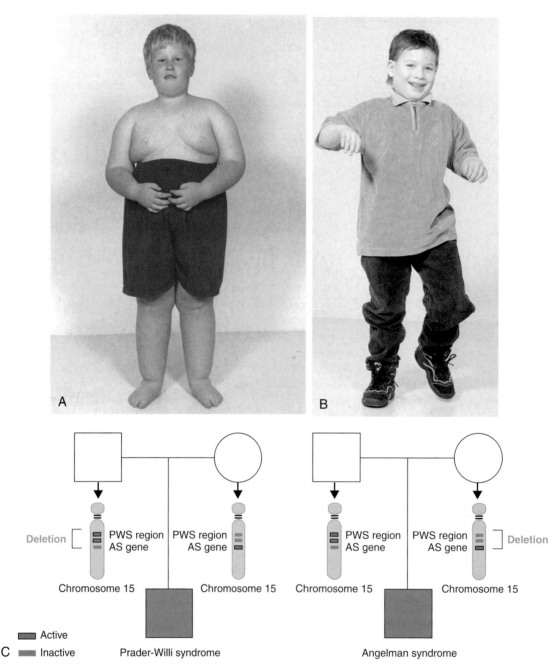

**FIGURE 5-11**

Illustration of the effect of imprinting on chromosome 15 deletions. A, Inheritance of the deletion from the father produces Prader–Willi syndrome (PWS) (note the inverted V-shaped upper lip, small hands, and truncal obesity). B, Inheritance of the deletion from the mother produces Angelman syndrome (note the characteristic posture). C, Pedigrees illustrating the inheritance pattern of this deletion and the activation status of genes in the critical region.

mother, the child develops Angelman syndrome, which is characterized by severe mental retardation, seizures, and an ataxic gait (Fig. 5-11,B). Both diseases are seen in about 1 of every 15,000 persons, and chromosome deletions are responsible for about 70% of cases of both diseases. The deletions that cause PWS and Angelman syndrome are microscopically indistinguishable and affect the same group of genes.

For some time, it was unclear how the same chromosomal deletion could produce such disparate results in different patients. Further analysis showed that the 4-Mb deletion

(the critical region) contains several genes that normally are transcribed only on the chromosome inherited from the father. These genes are transcriptionally inactive (imprinted) on the copy of chromosome 15 inherited from the mother. Similarly, other genes in the critical region are active only on the chromosome inherited from the mother and are inactive on the chromosome inherited from the father. Thus, several genes in this region are normally active on only one chromosome (Fig. 5-11,C). If the single active copy of one of these genes is lost through a chromosome deletion, then no gene product is produced at all, and disease results.

Molecular analysis using many of the tools and techniques outlined in Chapter 3 (microsatellite polymorphisms, cloning, and DNA sequencing) has identified several specific genes in the critical region of chromosome 15. The gene responsible for Angelman syndrome encodes a protein involved in ubiquitin-mediated protein degradation during brain development (consistent with the mental retardation and ataxia observed in this disorder). In brain tissue, this gene is active only on the chromosome inherited from the mother; thus, a maternally transmitted deletion removes the single active copy of this gene. Several genes in the critical region are involved in PWS, and they are transcribed only on the chromosome transmitted by the father.

Several mechanisms in addition to chromosome deletions can cause PWS and Angelman syndrome. One of these is **uniparental disomy**, a condition in which the person inherits two copies of a chromosome from one parent and none from the other (see Chapter 6 for further discussion). When two copies of the maternal chromosome 15 are inherited, PWS results because no active paternal genes are present in the critical region. Conversely, disomy of the paternal chromosome 15 produces Angelman syndrome. Point mutations in the identified Angelman syndrome gene can also produce disease. Finally, about 1% of cases of PWS result from a small deletion of the region that contains an imprinting control center on chromosome 15. This is the DNA sequence that apparently helps to set and reset the imprint itself. Box 5-2 presents clinical issues of PWS from the perspective of a patient's family.

### Beckwith–Wiedemann Syndrome

A second example of imprinting in the human genome is given by Beckwith–Wiedemann syndrome, an overgrowth condition accompanied by an increased predisposition to cancer. Beckwith–Wiedemann syndrome is usually recognizable at birth because of large size for gestational age, neonatal hypoglycemia, a large tongue, creases on the ear lobe, and omphalocele (an abdominal wall defect). Some children with Beckwith–Wiedemann syndrome also develop asymmetrical overgrowth of a limb or one side of the face or trunk (i.e., hemihyperplasia). Children with Beckwith–Wiedemann syndrome have an increased risk for developing Wilms tumor (a kidney cancer) and hepatoblastoma (a liver cancer). Both of these tumors can be treated effectively if they are detected early, so screening at regular intervals is an important part of management (see Chapter 15).

As with Angelman syndrome, a minority of Beckwith–Wiedemann syndrome cases (about 20% to 30%) are caused by the inheritance of two copies of a chromosome from the father and no copy of the chromosome from the mother (uniparental disomy, in this case affecting chromosome 11). Several genes on the short arm of chromosome 11 are imprinted on either the paternal or maternal chromosome (Fig. 5-12). These genes are found in two separate, differentially methylated regions (DMRs). In DMR1, the gene that encodes insulin-like growth factor 2 (*IGF2*) is normally inactive on the maternally transmitted chromosome and active on the paternally transmitted chromosome. Normally, then, a person has only one active copy of *IGF2*. When two copies of the paternal chromosome are inherited (i.e., paternal uniparental disomy) or there is loss of imprinting on the maternal copy of *IGF2*, an active *IGF2* gene is present in double dose. This results in increased levels of IGF2 during fetal development, contributing to the overgrowth features of Beckwith–Wiedemann syndrome. (Note that, in contrast to PWS and

---

**BOX 5-2**
## A Mother's Perspective of Prader–Willi syndrome

We have a 3½-year-old son, John, who has Prader–Willi syndrome. Months before John was born, we were concerned about his well-being because he wasn't as active in utero as his older siblings had been. At the first sight of John, the doctors suspected that things "weren't quite right." John opened his eyes but made no other movements. He couldn't adequately suck, he required supplemental oxygen, and he was "puffy." He remained hospitalized for nearly 3 weeks. The next 3 years were filled with visits to occupational therapists, physical therapists, home health care aides, early childhood service providers, and speech therapists.

From the day John was born, we searched diligently for a diagnosis. His father insisted that we need only love and help him. However, I wanted specifics on how to help him and knowledge from other parents who might have traveled a similar path. After extensive testing and three "chromosome checks," John's problem was diagnosed as Prader–Willi syndrome (PWS). We were glad to be provided with some direction and decided that we would deal with further challenges as they came upon us. We used what we learned about PWS to get started helping John reach his potential. We were not going to worry about all the potential problems John could have because of his PWS.

John attends a special education preschool at the local elementary school 4 days a week. The bus ride takes about 5 minutes, but it is long enough for John to very much anticipate it each day. If he is ill, we have to tell him that the bus is broken. He attends a Sunday school class with children of a similar age. He misbehaves by saying "hi" and "bye" very loudly to each participant. He receives speech therapy once a week, and I spend at least 30 minutes each day with John practicing speech, cognitive, and play skills. John has not yet experienced the feeding difficulties commonly observed in children with PWS. However, excessive eating and weight gain are more common in older children with PWS.

Compared with other 3-year-old children, John struggles with speech and motor developmental milestones. Yet, he loves to play with his siblings and their friends and to look at books. In fact, we struggle to keep people from doing too many things for John because they might prevent him from attaining the same goal independently. We feel very privileged to have him in our family.

Our expectations for John are that he achieves everything that is possible for him plus a little bit more. Indeed, some of his care providers are already impressed with his capabilities. I hope that his success is partly a result of the care and support that we have given to him. Moreover, I hope that John continues to overcome the daily challenges that face him.

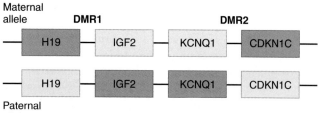

**FIGURE 5-12**
Schematic of the organization of several imprinted genes on chromosome 11p15.5 that are involved in the pathogenesis of Beckwith–Wiedemann syndrome and Russell–Silver syndrome. Beckwith–Wiedemann syndrome can arise from loss of imprinting of the growth promoting gene, *IGF2*, on the maternally transmitted chromosome, two copies of the paternal allele with an active *IGF2* as a consequence of uniparental disomy, or imprinting of the growth-suppressing gene, *CDKN1C*, on the maternally transmitted chromosome. Imprinting defects that lead to down-regulation of *IGF2* on the paternal allele cause some cases of Russell–Silver syndrome. *DMR*, Differentially methylated region; *red*, genes that are not methylated and therefore expressed; *green*, genes that are methylated and therefore silenced.

Angelman syndromes, which are produced by a *missing* gene product, Beckwith–Wiedemann syndrome is caused, at least in part, by *overexpression* of a gene product.)

In 50% to 60% of cases, Beckwith–Wiedemann syndrome is caused by a loss of the paternal imprint of DMR2, the region that contains several genes, including *KCNQ1* and *CDKN1C*. This is thought to result in silencing of growth inhibitors and thus overgrowth and increased predisposition to cancer, although the specific mechanism remains to be elucidated.

### Russell–Silver Syndrome

Russell–Silver syndrome is a clinically heterogeneous group of disorders characterized by growth retardation, proportionate short stature, leg length discrepancy, and a small, triangular-shaped face. About a third of Russell–Silver syndrome cases are caused by imprinting abnormalities of chromosome 11p15.5 that lead to down-regulation of IGF2 and diminished growth. Another 10% of cases of Russell–Silver syndrome are caused by maternal uniparental disomy. Thus, whereas up-regulation or extra copies of active IGF2 cause overgrowth in Beckwith–Wiedemann syndrome, down-regulation of IGF2 causes diminished growth in Russell–Silver syndrome.

> Some disease genes may be expressed differently when inherited from one sex versus the other. This is genomic imprinting. It is typically associated with methylation of DNA and chromatin condensation, which limit the action of transcription factors and decrease gene expression.

### Anticipation and Repeat Expansion

Since the early part of the 20th century, it has been observed that some genetic diseases seem to display an earlier age of onset and/or more severe expression in the more recent generations of a pedigree. This pattern is termed **anticipation**, and it has been the subject of considerable controversy and speculation. Many researchers believed that it was an artifact of better observation and clinical diagnosis in more recent times: A disorder that previously might have remained undiagnosed until age 60 years might now be diagnosed at age 40 years simply because of better diagnostic tools. Others, however, believed that anticipation could be a real biological phenomenon, although evidence for the actual mechanism remained elusive.

Molecular genetics has now provided good evidence that anticipation does have a biological basis. This evidence has come, in part, from studies of myotonic dystrophy, an autosomal dominant disease that involves progressive muscle deterioration and myotonia (inability to relax muscles after contraction) (Fig. 5-13). Seen in approximately 1 in 8000 persons, myotonic dystrophy is the most common muscular dystrophy that affects adults. This disorder is also typically characterized by cardiac arrhythmias (abnormal heart rhythms), testicular atrophy, insulin resistance, and cataracts. Most cases of myotonic dystrophy are caused by mutations in *DMPK*, a protein kinase gene located on chromosome 19.

**FIGURE 5-13**
A three-generation family affected with myotonic dystrophy. The degree of severity increases in each generation. The grandmother (*right*) is only slightly affected, but the mother (*left*) has a characteristic narrow face and somewhat limited facial expression. The baby is more severely affected and has the facial features of children with neonatal-onset myotonic dystrophy, including an open, triangle-shaped mouth. The infant has more than 1000 copies of the trinucleotide repeat, whereas the mother and grandmother each have approximately 100 repeats.

Analysis of *DMPK* has shown that the disease-causing mutation is an expanded CTG trinucleotide repeat (see Chapter 3) that lies in the 3′ untranslated portion of the gene (i.e., a region transcribed into mRNA but not translated into protein). The number of these repeats is strongly correlated with severity of the disease. Unaffected persons typically have 5 to 37 copies of the repeat. Those with 50 to 100 copies may be mildly affected or have no symptoms. Those with full-blown myotonic dystrophy have anywhere from 100 to several thousand copies of the repeated sequence. Expansion to large repeat numbers can produce congenital myotonic dystrophy; for

reasons that are not well understood, these large expansions are transmitted almost exclusively by females. The number of repeats often increases with succeeding generations: A mildly affected parent with 80 repeats might produce a severely affected offspring who has more than 1000 repeats (Fig. 5-14). As the number of repeats increases through successive generations, the age of onset decreases and severity often increases. There is thus strong evidence that expansion of this trinucleotide repeat is the cause of anticipation in myotonic dystrophy.

How does a mutation in the 3′ untranslated portion of *DMPK* produce the many disease features of myotonic

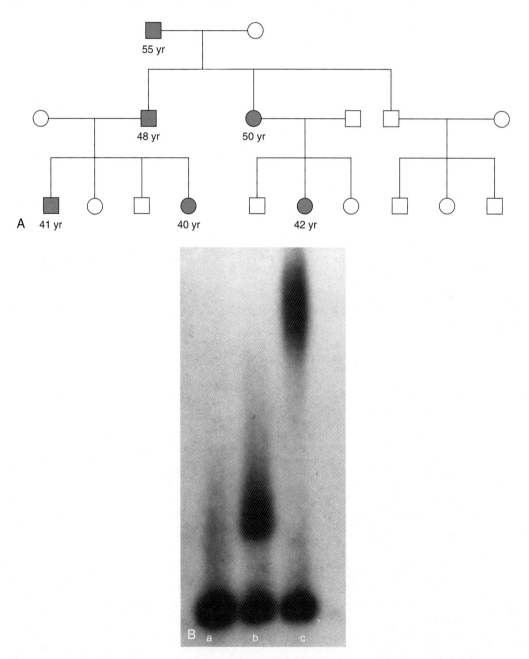

**FIGURE 5-14**
**A,** Myotonic dystrophy pedigree illustrating anticipation. In this case, the age of onset for family members affected with an autosomal dominant disease is lower in more recent generations. **B,** An autoradiogram from a Southern blot analysis of the myotonic dystrophy gene in three individuals. Individual **a** is homozygous for a 4- to 5-repeat allele and is normal. Individual **b** has one normal allele and one disease allele of 175 repeats; this individual has myotonic dystrophy. Individual **c** is also affected with myotonic dystrophy and has one normal allele and a disease-causing allele of approximately 900 repeats.
*(B courtesy of Dr. Kenneth Ward and Dr. Elaine Lyon, University of Utah Health Sciences Center.)*

dystrophy? There is now considerable evidence that the expanded repeat produces an mRNA product that remains in the nucleus of the cell and produces toxic gain-of-function effects. The abnormal mRNA interacts with proteins that normally bind other RNA products to regulate their splicing. As a result, several proteins, including several that are expressed in heart and skeletal muscle, are abnormally formed, giving rise to some of the pleiotropic features of the disease phenotype.

Recently, a gene on chromosome 3 was discovered in which a 4-bp (CCTG) expanded repeat can also cause myotonic dystrophy. Again, the repeat is located in the 3′ untranslated region of the gene. The phenotype associated with the chromosome 3 mutation is similar to that of the chromosome 19 mutation, although it sometimes is less severe. There is again evidence that this mutation produces a toxic mRNA that interferes with the normal function of RNA-binding proteins. Myotonic dystrophy thus illustrates several important genetic principles: anticipation, pleiotropy, and locus heterogeneity.

Repeat expansions have now been identified as a cause of more than 20 genetic diseases (Table 5-3), which can be assigned to three broad categories. The first includes neurological diseases, such as Huntington disease and most of the spinocerebellar ataxias, that are caused by a CAG or CTG repeat expansion in a protein-coding portion of the gene. The repeats generally expand in number from a normal range of 10 to 35 to a disease-causing range of approximately 50 to 100. Expansions tend to be larger when transmitted through the father than through the mother, and the mutations often have a gain-of-function effect. The second group consists of phenotypically more diverse diseases in which the expansions are again small in magnitude and are found in exons. The repeat sequence is heterogeneous, however, and anticipation is not a typical feature. The third category includes fragile X syndrome, myotonic dystrophy, two of the spinocerebellar ataxias, juvenile myoclonic epilepsy, and Friedreich ataxia. The repeat expansions are typically much larger than in the first two categories: The normal range is generally 5 to 50 trinucleotides, but the disease-causing range can vary from 100 to several thousand trinucleotides. The repeats are located outside the protein-coding regions of the gene in all of these disorders, and in some cases (e.g., myotonic dystrophy) the mutation produces a harmful RNA product rather than an abnormal or absent protein product. Repeat expansions are often larger when they are transmitted through the mother. Anticipation is seen in most of the diseases in the first and third categories.

**TABLE 5-3**
## Diseases Associated with Repeat Expansions

| Disease | Description | Repeat Sequence | Normal Range; Disease Range | Parent in Whom Expansion Usually Occurs | Location of Expansion |
|---------|-------------|-----------------|-----------------------------|-----------------------------------------|------------------------|
| **Category 1** | | | | | |
| Huntington disease | Loss of motor control, dementia, affective disorder | CAG | 6-34; 36-121 | More often through father | Exon |
| Spinal and bulbar muscular atrophy | Adult-onset motor-neuron disease associated with androgen insensitivity | CAG | 9-36; 38-62 | More often through father | Exon |
| Spinocerebellar ataxia type 1 | Progressive ataxia, dysarthria, dysmetria | CAG | 6-39; 40-82 | More often through father | Exon |
| Spinocerebellar ataxia type 2 | Progressive ataxia, dysarthria | CAG | 15-24; 32-200 | — | Exon |
| Spinocerebellar ataxia type 3 (Machado-Joseph disease) | Dystonia, distal muscular atrophy, ataxia, external ophthalmoplegia | CAG | 13-36; 61-84 | More often through father | Exon |
| Spinocerebellar ataxia type 6 | Progressive ataxia, dysarthria, nystagmus | CAG | 4-19; 20-33 | — | Exon |
| Spinocerebellar ataxia type 7 | Progressive ataxia, dysarthria, retinal degeneration | CAG | 4-35; 37-306 | More often through father | — |
| Spinocerebellar ataxia type 17 | Progressive ataxia, dementia, bradykinesia, dysmetria | CAG | 25-42; 47-63 | — | Exon |
| Dentatorubral-pallidoluysian atrophy (Haw River syndrome) | Cerebellar atrophy, ataxia, myoclonic epilepsy, choreoathetosis, dementia | CAG | 7-34; 49-88 | More often through father | Exon |

**TABLE 5-3**
## Diseases Associated with Repeat Expansions—cont'd

| Disease | Description | Repeat Sequence | Normal Range; Disease Range | Parent in Whom Expansion Usually Occurs | Location of Expansion |
|---|---|---|---|---|---|
| Huntington disease-like 2 | Features very similar to those of Huntington disease | CTG | 7-28; 66-78 | — | Exon |
| **Category 2** | | | | | |
| Pseudoachondroplasia, multiple epiphyseal dysplasia | Short stature, joint laxity, degenerative joint disease | GAC | 5; 6-7 | — | Exon |
| Oculopharyngeal muscular dystrophy | Proximal limb weakness, dysphagia, ptosis | GCG | 6; 7-13 | — | Exon |
| Cleidocranial dysplasia | Short stature, open skull sutures with bulging calvaria, clavicular hypoplasia, shortened fingers, dental anomalies | GCG, GCT, GCA | 17; 27 (expansion observed in one family) | — | Exon |
| Synpolydactyly | Polydactyly and syndactyly | GCG, GCT, GCA | 15; 22-25 | — | Exon |
| **Category 3** | | | | | |
| Myotonic dystrophy (DM1; chromosome 19) | Muscle loss, cardiac arrhythmia, cataracts, frontal balding | CTG | 5-37; 50 to several thousand | Either parent, but expansion to congenital region form through mother | 3′ untranslated |
| Myotonic dystrophy (DM2; chromosome 3) | Muscle loss, cardiac arrhythmia, cataracts, frontal balding | CCTG | 10-26; 75-11,000 | — | 3′ untranslated region |
| Friedreich ataxia | Progressive limb ataxia, dysarthria, hypertrophic cardiomyopathy, pyramidal weakness in legs | GAA | 6-32; 200-1700 | Disorder is autosomal recessive, so disease alleles are inherited from both parents | Intron |
| Fragile X syndrome (FRAXA) | Mental retardation, large ears and jaws, macroorchidism in males | CGG | 4-39; 200-900 | Exclusively through mother | 5′ untranslated region |
| Fragile site (FRAXE) | Mild mental retardation | GCC | 6-35; >200 | More often through mother | 5′ untranslated region |
| Spinocerebellar ataxia type 8 | Adult-onset ataxia, dysarthria, nystagmus | CTG | 16-34; >74 | More often through mother | 3′ untranslated region |
| Spinocerebellar ataxia type 10 | Ataxia and seizures | ATTCT | 10-20; 500-4500 | More often through father | Intron |
| Spinocerebellar ataxia type 12 | Ataxia, eye movement disorders; variable age at onset | CAG | 7-45; 55-78 | — | 5′ untranslated region |
| Progressive myoclonic epilepsy type 1 | Juvenile-onset convulsions, myoclonus, dementia | 12-bp repeat motif | 2-3; 30-75 | Autosomal recessive inheritance, so transmitted by both parents | 5′ untranslated region |

Anticipation refers to progressively earlier or more severe expression of a disease in more recent generations. Expansion of DNA repeats has been shown to cause anticipation in some genetic diseases. These diseases can be divided into three major categories, depending on the size of the expansion, the location of the repeat, the phenotypic consequences of the expansion, the effect of the mutation, and the parent in whom large expansions typically occur.

## The Fragile X Story: Molecular Genetics Explains a Puzzling Pattern of Inheritance

Since the 19th century, it has been observed that there is an approximate 25% excess of males among persons with mental retardation. This excess is partly explained by several X-linked conditions that cause mental retardation, of which the fragile X syndrome is the most common. In addition to mental retardation, fragile X syndrome is characterized by a distinctive facial appearance, with large ears and long face (Fig. 5-15), hypermobile joints, and macroorchidism (increased testicular volume) in postpubertal males. The degree of mental retardation tends to be milder and more variable in females than in males. The syndrome is termed "fragile X" because the X chromosomes of affected persons, when cultured in a medium that is deficient in folic acid, sometimes exhibit breaks and gaps near the tip of the long arm (Fig. 5-16).

Although the presence of a single fragile X mutation is sufficient to cause disease in either males or females, the prevalence of this condition is higher in males (1/4000) than in females (1/8000). The lower degree of penetrance in females, as well as variability in expression, reflects variation

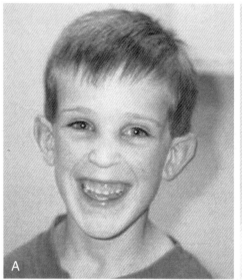
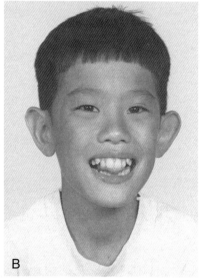
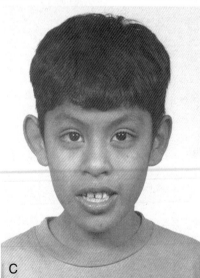

**FIGURE 5-15**
Boys with fragile X syndrome. Note the long faces, prominent jaws, and large ears and the similar characteristics of children from different ethnic groups: European (A), Asian (B), and Latin American (C).

**FIGURE 5-16**
An X chromosome from a male with fragile X syndrome, showing an elongated, condensed region near the tip of the long arm.
*(From Stein CK: Applications of cytogenetics in modern pathology. In McPherson RA, Pincus MR (eds): Henry's Clinical Diagnosis and Management by Laboratory Methods, 21st ed. Philadelphia: Saunders, 2006.)*

in patterns of X-inactivation (i.e., the percentage of active X chromosomes that carry the disease-causing mutation). Males who have affected descendants but are not affected themselves are termed *normal transmitting males.* In the mid-1980s, studies of fragile X syndrome pedigrees revealed a perplexing pattern: The mothers of transmitting males had a much lower percentage of affected sons than did the daughters of these males (Fig. 5-17). Because the mothers and daughters of normal transmitting males are both obligate carriers of the X-linked mutation, they should have equal risks of producing affected sons. Daughters of normal transmitting males were never affected with the disorder, but these women's sons could be affected. This pattern, dubbed the *Sherman paradox,* appeared to be inconsistent with the rules of X-linked inheritance.

Many mechanisms were proposed to explain this pattern, including autosomal and mitochondrial modifier loci. Resolution of the Sherman paradox came only with the cloning of the disease's gene, labeled *FMR1.* DNA sequence analysis showed that the 5' untranslated region of the gene contains a CGG repeat unit that is present in 6 to 50 copies in normal persons. Those with fragile X syndrome have 200 to 1000 or more CGG repeats (a full mutation). An intermediate number of repeats, ranging approximately from 50 to 200 copies, is seen in normal transmitting males and their female offspring. When these female offspring transmit the gene to their offspring, there is sometimes an expansion from the premutation of 50 to 200 repeats to the full mutation of more than 200 repeats. These expansions do not occur in male transmission. Furthermore, premutations tend to become larger in successive generations, and larger premutations are more likely to expand to a full mutation. These findings explain the Sherman paradox. Males with the premutation do not have daughters with fragile X syndrome because repeat expansion occurs in female transmission. Grandsons and great-grandsons of transmitting males are more likely to be affected by the disorder than are brothers of transmitting males because of progressive repeat expansion through successive generations of female permutation carriers.

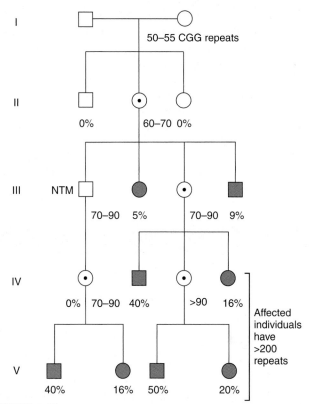

**FIGURE 5-17**
A pedigree showing the inheritance of the fragile X syndrome. Females who carry a premutation (50 to 230 CGG repeats) are shown by *dotted* lines. Affected individuals are represented by *solid* symbols. A normal transmitting male, who carries a premutation of 70 to 90 repeat units, is designated NTM. Note that the number of repeats increases each time the mutation is passed through another female. Also, only 5% of the NTM's sisters are affected, and only 9% of his brothers are affected, but 40% of his grandsons and 16% of his granddaughters are affected. This is the Sherman paradox.

Measurement of mRNA transcribed from *FMR1* has shown that the highest mRNA expression levels are in the brain, as would be expected. Persons with normal and premutation *FMR1* genes both produce mRNA. In fact, mRNA production is elevated in those with premutations, and it has been shown that this mRNA accumulates in the nucleus and has toxic effects, as do mutated mRNAs in myotonic dystrophy. Consequently, about one third of males with premutations develop a neurological disease characterized by ataxia and tremors in later life (after age 50 years). Approximately 20% of females with *FMR1* premutations experience premature ovarian failure (amenorrhea before age 40 years), again because of toxic mRNA effects. In contrast, those with full mutations have no *FMR1* mRNA in their cells, indicating that transcription of the gene has been eliminated. The CGG repeat is heavily methylated in those with the full mutation, as is a series of CG sequences 5' of the gene. The degree of methylation, which is likely to influence transcription of *FMR1*, is correlated with severity of expression of the disorder. A small percentage of persons with fragile X syndrome (<5%) do not have an expansion of the CGG repeat but instead have other loss-of-function point mutations in *FMR1*.

The protein product of *FMR1*, FMRP, binds to RNA and is capable of shuttling back and forth between the nucleus and the cytoplasm. It appears that FMRP may be involved in transporting mRNA from the nucleus to the cytoplasm, and it might also be involved in regulating the translation of mRNA.

Further study of this region has revealed the presence of another fragile site distal to the fragile X site. This site, termed FRAXE, is also associated with an expansion of a CGG trinucleotide repeat in the 5' region of a gene (labeled *FMR2*), subsequent hypermethylation, and a phenotype that includes mental retardation. Unlike fragile X syndrome, the CGG repeat at this locus can expand when transmitted through either males or females.

Cloning of the *FMR1* gene and the identification of a repeat expansion have already explained a great deal about the inheritance and expression of the fragile X syndrome. These advances have also improved diagnostic accuracy for the condition, because cytogenetic analysis of chromosomes often fails to identify fragile X heterozygotes. In contrast, DNA diagnosis, which consists of measurement of the length of the CGG repeat sequence and the degree of methylation of *FMR1*, is much more accurate.

## Study Questions

1. Females have been observed with five X chromosomes in each somatic cell. How many Barr bodies would such females have?

2. Explain why 8% to 10% of female carriers of the *DMD* gene have muscle weakness.

3. For X-linked recessive disorders, the ratio of affected males to affected females in populations increases as the disease frequency decreases. Explain this in terms of gene and genotype frequencies.

4. Figure 5-18 shows the inheritance of hemophilia A in a family. What is the risk that the male in generation IV is affected with hemophilia A? What is the risk that the female in generation IV is a heterozygous carrier? What is the risk that she is affected by the disorder?

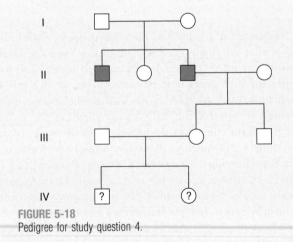

**FIGURE 5-18**
Pedigree for study question 4.

5. Pedigrees for autosomal dominant and X-linked dominant diseases are sometimes difficult to distinguish. Name as many features as you can that would help to tell them apart.

6. How would you distinguish mitochondrial inheritance from other modes of inheritance?

7. A man with Becker muscular dystrophy marries a phenotypically normal woman. On average, what percentage of their male offspring will be affected and what percentage of their female offspring will be affected by this disorder?

8. A female carrier of a Duchenne muscular dystrophy mutation marries a phenotypically normal man. On average, what percentage of their male offspring will be affected and what percentage of their female offspring will be affected with the disorder?

9. A boy and his brother both have hemophilia A. If there is no family history of hemophilia A in previous generations, what is the probability that the boys' aunt (the mother's sister) is a heterozygous carrier of the disease gene?

10. It is possible to create a zygote from two copies of the maternal genome alone. In amphibians, the zygote will develop and mature into an adult without fertilization by a sperm cell (this process is known as **parthenogenesis**). The same experiment has been attempted in mice, but it always results in early prenatal death. Explain this.

## Suggested Readings

Bolton-Maggs PHB, Pasi KJ. Haemophilias A and B. Lancet 2003;361:1801–9.

Carrel L, Willard HF. X-inactivation profile reveals extensive variability in X-linked gene expression in females. Nature 2005;434:400–4.

Chahrour M, Zoghbi HY. The story of Rett syndrome: From clinic to neurobiology. Neuron 2007;56:422–37.

Clayton-Smith J, Laan L. Angelman syndrome: A review of the clinical and genetic aspects. J Med Genet 2003;40: 87–95.

Clerc P, Avner P. Random X-chromosome inactivation: Skewing lessons for mice and men. Curr Opin Genet Dev 2006;16:246–53.

Dalkilic I, Kunkel LM. Muscular dystrophies: Genes to pathogenesis. Curr Opin Genet Dev 2003;13:231–38.

Deeb SS. The molecular basis of variation in human color vision. Clin Genet 2005;67:369–77.

Emery AE. The muscular dystrophies. Lancet 2002;359: 687–95.

Garber K, Smith KT, Reines D, Warren ST. Transcription, translation and fragile X syndrome. Curr Opin Genet Dev 2006;16:270–75.

Garber KB, Visootsak J, Warren ST. Fragile X syndrome. Eur J Hum Genet 2008;16:666–72.

Graw J, Brackmann HH, Oldenburg J, et al. Haemophilia A: From mutation analysis to new therapies. Nat Rev Genet 2005;6:488–501.

Horsthemke B, Buiting K. Imprinting defects on human chromosome 15. Cytogenet Genome Res 2006;113: 292–9.

Jiang Y-H, Bressler J, Beaudet AL. Epigenetics and human disease. Annu Rev Genomics Hum Genet 2004;5:479–510.

Orr HT, Zoghbi HY. Trinucleotide repeat disorders. Annu Rev Neurosci 2007;30:575–621.

Ranum LP, Cooper TA. RNA-mediated neuromuscular disorders. Annu Rev Neurosci 2006;29:259–77.

Ross MT, Grafham DV, Coffey AJ, et al. The DNA sequence of the human X chromosome. Nature 2005; 434:325–37.

Schapira AHV. Mitochondrial disease. Lancet 2006; 368: 70–82.

Straub T, Becker PB. Dosage compensation: The beginning and end of generalization. Nat Rev Genet 2007;8: 47–57.

Terracciano A, Chiurazzi P, Neri G. Fragile X syndrome. Am J Med Genet C Semin Med Genet 2005;137:32–7.

Wallace DC. A mitochondrial paradigm of metabolic and degenerative diseases, aging, and cancer: A dawn for evolutionary medicine. Annu Rev Genet 2005;39: 359–407.

Wattendorf DJ, Muenke M. Prader–Willi syndrome. Am Fam Physician 2005;72:827–30.

Weksberg R, Shuman C, Smith AC. Beckwith–Wiedemann syndrome. Am J Med Genet C Semin Med Genet 2005;137:12–23.

Wutz A, Gribnau J. X inactivation Xplained. Curr Opin Genet Dev 2007;17:387–93.

## Internet Resources

MITOMAP (extensive information about the mitochondrial genome and its role in disease) *http://www.mitomap.org/*

Muscular Dystrophy Association (information about various types of muscular dystrophy with links to other websites) *http://www.mdausa.org/*

National Hemophilia Foundation (information about hemophilia and links to other websites) *http://www.hemophilia. org/home.htm*

## Chapter 6

# CLINICAL CYTOGENETICS: THE CHROMOSOMAL BASIS OF HUMAN DISEASE

The previous two chapters have dealt with single-gene diseases. We turn now to diseases caused by alterations in the number or structure of chromosomes. The study of chromosomes and their abnormalities is called **cytogenetics**.

**Chromosome abnormalities** are responsible for a significant fraction of genetic diseases, occurring in approximately 1 of every 150 live births. They are the leading known cause of mental retardation and pregnancy loss. Chromosome abnormalities are seen in 50% of first-trimester and 20% of second-trimester spontaneous abortions. Thus, they are an important cause of morbidity and mortality.

As in other areas of medical genetics, advances in molecular genetics have contributed many new insights in the field of cytogenetics. For example, molecular techniques have permitted the identification of chromosome abnormalities, such as deletions, that affect very small regions. In some cases, specific genes that contribute to cytogenetic syndromes are being pinpointed. In addition, the ability to identify DNA polymorphisms in parents and offspring has enabled researchers to specify whether an abnormal chromosome is derived from the mother or from the father. This has increased our understanding of the biological basis of meiotic errors and chromosome abnormalities.

In this chapter we discuss abnormalities of chromosome number and structure. The genetic basis of sex determination is reviewed, the role of chromosome alterations in cancer is examined, and several diseases caused by chromosomal instability are discussed. Emphasis is given to the new contributions of molecular genetics to cytogenetics.

## CYTOGENETIC TECHNOLOGY AND NOMENCLATURE

Although it was possible to visualize chromosomes under microscopes as early as the mid-1800s, it was quite difficult to observe individual chromosomes. It was thus hard to count the number of chromosomes in a cell or to examine structural abnormalities. Beginning in the 1950s, several techniques were developed that improved our ability to observe chromosomes. These included the use of spindle poisons, such as colchicine and colcemid, that arrest dividing somatic cells in metaphase, when chromosomes are maximally condensed and easiest to see; the use of a hypotonic (low-salt) solution, which causes swelling of cells, rupture of the nucleus, and better separation of individual

chromosomes; and the use of staining materials that are absorbed differently by different parts of chromosomes, thus producing the characteristic light and dark bands that help to identify individual chromosomes.

> Our ability to study chromosomes has been improved by the visualization of chromosomes in metaphase, by hypotonic solutions that cause nuclear swelling, and by staining techniques that label chromosome bands.

Chromosomes are analyzed by collecting a living tissue (usually blood), culturing the tissue for the appropriate amount of time (usually 48 to 72 hours for peripheral lymphocytes), adding colcemid to produce metaphase arrest, harvesting the cells, placing the cell sediment on a slide, rupturing the cell nucleus with a hypotonic saline solution, staining with a designated nuclear stain, and photographing the metaphase spreads of chromosomes on the slide. The images of the 22 pairs of autosomes are arranged according to length, with the sex chromosomes in the right-hand corner. This ordered display of chromosomes is termed a **karyogram** or **karyotype** (Fig. 6-1). (The term *karyotype* refers to the number and type of chromosomes present in an individual, and *karyogram* is now often used to designate the printed display of chromosomes.) Currently, computerized image analyzers are usually used to display chromosomes.

After sorting by size, chromosomes are further classified according to the position of the centromere. If the centromere occurs near the middle of the chromosome, the chromosome is said to be **metacentric** (Fig. 6-2). An **acrocentric** chromosome has its centromere near the tip, and **submetacentric** chromosomes have centromeres somewhere between the middle and the tip. The tip of each chromosome is the **telomere**. The short arm of a chromosome is labeled $p$ (for *petite*), and the long arm is labeled $q$. In metacentric chromosomes, where the arms are of roughly equal length, the p and q arms are designated by convention.

> A karyotype, or karyogram, is a display of chromosomes ordered according to length. Depending on the position of the centromere, a chromosome may be acrocentric, submetacentric, or metacentric.

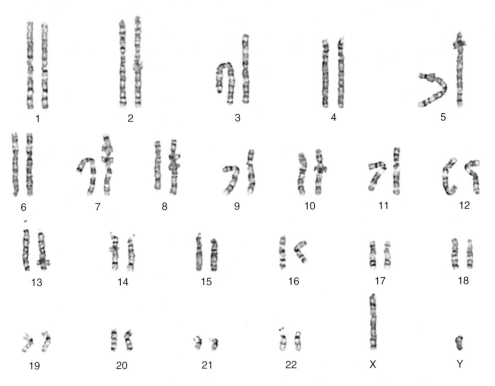

FIGURE 6-1
A banded (karyogram) of a normal female. The banded metaphase chromosomes are arranged from largest to smallest.

A normal female karyotype is designated 46,XX; a normal male karyotype is designated 46,XY. The nomenclature for various chromosome abnormalities is summarized in Table 6-1 and is indicated for each condition discussed in this chapter.

## Chromosome Banding

Early karyotypes were useful in counting the number of chromosomes, but structural abnormalities, such as balanced rearrangements or small chromosomal deletions, were often undetectable.

Staining techniques were developed in the 1970s to produce the chromosome bands characteristic of modern karyotypes. Chromosome banding helps greatly in the detection of deletions, duplications, and other structural abnormalities, and it facilitates the correct identification of individual chromosomes. The major bands on each chromosome are systematically numbered (Fig. 6-3). For example, 14q32 refers to the second band in the third region of the long arm of chromosome 14. Sub-bands are designated by decimal points following the band number (e.g., 14q32.3 is the third sub-band of band 2).

**TABLE 6-1**
## Standard Nomenclature for Chromosome Karyotypes

| Karyotype | Description |
|---|---|
| 46,XY | Normal male chromosome constitution |
| 47,XX,+21 | Female with trisomy 21, Down syndrome |
| 47,XY,+21[10]/46,XY[10] | Male who is a mosaic of trisomy 21 cells and normal cells (10 cells scored for each karyotype) |
| 46,XY,del(4)(p14) | Male with distal and terminal deletion of the short arm of chromosome 4 from band p14 to terminus |
| 46,XX,dup(5)(p14p15.3) | Female with a duplication within the short arm of chromosome 5 from bands p14 to p15.3 |
| 45,XY,der(13;14)(q10;q10) | A male with a balanced Robertsonian translocation of chromosomes 13 and 14. Karyotype shows that one normal 13 and one normal 14 are missing and replaced with a derivative chromosome composed of the long arms of chromosomes 13 and 14 |
| 46,XY,t(11;22)(q23;q22) | A male with a balanced reciprocal translocation between chromosomes 11 and 22. The breakpoints are at 11q23 and 22q22 |
| 46,XX,inv(3)(p21q13) | An inversion on chromosome 3 that extends from p21 to q13; because it includes the centromere, this is a pericentric inversion |
| 46,X,r(X)(p22.3q28) | A female with one normal X chromosome and one ring X chromosome formed by breakage at bands p22.3 and q28 with subsequent fusion |
| 46,X,i(Xq) | A female with one normal X chromosome and an isochromosome of the long arm of the X chromosome |

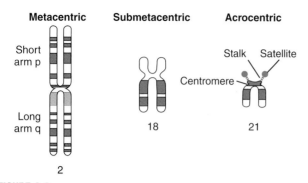

**FIGURE 6-2**
Metacentric, submetacentric, and acrocentric chromosomes. Note the stalks and satellites present on the short arms of the acrocentric chromosomes.

Several chromosome-banding techniques are used in cytogenetics laboratories. **Quinacrine** banding (Q-banding) was the first staining method used to produce specific banding patterns. This method requires a fluorescence microscope and is no longer as widely used as **Giemsa banding** (G-banding). To produce G-bands, a Giemsa stain is applied after the chromosomal proteins are partially digested by trypsin. **Reverse banding** (R-banding) requires heat treatment and reverses the usual white and black pattern that is seen in G-bands and Q-bands. This method is particularly helpful for staining the distal ends of chromosomes. Other staining techniques include **C-banding** and nucleolar organizing region stains (**NOR stains**). These latter methods specifically stain certain portions of the chromosome. C-banding stains the **constitutive heterochromatin**, which usually lies near the centromere, and NOR staining highlights the satellites and stalks of acrocentric chromosomes (see Fig. 6-2).

**High-resolution banding** involves staining chromosomes during prophase or early metaphase (prometaphase), before they reach maximal condensation. Because prophase and prometaphase chromosomes are more extended than metaphase chromosomes, the number of bands observable for all chromosomes increases from about 300 to 450 (as in Fig. 6-3) to as many as 800. This allows the detection of less obvious abnormalities usually not seen with conventional banding.

> Chromosome bands help to identify individual chromosomes and structural abnormalities in chromosomes. Banding techniques include quinacrine, Giemsa, reverse, C, and NOR banding. High-resolution banding, using prophase or prometaphase chromosomes, increases the number of observable bands.

### Fluorescence in situ Hybridization

In the widely used **fluorescence in situ hybridization (FISH)** technique, a labeled single-stranded DNA segment (probe) is exposed to denatured metaphase, prophase, or interphase

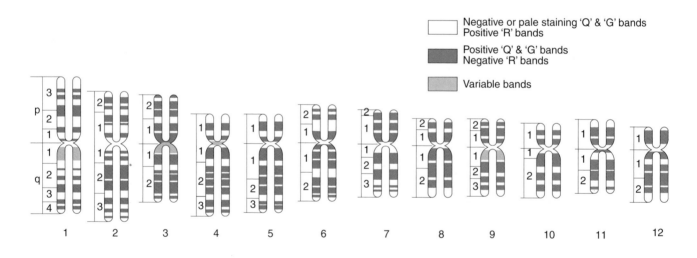

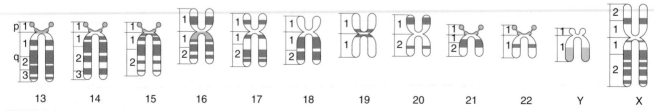

**FIGURE 6-3**
Schematic representation of the banding pattern of a G-banded karyotype; 300 bands are represented in this ideogram. The short and long arms of the chromosomes are designated, and the segments are numbered according to the standard nomenclature adopted at the Paris conference in 1971. In this illustration, both sister chromatids are shown for each chromosome.

chromosomes. The probe undergoes complementary base pairing (hybridization) only with the complementary DNA sequence at a specific location on one of the denatured chromosomes. Because the probe is labeled with a fluorescent dye, the location at which it hybridizes with the patient's chromosomes can be visualized under a fluorescence microscope. A common use of FISH is to determine whether a portion of a chromosome is deleted in a patient. In a normal person, a probe hybridizes in two places, reflecting the presence of two homologous chromosomes in a somatic cell nucleus. If a probe from the chromosome segment in question hybridizes to only one of the patient's chromosomes, then the patient probably has the deletion on the copy of the chromosome to which the probe fails to hybridize. FISH provides considerably better resolution than high-resolution banding approaches; it can typically detect deletions as small as 1 million base pairs (1 Mb) in size.

Extra copies of a chromosome region can also be detected using FISH. In this case, the probe hybridizes in three or more places instead of two. Combinations of FISH probes can also be used to detect chromosome rearrangements such as translocations (see later discussion).

Figure 6-4A illustrates a FISH result for a child who is missing a small piece of the short arm of chromosome 17. Although a centromere probe (used as a control) hybridizes to both copies of chromosome 17, a probe corresponding to a specific region of 17p hybridizes to only one chromosome. This demonstrates the deletion that causes the Smith–Magenis syndrome (Fig. 6-4B; also see Table 6-3 below).

Because FISH detection of missing or extra chromosomes can be carried out with interphase chromosomes, it is not necessary to stimulate cells to divide in order to obtain metaphase chromosomes (a time-consuming procedure required for traditional microscopic approaches). This allows analyses and diagnoses to be completed more rapidly. FISH analysis of interphase chromosomes is commonly used in the prenatal detection of fetal chromosome abnormalities and in the analysis of chromosome rearrangements in tumor cells.

The FISH technique has been extended by using multiple probes, each labeled with a different color, so that several of the most common numerical abnormalities (e.g., those of chromosomes 13, 18, 21, X, and Y) can be tested simultaneously in the same cell. In addition, techniques such as **spectral karyotyping** use varying combinations of five different fluorescent probes in conjunction with special cameras and image-processing software so that each chromosome is uniquely colored (painted along its entire length with a series of probes) for ready identification. Such images are especially useful for identifying small chromosome rearrangements (Fig. 6-5).

> FISH is a technique in which a labeled probe is hybridized to metaphase, prophase, or interphase chromosomes. FISH can be used to test for missing or additional chromosomal material as well as chromosome rearrangements. The FISH technique can be extended with multiple colors to detect several possible alterations of chromosome number simultaneously. Multiple probes can be used to paint each chromosome with a unique color, facilitating the detection of structural rearrangements.

## Comparative Genomic Hybridization

Losses or duplications of whole chromosomes or specific chromosome regions can be detected by a technique known as **comparative genomic hybridization (CGH)** (Fig. 6-6).

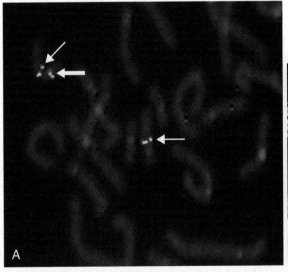

**FIGURE 6-4**
A, A fluorescence in situ hybridization (FISH) result. The *thinner arrows* point to a centromere-hybridizing probe for chromosome 17, and the *thicker arrow* points to a probe that hybridizes to 17p. The latter probe reveals only one spot in this individual, who has a deletion of 17p, producing the Smith–Magenis syndrome.
*(Courtesy of Dr. Arthur Brothman, University of Utah Health Sciences Center.)*
B, Face of an infant girl with Smith–Magenis syndrome. Note the broad forehead and relatively flat face.
*(Courtesy of Dr. Marilyn C. Jones, Children's Hospital, San Diego.)*

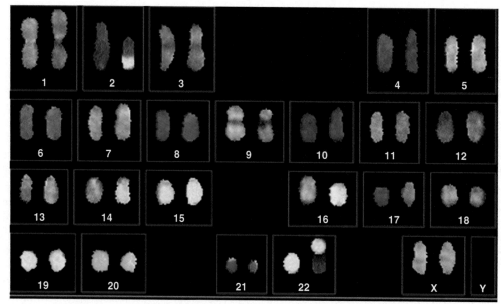

**FIGURE 6-5**
Spectral karyotype. The power of spectral karyotyping is demonstrated by the identification of a rearrangement between chromosomes 2 and 22. Note that a portion of chromosome 2 (*purple*) has exchanged places with a portion of chromosome 22 (*yellow*).
*(Courtesy of Dr. Art Brothman, University of Utah Health Sciences Center.)*

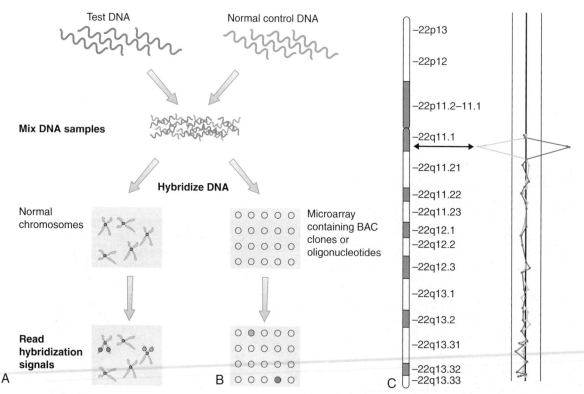

**FIGURE 6-6**
The comparative genomic hybridization (CGH) technique. **A**, Red-labeled test DNA (from a tumor sample in this case) and green-labeled reference DNA (from normal cells) are both denatured and hybridized to normal chromosomes. The ratio of green to red signal on the hybridized chromosomes indicates the location of duplications (*red signal*) or deletions (*green signal*) in the tumor chromosomes. **B**, Array CGH: Test and normal DNA are hybridized to probes embedded in a microarray. Duplications are indicated by hybridization of more red-labeled DNA to a probe containing a DNA sequence that is complementary to the duplicated region. Conversely, hybridization of only green-labeled DNA (reference DNA) indicates a deletion of the corresponding region. **C**, In a patient with DiGeorge sequence, a CGH test was performed in which the patient's DNA was labeled green and the control DNA was labeled red. The figure demonstrates a lack of green signal and excess red signal, signifying a deletion of chromosome 22q11.

DNA is extracted from a test source, such as cells from a tumor or whole-blood cells from a patient. The DNA is then labeled with a substance that exhibits one color (e.g., red) under a fluorescence microscope; DNA taken from normal control cells is labeled with a second color (e.g., green). In the earliest version of CGH, both sets of DNA are then hybridized to normal metaphase chromosomes on a slide. If any chromosome region is duplicated in the tumor cell, then the corresponding region of the metaphase chromosome will hybridize with excess quantities of the red-labeled DNA. This region will appear red under the microscope. Conversely, if a region is deleted in the tumor cell, then the corresponding region of the metaphase chromosome will hybridize only with green-labeled control DNA, and the region will appear green under the microscope. CGH is especially useful in scanning for deletions and duplications of chromosome material in cancer cells, where detection of such alterations can help to predict the type and/or severity of the cancer.

A serious limitation of CGH when used with metaphase chromosomes is that deletions or duplications smaller than 5 to 10 Mb cannot be detected microscopically. Higher resolution is offered by **array CGH**, in which test and control DNA is hybridized with a microarray (see Chapter 3) that contains probes whose DNA sequences correspond to specific regions of the genome. A commonly used CGH microarray contains approximately 3000 bacterial artificial chromosome (BAC, see Chapter 3) probes that each contain DNA inserts of about 150 kb. Because the BAC probes are on average about 1 Mb apart, this version of array CGH can detect duplications or deletions that are approximately 1 Mb in size (about 5 to 10 times better resolution than CGH with metaphase chromosomes). Other BAC microarrays are targeted so that higher resolution is achieved in regions where duplications or deletions are known to be associated with specific diseases (e.g., microdeletions such as Williams syndrome [see later]). A limitation of BAC microarrays is that alterations smaller than the BAC insert itself (about 150 kb) cannot easily be detected.

Still more recently, CGH has been developed using microarrays that contain hundreds of thousands of small oligonucleotide probes (see Chapter 3). These microarrays provide resolution to 50 to 100 kb or even less, allowing the detection of duplications and deletions that may affect only a single gene.

In addition to better resolution, array CGH offers a number of other advantages over traditional karyotype analysis. The process is highly automated, requiring less time from laboratory personnel. There is no need for dividing cells (in contrast to the analysis of metaphase chromosomes), and a minute amount of DNA is sufficient for analysis of the entire genome. For these reasons, array CGH is rapidly becoming one of the most commonly used techniques in cytogenetics laboratories. The primary disadvantage of CGH is that it cannot detect balanced rearrangements of chromosomes (i.e., reciprocal translocations or inversions).

> The CGH technique, in which differentially labeled DNA from test and control sources is hybridized to normal metaphase chromosomes or probes in microarrays, allows the detection of chromosome duplications and deletions but not balanced rearrangements. Array CGH can detect deletions and duplications shorter than 100 kb and requires only small amounts of DNA.

## ABNORMALITIES OF CHROMOSOME NUMBER

### Polyploidy

A cell that contains a multiple of 23 chromosomes in its nucleus is said to be **euploid** (Greek, *eu* = "good," *ploid* = "set"). Thus, haploid gametes and diploid somatic cells are euploid. **Polyploidy**, the presence of a complete set of extra chromosomes in a cell, is seen commonly in plants and often improves their agricultural value. Polyploidy also occurs in humans, although much less frequently. The polyploid conditions that have been observed in humans are **triploidy** (69 chromosomes in the nucleus of each cell) and **tetraploidy** (92 chromosomes in each cell nucleus). The karyotypes for these two conditions are designated 69,XXX and 92,XXXX, respectively (assuming that all of the sex chromosomes were X; other combinations of the X and Y chromosomes may be seen). Because the number of chromosomes present in each of these conditions is a multiple of 23, the cells are euploid in each case. However, the additional chromosomes encode a large amount of surplus gene product, causing multiple anomalies such as defects of the heart and central nervous system.

Triploidy is seen in only about 1 in 10,000 live births, but it accounts for an estimated 15% of the chromosome abnormalities occurring at conception. Thus, the vast majority of triploid conceptions are spontaneously aborted, and this condition is one of the most common causes of fetal loss in the first two trimesters of pregnancy. Triploid fetuses that survive to term typically die shortly after birth. The most common cause of triploidy is the fertilization of an egg by two sperm (**dispermy**). The resulting zygote receives 23 chromosomes from the egg and 23 chromosomes from each of the two sperm cells. Triploidy can also be caused by the fusion of an ovum and a polar body, each containing 23 chromosomes, and subsequent fertilization by a sperm cell. **Meiotic failure**, in which a diploid sperm or egg cell is produced, can also produce a triploid zygote.

Tetraploidy is much rarer than triploidy, both at conception and among live births. It has been recorded in only a few live births, and those infants survived for only a short period. Tetraploidy can be caused by a mitotic failure in the early embryo: All of the duplicated chromosomes migrate to one of the two daughter cells. It can also result from the fusion of two diploid zygotes.

> Cells that have a multiple of 23 chromosomes are said to be euploid. Triploidy (69 chromosomes) and tetraploidy (92 chromosomes) are polyploid conditions found in humans. Most polyploid conceptions are spontaneously aborted, and all are incompatible with long-term survival.

## Autosomal Aneuploidy

Cells that contain missing or additional individual chromosomes are termed **aneuploid** (not a multiple of 23 chromosomes). Usually only one chromosome is affected, but it is possible for more than one chromosome to be missing or duplicated. Aneuploidies of the autosomes are among the most clinically important of the chromosome abnormalities. They consist primarily of **monosomy** (the presence of only one copy of a chromosome in an otherwise diploid cell) and **trisomy** (three copies of a chromosome). Autosomal monosomies are nearly always incompatible with survival to term, so only a small number of them have been observed among live-born individuals. In contrast, some trisomies are seen with appreciable frequencies among live births. The fact that trisomies produce less-severe consequences than monosomies illustrates an important principle: *the body can tolerate excess genetic material more readily than it can tolerate a deficit of genetic material.*

The most common cause of aneuploidy is **nondisjunction**, the failure of chromosomes to disjoin normally during meiosis (Fig. 6-7). Nondisjunction can occur during meiosis I or meiosis II. The resulting gamete either lacks a chromosome or has two copies of it, producing a monosomic or trisomic zygote, respectively.

> Aneuploid conditions consist primarily of monosomies and trisomies. They are usually caused by nondisjunction. Autosomal monosomies are almost always lethal, but some autosomal trisomies are compatible with survival.

## Trisomy 21

**Trisomy 21** (karyotype 47,XY,+21 or 47,XX,+21)* is seen in approximately 1 of every 800 to 1000 live births, making it the most common autosomal aneuploid condition compatible with survival to term. This trisomy produces Down syndrome, a phenotype originally described by John Langdon Down in 1866. Almost 100 years elapsed between Down's description of this syndrome and the discovery (in 1959) that it is caused by the presence of an extra copy of chromosome 21.

Although there is considerable variation in the appearance of persons with Down syndrome, they present a constellation of features that help the clinician to make a diagnosis. The facial features include a low nasal root, upslanting palpebral fissures, measurably small and sometimes overfolded ears, and a flattened maxillary and malar region, giving the face a characteristic appearance (Fig. 6-8). Some of these features led to the use of the term "mongolism" in earlier literature, but this term is inappropriate and is no longer used. The cheeks are round, and the corners of the mouth are sometimes downturned. The neck is short, and the skin is redundant at the nape of the neck, especially in newborns. The occiput is flat, and the hands and feet tend to be rather broad and short. Approximately 50% of persons with Down syndrome have a deep flexion crease across their palms (formerly termed a "simian crease,"

---

*For brevity, the remainder of the karyotype designations for abnormalities not involving the sex chromosomes will indicate an affected male.

**FIGURE 6-7**
In meiotic nondisjunction, two chromosome homologs migrate to the same daughter cell instead of disjoining normally and migrating to different daughter cells. This produces monosomic and trisomic offspring.

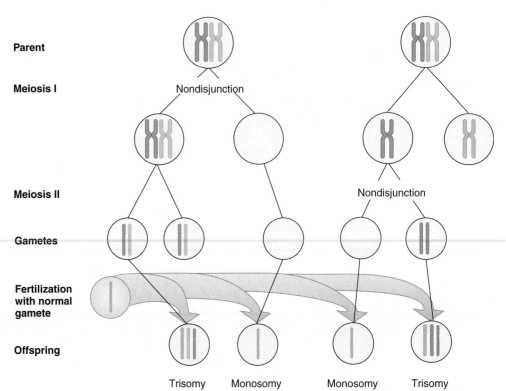

Parent

Meiosis I — Nondisjunction

Meiosis II — Nondisjunction

Gametes

Fertilization with normal gamete

Offspring

Trisomy    Monosomy    Monosomy    Trisomy

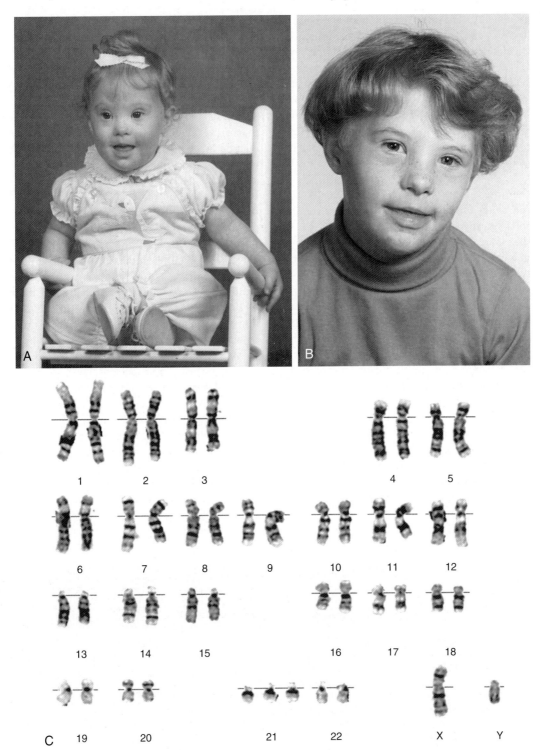

**FIGURE 6-8**
**A,** An infant with Down syndrome, illustrating typical features of this disorder: upslanting palpebral fissures, redundant skin of the inner eyelid (epicanthic fold), protruding tongue, and low nasal bridge. **B,** Same girl as in A, 7 years later. Note that the typical features are present but less obviously expressed. **C,** A karyogram of a male with trisomy 21.

though this term is also now considered inappropriate). Decreased muscle tone (hypotonia) is a highly consistent feature that is helpful in making a diagnosis.

Several medically significant problems occur with increased frequency among infants and children with Down syndrome. About 3% of these infants develop an obstruction of the duodenum or atresia (closure or absence) of the esophagus, duodenum, or anus. Respiratory infections are quite common, and the risk of developing leukemia is 15 to 20 times higher in patients with Down syndrome than in the general population. The most significant medical problem is that approximately 40% of these persons are born with structural heart defects. The most common of these is an atrioventricular (AV) canal, a defect in which the interatrial and interventricular septa fail to fuse normally during fetal development. The result is blood flow from the left side of the heart to the right side and then to the pulmonary vasculature, producing pulmonary hypertension. Ventricular septal defects (VSDs) are also common. Moderate to severe mental retardation (IQ ranging from 25 to 60) is seen in most persons with Down syndrome, and this condition alone accounts for approximately 10% of all cases of mental retardation in the United States.

Several other medical problems occur in infants and young children with Down syndrome. Conductive and sometimes neural hearing loss, hypothyroidism, and various eye abnormalities are the most important and common. Clinical Commentary 6-1 outlines a plan for routine medical care of infants and children with Down syndrome.

The medical problems seen in children with Down syndrome result in decreased survival rates. Congenital heart defects are the most important single cause of early mortality. In the early 1960s, only about half of children with this disorder survived as long as 5 years. As a result of improvements in corrective surgery, antibiotic treatment, and management of leukemia, the survival rate has increased considerably in the past 40 years. Currently, it is estimated that approximately 80% of children with Down syndrome will survive to age 10 years, and about half survive to 50 years of age. There is convincing evidence that enriched environments and educational interventions can produce significant improvements in intellectual function.

Males with Down syndrome are nearly always sterile, with only a few reported cases of reproduction. Many females with Down syndrome can reproduce, although approximately 40% fail to ovulate. A female with Down syndrome has a 50% risk of producing a gamete with two copies of chromosome 21 (which would then produce a trisomic zygote). However, because approximately 75% of trisomy 21 conceptions are spontaneously aborted, the risk of producing affected live-born offspring is considerably lower than 50% in women with Down syndrome. Because reproduction is uncommon, nearly all cases of trisomy 21 can be regarded as new mutations.

Approximately 95% of Down syndrome cases are caused by nondisjunction, and most of the remainder are caused by chromosome translocations (see later discussion). Comparisons of chromosome 21 polymorphisms in parents and offspring have demonstrated that the extra chromosome is contributed by the mother in 90% to 95% of trisomy 21 cases. About 75% of these maternal nondisjunctions occur during meiosis I, and the remainder occur during meiosis II. As discussed in greater detail later, there is a strong correlation between maternal age and the risk of producing a child with Down syndrome.

Mosaicism is seen in approximately 2% to 4% of trisomy 21 live births. These persons have some normal somatic cells and some cells with trisomy 21. This type of mosaicism in a male is designated 47,XY,+21[10]/46,XY[10], with the numbers inside the brackets indicating the number of cells scored with each karyotype. The most common cause of mosaicism in trisomy is a trisomic conception followed by loss of the extra chromosome during mitosis in some embryonic cells. Mosaicism often results in milder clinical expression of the phenotype associated with a chromosome abnormality.

## CLINICAL COMMENTARY 6-1
### Anticipatory Guidance and Health Supervision in Children with Down Syndrome

In recent years, an approach termed *health supervision and anticipatory guidance* has evolved for the care and treatment of persons with genetic syndromes and chronic diseases. After a thorough study of the disease in question (including an extensive literature review), basic guidelines are established for the screening, evaluation, and management of patients. If followed by the primary care practitioner or the specialist, these guidelines should help to prevent further disability or illness. We illustrate the health supervision and anticipatory guidance approach with the current guidelines for care of children with Down syndrome.

- As mentioned in the text, AV canals are the most common congenital heart defect seen in newborns with Down syndrome. Surgical correction of this condition is appropriate if it is detected before 1 year of age; after this time, pulmonary hypertension has been present too long for surgery to be successful. Accordingly, it is now recommended that an echocardiogram be performed during the newborn period.
- Because Down syndrome patients often have strabismus (deviation of the eye from its normal visual axis) and other eye problems, they should be examined regularly by their physician. If any symptoms or signs are observed, the patient is referred to an ophthalmologist familiar with Down syndrome. In asymptomatic children, an ophthalmological examination before the age of 4 years is recommended to evaluate visual acuity.
- Hypothyroidism is common, especially during adolescence. Therefore, thyroid hormone levels should be measured annually.
- Sensorineural and conductive hearing loss are both seen in children with Down syndrome. The routine follow-up should include a hearing test at birth and every 6 months until 2 years of age, with subsequent testing as needed.
- Instability of the first and second vertebrae has led to spinal cord injuries in some older Down syndrome patients. It is thus suggested that imaging studies be carried out in children with neurological symptoms and in those planning to participate in athletic activities.
- Referral of infants and children with Down syndrome to preschool programs to provide intervention for developmental disabilities is an important component of routine care.

Similar series of guidelines have been developed for children with trisomy 18, Williams syndrome, and Turner syndrome. In principle, the anticipatory guidance and health supervision approach can be applied to any genetic disease for which there is sufficient knowledge.

Depending on the timing and the way the mosaicism originated, some persons have **tissue-specific mosaicism**. As the term suggests, this type of mosaicism is confined only to certain tissues. This can complicate diagnosis, because karyotypes are usually made from a limited number of tissue types (usually circulating lymphocytes derived from a blood sample, or, less commonly, fibroblasts derived from a skin biopsy). Mosaicism affecting primarily the germline of a parent can lead to multiple recurrences of Down syndrome in the offspring. This factor helps to account for the fact that the recurrence risk for Down syndrome among mothers younger than 30 years is about 1% (i.e., 10 times higher than the population risk for this age group).

Because of the prevalence and clinical importance of Down syndrome, considerable effort has been devoted to defining the specific genes responsible for this disorder. Molecular approaches, such as cloning and sequencing, are being used to identify specific genes in this region that are responsible for the Down syndrome phenotype, and the process has been made easier with the availability of a complete DNA sequence for chromosome 21. A candidate for mental retardation in Down syndrome is *DYRK1A*, a kinase gene that causes learning and memory defects when it is over-expressed in mice. Another gene located in the critical region, *APP*, encodes the amyloid β precursor protein. A third copy of *APP* is likely to account for the occurrence of Alzheimer disease features in nearly all Down syndrome patients by 40 years of age. Mutations of *APP* cause a small percentage of Alzheimer disease cases (see Chapter 12), and Down syndrome individuals with partial trisomies that do not include the *APP* gene do not develop Alzheimer disease features.

> Trisomy 21, which causes Down syndrome, is the most common autosomal aneuploidy seen among live births. The most significant problems include mental retardation, gastrointestinal tract obstruction, congenital heart defects, and respiratory infections. The extra 21st chromosome is contributed by the mother in approximately 90% of cases. Mosaicism is seen in 2% to 4% of Down syndrome cases, and it often accompanies a milder phenotype. Specific genes contributing to the Down syndrome phenotype are being identified.

## Trisomy 18

Trisomy 18 (47,XY,+18), also known as Edwards syndrome, is the second most common autosomal trisomy, with a prevalence of about 1 per 6000 live births. It is, however, much more common at conception and is the most common chromosome abnormality among stillborns with congenital malformations. It is estimated that fewer than 5% of trisomy 18 conceptions survive to term.

The Edwards syndrome phenotype is as discernable as Down syndrome, but because it is less common, it is less likely to be recognized clinically. Infants with trisomy 18 have prenatal growth deficiency (weight that is low for gestational age), characteristic facial features, and a distinctive

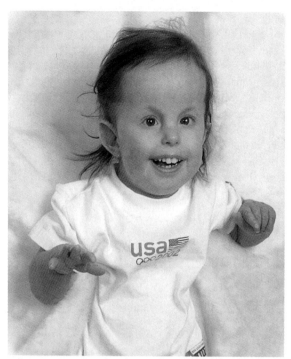

**FIGURE 6-9**
A 3-year-old girl with trisomy 18 (Edwards syndrome) with typical facial features including a narrow head, short palpebral fissures, and malformed external ears as well as characteristic overlapping of the index finger on top of the middle finger.

hand abnormality that often helps the clinician to make the initial diagnosis (Fig. 6-9). Minor anomalies of diagnostic importance include small ears with unraveled helices, a small mouth that is often hard to open, a short sternum, and short halluces (first toes). Most babies with trisomy 18 have major malformations. Congenital heart defects, particularly VSDs, are the most common and occur in 90% of children. Other medically significant congenital malformations include omphalocele (protrusion of the bowel into the umbilical cord), radial aplasia (missing radius bone), diaphragmatic hernia, and, occasionally, spina bifida.

About 50% of infants with trisomy 18 die within the first several weeks of life, and only about 5% survive to 12 months of age. A combination of factors, including aspiration pneumonia, predisposition to infections and apnea, and congenital heart defects, accounts for the high mortality rate.

Marked developmental disabilities are seen among those trisomy 18 patients who survive infancy. The degree of delay is much more significant than in Down syndrome, and most children are not able to walk independently. However, children with trisomy 18 do progress somewhat in their milestones, and older children learn some communication skills.

More than 95% of patients with Edwards syndrome have complete trisomy 18; only a small percentage have mosaicism. As in trisomy 21, there is a significant maternal age effect. Molecular analyses indicate that, as in trisomy 21, approximately 90% of trisomy 18 cases are the result of an extra chromosome transmitted by the mother.

## Trisomy 13

Trisomy 13 (47,XY,+13), also termed Patau syndrome, is seen in about 1 of every 10,000 births. The malformation pattern is quite distinctive and usually permits clinical recognition. It consists primarily of oral–facial clefts, microphthalmia (small, abnormally formed eyes), and postaxial polydactyly (Fig. 6-10). Malformations of the central nervous system are often seen, as are heart defects and renal abnormalities. Cutis aplasia (a scalp defect on the posterior occiput) can also occur.

The survival rate is very similar to that of trisomy 18, and 95% of live-born infants die during the first year of life.

Children who survive infancy have significant developmental retardation, with skills seldom progressing beyond those of a child of 2 years. However, as in trisomy 18, children with trisomy 13 do progress somewhat in their development and are able to communicate with their families to some degree.

About 80% of patients with Patau syndrome have full trisomy 13. Most of the remaining patients have trisomy of the long arm of chromosome 13 due to a translocation (see later discussion). As in trisomies 18 and 21, the risk of bearing a child with this condition increases with advanced maternal age. It is estimated that 95% or more of trisomy 13 conceptions are spontaneously lost during pregnancy.

> Trisomies of the 13th and 18th chromosomes are sometimes compatible with survival to term, although 95% or more of affected fetuses are spontaneously aborted. These trisomies are much less common at birth than is trisomy 21, and they produce severer disease features, with 95% mortality during the first year of life. As in trisomy 21, there is a maternal age effect, and the mother contributes the extra chromosome in approximately 90% of cases.

### Trisomies, Nondisjunction, and Maternal Age

The prevalence of Down syndrome among offspring of mothers of different ages is shown in Figure 6-11. Among mothers younger than 30 years of age, the risk is less than 1/1000. It increases to approximately 1/400 at age 35 years, 1/100 at age 40 years, and approximately 1/25 after age 45 years. Most other trisomies, including those in which the fetus does not survive to term, also increase in prevalence as maternal age increases. This risk is one of the primary indications for prenatal diagnosis among women older than 35 years (see Chapter 13).

Several hypotheses have been advanced to account for this increase, including the idea that a trisomic pregnancy is less

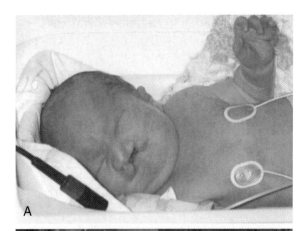
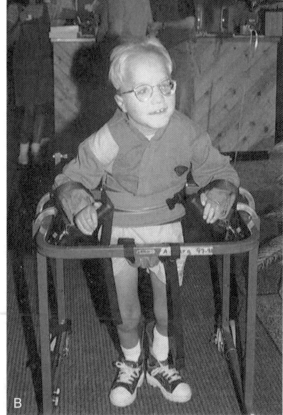

**FIGURE 6-10**

**A,** Newborn boy with full trisomy 13 (Patau syndrome). This baby has a cleft palate, atrial septal defect, inguinal hernia, and postaxial polydactyly of the left hand. **B,** A boy with full trisomy 13 at age 7 years (survival beyond the first year is uncommon). He has significant visual and auditory impairments.

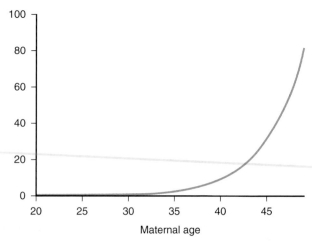

NUMBER OF DOWN SYNDROME
CASES PER 1,000 BIRTHS

Maternal age

**FIGURE 6-11**

The prevalence of Down syndrome among live births in relation to age of the mother. The prevalence increases with maternal age and becomes especially notable after the age of 35 years.

*(Data from Hook EB, Chambers GM: Birth defects. 1977;23[3A]:123-141.)*

likely to be spontaneously aborted in older women. Direct studies of the frequency of chromosome abnormalities in sperm and egg cells indicate that the pattern is instead due to an increase in nondisjunction among older mothers. Recall that nearly all of a female's oocytes are formed during her embryonic development. (There is recent evidence that a small number of oocytes may be produced later in life.) They remain suspended in prophase I until they are shed during ovulation. Thus, an ovum produced by a 45-year-old woman is itself about 45 years old. This long period of suspension in prophase I might impair normal chromosome disjunction, although the precise nature of this mechanism is not understood.

Many factors have been examined to determine whether they can affect the frequency of nondisjunction in women. These include hormone levels, cigarette smoking, autoimmune thyroid disease, alcohol consumption, and radiation (the latter does increase nondisjunction when administered in very large doses to experimental animals). None of these factors has shown consistent correlations with nondisjunction in humans, however, and maternal age remains the only known correlated factor.

Although maternal age is strongly correlated with the risk of Down syndrome, approximately three fourths of Down syndrome children are born to mothers younger than 35 years of age. This is because the great majority of children (more than 90%) are borne by women in this age group.

Numerous studies, including direct analysis of sperm cells, have tested the hypothesis of a paternal age effect for trisomies. The consensus is that such an effect, if it exists at all, is minor. This might reflect the fact that spermatocytes, unlike oocytes, are generated throughout the life of the male.

> Nearly all autosomal trisomies increase with maternal age as a result of nondisjunction in older mothers. There is little evidence for a paternal age effect on nondisjunction in males.

### Sex Chromosome Aneuploidy

Among live-born infants, about 1 in 400 males and 1 in 650 females have some form of sex chromosome aneuploidy. Primarily because of X inactivation, the consequences of this class of aneuploidy are less severe than those of autosomal aneuploidy. With the exception of the absence of an X chromosome, all of the sex chromosome aneuploidies are compatible with survival in at least some cases.

## Monosomy of the X Chromosome (Turner Syndrome)

The phenotype associated with a single X chromosome (45, X) was described by Henry Turner in 1938. (An earlier description was given by Otto Ullrich in 1930.) Persons with Turner syndrome are female and usually have a characteristic phenotype, including the variable presence of proportionate short stature, sexual infantilism and ovarian dysgenesis, and a pattern of major and minor malformations. The physical features can include a triangle-shaped face; posteriorly rotated external ears; and a broad, "webbed neck" (Fig. 6-12). In addition, the chest is broad and shield-like in shape. Lymphedema of the hands and feet is observable at birth. Many patients with Turner syndrome have congenital heart

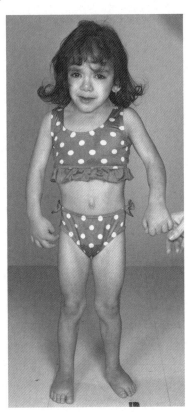

**FIGURE 6-12**
A girl with Turner syndrome (45,X). Note the characteristically broad, webbed neck. Stature is reduced, and swelling (lymphedema) is seen in the ankles and wrists.

defects, most commonly obstructive lesions of the left side of the heart (bicuspid aortic valve in 50% of patients and coarctation [narrowing] of the aorta in 15% to 30%). Severe obstructions should be surgically repaired. About 50% of persons with Turner syndrome have structural kidney defects, but they usually do not cause medical problems. There is typically some diminution in spatial perceptual ability, but intelligence is usually normal.

Girls with Turner syndrome exhibit proportionate short stature and do not undergo an adolescent growth spurt. Mature height is reduced by approximately 20 cm, on average. Growth hormone administration produces increased height somewhat in these girls, and families now commonly choose this therapy. In most persons with Turner syndrome, streaks of connective tissue, rather than ovaries, are seen (gonadal dysgenesis). Lacking normal ovaries, they do not usually develop secondary sexual characteristics, and most women with this condition are infertile (about 5% to 10% have sufficient ovarian development to undergo menarche, and a small number have borne children). Teenagers with Turner syndrome are typically treated with estrogen to promote the development of secondary sexual characteristics. The dose is then continued at a reduced level to maintain these characteristics and to help prevent osteoporosis.

The diagnosis of Turner syndrome is often made in the newborn infant, especially if there is a noticeable webbing of the neck coupled with a heart defect. The facial features are more subtle than in the autosomal abnormalities described previously, but the experienced clinician can often diagnose Turner syndrome on the basis of one or more of the listed clues. If Turner syndrome is not recognized in infancy or childhood, it is often diagnosed later because of short stature and/or amenorrhea.

The chromosome abnormalities in persons with Turner syndrome are quite variable. About 50% of these patients have a 45,X karyotype in their peripheral lymphocytes. At least 30% to 40% have mosaicism, most commonly 45,X/46,XX and less commonly 45,X/46,XY. Mosaics who have Y chromosomes in some cells are predisposed to neoplasms (gonadoblastomas) in the gonadal streak tissue. About 10% to 20% of patients with Turner syndrome have structural X chromosome abnormalities involving a deletion of some or all of Xp. This variation in chromosome abnormality helps to explain the considerable phenotypic variation seen in this syndrome.

Approximately 60% to 80% of monosomy X cases are caused by the absence of a paternally derived sex chromosome, occurring either during early mitosis in the embryo or during meiosis in the father (i.e., the offspring receives an X chromosome only from the mother). The 45,X karyotype is estimated to occur in 1% to 2% of conceptions, but Turner syndrome is seen in only about 1/2000 to 1/3000 live-born girls. Thus, the great majority (more than 99%) of 45,X conceptions are lost prenatally. Among those that do survive to term, many are chromosomal mosaics, and mosaicism of the placenta alone (**confined placental mosaicism**) is especially common. It is likely that the presence of some normal cells in mosaic fetuses enhances fetal survival.

Molecular analysis has begun to pinpoint specific genes involved in the Turner syndrome phenotype. For example, mutations in the *SHOX* gene, which encodes a transcription factor expressed in embryonic limbs, produce short stature. This gene is located on the distal tip of the X and Y short arms (in a region of the X chromosome that escapes inactivation; see Clinical Commentary 6-2). Thus, it is normally transcribed in two copies in both males and females. In females with Turner syndrome, this gene would be present in only one active copy, and the resulting haploinsufficiency contributes to short stature.

## CLINICAL COMMENTARY 6-2
### XX Males, XY Females, and the Genetic Basis of Sex Determination

During normal meiosis in the male, crossover occurs between the tip of the short arm of the Y chromosome and the tip of the short arm of the X chromosome. These regions of the X and Y chromosomes contain highly similar DNA sequences. Because this resembles the behavior of autosomes during meiosis, the distal portion of the Y chromosome is known as the **pseudoautosomal** region. It spans approximately 2.5 Mb.

Just centromeric of the pseudoautosomal region lies a gene known as *SRY* (sex-determining region on the Y). This gene, which is expressed in embryonic development, encodes a product that interacts with other genes to initiate the development of the undifferentiated embryo into a male (including Sertoli cell differentiation and secretion of müllerian-inhibiting substance). The *SRY* gene product is a member of the high-mobility group (HMG) family of DNA-bending transcription factors. By bending DNA, the protein is thought to promote DNA–DNA interactions that trigger events in the developmental cascade leading to male differentiation. In particular, the protein encoded by *SRY* interacts antagonistically with that of *DAX1*, which represses genes that promote differentiation of the embryo into a male. In the absence of *SRY*, *DAX1* continues to repress these genes, and a female embryo is created.

Loss-of-function mutations of *SRY* can produce individuals with an XY karyotype but a female phenotype. In addition, when the mouse *Sry* gene is inserted experimentally into a female mouse embryo, a male mouse is produced. Thus, there is very good evidence that the *SRY* gene is the initiator of male sexual differentiation in the embryo. Mutations in another member of this gene family, *SOX9*, can produce sex reversal (XY females) and campomelic dysplasia (malformations of bone and cartilage).

Approximately 1 of every 20,000 males presents with a phenotype similar to Klinefelter syndrome (without increased height), but chromosome evaluation shows that they have a normal *female* karyotype (46,XX). It has been demonstrated that these XX males have an X chromosome that includes the *SRY* gene. This is explained as a result of a faulty crossover between the X and Y chromosomes during male meiosis, such that the *SRY* gene, instead of remaining on the Y chromosome, is transferred to the X chromosome. The offspring who inherits this X chromosome from his father consequently has a male phenotype. Conversely, it should be apparent that an offspring who inherits a Y chromosome that lacks the *SRY* gene would be an XY female. These females have gonadal streaks rather than ovaries and have poorly developed secondary sexual characteristics.

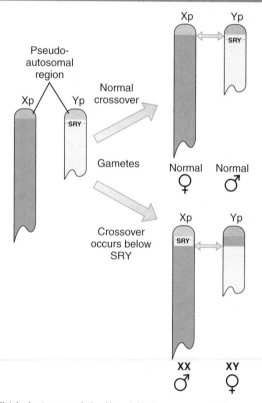

The distal short arms of the X and Y chromosomes exchange material during meiosis in the male. The region of the Y chromosome in which this crossover occurs is called the pseudoautosomal region. The *SRY* gene, which triggers the process leading to male gonadal differentiation, is located just outside the pseudoautosomal region. Occasionally, the crossover occurs on the centromeric side of the *SRY* gene, causing it to lie on an X chromosome instead of on the Y chromosome. An offspring receiving this X chromosome will be an XX male, and an offspring receiving the Y chromosome will be an XY female.

> Most females with Turner syndrome have a 45,X karyotype. Although this disorder is common at conception, it is relatively rare among live births, reflecting a very high rate of spontaneous abortion. Mosaicism, including confined placental mosaicism, appears to increase the probability of survival to term.

## Klinefelter Syndrome

Like Down and Turner syndromes, the syndrome associated with a 47,XXY karyotype was identified before the underlying chromosomal abnormality was understood. Described in 1942 by Harry Klinefelter, the syndrome that bears his name is seen in approximately 1/500 to 1/1000 male births. Although Klinefelter syndrome is a common cause of primary hypogonadism in males, the phenotype is less striking than those of the syndromes described thus far. Patients with Klinefelter syndrome tend to be taller than average, with disproportionately long arms and legs (Fig. 6-13). Clinical examination of postpubertal patients reveals small testes (less than 10 mL in volume), and most patients with Klinefelter syndrome are sterile as a result of atrophy of the seminiferous tubules. Testosterone levels in adolescents and adults are low. Gynecomastia (breast development) is seen in approximately one third of affected persons and leads to an increased risk of breast cancer, which can be reduced by mastectomy (breast removal). Body hair is typically sparse after puberty, and muscle mass tends to be reduced. In addition, there is a predisposition for learning disabilities and a reduction in verbal IQ. Although intelligence is usually in the normal range, the IQ is on average 10 to 15 points lower than that of the affected person's siblings. Because of the subtlety of this disorder, Klinefelter syndrome is often not diagnosed until after puberty, and the condition is sometimes first ascertained in fertility clinics.

The extra X chromosome is derived maternally in about 50% of Klinefelter cases, and the syndrome increases in incidence with advanced maternal age. Mosaicism, which is seen in about 15% of patients, increases the likelihood of viable sperm production. Individuals with the 48,XXXY and 49,XXXXY karyotypes have also been reported. Because they have a Y chromosome, they have a male phenotype, but the degree of mental deficiency and physical abnormality increases with each additional X chromosome.

Testosterone therapy, beginning in midadolescence, can enhance secondary sex characteristics, and it helps to decrease the risk of osteoporosis. There is some evidence that this treatment also improves psychological well-being.

> Males with Klinefelter syndrome (47,XXY) are taller than average, might have reduced IQ, and are usually sterile. Testosterone therapy and mastectomy for gynecomastia are sometimes recommended.

## Trisomy X

The 47,XXX karyotype occurs in approximately 1/1000 females and usually has benign consequences. Overt physical abnormalities are rarely seen, but these females sometimes suffer from sterility, menstrual irregularity, or mild mental retardation. As in Klinefelter syndrome, the 47,XXX karyotype is often first ascertained in fertility clinics. Approximately 90% of cases are the result of nondisjunction in the mother, and, as with other trisomies, the incidence increases among the offspring of older mothers.

Females have also been seen with four, five, or even more X chromosomes. Each additional X chromosome is accompanied by increased mental retardation and physical abnormality.

## 47,XYY Syndrome

The final sex chromosome aneuploidy to be discussed is the 47,XYY karyotype. Males with this karyotype tend to be taller than average, and they have a 10- to 15-point reduction in average IQ. This condition, which causes few physical problems, achieved great notoriety when its incidence in the male prison population was discovered to be as high as 1/30, compared with 1/1000 in the general male population. This led to the suggestion that this karyotype might confer a predisposition to violent, criminal behavior. A number of studies have addressed this issue and have shown that XYY males are not inclined to commit violent crimes. There is, however, evidence of an increased incidence of minor behavioral disorders, such as hyperactivity, attention deficit disorder, and learning disabilities.

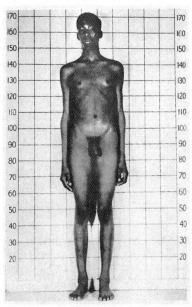

**FIGURE 6-13**
A male with Klinefelter syndrome (47,XXY). Stature is increased, gynecomastia may be present, and body shape may be somewhat feminine. *(From McKusick VA: J Chronic Dis 1960; 12-1-202.)*

> The 47,XXX and 47,XYY karyotypes are seen in about 1/1000 females and males, respectively. Each involves a slight degree of reduction in IQ but few physical problems.

## CHROMOSOME ABNORMALITIES AND PREGNANCY LOSS

For a long time it was difficult to detect the early stages of pregnancy with certainty. It was thus possible for a woman to become pregnant but to lose the embryo before knowing of the pregnancy. Sensitive urinary assays of chorionic gonadotropin levels, which increase when the embryo implants in the uterine wall, have allowed researchers to pinpoint accurately the occurrence of pregnancy at this early stage. Follow-up of women in whom implantation was verified in this way has shown that about one third of pregnancies are lost after implantation (the number lost before implantation is unknown). Therefore, spontaneous pregnancy loss is common in humans.

As mentioned earlier, chromosome abnormalities are the leading known cause of pregnancy loss. It is estimated that a minimum of 10% to 20% of conceptions have a chromosome abnormality, and at least 95% of these chromosomally abnormal conceptions are lost before term. Karyotype studies of miscarriages indicate that about 50% of the chromosome abnormalities are trisomies, 20% are monosomies, 15% are triploids, and the remainder consists of tetraploids and structural abnormalities. Some chromosome abnormalities that are common at conception seldom or never survive to term. For example, trisomy 16 is thought to be the most common trisomy at conception, but it is never seen in live births.

It is possible to study chromosome abnormalities directly in sperm and egg cells. Oocytes are typically obtained from unused material in in vitro fertilization studies. Karyotypes obtained from these cells indicate that 20% to 25% of oocytes have missing or extra chromosomes. Human sperm cells can be studied by FISH analysis or after fusing them with hamster oocytes so that their DNA begins mitosis and condenses, allowing easier visualization. The frequency of aneuploidy in these sperm cells is about 3% to 4%. Structural abnormalities (see later discussion) are seen in about 1% of oocytes and 5% of sperm cells, and the incidence increases with advanced paternal age. Undoubtedly, this high rate of chromosome abnormality contributes importantly to later pregnancy loss.

These approaches, while informative, can involve some biases. For example, mothers in whom in vitro fertilization is performed are not a representative sample of the population. In addition, their oocytes have been stimulated artificially, and only those oocytes that could not be fertilized by sperm cells are studied. Thus, the oocytes themselves might not be a representative sample. The sperm cells studied in human–hamster hybrids represent only those that are capable of penetrating the hamster oocyte and again might not be a representative sample.

FISH analysis of aneuploidy can evaluate thousands of cells fairly rapidly, giving it an important advantage over the human–hamster technique. In general, the FISH studies have yielded results similar to those of the human–hamster studies, showing that on average, the frequency of disomy is approximately 0.15% for each autosome and 0.26% for the sex chromosomes. These studies have also confirmed a tendency for elevated frequencies of nondisjunction of the sex chromosomes and some of the acrocentric chromosomes, including chromosome 21, in sperm cells.

> Pregnancy loss is common in humans, with approximately one third of pregnancies lost spontaneously after implantation. Chromosome abnormalities, which have been studied in sperm and egg cells and in miscarriages and stillbirths, are an important cause of pregnancy loss.

## ABNORMALITIES OF CHROMOSOME STRUCTURE

In addition to the loss or gain of whole chromosomes, parts of chromosomes can be lost or duplicated as gametes are formed, and the arrangement of portions of chromosomes can be altered. Structural chromosome abnormalities may be **unbalanced** (the rearrangement causes a gain or loss of chromosomal material) or **balanced** (the rearrangement does not produce a loss or gain of chromosome material). Unlike aneuploidy and polyploidy, balanced structural abnormalities often do not produce serious health consequences. Nevertheless, abnormalities of chromosome structure, especially those that are unbalanced, can produce serious disease in individuals or their offspring.

Alterations of chromosome structure can occur when homologous chromosomes line up improperly during meiosis (e.g., unequal crossover, as described in Chapter 5). In addition, **chromosome breakage** can occur during meiosis or mitosis. Mechanisms exist to repair these breaks, and usually the break is repaired perfectly with no damage to the daughter cell. Sometimes, however, the breaks remain, or they heal in a fashion that alters the structure of the chromosome. The likelihood of chromosome breakage may be increased in the presence of certain harmful agents, called **clastogens**. Clastogens identified in experimental systems include ionizing radiation, some viral infections, and some chemicals.

### Translocations

A **translocation** is the interchange of genetic material between nonhomologous chromosomes. Balanced translocations represent one of the most common chromosomal aberrations in humans, occurring in 1 of every 500 to 1000 individuals (Table 6-2). There are two basic types of translocations, **reciprocal** and **Robertsonian**.

#### Reciprocal Translocations

Reciprocal translocations happen when breaks occur in two different chromosomes and the material is mutually exchanged. The resulting chromosomes are called **derivative chromosomes**. The carrier of a reciprocal translocation is usually unaffected because he or she has a normal complement of genetic material. However, the carrier's offspring can be normal, can carry the translocation, or can have duplications or deletions of genetic material.

An example of a reciprocal translocation between chromosomes 3 and 6 is shown in Figure 6-14. The distal half of the

**TABLE 6-2**

## Prevalence of Chromosomal Abnormalities among Newborns

| Abnormality | Prevalence at Birth |
|---|---|
| **Autosomal Syndromes** | |
| Trisomy 21 | 1/800 |
| Trisomy 18 | 1/6000 |
| Trisomy 13 | 1/10,000 |
| Unbalanced rearrangements | 1/17,000 |
| Balanced Rearrangements | |
|    Robertsonian translocations | 1/1000 |
|    Reciprocal translocations | 1/11,000 |
| **Sex Chromosome Abnormalities** | |
| 47,XXY | 1/1000 male births |
| 47,XYY | 1/1000 male births |
| 45,X* | 1/5000 female births |
| 47,XXX | 1/1000 female births |
| **All Chromosome Abnormalities** | |
| Autosomal disorders and unbalanced rearrangements | 1/230 |
| Balanced rearrangements | 1/500* |

*The 45,X karyotype accounts for about half of the cases of Turner syndrome.*

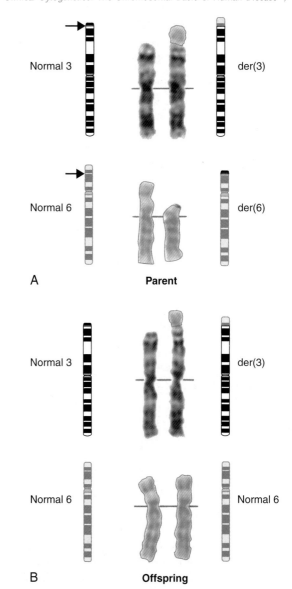

**FIGURE 6-14**

**A,** The parent has a reciprocal balanced translocation involving the short arms of chromosomes 6 and 3. The distal short arm of the 6 has been translocated to the very distal tip of the 3. A small piece of chromosome 3 is attached to the derivative 6. This person had a child whose chromosomes are depicted in **B**. The child received the derivative chromosome 3 (with part of the 6 short arm attached) and the normal 6; from the other parent, the child received a normal 3 and a normal 6. Therefore, the child had a partial trisomy of the 6 short arm and, presumably, a small deletion of the 3 short arm.

short arm of chromosome 6 is translocated to the short arm of chromosome 3, and a small piece of chromosome 3 is translocated to the short arm of chromosome 6. If the translocations occur at 3p13 and 6p14, the karyotype is designated 46,XX, t(3;6)(p13;p14). The offspring of this woman received the derivative chromosome 3, termed der(3), and the normal 6; thus, the child had a **partial trisomy** of the distal portion of chromosome 6 (i.e., 6p trisomy). This is a well-established but rather uncommon chromosomal syndrome.

> Reciprocal translocations are caused by two breaks on different chromosomes, with a subsequent exchange of material. Although carriers of balanced reciprocal translocations usually have normal phenotypes, their offspring might have a partial trisomy or a partial monosomy and an abnormal phenotype.

## Robertsonian Translocations

In Robertsonian translocations, the short arms of two nonhomologous chromosomes are lost and the long arms fuse at the centromere to form a single chromosome (Fig. 6-15). This type of translocation is confined to the acrocentric chromosomes (13, 14, 15, 21, and 22), because the short arms of these chromosomes are very small and contain no essential genetic material. Because the carriers of Robertsonian translocations lose no essential genetic material, they are phenotypically normal but have only 45 chromosomes in each cell. Their offspring, however, may inherit an extra or missing long arm of an acrocentric chromosome.

A common Robertsonian translocation involves fusion of the long arms of chromosomes 14 and 21. The karyotype of

**Robertsonian translocation**

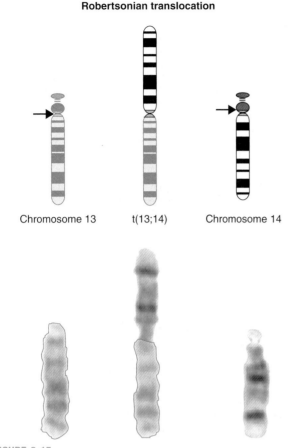

Chromosome 13          t(13;14)          Chromosome 14

**FIGURE 6-15**
In a Robertsonian translocation, shown here, the long arms of two acrocentric chromosomes (13 and 14) fuse, forming a single chromosome.

a male carrier of this translocation is 45,XY,der(14;21)(q10; q10). This person lacks one normal 14 and one normal 21 and instead has a chromosome derived from a translocation of the entire long arms of chromosomes 14 and 21. During meiosis in this person, the translocation chromosome must still pair with its homologs. Figure 6-16 illustrates the ways these chromosomes can segregate in the gametes formed by the translocation carrier. If **alternate segregation** occurs, then the offspring are either chromosomally normal or have a balanced translocation with a normal phenotype. If one of the **adjacent segregation** patterns occurs, then the gametes are unbalanced and the offspring may have trisomy 14, monosomy 14, monosomy 21, or trisomy 21. (Note that these trisomies and monosomies are genetically the same as trisomies and monosomies produced by nondisjunction, because only the long arms of these chromosomes contain genetically significant material.) Fetuses with the first three possibilities do not survive to term, and the last translocation results in an infant with three copies of the long arm of chromosome 21 and a Down syndrome phenotype. Robertsonian translocations are responsible for approximately 5% of Down syndrome cases.

It is expected that the three types of conceptions compatible with survival would occur in equal frequencies: One third would be completely normal, one third would carry the translocation but be phenotypically normal, and one third would have Down syndrome. In part because of prenatal loss, the actual fraction of live-born offspring with Down syndrome is less than one third (about 10% to 15% for mothers who carry the translocation, and only 1% to 2% for fathers who carry it). This recurrence risk, however, is greater than the risk for parents of a child who has the nondisjunction type of Down syndrome (1% for mothers younger than 30 years of age). This difference in recurrence risks demonstrates why it is critical to order a chromosome study whenever a condition such as Down syndrome is suspected.

> Robertsonian translocations occur when the long arms of two acrocentric chromosomes fuse at the centromere. The carrier of a Robertsonian translocation can produce conceptions with monosomy or trisomy of the long arms of acrocentric chromosomes.

### Deletions

A **deletion** is caused by a chromosome break and subsequent loss of genetic material. A single break leading to a loss that includes the chromosome's tip is called a **terminal deletion**. An **interstitial deletion** results when two breaks occur and the material between the breaks is lost. For example, a chromosome segment with normal DNA can be symbolized ABCDEFG. An interstitial deletion could produce the sequence ABEFG, and a terminal deletion could produce ABCDE.

Usually, a gamete containing a chromosome with a deletion unites with a normal gamete to form a zygote. The zygote then has one normal chromosome and a homolog with the deletion. Microscopically visible deletions generally involve multiple genes, and the consequences of losing this much genetic material from even one member of the chromosome pair can be severe. After the three autosomal aneuploidies described earlier, the autosomal deletion syndromes are the most common group of clinically significant chromosome abnormalities.

A well-known example of a chromosome deletion syndrome is the *cri-du-chat* syndrome. This term (French, "cry of the cat") describes the distinctive cry of the child. This cry usually becomes less obvious as the child ages, making a clinical diagnosis more difficult after 2 years of age. *Cri-du-chat* syndrome is caused by a deletion of the distal short arm of chromosome 5; the karyotype is 46,XY,del(5p). Seen in approximately 1 in 50,000 live births, it is also characterized by mental retardation (average IQ about 35), microcephaly (small head), and a characteristic but not distinctive facial appearance. Although mortality rates are increased, many persons with cri-du-chat syndrome now survive to adulthood.

Wolf–Hirschhorn syndrome (Fig. 6-17), caused by a deletion of the distal short arm of chromosome 4, is another well-characterized deletion syndrome. Other well-known deletions include those of 18p, 18q, and 13q. With the exception of the 18p deletion syndrome, each of these disorders is relatively distinctive, and the diagnosis can often be made before the karyotype is obtained. The features of the

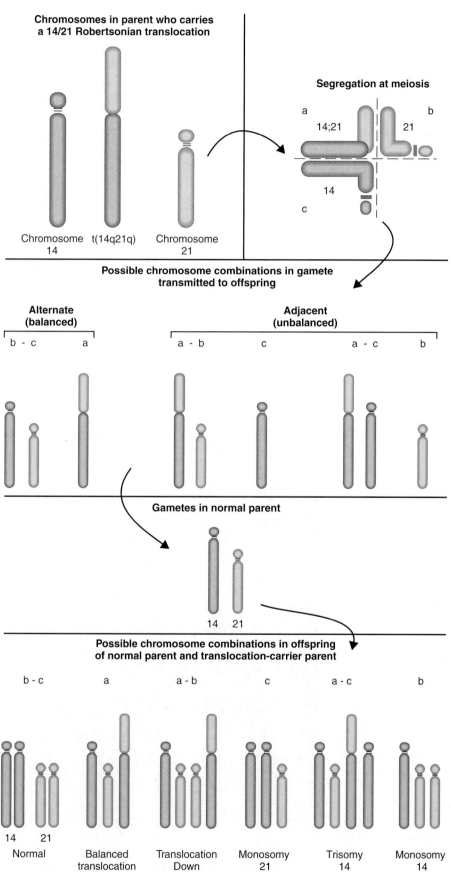

**Chromosomes in parent who carries a 14/21 Robertsonian translocation**

Chromosome 14    t(14q21q)    Chromosome 21

**Segregation at meiosis**

a    14;21    b    21

14    c

**Possible chromosome combinations in gamete transmitted to offspring**

**Alternate (balanced)**

b - c    a

**Adjacent (unbalanced)**

a - b    c    a - c    b

**Gametes in normal parent**

14    21

**Possible chromosome combinations in offspring of normal parent and translocation-carrier parent**

b - c    a    a - b    c    a - c    b

14    21
Normal

Balanced translocation t(14q21q)

Translocation Down syndrome

Monosomy 21

Trisomy 14

Monosomy 14

**FIGURE 6-16**
The possible segregation patterns for the gametes formed by a carrier of a Robertsonian translocation. Alternate segregation (quadrant *a* alone, or quadrant *b* with quadrant *c*) produces either a normal chromosome constitution or a translocation carrier with a normal phenotype. Adjacent segregation (quadrant *a* with *b*, quadrant *c* alone, quadrant *a* with *c*, or quadrant *b* alone) produces unbalanced gametes and results in conceptions with translocation Down syndrome, monosomy 21, trisomy 14, or monosomy 14, respectively. For example, monosomy 14 is produced when the parent who carries the translocation transmits a copy of chromosome 21 but does not transmit a copy of chromosome 14 (as in the *lower right corner*).

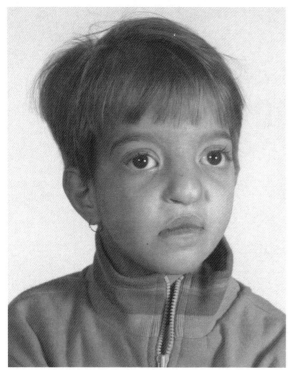

**FIGURE 6-17**
Child with Wolf–Hirschhorn syndrome (46,XX,del[4p]). Note the wide-spaced eyes and repaired cleft lip.

18p deletion syndrome are more subtle, and usually this condition is recognized when chromosome analysis is performed for evaluation of developmental disability.

> Microscopically observable chromosome deletions, which may be either terminal or interstitial, usually affect a fairly large number of genes and produce recognizable syndromes.

### Microdeletion Syndromes

The deletions described thus far all involve relatively large segments of chromosomes, and many of them were described before chromosome banding techniques were developed. With the advent of high-resolution banding, it has become possible to identify microscopically a large number of deletions that were previously too small for detection. In addition, advances in molecular genetics, particularly the FISH technique (see Fig. 6-4) and the development of large numbers of easily identified polymorphisms, have permitted the detection of deletions that are often too small to be observed microscopically (i.e., <5 Mb).

Prader–Willi syndrome, a disorder discussed in Chapter 5, is a good example of a **microdeletion syndrome**. Although this condition was described in the 1950s,* it was not until 1981 that advanced banding techniques detected a small deletion

of chromosome bands 15q11-q13 in about 50% of these patients. With the use of molecular techniques, deletions that were too small to be detected cytogenetically were also discovered. In total, about 70% of Prader–Willi cases are caused by microdeletions of 15q.

Because of imprinting, the inheritance of a microdeletion of the paternal chromosome 15 material produces Prader–Willi syndrome, while a microdeletion of the maternally derived chromosome 15 produces the phenotypically distinct Angelman syndrome (see Chapter 5).

Williams syndrome, which is characterized by mental retardation, supravalvular aortic stenosis (SVAS), multiple peripheral pulmonary arterial stenoses, characteristic facial features, dental malformations, and hypercalcemia, is another example of a microdeletion syndrome (Fig. 6-18).

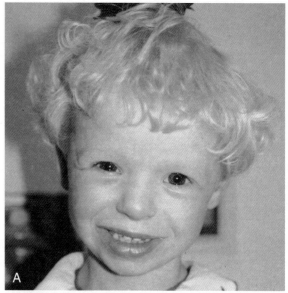

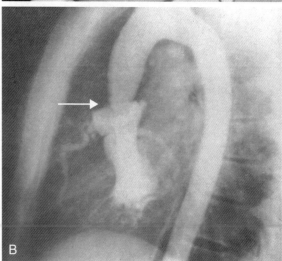

**FIGURE 6-18**
**A,** Girl with Williams syndrome, illustrating typical facial features: broad forehead, short palpebral fissures, low nasal bridge, anteverted nostrils, long philtrum, full cheeks, and relatively large mouth with full lips. **B,** Angiogram illustrating supravalvular aortic stenosis (narrowing of the ascending aorta) *(arrow).*
*(Courtesy Dr. Mark Keating, Harvard University.)*

---

*Although Prader is commonly credited with the first complete description of Prader–Willi syndrome in 1956, John Langdon Down (of Down syndrome fame) published a fairly complete description of the disorder in 1887.

A series of molecular analyses have identified some of the individual genes responsible for the Williams syndrome phenotype. The gene that encodes elastin, *ELN*, for example, is located in the Williams syndrome critical region and is expressed in blood vessels. Elastin is an important component of the aortic wall (microfibrils, which were discussed in Chapter 4 in the context of Marfan syndrome, are another component). Mutations or deletions of elastin alone result in isolated SVAS without the other features of Williams syndrome. Larger deletions, encompassing additional genes, produce the complete Williams syndrome phenotype. A second gene in the critical region, *LIMK1*, encodes a brain-expressed kinase that may be involved in the visual–spatial cognition defects observed in patients with Williams syndrome. This is supported by observation of patients with partial deletions of the critical region affecting only the *ELN* and *LIMK1* genes. These persons have SVAS and visual–spatial cognitive deficiency but none of the other features of Williams syndrome.

High-resolution banding and molecular genetic techniques have often led to a more precise specification of the critical chromosome region that must be deleted to cause a given syndrome. Wolf–Hirschhorn syndrome, for instance, can be produced by the deletion of only a very small telomeric segment of 4p. In some instances, specific genes responsible for chromosome abnormality syndromes can be pinpointed. For example, persons with a deletion of 11p can present with a series of features including *W*ilms tumor (a kidney tumor), *a*niridia (absence of the iris), *g*enitourinary abnormalities,* and mental *r*etardation (sometimes termed the WAGR syndrome). The genes responsible for the kidney tumor and for aniridia have now each been identified and cloned. Because the WAGR syndrome involves the deletion of a series of adjacent genes, it is sometimes referred to as an example of a **contiguous gene syndrome**. In addition to microdeletions, microduplications can also produce contiguous gene syndromes.

Some of the microdeletion syndromes, such as the Prader–Willi and Williams syndromes, manifest deletions of a critical region of very consistent size (e.g., 4 Mb for Prader–Willi syndrome). Recent studies show that this is caused by the presence of multiple repeated sequences, termed **low-copy repeats** (see Chapter 2), at the deletion boundaries. These repeated sequences promote unequal crossing over (see Chapter 5), which then produces duplications and deletions of the region bounded by the repeat elements.

Several additional examples of microdeletions are given in Table 6-3. Many of these conditions, including the Prader–Willi, Miller–Dieker, Williams, and velocardiofacial syndromes (Clinical Commentary 6-3), are now diagnosed using the FISH or CGH techniques.

---

*Because individuals with the WAGR syndrome also have gonadoblastomas (gonadal tumors), some authorities believe that the "G" should stand for gonadoblastoma rather than genitourinary abnormality.

**TABLE 6-3**
## Microdeletion Syndromes*

| Syndrome | Clinical Features | Chromosomal Deletion |
|---|---|---|
| Prader–Willi | Mental retardation, short stature, obesity, hypotonia, characteristic facies, small feet | 15q11-13 |
| Angelman | Mental retardation, ataxia, uncontrolled laughter, seizures | 15q11-13 |
| Langer–Giedion | Characteristic facies, sparse hair, exostosis, variable mental retardation | 8q24 |
| Miller–Dieker | Lissencephaly, characteristic facies | 17p13.3 |
| Velocardiofacial/ DiGeorge | Characteristic facies, cleft palate, heart defects, poorly developed thymus | 22q11 |
| Smith–Magenis | Mental retardation, hyperactivity, dysmorphic features, self-destructive behavior | 17p11.2 |
| Williams | Developmental disability, characteristic facies, supravalvular aortic stenosis | 7q1 |
| Aniridia, Wilms tumor | Mental retardation, aniridia, predisposition to Wilms tumor, genital defects | 11p13 |
| Deletion 1p36 | Mental retardation, seizures, hearing loss, heart defects, growth failure, distinctive facial features | 1p36 |
| Rubinstein–Taybi | Mental retardation, broad thumbs and great toes, characteristic facial features, vertebral and sternal abnormalities, pulmonary stenosis | 16p13.3 |
| Alagille | Neonatal jaundice, "butterfly" vertebrae, pulmonic valvular stenosis, characteristic facial features | 20p12 |

*For most of these conditions, only some cases are caused by the listed microdeletion; other cases may be caused by single-gene mutations within the same region.

## CLINICAL COMMENTARY 6-3
### DiGeorge Sequence, Velocardiofacial Syndrome, and Microdeletions of Chromosome 22

The DiGeorge sequence* is characterized by structural or functional defects of the thymus, conotruncal heart defects, hypoparathyroidism (reduced parathyroid function), and secondary hypocalcemia (decreased serum calcium). This pattern of malformations is caused by an alteration of the embryonic migration of neural crest cells to the developing structures of the neck. In the 1980s some children with DiGeorge sequence were found to have a deletion of part of the long arm of chromosome 22, often related to an unbalanced translocation between this and another chromosome. This led to the hypothesis that genes on chromosome 22 were responsible for this condition.

Independent of this work, a condition called the velocardiofacial (VCF) syndrome, or Shprintzen syndrome, was described in the late 1970s. This syndrome includes palatal (velum) abnormalities (including cleft palate), a characteristic facial appearance (see figure), and, in some cases, heart malformations. In addition, these patients have learning disabilities and developmental delay. Later it was discovered that some persons with VCF have dysfunctional T cells (these cells mature in the thymus), and some have all features of DiGeorge sequence. This suggested that DiGeorge sequence was somehow related to VCF syndrome.

The resemblance between the DiGeorge sequence and VCF syndrome led to the hypothesis that both conditions were caused by abnormalities on chromosome 22. High-resolution chromosome studies, including FISH, of patients with DiGeorge sequence and patients with VCF syndrome revealed small deletions of chromosome 22 in both groups. These analyses also helped to narrow the critical region that causes both conditions. Approximately 80% to 90% of infants with DiGeorge sequence have a 3-Mb microdeletion of the 22q11.2 region, and 80% to 100% of VCF patients have the same microdeletion. In addition, 15% to 20% of persons with isolated conotruncal defects exhibit this deletion. Thus, most persons with either DiGeorge sequence or VCF syndrome have a microdeletion of 22q11.2 and are collectively described as having 22q11.2 deletion syndrome. With a prevalence of 1 in 3000 to 4000 live births, this is the most common human microdeletion syndrome.

About 90% of persons with 22q11.2 microdeletions lack the same 3-Mb region, containing about 35 genes. Another 8% have a smaller, 1.5-Mb deletion located within the 3-Mb region. No consistent phenotypic differences are seen between these two groups of patients. Both the 1.5- and 3-Mb regions are flanked by low-copy repeats that are thought to promote unequal crossing over, and thus deletion, in this region. One of the genes located in the deleted region, *TBX1*, encodes a transcription factor that helps to regulate the migration of neural crest cells and the development of facial structures, the thymus, the parathyroid, and the heart. In mouse models, *Txb1* haploinsufficiency produces many of the features of DiGeorge sequence and VCF syndromes.

This example illustrates how cytogenetic studies can demonstrate potential biological relationships between genetic syndromes. Further studies are under way to characterize the individual genes in this region and how they contribute to the phenotypic variation seen in DiGeorge sequence and VCF syndromes.

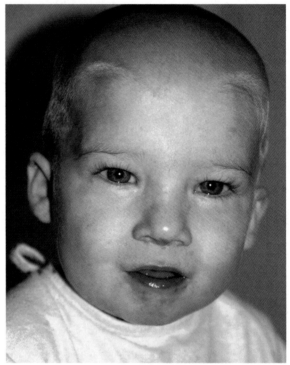

Face of a boy with 22q11 deletion syndrome. Note the narrow, tall nasal root and nasal bridge and the somewhat smooth philtrum.
*(Courtesy of Dr. Lynne M. Bird, Children's Hospital, San Diego.)*

---

*A **sequence** is defined as a series of alterations that are all due to a single, primary defect (see Chapter 15 for further discussion). In DiGeorge sequence, the primary defect is an abnormality in the migration of neural crest cells.

---

> Microdeletions are a subtype of chromosome deletion that can be observed only in banded chromosomes or, in some cases, using molecular genetic approaches. Syndromes caused by the deletion of a series of adjacent genes are sometimes called *contiguous gene syndromes*.

### Subtelomeric Rearrangements

The regions near telomeres of chromosomes tend to have a high density of genes. Consequently, rearrangements of genetic material (e.g., deletions, duplications) in these regions often result in genetic disease. It is estimated that at least 5% of unexplained cases of mental retardation are caused by subtelomeric rearrangements. The most common of these rearrangements is a deletion of several thousand bases of chromosome 1p36, which is seen in approximately one in 5000 live births. This condition, termed monosomy 1p36 syndrome, is associated with mental retardation, developmental delay, seizures, hearing impairment, heart defects, hypotonia, and characteristic facial features (Fig. 6-19).

Collections of probes have been designed so that FISH analysis of metaphase chromosomes can be undertaken to determine whether a subtelomeric deletion or duplication has occurred in a patient. Increasingly, comparative genomic hybridization (CGH, discussed earlier) is being carried out, hybridizing differentially labeled patient and control DNA samples to microarrays that contain probes corresponding to all human subtelomeric regions. If a subtelomeric region is duplicated or deleted, the patient's DNA will show either excessive or deficient hybridization to the probe corresponding to that region.

> Subtelomeric rearrangements involve deletions or duplications of DNA in the gene-rich regions near telomeres. They can be detected by hybridizing specifically designed FISH probes to metaphase chromosomes or by comparative genomic hybridization of patient and control DNA to microarrays containing subtelomeric probes.

### Uniparental Disomy

As discussed earlier, about 70% of Prader–Willi cases are caused by microdeletions. Most of the remaining cases involve **uniparental disomy** (*di* = "two"), a condition in which one parent has contributed two copies of a chromosome and the other parent has contributed no copies (Fig. 6-20). If the parent has contributed two copies of one homolog, the condition is termed **isodisomy**. If the parent

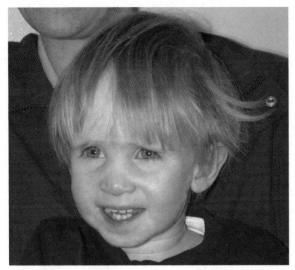

**FIGURE 6-19**
Face of young boy with 1q36 deletion syndrome. Note the horizontal eyebrows, deep-set eyes, broad nasal root, and pointed chin.

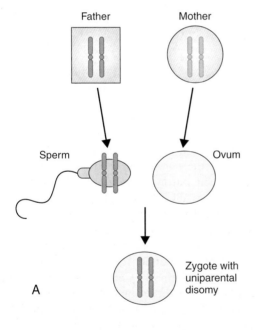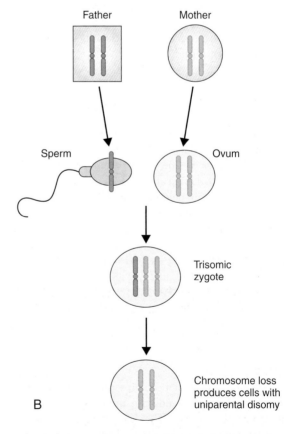

**FIGURE 6-20**
Two mechanisms that can produce uniparental disomy. **A,** Parental nondisjunction produces a sperm cell with two copies of a specific chromosome, and maternal nondisjunction produces an ovum with no copies of the same chromosome. The resulting zygote has two copies of the father's chromosome and no copies of the mother's chromosome (in this example the father contributes both chromosomes, but it is also possible that the mother could contribute both chromosomes). **B,** Nondisjunction (in the mother, in this example) results in a trisomic zygote. Loss of the paternal chromosome during mitosis produces embryonic cells that have two copies of the mother's chromosome.

has contributed one copy of each homolog, it is termed **heterodisomy**. Isodisomy or heterodisomy of an imprinted chromosome can cause diseases such as Prader–Willi syndrome (i.e., the inheritance of two copies from the mother and none from the father means that the offspring receives no active paternal genes in the imprinted region; see Chapter 5). Isodisomy can result in autosomal recessive disease in the offspring of a heterozygous parent if the parent contributes two copies of the chromosome homolog that contains the disease-causing mutation (see Fig. 6-20). The first documented case of uniparental disomy in a human was seen in a person with cystic fibrosis whose heterozygous carrier parent had transmitted two copies of chromosome 7 containing a mutated *CFTR* gene, while the other parent transmitted neither copy of chromosome 7.

Uniparental disomy can arise in a number of ways. A trisomic conception can lose one of the extra chromosomes, resulting in an embryo that has two copies of the chromosome contributed by one parent. Disomy can also result from the union of a gamete that contains two copies of a specific chromosome with a gamete that contains no copies of that chromosome (see Fig. 6-20). In the early embryo, cells with uniparental disomy can be produced by mitotic errors, such as chromosome loss with subsequent duplication of the homologous chromosome. In addition to the Prader–Willi and Angelman syndromes and cystic fibrosis, uniparental disomy has been seen in cases of Russell–Silver syndrome, hemophilia A (see Chapter 5), and Beckwith–Wiedemann syndrome (see Chapters 5 and 15).

### Duplications

A partial trisomy, or **duplication**, of genetic material may be seen in the offspring of persons who carry a reciprocal translocation. Duplications can also be caused by unequal crossover during meiosis, as described for the X-linked color vision loci (see Chapter 5) and for Charcot–Marie–Tooth disease (see Chapter 3). Duplications tend to produce less-serious consequences than deletions, again illustrating the principle that a loss of genetic material is more serious than an excess of genetic material.

> Duplications can arise from unequal crossover, or they can occur among the offspring of reciprocal translocation carriers. Duplications generally produce less-serious consequences than do deletions of the same region.

### Ring Chromosomes

Deletions sometimes occur at both tips of a chromosome. The remaining chromosome ends can then fuse, forming a **ring chromosome** (Fig. 6-21). The karyotype of a female with a ring X chromosome is 46,X,r(X). If the ring chromosome includes a centromere, it can often proceed through cell division, but its structure can create difficulties. Ring chromosomes are often lost, resulting in monosomy for the chromosome in at least some cells (i.e., mosaicism for the

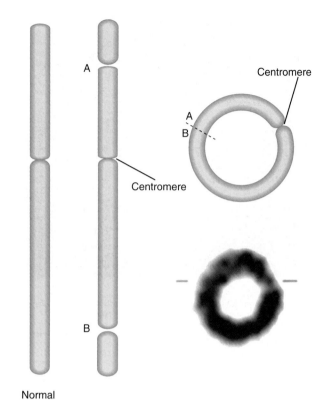

**FIGURE 6-21**
Both tips of a chromosome can be lost, leaving sticky ends that attach to each other, forming a ring chromosome. A chromosome 12 ring is shown here.

ring chromosome may be seen). Ring chromosomes have been described in at least one case for each of the human autosomes.

### Inversions

An **inversion** is the result of two breaks on a chromosome followed by the reinsertion of the intervening fragment at its original site but in inverted order. Thus, a chromosome symbolized as ABCDEFG might become ABEDCFG after an inversion. If the inversion includes the centromere, it is called a **pericentric inversion**. Inversions that do not involve the centromere are termed **paracentric inversions**.

Like reciprocal translocations, inversions are a balanced structural rearrangement. Consequently, they seldom produce disease in the inversion carrier (recall from Chapter 5, though, that an inversion that interrupts the factor VIII gene produces severe hemophilia A). Inversions can interfere with meiosis, however, producing chromosome abnormalities in the offspring of inversion carriers. Because chromosomes must line up in perfect order during prophase I, a chromosome with an inversion must form a loop to line up with its normal homolog (Fig. 6-22). Crossing over within this loop can result in duplications or deletions in the chromosomes of daughter cells. Thus, the offspring of persons who carry inversions often have chromosome deletions or duplications. It is estimated that about 1 in 1000 people carries an inversion and is therefore at risk for producing gametes with duplications or deletions.

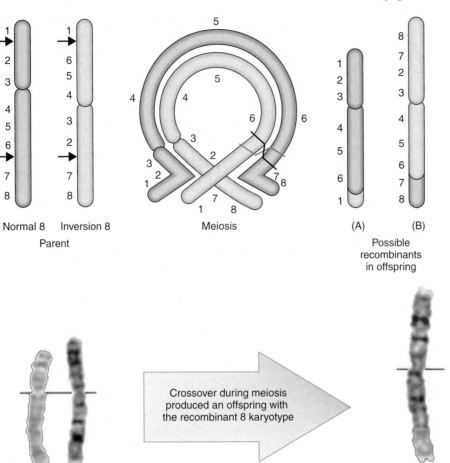

Normal 8    Inversion 8
Parent

Meiosis

(A)    (B)

Possible
recombinants
in offspring

**FIGURE 6-22**
A pericentric inversion in chromosome 8 causes the formation of a loop during the alignment of homologous chromosomes in meiosis. Crossing over in this loop can produce duplications or deletions of chromosome material in the resulting gamete. The offspring in the lower right received one of the recombinant 8 chromosomes from this parent.

Crossover during meiosis produced an offspring with the recombinant 8 karyotype

(B)

Parent                                    Offspring

Figure 6-22 gives an example of a pericentric inversion on chromosome 8 (46,XX,inv[8]). About 5% of the offspring of persons who carry this inversion receive a deletion or duplication of the distal portion of 8q. This combination results in the recombinant 8 syndrome, which is characterized by mental retardation, heart defects, seizures, and a characteristic facial appearance.

> Chromosome inversions are relatively common structural abnormalities and may be either pericentric (including the centromere) or paracentric (not including the centromere). Parents with inversions are usually normal in phenotype but can produce offspring with deletions or duplications.

### Isochromosomes

Sometimes a chromosome divides along the axis perpendicular to its usual axis of division (Fig. 6-23). The result is an **isochromosome**, a chromosome that has two copies of one arm and no copies of the other. Because the genetic material is substantially altered, isochromosomes of most autosomes are lethal. Most isochromosomes observed in live births involve the X chromosome, and babies with isochromosome

Xq (46,X,i[Xq]) usually have features of Turner syndrome. Isochromosome 18q, which produces an extra copy of the long arm of chromosome 18, has been observed in infants with Edwards syndrome. Although most isochromosomes appear to be formed by faulty division, they can also be created by Robertsonian translocations of homologous acrocentric chromosomes (e.g., a Robertsonian translocation of the two long arms of chromosome 21).

### CHROMOSOME ABNORMALITIES AND CLINICAL PHENOTYPES

As we have seen, most autosomal aberrations induce consistent patterns of multiple malformations, minor anomalies, and phenotypes with variable degrees of developmental retardation. Although the individual features are usually nonspecific (e.g., palmar creases can be seen in both Down syndrome and trisomy 18), the general pattern of features is usually distinctive enough to permit a clinical diagnosis. This is especially true of the well-known chromosome syndromes: Down syndrome, Edwards syndrome, Patau syndrome, and Turner syndrome. However, there is considerable phenotypic variability even within these syndromes. No one patient has every feature; most congenital malformations (e.g., heart defects) are seen in only some affected individuals. This

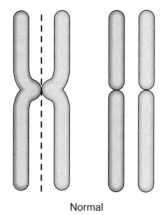

Normal

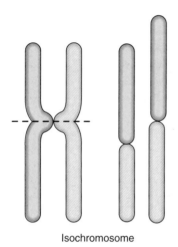

Isochromosome

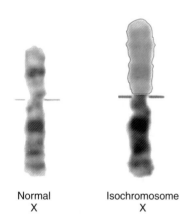

Normal                Isochromosome
X                     X

**FIGURE 6-23**

*Top*, Normal chromosome division. *Center*, An isochromosome is formed when a chromosome divides along an axis perpendicular to its usual axis of division. This produces one chromosome with only the short arms and another with only the long arms. *Bottom*, A normal X chromosome is compared with an isochromosome of Xq.

phenotypic variability, and the attendant potential for misdiagnosis, underscores the need to order a karyotype whenever clinical features suggest a chromosome abnormality.

Usually the biological basis for phenotypic variability is unknown, although mechanisms such as mosaicism, which often leads to milder expression, are being uncovered. The

basis of variable expression of chromosome syndromes will be better understood as the individual genes involved in these abnormalities are identified and characterized.

In spite of the variability of chromosome syndromes, it is possible to make several generalizations:

- Most chromosome abnormalities (especially those involving autosomes) are associated with developmental delay in children and mental retardation in older persons. This reflects the fact that a large number of human genes, perhaps one third or more of the total, participate in the development of the central nervous system. Therefore, a chromosome abnormality, which typically can affect hundreds of genes, is very likely to involve genes that affect nervous system development.
- Most chromosome syndromes involve alterations of facial morphogenesis that produce characteristic facial features. For this reason, the patient often resembles other persons with the same disorder more than members of his or her own family. Usually the facial features and minor anomalies of the head and limbs are the best aids to diagnosis (see Chapter 15).
- Growth delay (short stature and/or poor weight gain in infancy) is commonly seen in autosomal syndromes.
- Congenital malformations, especially congenital heart defects, occur with increased frequency in most autosomal chromosome disorders. These defects occur in specific patterns. For example, AV canals and VSDs are common in children with Down syndrome. Other congenital heart defects, such as aortic coarctation or hypoplastic (underdeveloped) left ventricle, are seldom seen in these children but may be seen in those with Turner syndrome.

The most common clinical indications for a chromosome analysis are a newborn with multiple congenital malformations or a child with developmental retardation. A summary of clinical situations in which a chromosome evaluation should be considered is given in Box 6-1.

---

**BOX 6-1**
**Indications for Performing Chromosome Analysis**

- Persons with a suspected recognizable chromosome syndrome (e.g., Down syndrome)
- Persons with an unrecognizable pattern of two or more malformations
- Persons with ambiguous genitalia
- Mental retardation or developmental delay in children who have multiple physical abnormalities
- Parents and children of persons with chromosomal translocations, deletions, or duplications
- Stillborn infants with malformation or with no recognizable reason for fetal death
- Females with proportionate short stature and primary amenorrhea
- Males with small testes or significant gynecomastia

Chromosome abnormalities typically result in developmental delay, mental retardation, characteristic facial features, and various types of congenital malformations. Despite some overlap of phenotypic features, many chromosome abnormalities can be recognized by clinical examination.

## CANCER CYTOGENETICS

Most of the chromosome abnormality syndromes discussed thus far are caused by errors that occur in the meiotic process leading to gamete formation. Chromosome rearrangements can also take place in somatic cells; these are responsible for a number of important cancers in humans. The first of these to be recognized was a chromosome alteration seen consistently in patients with chronic myelogenous leukemia (CML). Initially, it was suggested that the chromosome alteration was a deletion of the long arm of either chromosome 21 or chromosome 22. With the subsequent development of chromosome banding techniques, the abnormality was identified as a reciprocal translocation between chromosomes 9 and 22. The **Philadelphia chromosome**, as this translocation is commonly known, consists of a translocation of most of chromosome 22 onto the long arm of chromosome 9. A small distal portion of 9q in turn is translocated to chromosome 22. The net effect is a smaller chromosome 22, which explains why the Philadelphia chromosome was initially thought to be a deletion. This translocation (Fig. 6-24) is seen in most cases of CML.

Much has been learned about the effects of this translocation by isolating the genes that are located near the translocation **breakpoints** (i.e., the locations on the chromosomes at which the breaks occur preceding translocation). A proto-oncogene (see Chapter 11) called *ABL* is moved from its normal position on 9q to 22q. This alters the *ABL* gene product, causing increased tyrosine kinase activity, which leads to malignancy in hematopoietic cells (i.e., cells that form blood cells such as lymphocytes). Drugs have now been developed to inhibit the tyrosine kinase encoded by this gene, offering much more effective treatment for CML.

A second example of a translocation that produces cancer is given by Burkitt lymphoma, a childhood jaw tumor. In this case, a reciprocal translocation involving chromosomes 8 and 14 moves the *MYC* proto-oncogene from 8q24 to 14q32, near the immunoglobulin heavy chain loci (see Chapter 9). Transcription regulation sequences near the immunoglobulin genes then activate *MYC*, causing malignancies to form.

More than 100 different rearrangements, involving nearly every chromosome, have been observed in more than 40 different types of cancer. Some of these are summarized in Table 6-4. Increasingly, these translocations are identified using spectral karyotypes. In some cases, identification of the chromosome rearrangement leads to a more accurate prognosis and better therapy. Hence, the cytogenetic evaluation of bone marrow cells from leukemia patients

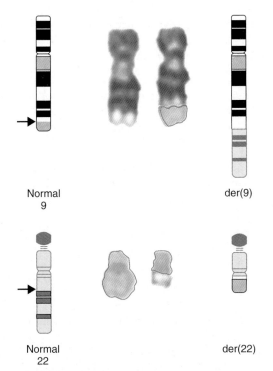

**Normal 9**          **der(9)**

**Normal 22**          **der(22)**

**FIGURE 6-24**
Reciprocal translocation between chromosome 22 and the long arm of chromosome 9 (the Philadelphia chromosome). The occurrence of this translocation in hematopoietic cells can produce chronic myelogenous leukemia.

**TABLE 6-4**
## Specific Cytogenetic Changes Observed in Selected Leukemias and Solid Tumors

| Type | Most Common Chromosome Aberration |
|------|-----------------------------------|
| **Leukemias** | |
| Chronic myelogenous leukemia | t(9;22)(q34;q11) |
| Acute myeloblastic leukemia | t(8;21)(q22;q22) |
| Acute promyelocytic leukemia | t(15;17)(q22;q11-12) |
| Acute myeloid leukemia | +8,-7,-5,del(5q),del(20q) |
| Acute lymphocytic leukemia | t(12;21)(p13;q22) |
| **Solid Tumors** | |
| Burkitt lymphoma | t(8;14)(9q24;q32) |
| Ewing sarcoma | t(11;22)(q24;q12) |
| Meningioma | Monosomy 22 |
| Retinoblastoma | del(13)(q14) |
| Wilms tumor | del(11)(p13) |
| Neuroblastoma | *N-MYC* amplification |
| Breast cancer | *HER2/NEU* amplification |

is a routine part of diagnosis. Furthermore, identification and characterization of the genes that are altered in translocation syndromes are leading to a better understanding of carcinogenesis in general.

▶ Balanced translocations in somatic cells can sometimes cause malignancies by interrupting or altering genes or their regulatory sequences.

## CHROMOSOME INSTABILITY SYNDROMES

Several autosomal recessive disease conditions exhibit an increased incidence of chromosome breaks under specific laboratory conditions. These conditions, which are termed **chromosome instability syndromes**, include ataxia–telangiectasia,

Bloom syndrome, Fanconi anemia, and xeroderma pigmentosum (see Chapter 2). Among patients with Fanconi anemia, the frequency of breaks can be increased further if the chromosomes are exposed to certain alkylating agents. Patients with Bloom syndrome also have a high incidence of somatic cell sister chromatid exchange (exchange of chromosome material between sister chromatids; see Chapter 2). Each of these syndromes is associated with a significant increase in cancer risk. All are thought to be the result of faulty DNA replication or repair, as discussed in Chapter 2.

▶ The chromosome instability syndromes all involve increased frequencies of chromosome breakage and an increased risk of malignancy. All are associated with defects in DNA replication or repair.

## Study Questions

1. Distinguish among haploidy, diploidy, polyploidy, euploidy, and aneuploidy.

2. Explain the uses and relative advantages of FISH, spectral karyotyping, and comparative genomic hybridization (CGH).

3. Describe three ways in which triploidy could arise.

4. Studies of karyotypes obtained by prenatal diagnosis at 10 weeks' gestation (chorionic villus sampling; see Chapter 13) reveal prevalence rates of chromosome abnormalities that differ from those obtained in karyotypes at 16 weeks' gestation (amniocentesis; see Chapter 13). Explain this.

5. Even though conditions such as Down syndrome and Edwards syndrome can usually be diagnosed accurately by clinical examination alone, a karyotype is always recommended. Why?

6. Rank the following, from lowest to highest, in terms of the risk of producing a child with Down syndrome:
   45-year-old woman with no previous family history of Down syndrome

   25-year-old woman who has had one previous child with Down syndrome
   25-year-old male carrier of a 21/14 Robertsonian translocation
   25-year-old female carrier of a 21/14 Robertsonian translocation

7. Females with the 49,XXXXX karyotype have been reported. Explain how this karyotype could occur.

8. A man with hemophilia A and a normal woman produce a child with Turner syndrome (45,X). The child has normal factor VIII activity. In which parent did the meiotic error occur?

9. A cytogenetics laboratory reports a karyotype of 46,XY,del(8)(p11) for one patient and a karyotype of 46,XY,dup(8)(p11) for another patient. Based on this information alone, which patient is expected to be more severely affected?

10. Why do translocations in somatic cells sometimes lead to cancer?

## Suggested Readings

Albertson DG, Collins C, McCormick F, Gray JW. Chromosome aberrations in solid tumors. Nat Genet 2003;34(4):369–76.

American Academy of Pediatrics. Committee on Genetics. Health supervision for children with Down syndrome. Pediatrics 2001;107(2):442–9.

Antonarakis SE, Epstein CJ. The challenge of Down syndrome. Trends Mol Med 2006;12(10):473–9.

Aradhya S, Cherry AM. Array-based comparative genomic hybridization. Clinical contexts for targeted and whole-genome designs. Genet Med 2007;9(9):553–9.

Battaglia A. Del 1p36 syndrome: A newly emerging clinical entity. Brain and Development 2005;27(5):358–61.

Blaschke RJ, Rappold G. The pseudoautosomal regions, SHOX and disease. Curr Opin Genet Dev 2006;16(3):233–9.

Carey JC. Trisomy 18 and 13 syndromes. In Cassidy SB, Allanson JE (eds): Management of Genetic Syndromes. Hoboken, NJ: Wiley–Liss, 2005, pp. 555–68.

Emanuel BS, Saitta SC. From microscopes to microarrays. Dissecting recurrent chromosomal rearrangements. Nat Rev Genet 2007;8(11):869–83.

Fleming A, Vilain E. The endless quest for sex determination genes. Clin Genet 2005;67(1):15–25.

Frohling S, Dohner H. Chromosomal abnormalities in cancer. N Engl J Med 2008;359(7):722–34.

Gartler SM. The chromosome number in humans. A brief history. Nat Rev Genet 2006;7(8):655–60.

Gotz MJ, Johnstone EC, Ratcliffe SG. Criminality and antisocial behaviour in unselected men with sex chromosome abnormalities. Psychol Med 1999;29(4):953–62.

Graham GE, Allanson JE, Gerritsen JA. Sex chromosome abnormalities. In: Rimoin DL, Connor JM, Pyeritz RE, Korf BR (eds): Emery and Rimoin's Principles and Practice of Medical Genetics, 5th ed. London: Churchill Livingstone, 2007, pp. 1038–57.

Hall H, Hunt P, Hassold T. Meiosis and sex chromosome aneuploidy: How meiotic errors cause aneuploidy; how aneuploidy causes meiotic errors. Curr Opin Genet Dev 2006;16(3):323–9.

Kobrynski LJ, Sullivan KE. Velocardiofacial syndrome, DiGeorge syndrome: The chromosome 22q11.2 deletion syndromes. Lancet 2007;370(9596):1443–52.

Lanfranco F, Kamischke A, Zitzmann M, Nieschlag E. Klinefelter's syndrome. Lancet 2004;364(9430):273–83.

Lee C, Iafrate AJ, Brothman AR. Copy number variations and clinical cytogenetic diagnosis of constitutional disorders. Nat Genet 2007;39(Suppl 7):S48–S54.

Loscalzo ML. Turner syndrome. Pediatr Rev 2008;29(7): 219–27.

Patterson D, Costa AC. Down syndrome and genetics—a case of linked histories. Nat Rev Genet 2005;6(2): 137–47.

Pober BR, Morris CA. Diagnosis and management of medical problems in adults with Williams–Beuren syndrome. Am J Med Genet C Semin Med Genet 2007;145(3):280–90.

Robinson WP. Mechanisms leading to uniparental disomy and their clinical consequences. Bioessays 2000;22(5):452–9.

Roizen NJ, Patterson D. Down's syndrome. Lancet 2003;361 (9365):1281–9.

Schinzel A. Catalogue of Unbalanced Chromosome Aberrations. Berlin: Walter de Gruyter, 2001.

Shaffer LG, Tommerup N (eds): An International System for Human Cytogenetic Nomenclature. Basel: S. Karger, 2005.

Speicher MR, Carter NP. The new cytogenetics. Blurring the boundaries with molecular biology. Nat Rev Genet 2005; 6(10):782–92.

Spinner NB, Saitta SC, Emanuel BS. Deletions and other structural abnormalities of the autosomes. In: Rimoin DL, Connor JM, Pyeritz RE, Korf BR (eds): Emery and Rimoin's Principles and Practice of Medical Genetics, 5th ed. London: Churchill Livingstone, 2007, pp. 1058–82.

Stankiewicz P, Beaudet AL. Use of array CGH in the evaluation of dysmorphology, malformations, developmental delay, and idiopathic mental retardation. Curr Opin Genet Dev 2007;17(3):182–92.

Sybert VP, McCauley E. Turner's syndrome. N Engl J Med 2004;351(12):1227–38.

Tolmie JL, MacFayden U. Clinical genetics of common autosomal trisomies. In: Rimoin DL, Connor JM, Pyeritz RE, Korf BR (eds): Emery and Rimoin's Principles and Practice of Medical Genetics, 5th ed. London: Churchill Livingstone, 2007, pp. 1015–37.

Wattendorf DJ, Muenke M. Klinefelter syndrome. Am Fam Physician 2005;72(11):2259–62.

## Internet Resources

European Cytogenetics Association (a series of URLs for various cytogenetics websites) *http://www.biologia.uniba.it/eca/*

Mitelman Database of Chromosome Aberrations in Cancer *http://cgap.nci.nih.gov/Chromosomes/Mitelman*

National Association for Down Syndrome (contains many URLs for Down syndrome websites) *http://www.nads.org/*

Support Organization for Trisomy 18, 13, and Related Disorders (S.O.F.T.) *http://www.trisomy.org/*

# Chapter 7

# BIOCHEMICAL GENETICS: DISORDERS OF METABOLISM

Each of us is composed of a large number of complex molecules that are hierarchically arranged in space to form cells, tissues, organs, and ultimately a complete human being. These molecules are constructed from individual elements that may be synthesized endogenously or obtained from the environment. Once created, these molecules are not static. In fact, they are perpetually being synthesized, degraded, excreted, and sometimes recycled in a tightly choreographed metabolic dance.

Each metabolic process consists of a sequence of catalytic steps mediated by enzymes encoded by genes. Usually, these genes are replicated with high fidelity, and enzymatic systems continue to work effectively from generation to generation. Occasionally, mutations reduce the efficiency of encoded enzymes to a level at which normal metabolism cannot occur. Such variants of metabolism were recognized by Sir Archibald Garrod at the beginning of the 20th century, based partly on his studies of alkaptonuria (AKU). Garrod recognized that these variants illustrated "chemical individualities," and he called these disorders "inborn errors of metabolism," thus setting the cornerstone for contemporary biochemical genetics.

AKU is a rare disorder in which homogentisic acid (HGA), an intermediate metabolite in phenylalanine and tyrosine metabolism (Fig. 7-1), is excreted in large quantities in urine, causing it to darken on standing. Hence, AKU was classically referred to as "black urine disease." Additionally, an oxidation product of HGA is directly deposited in connective tissues, resulting in abnormal pigmentation and debilitating arthritis.

Garrod proposed in 1902 that AKU was caused by a deficiency of the enzyme that normally splits the aromatic ring of HGA. Fifty years later, it was established that AKU is produced by a failure to synthesize homogentisate 1,2-dioxygenase (HGO). However, it was not until 1996 that scientists identified the gene that is altered in AKU, based on homology to a gene encoding an HGO enzyme that was isolated from a fungus species. The coding region of *HGO* comprises 14 exons distributed over 60 kb of DNA. Many of the mutations identified in *HGO* encode proteins that show no HGO activity when expressed in vitro. This indicates that AKU is caused by a loss-of-function mutation, confirming the hypothesis put forth by Garrod more than a century ago.

Almost all biochemical processes of human metabolism are catalyzed by enzymes. Variations of enzymatic activity among humans are common, and a minority of these variants cause disease. These concepts were introduced by Archibald Garrod and exemplified by his studies of alkaptonuria.

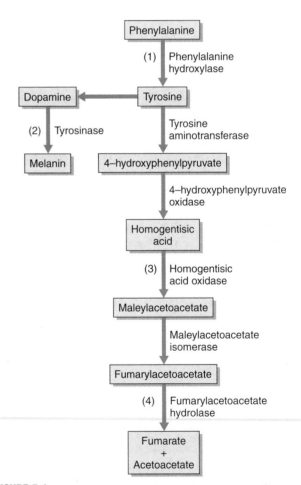

**FIGURE 7-1**

Major pathway of phenylalanine metabolism. Different enzymatic defects in this pathway cause classic PKU *(1)*, tyrosinase-negative oculocutaneous albinism *(2)*, AKU *(3)*, and tyrosinemias *(4)*.

## VARIANTS OF METABOLISM
### Prevalence of Metabolic Disease

Hundreds of different inborn errors of metabolism have been described to date, and most of these are rare. Taken together, however, metabolic disorders account for a substantial percentage of the morbidity and mortality directly attributable to genetic disease (Table 7-1). One survey conservatively estimated that the incidence of metabolic disorders is approximately 1 in every 2500 births, or 10% of all monogenic conditions in children. Moreover, we are beginning to understand that different alleles of genes that encode enzymes can alter one's risk for many common diseases, such as diabetes, heart disease, stroke, and cancer.

The diagnosis of a metabolic disorder can be challenging (Clinical Commentary 7-1); thus, the morbidity associated with metabolic defects is probably underestimated. In the

**TABLE 7-1**
## Disorders of Metabolism

| Name | Prevalence | Mutant Gene Product | Chromosomal Location |
|---|---|---|---|
| **Carbohydrate Disorders** | | | |
| Classic galactosemia | 1/35,000 to 1/60,000 | Galactose-1-phosphate uridyl transferase | 9p13 |
| Hereditary fructose intolerance | 1/20,000 | Fructose 1,6-bisphosphate aldolase | 9q13-q32 |
| Fructosuria | ~1/100,000 | Fructokinase | 2p23 |
| Hypolactasia (adult) | Common | Lactase | 2q21 |
| Diabetes mellitus type 1 | 1/400 (Europeans) | Multiple | Polygenic |
| Diabetes mellitus type 2 | 1/20 | Multiple | Polygenic |
| Maturity-onset diabetes of the young (MODY) | ~1/400 | Multiple | Multiple loci |
| **Amino Acid Disorders** | | | |
| Phenylketonuria | 1/10,000 | Phenylalanine hydroxylase | 12q24 |
| Tyrosinemia (type 1) | 1/100,000 | Fumarylacetoacetate hydrolase | 15q23-25 |
| Maple syrup urine disease | 1/180,000 | Branched-chain α-ketoacid dehydrogenase (multiple subunits) | Multiple loci |
| Alkaptonuria | 1/250,000 | Homogentisic acid oxidase | 3q2 |
| Homocystinuria | 1/340,000 | Cystathionine β-synthase | 21q2 |
| Oculocutaneous albinism | 1/35,000 | Tyrosinase | 11q |
| Cystinosis | 1/100,000 | CTNS | 17p13 |
| Cystinuria | 1/7000 | SLC3A1 (type 1) | 2p |
| | | SLC7A9 (types II & III) | 19q13 |
| **Lipid Disorders** | | | |
| MCAD | 1/20,000 | Medium-chain acyl-CoA dehydrogenase | 1p31 |
| LCAD | Rare | Long-chain acyl-CoA dehydrogenase | 2q34-q35 |
| SLO | 1/10,000 | Δ7-sterol reductase | 11q12-q13 |
| **Organic Acid Disorders** | | | |
| Methylmalonic acidemia | 1/20,000 | Methylmalonyl-CoA mutase | 6p |
| Propionic acidemia | Rare | Propionyl-CoA carboxylase | 13q32; 3q |

*Continued*

**TABLE 7-1**
Disorders of Metabolism—cont'd

| Name | Prevalence | Mutant Gene Product | Chromosomal Location |
|---|---|---|---|
| **Urea Cycle Defects** | | | |
| Ornithine transcarbamylase deficiency | 1/70,000 to 1/100,000 | Ornithine carbamyl transferase | Xp21 |
| Carbamyl phosphate synthetase deficiency | 1/70,000 to 1/100,000 | Carbamyl phosphate synthetase I | 2p |
| Argininosuccinic acid synthetase deficiency | 1/70,000 to 1/100,000 | Argininosuccinic acid synthetase | 9q34 |
| **Energy Production Defects** | | | |
| Cytochrome c oxidase deficiency | Rare | Cytochrome oxidase peptides | Multiple loci |
| Pyruvate carboxylase deficiency | Rare | Pyruvate carboxylase | 11q |
| Pyruvate dehydrogenase complex ($E_1$) deficiency | Rare | Pyruvate decarboxylase, $E_1\alpha$ | Xp22 |
| NADH-CoQ reductase deficiency | Rare | Multiple nuclear genes | Multiple loci |
| **Heavy Metal Transport Defects** | | | |
| Wilson disease | 1/50,000 | ATP7B | 13q14 |
| Menkes disease | 1/250,000 | ATP7A | Xq13 |
| Hemochromatosis | 1/200 to 1/400 (Europeans) | HFE | 6p21 |
| Acrodermatitis enteropathica | Rare | SLC39A4 | 8q24 |

*LCAD, long-chain acyl-CoA dehydrogenase; MCAD, medium-chain acyl-coenzyme A dehydrogenase; SLO, Smith–Lemli–Opitz syndrome.*

1970s, an often fatal acute metabolic encephalopathy called Reye syndrome was diagnosed in many children. In the following decades, we learned that some children with an encephalopathy indistinguishable from Reye syndrome had a urea cycle defect that produced hyperammonemia (increased levels of circulating ammonia) and death. Recognition of Reye syndrome as a **phenocopy** of a urea cycle defect is important because, in addition to supportive care,* the children can now receive direct treatment for the urea cycle defects. Similarly, on postmortem examination, some children who have died from sudden infant death syndrome (SIDS) have been found to have a defect of fatty acid metabolism. These are also treatable disorders, and life-threatening episodes can be avoided with appropriate care.

▶ Although individual metabolic disorders are rare, their overall direct and indirect contribution to morbidity and mortality is substantial.

### Inheritance of Metabolic Defects

Most metabolic disorders are inherited in an autosomal recessive pattern: Only individuals having two mutant alleles are affected. Although a mutant allele produces reduced or no enzyme activity (loss of function), it usually does not alter the health of a heterozygous carrier. Because many of the genes encoding disease-related enzymes have been cloned and their mutations characterized, carrier testing and prenatal diagnosis are available for many metabolic disorders. However, testing samples of dried blood for elevated levels of metabolites in the newborn period (e.g., for phenylketonuria and galactosemia; see Chapter 13) remains the most commonly used population-based screening test for metabolic disorders. Expanded newborn screening that tests for dozens of different disorders by checking for the presence of abnormal metabolites in blood is becoming increasingly more common. As the technology for rapid and efficient DNA testing of mutant alleles progresses, population-based screening for additional disorders is likely to be incorporated.

▶ Most inborn errors of metabolism are inherited in an autosomal recessive pattern. The carrier state usually is not associated with morbidity. Carrier and diagnostic testing are becoming widely available for many disorders.

### Types of Metabolic Processes

Metabolic disorders have been classified in many different ways, based on the pathological effects of the pathway

---

*Supportive care is care that supports the elementary functions of the body, such as maintenance of fluid balance, oxygenation, and blood pressure, but is not aimed at treating the disease process directly.

## CLINICAL COMMENTARY 7-1
### *Diagnosis of a Metabolic Disorder*

The presentations of persons with inborn errors of metabolism are highly variable. During gestation, the maternal–placental unit usually provides essential nutrients and prevents the accumulation of toxic substrates. Thus, a fetus is infrequently symptomatic. However, after birth, persons with metabolic disorders can present at ages ranging from the first 24 hours of life to adulthood. The presentation may be precipitous and characterized by dramatic alterations in homeostasis and even death. In contrast, the disorder may be insidious, with only subtle changes in function over long periods. For most metabolic disorders, the presymptomatic period and onset of symptoms lie somewhere between these two extremes. The following case illustrates this point.

Anthony is a 9-month-old Latin American boy who comes to the emergency department accompanied by his parents. His parents complain that he has been irritable and vomiting for the last 36 hours, and over the past 12 hours he has become increasingly sleepy. They sought medical attention because it was difficult to awaken Anthony to breast-feed him. Anthony's medical history is unremarkable. He has a healthy 8-year-old sister and had a brother who died in his crib at 7 months of age. An investigation of the brother's death and an autopsy were performed. The findings were consistent with sudden infant death syndrome (SIDS).

Anthony is hospitalized and is noted to be hypoglycemic (low serum glucose level), slightly acidemic (serum pH <7.4), and hyperammonemic (elevated plasma ammonia). Intravenous infusion of glucose transiently improves his level of alertness, but he becomes comatose and dies 5 days later. An autopsy reveals marked cerebral edema (swelling of the brain) and fatty infiltration of the liver consistent with a diagnosis of Reye syndrome. Anthony's mother is concerned that the boys' deaths are related to each other, especially since she is pregnant again. She is counseled that the causes of death are unrelated and neither disorder is likely to recur in her family.

One year later her 6-month-old daughter, Maria, is hospitalized for the third time because of lethargy and weakness. Laboratory studies reveal moderate hypoglycemia, hyperammonemia, and ketonuria (ketones in the urine). Additional studies, including measurement of urine organic acids,* serum amino acids, and free and esterified plasma carnitines, suggest that Maria has a defect of fatty acid oxidation. Therapy is initiated with intravenous glucose, oral carnitine, and the restriction of fats to no more than 20% of her caloric requirements. More specific biochemical and molecular studies confirm that Maria has medium-chain acyl-CoA dehydrogenase (MCAD) deficiency. Molecular studies from preserved tissues that had been collected at autopsy from Maria's deceased brothers indicate that they also had MCAD deficiency. Maria's asymptomatic older sister is similarly affected. Both girls are healthy 2 years later, eating a low-fat diet and using a carnitine supplement. They have a new baby brother who underwent prenatal testing for MCAD and was found to be unaffected.

The disparate presentations of MCAD deficiency in this family (sudden death, acute illness, chronic illness, and asymptomatic) illustrate the phenotypic variability often observed in persons with inborn errors of metabolism, even those sharing an identical mutation. Thus, there might not be a disease-specific pattern of symptoms and findings. Often it is the heightened index of suspicion of care providers that leads to the testing necessary to identify a metabolic disorder. Supportive therapy can be lifesaving and should be initiated before making a diagnosis. Nevertheless, it is imperative that prudent attempts be made to make a specific diagnosis, because it can have important implications for the family (e.g., prenatal testing, presymptomatic therapy). The treatment of MCAD deficiency is completely effective in preventing early death from the toxic effects of accumulated fatty acid intermediates.

---

*Organic acids are carbon-based acids that are products of intermediate metabolism and normally do not accumulate in plasma or urine beyond the buffering capacities of these fluids.*

---

blocked (e.g., absence of end product, accumulation of substrate); different functional classes of proteins (e.g., receptors, hormones); associated cofactors (e.g., metals, vitamins); and pathways affected (e.g., glycolysis, citric acid cycle). Each of these has advantages and disadvantages, and none of them encompasses all metabolic disorders. However, the classification that most completely integrates our knowledge of cell biology, physiology, and pathology with metabolic disorders categorizes defects of metabolism by the types of processes that are disturbed.

## DEFECTS OF METABOLIC PROCESSES

Almost all biochemical reactions in the human body are controlled by enzymes, which act as **catalysts**. The catalytic properties of enzymes typically increase reaction rates by more than a million-fold. These reactions mediate the synthesis, transfer, use, and degradation of biomolecules to build and maintain the internal structures of cells, tissues, and organs. Biomolecules can be categorized into four primary groups: nucleic acids, proteins, carbohydrates, and lipids. The major metabolic pathways that metabolize these molecules include glycolysis, citric acid cycle, pentose phosphate shunt, gluconeogenesis, glycogen and fatty acid synthesis and storage, degradative pathways, energy production, and

transport systems. We now discuss how defects in each of these metabolic pathways can cause human disease.

### Carbohydrate Metabolism

Because of the many different applications that they serve in all organisms, carbohydrates are the most abundant organic substance on Earth. Carbohydrates function as substrates for energy production and storage, as intermediates of metabolic pathways, and as the structural framework of DNA and RNA. Consequently, carbohydrates account for a major portion of the human diet and are metabolized into three principal monosaccharides: glucose, galactose, and fructose. Galactose and fructose are converted to glucose before glycolysis. The failure to effectively use these sugars accounts for the majority of the inborn errors of human carbohydrate metabolism.

### Galactose

The most common monogenic disorder of carbohydrate metabolism, transferase deficiency galactosemia (classic galactosemia) affects 1 in every 30,000 newborns. It is most commonly caused by mutations in the gene encoding galactose-1-phosphate uridyl transferase (GAL-1-P uridyl transferase) (Fig. 7-2). This gene is composed of 11 exons distributed across 4 kb of DNA, and approximately 70% of

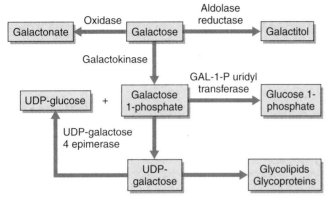

**FIGURE 7-2**
Major pathways of galactose metabolism. The most common enzymatic abnormality producing galactosemia is a defect of GAL-1-P uridyl transferase. Defects of galactokinase or of UDP-galactose 4-epimerase are much less common causes of galactosemia. GAL, galactose; UDP, uridine diphosphate.

galactosemia-causing alleles in people of Western European origin have a single missense mutation in exon 6. As a result of diminished GAL-1-P uridyl transferase activity, affected persons cannot effectively convert galactose to glucose; consequently, galactose is alternatively metabolized to galactitol and galactonate (see Fig. 7-2). Although galactose and its metabolites accumulate in many tissues, the pathophysiology of classic galactosemia is not well understood.

Classic galactosemia typically manifests in the newborn period with poor sucking, failure to thrive, and jaundice. If galactosemia is left untreated, sepsis, hyperammonemia, and shock leading to death usually follow. Cataracts (opacification of the lens of the eye) are found in about 10% of infants. Newborn screening for galactosemia is widespread, and most affected persons are identified as they begin to develop symptoms. Early identification affords prompt treatment, which consists largely of eliminating dietary galactose. This substantially reduces the morbidity associated with the acute effects of elevated levels of galactose metabolites. Long-term disabilities include poor growth, developmental delay,* mental retardation, and ovarian failure in females. These sequelae are thought to be caused by endogenous production of galactose. The effects of prospective dietary therapy on the prevalence of these long-term sequelae are less clear. Early studies suggested that there was no effect, but as more longitudinal data become available, it appears that patients treated early in life with dietary therapy may have a better cognitive outcome.

Galactosemia can also be caused by mutations in the genes encoding galactokinase or uridine diphosphate galactose-4-epimerase (UDP-galactose-4-epimerase) (see Fig. 7-2). Deficiency of galactokinase is associated with the formation of cataracts but does not cause growth failure, mental retardation, or hepatic disease. Dietary restriction of galactose is also the treatment for galactokinase deficiency. Deficiency of UDP-galactose-4-epimerase can be limited to red blood cells and leukocytes, causing no ill effects, or it can be systemic and produce symptoms similar to those of classic galactosemia. Treatment is aimed at reducing the dietary intake of galactose, but not as severely as in patients with classic galactosemia, because some galactose must be provided to produce UDP-galactose for the synthesis of some complex carbohydrates.

> Galactosemia is one of the most common inherited disorders of carbohydrate metabolism. Newborn screening for galactosemia is widespread. Early identification allows prompt treatment, which consists largely of eliminating dietary galactose. Mutations in the gene that encodes GAL-1-P uridyl transferase are the most common cause of galactosemia.

### Fructose

Three autosomal recessive defects of fructose metabolism have been described. The most common is caused by mutations in the gene encoding hepatic fructokinase. This enzyme catalyzes the first step in the metabolism of dietary fructose, the conversion to fructose-1-phosphate (Fig. 7-3). Inactivation of hepatic fructokinase results in asymptomatic fructosuria (presence of fructose in the urine).

In contrast, hereditary fructose intolerance (HFI) results in poor feeding, failure to thrive, hepatic and renal insufficiency, and death. HFI is caused by a deficiency of fructose 1,6-bisphosphate aldolase in the liver, kidney cortex, and small intestine. Infants and adults with HFI are asymptomatic unless they ingest fructose or sucrose (a sugar composed of fructose and glucose). Infants who are breast-fed become symptomatic after weaning, when fruits and vegetables are added to their diet. Affected infants may survive into childhood because they avoid foods they consider noxious, thereby self-limiting their intake of fructose. The prevalence of HFI may be as high as 1 in 20,000 births, and, since the cloning of the gene encoding fructose-1-phosphate aldolase, differences in the geographic distribution of mutant alleles have been found.

Deficiency of hepatic fructose 1,6-bisphosphatase (FBPase) causes impaired gluconeogenesis, hypoglycemia (reduced level of circulating glucose), and severe metabolic acidemia (serum pH <7.4). Affected infants are commonly presented for treatment shortly after birth, although cases diagnosed later in childhood have been reported. If patients are adequately supported beyond childhood, growth and development appear to be normal. A handful of mutations have been found in the gene encoding FBPase, some of which encode mutant proteins that are inactive in vitro.

> Asymptomatic deficiency of fructokinase is the most common defect of fructose metabolism. Hereditary fructose intolerance is less prevalent but is associated with much more severe problems.

### Glucose

Abnormalities of glucose metabolism are the most common errors of carbohydrate metabolism. However, the causes of

---

*Developmental delay is the delayed attainment of age-appropriate motor, speech, and/or cognitive milestones; the outcomes of developmental delay range from normal to profound mental retardation.

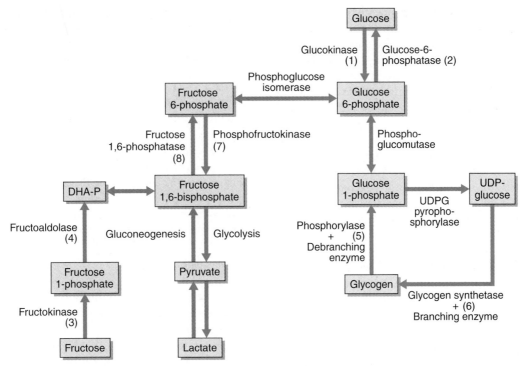

**FIGURE 7-3**
Summary of glucose, fructose, and glycogen metabolism. Enzymatic defects in this pathway cause hyperglycemia *(1)*, Von Gierke disease *(2)*, fructosuria *(3)*, hereditary fructose intolerance *(4)*, Cori disease *(5)*, Anderson disease *(6)*, Tarui disease *(7)*, and fructose 1,6-bisphosphatase (FBPase) deficiency *(8)*. UDP, uridine diphosphate.

these disorders are heterogeneous and include both environmental and genetic factors. Historically, disorders associated with elevated levels of plasma glucose (hyperglycemia) have been classified into three categories: type 1 diabetes mellitus (T1DM), T2DM, and maturity-onset diabetes of the young (MODY), of which there are many subtypes. T1DM is associated with reduced or absent levels of plasma insulin and usually manifests in childhood. DM type 2 is characterized by insulin resistance and, most commonly, adult onset. A more detailed discussion of T1DM and T2DM can be found in Chapter 12.

Substantial advances in understanding of the pathophysiology of common diabetes have been achieved by identifying mutations that cause rare monogenic forms of hyperglycemia. Mutations in the insulin receptor gene have been associated with a disorder characterized by insulin resistance and acanthosis nigricans (hypertrophic skin with a corrugated appearance). These mutations can decrease the number of insulin receptors on a cell's surface, or they can decrease the insulin-binding activity level or the insulin-stimulated tyrosine kinase activity level. Mutations in mitochondrial DNA and in the genes encoding insulin and glucokinase have also been associated with hyperglycemic disorders.

▶ Studies of rare, monogenic forms of diabetes define the pathways that may be disturbed in the more common forms of diabetes mellitus.

## Lactose
The ability to metabolize lactose (a sugar composed of glucose and galactose) depends, in part, on the activity of an intestinal brush-border enzyme called lactase-phlorizin hydrolase

(LPH). In most mammals, LPH activity diminishes after infants are weaned from maternal milk. However, the persistence of intestinal LPH activity is a common autosomal recessive trait in human populations, with an incidence ranging from 5% to 90%. The geographic distribution of lactase persistence is concordant with areas of high milk intake, such as northwestern Europe and certain parts of Africa. The persistent ability for adults to use dairy products as a source of vitamin D had a selective advantage in these populations.

Lactase nonpersistence (i.e., adult-type hypolactasia or lactose intolerance) is common in most tropical and subtropical countries. Persons with lactase nonpersistence can experience nausea, bloating, and diarrhea after ingesting lactose. Thus, in many regions where reduced lactase activity is prevalent, the lactose in dairy products is often partially metabolized (e.g., by lactobacilli in the preparation of yogurt) before consumption. The role of lactase nonpersistence as a cause of abdominal pain and symptoms of irritable bowel syndrome is controversial.

LPH is encoded by the lactase gene (*LCT*) on chromosome 2. In European populations, adult LPH expression is regulated by a polymorphism located in an upstream gene named *minichromosome maintenance 6* (*MCM6*). However, in sub-Saharan African populations in which lactase persistence is common, this polymorphism is found at low frequency in groups from West Africa and is absent in dairy-consuming populations from East Africa. Recently, lactase persistence in these populations was shown to be caused by different polymorphisms that appear to increase transcription of *LCT*. These polymorphisms appear to have arisen relatively recently in human evolution and have increased in incidence as a result of natural selection independently in populations in Europe and Africa. In each case, the selective force appears to have been a local adaptive response to the higher fitness afforded by the ability to consume dairy products.

Mutations that abolish lactase activity altogether cause congenital lactase deficiency and produce severe diarrhea and malnutrition in infancy. Such mutations are very rare.

## Glycogen

Carbohydrates are most commonly stored as glycogen in humans. Consequently, enzyme deficiencies that lead to impaired synthesis or degradation of glycogen are also considered disorders of carbohydrate metabolism. Defects of each of the proteins involved in glycogen metabolism have been identified (Table 7-2). These cause different forms of glycogen storage disorders and are classified numerically according to the chronological order in which their enzymatic basis was described. The two organs most severely affected by glycogen storage disorders are the liver and skeletal muscle. Glycogen storage disorders that affect the liver typically cause hepatomegaly (enlarged liver) and hypoglycemia (low plasma glucose level). Glycogen storage disorders that affect skeletal muscle cause exercise intolerance, progressive weakness, and cramping. Some glycogen storage disorders, such as Pompe disease, can also affect cardiac muscle, causing cardiomyopathy and early death. Treatment of some glycogen storage disorders by enzyme replacement (e.g., giving active forms of the enzyme intraveneously) can improve, and in some cases prevent, symptoms and therefore preserve function and prevent early death.

**TABLE 7-2**
## Glycogen Storage Disorders

| Type | Defect | Major Affected Tissues |
|------|--------|------------------------|
| Ia (Von Gierke) | Glucose-6-phosphatase | Liver, kidney, intestine |
| Ib | Microsomal glucose-6-phosphate transport | Liver, kidney, intestine, neutrophils |
| II (Pompe) | Lysosomal acid β-glucosidase | Muscle, heart |
| IIIa (Cori) | Glycogen debranching enzyme | Liver, muscle |
| IIIb | Glycogen debranching enzyme | Liver |
| IV (Anderson) | Branching enzyme | Liver, muscle |
| V (McArdle) | Muscle phosphorylase | Muscle |
| VI (Hers) | Liver phosphorylase | Liver |
| VII (Tarui) | Muscle phosphofructokinase | Muscle |

> Each disorder of glycogen metabolism is uncommon, yet taken together these defects account for substantial morbidity. Early intervention can prevent severe disabilities and early death.

## Amino Acid Metabolism

Proteins play the most diverse roles of the major biomolecules (e.g., providing mechanical support, coordinating immune responses, generating motion). Indeed, nearly all known enzymes are proteins. The fundamental structural units of proteins are amino acids. Some amino acids can be synthesized endogenously (nonessential), and others must be obtained from the environment (essential). Many defects of the metabolism of amino acids have been described.

## Phenylalanine

Defects in the metabolism of phenylalanine (an essential amino acid) cause the hyperphenylalaninemias, some of the most widely studied of all inborn errors of metabolism. These disorders are caused by mutations in genes that encode components of the phenylalanine hydroxylation pathway (see Fig. 7-1). Elevated levels of plasma phenylalanine disrupt essential cellular processes in the brain such as myelination and protein synthesis, eventually producing severe mental retardation. Most cases of hyperphenylalaninemia are caused by mutations of the gene that encodes phenylalanine hydroxylase (PAH), resulting in classic phenylketonuria (PKU). Hundreds of different mutations have been identified in PAH, including substitutions, insertions, and deletions. The prevalence of hyperphenylalaninemia varies widely among groups from different geographic regions; PKU ranges from 1 in every 10,000 people of European origin to 1 in every 90,000 in those of African ancestry. Less commonly, hyperphenylalaninemia is caused by defects in the synthesis of tetrahydrobiopterin, a cofactor necessary for the hydroxylation of phenylalanine, or by a deficiency of dihydropteridine reductase.

Treatment of most hyperphenylalaninemias is aimed at restoring normal blood phenylalanine levels by restricting dietary intake of phenylalanine-containing foods (Box 7-1). However, phenylalanine is an essential amino acid, and adequate supplies are necessary for normal growth and development. A complete lack of phenylalanine is fatal. Thus, a fine balance must be maintained between providing enough protein and phenylalanine for normal growth and preventing the serum phenylalanine level from rising too high. Persons with PKU clearly benefit from lifelong treatment. Thus, once PKU is diagnosed, the person must follow a phenylalanine-restricted diet for life.

Hyperphenylalaninemia in a pregnant woman results in elevated phenylalanine levels in the fetus. This can cause poor growth, birth defects, microcephaly, and mental retardation in the fetus (regardless of the fetus's genotype). Thus, it is important that women with PKU receive appropriate pregnancy counseling. Optimally, they should maintain a low-phenylalanine diet at conception and throughout pregnancy.

BOX 7-1
# Dietary Management of Inborn Errors of Metabolism

The most important component of therapy for many inborn errors of metabolism is manipulation of the diet. This commonly includes restriction of substrates that are toxic, such as carbohydrates (e.g., in galactosemia, diabetes mellitus), fats (e.g., in MCAD deficiency), and amino acids (e.g., in PKU, MSUD, urea cycle defects); avoidance of fasting; replacement of deficient cofactors (e.g., B vitamins, carnitine); or using alternative pathways of catabolism to eliminate toxic substances. Because many metabolic disorders are diagnosed in infants whose nutritional requirements change quickly (sometimes weekly), it is imperative to provide infants with diets that provide adequate calories and nutrients for normal growth and development. Thus, the responsibility for maintaining a special diet begins with the parents of an affected child and shifts to the child when he or she is capable of managing independently. For most persons with metabolic diseases, a special diet must be maintained for life. This results in many unique challenges that are often unforeseen by families and care providers alike. Consequently, it is important that support and guidance be provided by clinical dietitians, gastroenterologists, psychologists, genetic counselors, and biochemical geneticists.

For example, newborns with PKU are fed a low-phenylalanine diet to prevent the effects of hyperphenylalaninemia on the brain. Breast milk contains too much phenylalanine to be used as the only source of nutrients. Therefore, many infants are fed an expensive low-phenylalanine formula that is available only by prescription, a medical food.* Small quantities of breast milk can be mixed with the formula, although the breast milk must be pumped and carefully titrated to avoid giving the infant too much phenylalanine. Serum phenylalanine levels are measured frequently, and

*U.S. Public Law 100-290 defines a medical food as a food that is formulated to be consumed or administered enterally under the supervision of a physician and that is intended for the specific dietary management of a disease or condition for which distinctive nutritional requirements, based on recognized scientific principles, are established by medical evaluation.

adjustments must be made to the diet to compensate for an infant's growth and varying individual tolerances for phenylalanine. These interventions can disrupt mother–infant bonding and distort a family's social dynamics.

As a child with PKU grows older, low-protein food substitutes are introduced to supplement the formula (e.g., low-protein breads and pasta). To put this in perspective, consider that a 10-year-old child with PKU might ingest 300 to 500 mg of phenylalanine per day. Thus, three or four slices of regular bread would meet a child's nutritional needs and dietary phenylalanine limit, because of the relatively high protein content of grains. Low-protein foods make the diet more substantive and varied. Indeed, seven slices of low-protein bread contain the phenylalanine equivalent of one piece of regular bread. Yet, many of these foods have an odor, taste, texture, or appearance that differs distinctively from foods containing normal amounts of protein.

The intake of many fruits, fats, and carbohydrates is less restricted (see table below). However, phenylalanine is found unexpectedly in many food items (e.g., gelatin, beer). In fact, the FDA requires manufacturers to label products containing aspartame (a common artificial sweetener that contains phenylalanine) with a warning for persons with PKU.

Teenagers with PKU sometimes have difficulty socializing with their peers because their dietary restrictions limit food choices at restaurants, sporting events, and parties. Adults with PKU must consume more medical food than they did during childhood due to their size and protein requirements. Women with PKU must be on a severely restricted low-phenylalanine diet during pregnancy, because hyperphenylalaninemia is a known teratogen (see text).

Inborn errors of metabolism are chronic diseases that are often treated by substantially modifying the diet. This can require considerable changes in lifestyle, precipitating financial and emotional hardships of which health care providers need to be aware.

## Phenylalanine Content of Some Common Foods

| Food | Measure | Phenylalanine (mg) |
|---|---|---|
| Turkey, light meat | 1 cup | 1662 |
| Tuna, water-packed | 1 cup | 1534 |
| Baked beans | 1 cup | 726 |
| Lowfat milk (2% fat) | 1 cup | 393 |
| Soy milk | 1 ounce | 46 |
| Breast milk | 1 ounce | 14 |
| Broccoli (raw) | 3 tablespoons | 28 |
| Potato (baked) | 2 tablespoons | 14 |
| Watermelon | 1/2 cup | 12 |
| Grapefruit (fresh) | 1/4 fruit | 13 |
| Beer | 6 ounces | 11 |
| Gelatin dessert | 1/2 cup | 36 |

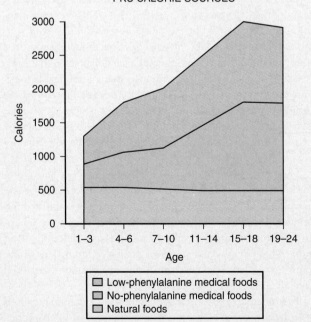

PKU CALORIE SOURCES

Legend:
☐ Low-phenylalanine medical foods
☐ No-phenylalanine medical foods
☐ Natural foods

Sources of calories of persons with PKU at different ages. The amount of no-protein medical foods and low-protein medical foods eaten increases with age as the need for energy and protein increases.
*(Courtesy of Kathleen Huntington and Diane Waggoner, Oregon Health and Science University.)*

Hyperphenylalaninemias are the most common defects of amino acid metabolism. Classic PKU is caused by mutations in the gene that encodes phenylalanine hydroxylase. Hyperphenylalaninemias are treated by restricting dietary intake of phenylalanine-containing foods.

## Tyrosine

The amino acid tyrosine is the starting point of the synthetic pathways leading to catecholamines, thyroid hormones, and melanin pigments, and it is incorporated into many proteins. Elevated levels of serum tyrosine can be acquired (e.g., severe hepatocellular dysfunction), or they can result from an inborn error of catabolism, of which there are several. Hereditary tyrosinemia type 1 (HT1) is the most common metabolic defect and is caused by a deficiency of fumarylacetoacetate hydrolase (FAH), which catalyzes the last step in tyrosine catabolism (see Fig. 7-1). Accumulation of FAH's substrate, fumarylacetoacetate, and its precursor, maleylacetoacetate, is thought to be mutagenic and toxic to the liver. Consequently, HT1 is characterized by dysfunction of the renal tubules, acute episodes of peripheral neuropathy, progressive liver disease leading to cirrhosis, and a high risk for developing liver cancer (hepatocellular carcinoma).

Management of HT1 includes supportive care, dietary restriction of phenylalanine and tyrosine, and administration of 2-(nitro-4-trifluoro-methylbenzoyl)-1,3-cyclohexanedione (NTBC) or nitisinone, an inhibitor of an enzyme upstream of FAH (4-hydroxyphenylpyruvate dioxygenase). Use of NTBC, combined with a low-tyrosine diet, has produced marked improvement in children with HT1. The long-term effects of NTBC are still unclear, but children who have been treated for more than 15 years appear to be doing well. Liver transplantation can be curative but is typically reserved for persons who fail to respond to NTBC or who develop a malignancy. In a mouse model of HT1, gene therapy (see Chapter 13) has been used to repopulate the liver with cells that exhibit stable long-term expression of FAH (FAH$^+$ hepatocytes). Because some FAH-hepatocytes remain in these livers, it is unclear whether the risk of hepatocellular carcinoma is reduced. Gene therapy for HT1 in humans might someday replace life-long dietary and pharmacologic treatment.

The gene that encodes FAH has been cloned, and mutations have been identified in many families. A splice-site mutation is quite common in French-Canadians, and its high frequency is probably the result of founder effect (see Chapter 3). This mutation results in an in-frame deletion of an exon, which removes a series of critically important amino acids from FAH. Missense and nonsense mutations of *FAH* have also been found in persons with HT1.

Tyrosinemia type 2 (oculocutaneous tyrosinemia) is caused by a deficiency of tyrosine aminotransferase. It is characterized by corneal erosions, thickening of the skin on the palms and the soles, and variable mental retardation. Tyrosinemia type 3 is associated with reduced activity of 4-hydroxyphenylpyruvate dioxygenase and neurological dysfunction. Only a few affected persons have been reported.

Deficiency of FAH causes HT1. Accumulation of the substrates of FAH leads to neurological, kidney, and liver dysfunction. Although liver transplantation has been the cornerstone of treatment for HT1, the use of drugs that block the production of FAH has proved effective.

## Branched-Chain Amino Acids

Approximately 40% of the preformed amino acids required by mammals are branched-chain amino acids (BCAAs) such as valine, leucine, and isoleucine. BCAAs can be used as a source of energy through an oxidative pathway that uses an $\alpha$-ketoacid as an intermediate. Decarboxylation of $\alpha$-ketoacids is mediated by a multimeric enzyme complex called branched-chain $\alpha$-ketoacid dehydrogenase (BCKAD). The BCKAD complex is composed of at least four catalytic components and two regulatory enzymes, which are encoded by six genes. A deficiency of any one of these six components produces a disorder known as maple syrup urine disease (MSUD), so named because the urine of affected persons has an odor reminiscent of maple syrup.

The prevalence of MSUD in the general population is low, but MSUD is relatively common in the Mennonite community of Lancaster County, Pennsylvania, where approximately 1 in every 7 persons is a heterozygous carrier. All of these carriers have the same disease-causing mutation of $E_1\alpha$, one of the loci encoding a catalytic component of BCKAD, and all are descendants of a couple who emigrated from Europe in the 18th century. This is another example of founder effect in a small, relatively isolated population (see Chapter 3).

Untreated patients with MSUD accumulate BCAAs and their associated ketoacids, leading to progressive neurodegeneration and death in the first few months of life. Treatment of MSUD consists of dietary restriction of BCAAs to the minimum required for normal growth. Despite treatment, episodic deterioration is common, and supportive care is required during these crises. Because increasing BCKAD activity by only a few percentage points can alter the course of disease substantially, therapy with thiamine, a cofactor of BCKAD, is used to treat these patients. Gene therapy for MSUD is also being investigated.

Maple syrup urine disease is caused by defects in branched-chain $\alpha$-ketoacid dehydrogenase. Accumulation of BCAAs causes progressive neurodegeneration and death. Treatment consists of restricting dietary intake of BCAAs to a minimal level.

Early detection of amino acid defects, coupled with prompt intervention, can prevent severe physical impairment and death. Moderate increases of enzyme activity can dramatically alter the course of some aminoacidopathies, making them good candidates for somatic cell gene therapy.

### Lipid Metabolism

Lipids (Greek, *lipos*, "fat") are a heterogeneous group of biomolecules that are insoluble in water and highly soluble in organic solvents (e.g., chloroform). They provide the backbone for phospholipids and sphingolipids, which are components of all biological membranes. Lipids such as cholesterol are also constituents of steroid hormones; they act as intracellular messengers and serve as an energy substrate. Elevated serum lipid levels (hyperlipidemia) are common and result from defective lipid transport mechanisms. Errors in the metabolism of fatty acids (hydrocarbon chains with a terminal carboxylate group) are much less common. However, characterizing hyperlipidemias (see Chapter 12), errors of fatty acid metabolism, and defects of cholesterol production and use has been a powerful approach for understanding the biochemical basis of lipid catabolism.

During fasting and prolonged aerobic exercise, fatty acids are mobilized from adipose tissue and become a major substrate for energy production in the liver, skeletal muscle, and cardiac muscle. Major steps in this pathway include the uptake and activation of fatty acids by cells, transport across the outer and inner mitochondrial membranes, and entry into the β-oxidation spiral in the mitochondrial matrix (Fig. 7-4). Defects in each of these steps have been described in humans, although defects of fatty acid oxidation (FAO) are the most common.

## Fatty Acids

The most common inborn error of fatty acid metabolism results from a deficiency of medium-chain acyl-coenzyme A dehydrogenase (MCAD). MCAD deficiency is characterized by episodic hypoglycemia, which is often provoked by fasting (see Clinical Commentary 7-1). Commonly, a child with MCAD deficiency presents with vomiting and lethargy after a period of diminished oral intake due to a minor illness (e.g., upper respiratory illness, gastroenteritis). Fasting results in the accumulation of fatty acid intermediates, a failure to produce ketones in sufficient quantities to meet tissue demands, and the exhaustion of glucose supplies. Cerebral edema and encephalopathy result from the indirect and direct effects of fatty acid intermediates in the central nervous system. Death often follows unless a usable energy source such as glucose is provided promptly. Between these episodes, children with MCAD deficiency often have normal examinations. Treatment consists of the avoidance of fasting, ensuring an adequate source of calories, and providing supportive care during periods of nutritional stress.

To date, most of the reported MCAD patients are of northwestern European origin, and the majority have an A-to-G missense mutation that results in the substitution of glutamate for lysine. Additional substitution, insertion, and deletion mutations have been identified but are much less common. The molecular characterization of MCAD has made it possible to offer direct DNA testing as a reliable and inexpensive diagnostic tool. Furthermore, because MCAD deficiency meets the criteria established for newborn screening (see Chapter 13), testing for it has been added to some existing newborn screening programs in the United States and elsewhere.

Long-chain acyl-CoA fatty acids are metabolized in a sequence of steps catalyzed by a handful of different enzymes. The first step is controlled by long-chain acyl-CoA dehydrogenase (LCAD). The second step is catalyzed by enzymes that are part of an enzyme complex called the mitochondrial trifunctional protein (TFP). One of the enzymes of the TFP is long-chain L-3-hydroxyacyl-CoA dehydrogenase (LCHAD).

LCHAD deficiency is one of the most severe of the FAO disorders. The first cases reported manifested with severe liver disease ranging from fulminant neonatal liver failure to a chronic, progressive destruction of the liver. Over the last 10 years, the phenotype has expanded to include cardiomyopathy, skeletal myopathy, retinal disease, peripheral neuropathy, and sudden death. Its clinical and biochemical characteristics clearly differentiate it from other FAO disorders.

In the last decade, a number of women pregnant with a fetus affected with LCHAD deficiency have developed a severe liver disease called acute fatty liver of pregnancy (AFLP) and HELLP syndrome (hemolysis, elevated liver function tests, low platelets). It is hypothesized that failure of the fetus to metabolize free fatty acids results in the accumulation of abnormal fatty acid metabolites in the maternal liver and placenta. Accumulation in the liver might cause the abnormalities observed in women with AFLP and HELLP. Accumulation in the placenta might cause intrauterine growth retardation and increase the probability of preterm delivery, both of which are common in children with LCHAD deficiency.

## Cholesterol

Elevated levels of plasma cholesterol have been associated with various conditions, most notably atherosclerotic heart disease. It has been demonstrated that substantially reduced levels of cholesterol can adversely affect growth and development. The final step of cholesterol biosynthesis is catalyzed by the

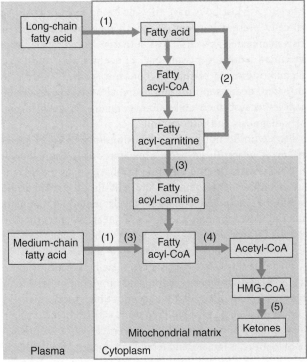

**FIGURE 7-4**
Summary of fatty acid metabolism: *1,* Fatty acid entry into a cell; *2,* activation and transesterification; *3,* mitochondrial uptake; *4,* oxidation via β-oxidation; *5,* formation of ketone bodies. Note that medium-chain fatty acids can traverse the mitochondrial membrane without carnitine-mediated transport. CoA, coenzyme A.

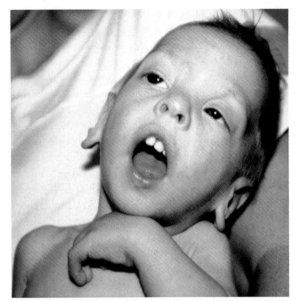

**FIGURE 7-5**
A child with Smith–Lemli–Opitz syndrome. Note the broad nasal root, upturned nasal tip, and inner epicanthal folds that are characteristic of this disorder.
*(From Jones K: Smith's Recognizable Patterns of Human Malformation, 6 ed, pp 116. Philadelphia: Saunders, 2006.)*

enzyme Δ7-sterol reductase (DHCR7). For years it has been noted that persons with an autosomal recessive disorder called Smith–Lemli–Opitz (SLO) syndrome have reduced levels of cholesterol and increased levels of 7-dehydrocholesterol (a precursor of DHCR7). SLO is characterized by various congenital anomalies of the brain, heart, genitalia, and hands (Fig. 7-5). It is unusual in this respect, because most inborn errors of metabolism do not cause congenital malformations.

In 1998, SLO was discovered to be caused by mutations in the DHCR7 gene, and to date more than 100 different mutations in DHCR7 have been found. Most of these are missense mutations that result in substitutions of a highly conserved residue of the protein. Population screens for mutation DHCR7 alleles suggests that the carrier frequency in populations of European ancestry is 3% to 4%. This high frequency suggests the incidence of SLO should be much higher than is commonly observed. One explanation is that some pregnancies with affected fetuses result in miscarriages or that SLO is undetected in some mildly affected patients. Supplementing the diet of SLO children with cholesterol can ameliorate their growth and feeding problems, although its effect on cognitive development is less clear.

> The clinical presentation of children with defects of lipid metabolism varies from slow deterioration to sudden death. MCAD deficiency is the most common of these disorders. Most affected persons can be diagnosed by biochemical analysis of a drop of dried blood at birth.

## Steroid Hormones

Cholesterol is the precursor for several major classes of steroid hormones including glucocorticoids (e.g., cortisol), mineralocorticoids (e.g., aldosterone), androgens (e.g., testosterone), and estrogens (Fig. 7-6). The actions of these steroid hormones are typically mediated by binding to an intracellular receptor, and these receptor–ligand complexes have myriad effects on a wide range of physiological processes. Defects in the synthesis of steroid hormones or their receptors can cause a wide variety of abnormalities.

Congenital adrenal hyperplasia (CAH) consists of a group of genetically heterogeneous autosomal recessive disorders of cortisol biosynthesis. Approximately 95% of cases of CAH are caused by mutations in *CYP21A2*, the gene that encodes 21-hydroxylase, and are characterized by cortisol deficiency, variable deficiency of aldosterone, and an excess of androgens. The overall incidence of 21-hydroxylase deficiency is about 1 in 15,000; therefore, the carrier frequency is about 1 in every 60 persons. However, the incidence of CAH varies widely among different ethnic groups. For example, among the Yupic of Alaska the incidence is 1 in 280.

The clinical severity of CAH varies widely and depends on the extent of residual 21-hydroxylase activity. The severe or classic form is typified by overproduction of cortisol precursors and thus excess adrenal androgens. In addition, aldosterone deficiency leads to loss of salt. In the milder forms, sufficient cortisol is produced, but there is still an overproduction of androgens. Female infants with CAH typically have ambiguous genitalia (Fig. 7-7) at birth due to in utero exposure to high concentrations of androgens, and CAH is the most common cause of ambiguous genitalia in 46,XX infants. Boys with CAH have normal genitalia at birth, so the age of diagnosis depends on the severity of aldosterone deficiency. Most boys have the salt-losing form of CAH and typically present between 7 and 14 days of age in adrenal crisis manifested as weight loss, lethargy, dehydration, hyponatremia (decreased plasma $Na^+$ concentration), and hyperkalemia (increased plasma $K^+$ concentration). If CAH is left untreated, death soon follows. Adrenal crisis is less common in girls because their ambiguous genitalia typically leads to early diagnosis and intervention.

Boys who do not have a salt-losing form of CAH present at 2 to 4 years of age with premature virilization. Persons with nonclassic or mild CAH do not have cortisol deficiency but manifest symptoms in late childhood or adulthood caused by increased androgen levels including premature pubertal development, hirsutism (increased hair growth in women in areas where hair is usually thin), amenorrhea or oligomenorrhea, polycystic ovaries, or acne.

Treatment of CAH consists of replacing cortisol, suppressing adrenal androgen secretion, and providing mineralocorticoids to return electrolyte concentrations to normal. The surgical management of children born with ambiguous genitalia is complex and somewhat controversial, but most girls with CAH identify as female, and feminizing surgery during the first year of life is standard. In pregnancies in which the fetus is at risk for classic CAH, steroids are administered to the mother to suppress the fetal overproduction of androgens and reduce the incidence of ambiguous genitalia in affected female infants.

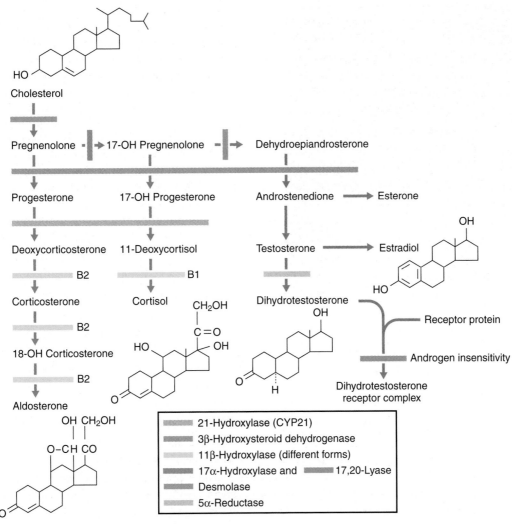

**FIGURE 7-6**
Summary of steroid synthesis. Enzymes involved in the production of cortisol, aldosterone, and testosterone are indicated.
*(Modified from Turnpenny P: Emery's Elements of Medical Genetics, 13th ed. Philadelphia: Churchill Livingstone, 2007.)*

*CYP21A2* is located on chromosome 6p21 within the major histocompatibility complex (MHC; see Chapter 9). About 90% of mutant *CYP21A2* alleles are caused by gene conversion* in which deleterious mutations are transferred to *CYP1A2*. These mutations result in a protein product that lacks normal 21-hydroxylase activity.

> Congenital adrenal hyperplasia is a relatively common defect of cortisol synthesis that causes masculinization of the genitalia in females and premature virilization in males. It is commonly caused by diminished activity of 21-hydroxylase.

*Gene conversion is a process in which two different DNA segments recombine in such a way that one segment is altered to become identical to the other.

### Steroid Hormone Receptors

Defects of most steroid hormone receptors are rare. For example, defects of the estrogen receptor have been found in a small number of persons in whom failed epiphyseal closure results in tall stature. Mutations in the gene that encodes the glucocorticoid receptor can produce hereditary resistance to the actions of cortisol.

In contrast, mutations in the X-linked gene that encodes the androgen receptor (AR) are relatively common. These mutations commonly result in complete or partial androgen insensitivity syndromes (CAIS or PAIS) in 46,XY persons. CAIS was formerly called "testicular feminization syndrome" and is characterized by typical female external genitalia at birth, absent or rudimentary müllerian structures (i.e., fallopian tubes, uterus, and cervix), a short vaginal vault, inguinal or labial testis, and reduced or absent secondary sexual characteristics. Persons with PAIS can have ambiguous external genitalia, varied positioning of the testes, and absent or present secondary sexual characteristics. Nearly all affected persons are infertile.

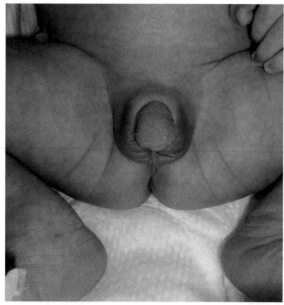

**FIGURE 7-7**
A 46,XX infant with congenital adrenal hyperplasia. The external genitalia are virilized, with rugation and partial fusion of the labial folds. No gonads were palpable, and a uterus and ovaries were visualized by ultrasonography. *(Courtesy of Melissa Parisi, University of Washington.)*

CAIS and PAIS are transmitted in an X-linked recessive pattern. More than 95% of persons with CAIS have mutations in the *AR* gene, and to date, hundreds of different mutations have been found. Most of these mutations are predicted to impair androgen binding or binding of the androgen receptor to DNA. Expansion of a polyglutamine tract in the androgen receptor causes a completely different genetic disorder called spinal bulbar muscular atrophy (see Chapter 5).

### Peroxisomal Enzymes

Peroxisomes are organelles that contain more than 50 enzymes that catalyze synthetic and degradative metabolic functions related mainly to lipid metabolism. Disorders of peroxisomes are typically divided into two groups: peroxisome biogenesis disorders (PBDs) and the single peroxisomal enzyme deficiencies (PEDs).

The PBD group comprises four disorders: Zellweger syndrome, neonatal adrenoleukodystrophy, infantile Refsum disease, and rhizomelic chondrodysplasia punctata type 1. Zellweger syndrome is the most severe of these disorders and manifests in newborns as severe hypotonia, progressive disease of the white matter of the brain, a distinctive facial appearance, and typically death in infancy. Children with neonatal adreno-leukodystrophy have similar but less-severe symptoms along with seizures. Infantile Refsum disease is less severe than Zellweger syndrome and neonatal adrenoleukodystrophy, yet affected children still have developmental delay, learning disabilities, hearing loss, and visual impairment.

PBDs are caused by mutations in genes that encode peroxins. These proteins are necessary both for peroxisome biogenesis and for importing proteins of the peroxisomal matrix and membrane. To date, mutations in more than a dozen different peroxin-encoding genes have been discovered, and for many of these genes, mutations can cause Zellweger syndrome, neonatal adrenoleukodystrophy, or infantile Refsum disease.

Nearly a dozen different PEDs have been described to date, and persons with these enzyme defects have widely disparate clinical characteristics depending on which peroxisomal function is primarily impaired. One of the most common is X-linked adrenoleukodystrophy (ALD, which involves faulty β-oxidation of very long chain fatty acids (VLCFA). ALD is subdivided into several disorders, depending in part on age of onset. The most common of these are childhood cerebral ALD or CCALD and adrenomyeloneuropathy (AMN). CCALD typically manifests between the ages of 3 and 10 years with progressive cognitive and behavioral deterioration that leads to profound disability. AMN causes similar but less-severe neurological symptoms but has a much later age of onset and slower rate of progression. Unlike many X-linked recessive disorders, 40% to 50% of women who are heterozygous for ALD develop AMN-like symptoms.

> Peroxisomal defects cause a wide variety of neurological problems in children and adults. These range from relatively mild and slowly progressive to severe and life threatening.

### Degradative Pathways

Most biomolecules are dynamic, perpetually being recycled as part of the normal metabolic state of a cell. Existing molecules are degraded into their constituents to produce substrates for building new molecules. The byproducts of energy production, substrate conversions, and anabolism also need to be processed and eliminated. Errors in these degradative pathways result in the accumulation of metabolites that would otherwise have been recycled or eliminated.

### Lysosomal Storage Disorders

The lysosomal storage disorders are the prototypical inborn errors of metabolism: Disease results from the accumulation of substrate. Enzymes within lysosomes catalyze the stepwise degradation of sphingolipids, glycosaminoglycans (mucopolysaccharides), glycoproteins, and glycolipids. Accumulation (storage) of undegraded molecules results in cell, tissue, and organ dysfunction. Most of the lysosomal disorders are caused by enzyme deficiencies, although some are caused by the inability to activate an enzyme or to transport an enzyme to a subcellular compartment where it can function properly. Many of the lysosomal storage disorders are found with uncommonly high prevalence in various ethnic populations as a result of founder effect, genetic drift, and possibly natural selection (see Chapter 3).

### Mucopolysaccharidoses

The mucopolysaccharidoses (MPS disorders) are a heterogeneous group of conditions caused by a reduced ability to degrade one or more glycosaminoglycans (e.g., dermatan sulfate, heparan sulfate, keratan sulfate, chondroitin sulfate).

These glycosaminoglycans are degradation products of proteo-glycans found in the extracellular matrix. Ten different enzyme deficiencies cause six different MPS disorders, which share many clinical features (Table 7-3), but these disorders can be distinguished from each other by clinical, biochemical, and molecular analyses. Assays that measure enzyme activity in fibroblasts, leukocytes, or serum are available for each MPS disorder, and prenatal testing after amniocentesis or chorionic villus sampling (see Chapter 13) is possible. Except for the X-linked recessive Hunter syndrome, all of the MPS disorders are inherited in an autosomal recessive fashion.

All MPS disorders are characterized by chronic and progressive multisystem deterioration, which causes hearing, vision, joint, and cardiovascular dysfunction (Fig. 7-8). Hurler, severe Hunter, and Sanfilippo syndromes are characterized by mental retardation, and normal cognition is observed in other MPS disorders.

Deficiency of iduronidase (MPS I) is the prototypic MPS disorder. It produces a spectrum of phenotypes that have been traditionally delimited into three groups—Hurler, Hurler–Scheie, and Scheie syndromes—which manifest severe, moderate, and mild disease, respectively. MPS I disorders cannot be distinguished from each other by measuring enzyme activity; therefore, the MPS I phenotype is usually assigned on the basis of clinical criteria. The iduronidase gene has been cloned, and eventually genotype–phenotype correlations may lead to earlier and more accurate classification.

Hunter syndrome (MPS II) is caused by a deficiency of iduronate sulfatase. It is categorized into mild and severe phenotypes based on clinical assessment. The onset of disease usually occurs between 2 and 4 years of age. Affected children develop coarse facial features, short stature, skeletal deformities, joint stiffness, and mental retardation. The gene that encodes iduronate sulfatase is composed of 9 exons spanning 24 kb. Twenty percent of all identified mutations are large deletions, and most of the remainder are missense and nonsense mutations.

Symptomatic treatment has been the standard of care for MPS for many years. More recently, restoration of endogenous enzyme activity has been accomplished by either bone marrow transplantation (BMT) or enzyme replacement with recombinant enzyme. BMT has become the mainstay of treatment for persons with Hurler syndrome and has been shown to improve the coarse facial features, upper airway obstruction, and cardiac disease. It also appears to mitigate neurological deterioration, although the long-term outcomes are still being investigated. BMT has been less successful for other MPS disorders. BMT is, in general, also associated with substantial morbidity and mortality. Enzyme replacement for Hurler syndrome was approved by the U.S. Food and Drug Administration in 2003 and has been shown to improve hepatosplenomegaly and respiratory disease. Early studies of enzyme replacement in Hunter syndrome (MPS II) and Maroteaux–Lamy syndrome (MPS VI) are promising.

> Defects of glycosaminoglycan degradation cause a heterogeneous group of disorders called mucopolysaccharidoses (MPS disorder). All MPS disorders are characterized by progressive multisystem deterioration causing hearing, vision, joint, and cardiovascular dysfunction. These disorders can be distinguished from each other by clinical, biochemical, and molecular studies.

**TABLE 7-3**
## Mucopolysaccharidoses*

| Name | Mutant Enzyme | Clinical Features |
|---|---|---|
| Hurler–Scheie | α-L-Iduronidase | Coarse face, hepatosplenomegaly, corneal clouding, dysostosis multiplex,[†] mental retardation |
| Hunter | Iduronate sulfatase | Coarse face, hepatosplenomegaly, dysostosis multiplex, mental retardation, behavioral problems |
| Sanfilippo A | Heparan-N-sulfamidase | Behavioral problems, dysostosis multiplex, mental retardation |
| Sanfilippo B | α-N-Acetylglucosaminidase | Behavioral problems, dysostosis multiplex, mental retardation |
| Sanfilippo C | Acetyl-CoA:α-glucosaminide N-acetyltransferase | Behavioral problems, dysostosis multiplex, mental retardation |
| Sanfilippo D | N-Acetylglucosamine-6-sulfatase | Behavioral problems, dysostosis multiplex, mental retardation |
| Morquio A | N-Acetylglucosamine-6-sulfatase | Short stature, bony dysplasia, hearing loss |
| Morquio B | β-Galactosidase | Short stature, bony dysplasia, hearing loss |
| Maroteaux–Lamy | Aryl sulfatase B | Short stature, corneal clouding, cardiac valvular disease, dysostosis multiplex |
| Sly | β-Glucuronidase | Coarse face, hepatosplenomegaly, corneal clouding, dysostosis multiplex |

*Hunter syndrome is an X-linked recessive disorder; the remaining MPS disorders are autosomal recessive.
[†]Dysostosis multiplex is a distinctive pattern of changes in bone, including a thickened skull, anterior thickening of the ribs, vertebral abnormalities, and shortened and thickened long bones.

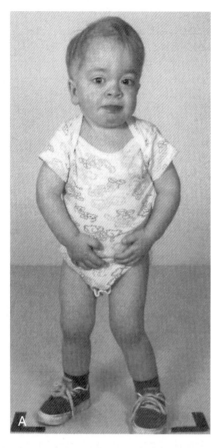

**FIGURE 7-8**
**A,** A boy with a mutation in α-ʟ-iduronidase, which causes Hurler syndrome. Note his coarse facial features, crouched stance, thickened digits, and protuberant abdomen. **B,** Transgenic mice with a targeted disruption of α-ʟ-iduronidase. Progressive coarsening of the face is apparent as 8-week-old mice *(left)* grow to become 52-week-old mice *(right)*. *(Courtesy of Dr. Lorne Clarke, University of British Columbia.)*

## Sphingolipidoses (Lipid Storage Diseases)

Defects in the degradation of sphingolipids (sphingolipidoses) result in their gradual accumulation, which leads to multiorgan dysfunction (Table 7-4). Deficiency of the lysosomal enzyme glucosylceramidase (also known as glucocerebrosidase or β-glucosidase), causes an accumulation of glucosylceramide that results in Gaucher disease. The most common metabolic storage disorder in humans, this condition is characterized by visceromegaly (enlarged visceral organs), multiorgan failure, and debilitating skeletal disease. Gaucher disease has traditionally been divided into three subtypes, which can be distinguished by their clinical features. Type 1 is most common and does not involve the central nervous system. Type 2 is the most severe, often leading to death within the first 2 years of life. Type 3 Gaucher disease is intermediate between the other two forms. In practice, the clinical phenotypes overlap, and the spectrum of Gaucher is so broad that it ranges from death in utero to persons who remain asymptomatic even in old age.

The extent to which specific organs are affected by Gaucher disease determines a person's clinical course. Splenomegaly, hepatomegaly, and pulmonary disease are shared among all three clinical types of Gaucher disease. Splenomegaly is associated with anemia, leukopenia, and thrombocytopenia, and splenic infarction can cause abdominal pain. Hepatomegaly can cause liver dysfunction, but cirrhosis and hepatic failure are uncommon.

Gaucher disease is caused by more than 200 different mutations in *GBA*, the gene that encodes glucosylceramidase. The frequency of alleles that cause type 1 Gaucher disease is particularly high in Ashkenazi Jews, in whom the five most common alleles account for 97% of all mutations. Persons with the same genotype can have very different clinical outcomes. However persons with at least one *N370S* allele, one of the most common alleles, do not develop primary neurological disease and tend to have a milder outcome in general.

Traditionally, treatment of persons with Gaucher disease was largely supportive (e.g., splenectomy for hypersplenism, blood transfusions for anemia). Enzyme replacement can reverse symptoms resulting from spleen and liver involvement. However, the effectiveness of enzyme replacement to treat neurological symptoms remains to be determined. Some persons with severe involvement, particularly chronic neurological conditions, benefit from BMT.

Enzymes that function in lysosomes are targeted and transported into the lysosomal space by specific pathways. Targeting is mediated by receptors that bind mannose-6-phosphate recognition markers attached to the enzyme (i.e., a posttranslational modification). The synthesis of these

**TABLE 7-4**
## Lysosomal Storage Disorders*

| Name | Mutant Enzyme | Clinical Features |
|---|---|---|
| Tay–Sachs | β-Hexosaminidase (A isoenzyme) | Hypotonia, spasticity, seizures, blindness |
| Gaucher (type 1; non-neuropathic) | β-Glucosidase | Splenomegaly, hepatomegaly, bone marrow infiltration, brain usually spared |
| Niemann–Pick, type 1A | Sphingomyelinase | Hepatomegaly, corneal opacities, brain deterioration |
| Fabry | α-Galactosidase | Paresthesia of the hands and feet, corneal dystrophy, hypertension, renal failure, cardiomyopathy |
| $G_{M1}$ gangliosidosis (infantile) | β-Galactosidase | Organomegaly, dysostosis multiplex,[†] cardiac failure |
| Krabbe | β-Galactosidase | Hypertonicity, blindness, deafness, seizures, (galactosylceramide–specific) atrophy of the brain |
| Metachromatic leukodystrophy | Aryl sulfatase A | Ataxia, weakness, blindness, brain atrophy (late-infantile) |
| Sandhoff | β-Hexosaminidase (total) | Optic atrophy, spasticity, seizures |
| Schindler | α-N-Acetylgalactosaminidase | Seizures, optic atrophy, retardation |
| Multiple sulfatase deficiency | Aryl sulfatase A, B, C | Retardation, coarse facial features, weakness, hepatosplenomegaly, dysostosis multiplex |

*Of the lysosomal storage disorders included in this table, Fabry syndrome is X-linked recessive and the remainder are autosomal recessive.

†Dysostosis multiplex is a distinctive pattern of changes in bone, including a thickened skull, anterior thickening of the ribs, vertebral abnormalities, and shortened and thickened long bones.

recognition markers is deficient in I-cell disease (mucolipidosis II), so named because the cytoplasm of fibroblasts from affected persons was found by light microscopy to contain inclusions. These inclusions represent partially degraded oligosaccharides, lipids, and glycosaminoglycans. As a consequence of recognition marker deficiency, newly synthesized lysosomal enzymes are secreted into the extracellular space instead of being correctly targeted to lysosomes. Persons with I-cell disease have coarse facial features, skeletal abnormalities, hepatomegaly, corneal opacities, mental retardation, and early death. There is no specific treatment for I-cell disease.

> Many different lysosomal enzymes catalyze the degradation of sphingolipids, glycosaminoglycans, glycoproteins, and glycolipids. Deficiency of a lysosomal enzyme causes accumulation of substrate, visceromegaly, organ dysfunction, and early death if untreated. Genes encoding many of the lysosomal enzymes have been cloned, and treatment with bone marrow transplantation and enzyme replacement is effective for some conditions.

## Urea Cycle Disorders

The primary role of the urea cycle is to prevent the accumulation of nitrogenous wastes by incorporating nitrogen into urea, which is subsequently excreted by the kidney. Additionally, the urea cycle is responsible for the de novo synthesis of arginine. The urea cycle consists of five major biochemical reactions (Fig. 7-9); defects in each of these steps have been described in humans.

Deficiencies of carbamyl phosphate synthetase (CPS), ornithine transcarbamylase (OTC), argininosuccinic acid synthetase (ASA), and argininosuccinase (AS) result in the accumulation of urea precursors such as ammonium and glutamine. Consequently, the clinical presentations of persons with CPS, OTC, ASA, and AS deficiencies are similar, producing progressive lethargy and coma and closely resembling the clinical presentation of Reye syndrome. Affected persons present in the neonatal period or at any time thereafter, and there is wide interfamilial variability in severity. In contrast, arginase deficiency causes a progressive spastic quadriplegia and mental retardation. Differential diagnosis of these disorders is based on biochemical testing.

Each of these disorders, except OTC deficiency, is inherited in an autosomal recessive pattern. Although OTC deficiency is an X-linked recessive disorder, women can be symptomatic carriers depending, in part, on the fraction of hepatocytes in which the normal allele is inactivated. The goal of therapy for each disorder is to provide sufficient calories and protein for normal growth and development while preventing hyperammonemia.

OTC deficiency, an X-linked condition, is the most prevalent of the urea cycle disorders and thus has been intensively studied. A variety of exon deletions and missense mutations have been described, and mutations that affect RNA processing have been observed.

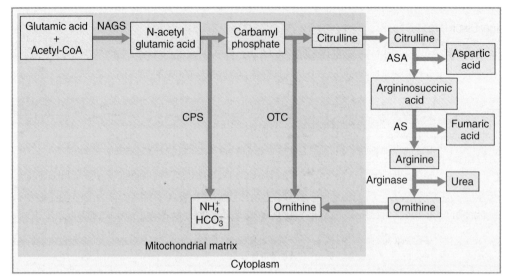

**FIGURE 7-9**
Schematic diagram of the urea cycle. AS, argininosuccinase; ASA, argininosuccinic acid synthetase; CoA, coenzme A; CPS, carbamyl phosphate synthetase; NAGS, *N*-acetylglutamate synthetase; OTC, ornithine transcarbamylase.

> The urea cycle consists of five major biochemical reactions that convert nitrogenous waste products to urea, which is subsequently excreted by the kidney. Enzymatic defects in this pathway lead to the accumulation of urea precursors, progressive neurological impairment, and death if untreated. The genes that cause most of these disorders have been cloned, including the most common observed defect, X-linked ornithine transcarbamylase (OTC) deficiency.

## Energy Production

Energy for cellular activities can be produced from many different substrates, including glucose, ketones, amino acids, and fatty acids. Catabolism of these substrates requires their stepwise cleavage into smaller molecules (via processes such as the citric acid cycle or β-oxidation), followed by passage of hydrogen ions through the oxidative phosphorylation (OXPHOS) system. Alternatively, some substrates are processed anaerobically.

The OXPHOS system consists of five multiprotein complexes that transfer electrons to $O_2$. These complexes are composed of more than 100 polypeptides and are located in the inner mitochondrial membrane. Thirteen of these polypeptides are encoded by the mitochondrial genome (see Fig. 5-9), and the remainder are encoded by nuclear genes. Thus, assembly and function of the OXPHOS system require ongoing signaling and transport between the nucleus and the mitochondrion. OXPHOS regulation is mediated by a wide variety of factors, including $O_2$ supply, hormone levels, and metabolite-induced transcription control.

More than 20 disorders characterized by OXPHOS defects are caused by substitutions, insertions, or deletions in the mitochondrial genome and are maternally inherited (see Chapter 5). In addition, nuclear genes that can cause mitochondrial DNA (mtDNA) deletions or depletion of mtDNA have been isolated, and these disorders are inherited in an autosomal recessive pattern. Mutations in genes affecting the OXPHOS system produce very complex phenotypes as a consequence of the varying metabolic requirements of different tissues and systems at different developmental stages.

Electron transfer flavoprotein (ETF) and ETF-ubiquinone oxidoreductase (ETF-QO) are nuclear-encoded proteins through which electrons can enter the OXPHOS system. Inherited defects in either of these proteins cause glutaric acidemia type II, which is characterized by hypotonia, hepatomegaly, hypoketotic or nonketotic hypoglycemia, and metabolic acidemia. Most affected persons present in the neonatal period or shortly thereafter, and despite aggressive therapy, affected children often die within months.

In most tissues, the metabolism of pyruvate proceeds through pyruvate dehydrogenase, the citric acid cycle, and the OXPHOS system. However, in tissues with high glycolytic activity and a reduced or absent OXPHOS capacity, the end products of metabolism are pyruvate and lactic acid (see Fig. 7-3). Lactate is produced by the reduction of pyruvate, and the bulk of circulating lactate is usually absorbed by the liver and converted to glucose. Defects in the pathways of pyruvate metabolism produce lactic acidemia. The most common of such disorders is a deficiency of the pyruvate dehydrogenase (PDH) complex. It may be caused by mutations in the genes encoding one of five components of the PDH complex: $E_1$, $E_2$, $E_3$, X-lipoate, or PDH phosphatase. These disorders are characterized by varying degrees of lactic acidemia, developmental delay, and abnormalities of the central nervous system. It has been suggested that the facial features of some children with PDH deficiency resemble those of children with fetal alcohol syndrome (see Clinical Commentary 15-5 in Chapter 15). It has also been proposed that acetaldehyde from the circulation of mothers with alcoholism inhibits PDH in the fetus, creating a phenocopy of PDH deficiency.

> The phenotype produced by defects of energy metabolism is complex because of the varying oxidative demands of different tissues and organs. Once the condition is diagnosed, the goal of treatment is to use alternative pathways of energy production.

### Transport Systems

The efficient movement of molecules between compartments (e.g., organelles, cells, the environment), and hence across a barrier, often requires a macromolecule that connects the compartments and mediates transport through the barrier. Abnormalities of these transport systems have myriad effects, depending on whether altered barrier integrity or the accumulation of substrate has a greater impact on normal physiology.

### Cystine

Cystine is the disulfide derivative of the amino acid cysteine. Abnormal cystine transport can produce two diseases: cystinuria and cystinosis. Both disorders are inherited in an autosomal recessive fashion.

Abnormal cystine transport between cells and the extracellular environment causes cystinuria, one of the most common inherited disorders of metabolism. Although cystinuria produces substantial morbidity, early death is uncommon. Cystinuria is a genetically heterogeneous disorder caused by a defect of dibasic amino acid transport affecting the epithelial cells of the gastrointestinal tract and renal tubules. As a result, cystine, lysine, arginine, and ornithine are excreted in urine in quantities higher than normal. Cystine is the most insoluble of the amino acids; therefore, elevated urinary cystine predisposes to the formation of renal calculi (kidney stones). Complications of chronic nephrolithiasis (presence of kidney stones) include infection, hypertension, and renal failure. Treatment of cystinuria consists largely of rendering cystine more soluble. This is accomplished by administering pharmacological amounts of water (4-6 L/day), alkalinizing the urine, and using chelating agents such as penicillamine.

Based on amino acid excretion studies, cystinuria has been divided into three phenotypes. Type I cystinuria has been associated with missense, nonsense, and deletion mutations in a gene termed *soluble carrier family 3, member 1 amino acid transporter* (*SLC3A1*). Types II and III cystinuria are caused by mutations in a gene called *SLC7A9*. *SLC3A1* and *SLC7A9* encode the heavy and light subunits of the amino

acid transporter $b^{0,+}$ located on the brush-border plasma membrane of epithelial cells in the proximal tubules of the kidney. In vitro studies of mutant *SLC3A1* and *SLC7A9* protein have demonstrated a marked reduction in transport activity, providing direct evidence for the role of these proteins in cystinuria.

Cystinosis is a rare disorder caused by a diminished ability to transport cystine across the lysosomal membrane. This produces an accumulation of cystine crystals in the lysosomes of most tissues. Affected persons are normal at birth but develop electrolyte disturbances, corneal crystals, rickets, and poor growth by the age of 1 year. Renal glomerular damage is severe enough to necessitate dialysis or transplantation within the first decade of life. Transplanted kidneys function normally, but chronic complications such as diabetes mellitus, pancreatic insufficiency, hypogonadism, myopathy, and blindness occur. Until recently, treatment was largely supportive, including renal transplantation. However, cysteine-depleting agents such as cysteamine have proved successful in slowing renal deterioration and improving growth. A novel gene encoding an integral lysosomal membrane protein was recently found to be mutated in persons with cystinosis.

> The phenotype of transport defects is partly contingent on the degree of barrier disruption as well as the compartments through which normal traffic is compromised. Abnormal cystine transport between cells and the extracellular environment causes cystinuria, renal disease, and hypertension. Cystinosis is produced by a defect of cystine efflux from the lysosome; it leads to severe chronic disabilities and, if untreated, death.

### Heavy Metals

Many of the enzymes controlling metabolic processes require additional factors (**cofactors**) to function properly and efficiently. These cofactors are commonly trace elements such as ions of heavy metals (metals that are more dense than those in the first two groups of the periodic table). At least 12 trace elements are essential to humans. For example, a zinc ion acts as a cofactor in carbonic anhydrase, placing a hydroxide ion next to carbon dioxide to facilitate the formation of bicarbonate. Although an adequate supply of trace elements is critical for normal metabolism, excessive amounts of circulating and/or stored heavy metals are highly toxic. Accordingly, a complex series of transport and storage proteins precisely control heavy metal homeostasis. Abnormalities of these proteins cause progressive dysfunction of various organs, often leading to premature death if untreated. Human disorders that disrupt the normal homeostasis of copper (Wilson disease, Menkes disease, occipital horn syndrome), iron (hereditary hemochromatosis; Clinical Commentary 7-2), and zinc (hereditary acrodermatitis enteropathica) have been described.

## CLINICAL COMMENTARY 7-2
### Hereditary Hemochromatosis

The term hemochromatosis refers to all disorders characterized by excessive iron storage. A subgroup of these disorders are hereditary and can be caused by mutations in one of several different genes. The most common form of hereditary hemochromatosis (HH) is an autosomal recessive disorder of iron metabolism in which excessive iron is absorbed in the small intestine and then accumulates in a variety of organs such as the liver, kidney, heart, joints, and pancreas. It was described by von Recklinghausen, the same physician who described neurofibromatosis 1, in 1889. Approximately 1 of every 8 northern Europeans is a carrier for HH, and 1 of every 200 to 400 persons is a homozygote. Although the penetrance of the disease-causing genotype is incomplete (as discussed later), HH is one of the most common genetic disorders observed in people of European ancestry. Its prevalence is much lower in Asian and African populations.

The most common symptom of HH is fatigue, although the clinical presentation of patients with HH can vary considerably. Additional findings include joint pain, diminished libido, diabetes, increased skin pigmentation, cardiomyopathy, liver enlargement, and cirrhosis. Abnormal serum iron parameters can identify most men at risk for iron overload, but HH is not detected in many premenopausal women. The most sensitive diagnostic test for HH is a liver biopsy accompanied by histochemical staining for hemosiderin (a form of stored iron).

As early as the 1970s, an increased frequency of the human leukocyte antigen HLA-A3 allele in HH patients indicated that an HH gene might be located near the major histocompatibility region (MHC) on chromosome 6p. Subsequent linkage studies confirmed this hypothesis in the late 1970s, but it was not until 1996 that the HH gene was cloned. The HH gene is a widely expressed HLA class I–like gene, *HFE*. The gene product is a cell-surface protein that binds to the transferrin receptor (transferrin carries iron molecules), overlapping the binding site for transferrin and inhibiting transferrin-mediated iron uptake. However, this does not directly affect iron transport from the small intestine. Instead, this interaction is thought to be involved in a cell's ability to sense iron levels. This function is disrupted in persons with mutations in *HFE*, resulting in excessive iron absorption from the small intestine and iron overload. Thus, hemochromatosis is not caused by a defect of an iron transport protein but rather by a defect in the regulation of transport.

A single missense mutation that results in the substitution of a tyrosine for cysteine in a $\beta_2$-microglobulin–binding domain accounts for 85% of all HH-causing mutations. A single ancestral haplotype predominates in Europeans, suggesting that there was a selective advantage conferred by having at least one copy of the HH gene. Because iron deficiency affects one third of the world's population and is significantly less common in HH heterozygotes, it is likely that this explains the higher incidence of HH in many populations.

Treatment of HH consists of reducing the accumulated iron in the body. This is accomplished by serial phlebotomy or by the use of an iron-chelating agent such as deferoxamine. Depending on the quantity of iron stored, return to a normal level of iron can take a few years. However, iron reduction prevents further liver damage, cures the cardiomyopathy, returns skin pigmentation to normal, and might improve the diabetes. Persons who have not developed irreversible liver damage have a nearly normal life expectancy.

The estimated penetrance of HH depends on a person's age, sex, and whether the presence of disease is measured by histological findings such as hepatic fibrosis or clinical symptoms. Most men who are homozygous for an HH-causing mutation do not develop clinical symptoms, and those who do are seldom symptomatic before the age of 40 years. An even smaller fraction of homozygous women develop clinical symptoms. If symptoms are seen, they typically occur 20 years or so later than in men because iron accumulation in women is tempered by iron losses during menstruation, gestation, and lactation.

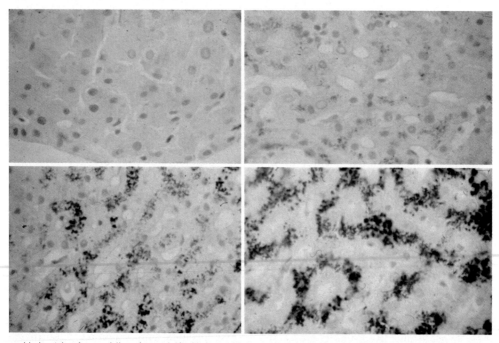

Comparison of hemosiderin stain of normal liver *(upper left)* with hemosiderin stain of livers from persons affected with hemochromatosis *(upper right, lower right,* and *lower left)*. Note the varying degree of increased deposition of hemosiderin livers of HH homozygotes. This damages the liver, impairs its function, and can lead to cirrhosis and liver cancer.

## Copper

Copper is absorbed by epithelial cells of the small intestine and is subsequently distributed by various chaperone proteins that shuttle it to different places in the cell (e.g., cytoplasmic enzymes that use copper as a cofactor, enzymes in the mitochondria). Some copper is transported to the liver to be incorporated into proteins that distribute it to other parts of the body (e.g., brain). Excess copper in hepatocytes is secreted into bile and excreted from the body.

Menkes disease (MND) is an X-linked recessive disorder described in 1962 by John Menkes, who studied five male siblings, all of whom died before 3 years of age. MND is characterized by mental retardation, seizures, hypothermia, twisted and hypopigmented hair (pili torti), loose skin, arterial rupture, and death in early childhood. In MND patients, copper can be absorbed by the gastrointestinal epithelium but cannot be exported effectively from these cells into the blood stream. Consequently, when the intestinal cells slough, the trapped copper is excreted from the body. The lack of available copper leads to an overall deficiency of copper.

Copper is a required cofactor in tyrosinase, lysyl oxidase, superoxide dismutase, cytochrome c oxidase, and dopamine β-hydroxylase. Reduced availability of copper leads to impaired enzyme function, explaining the major clinical features of MND. For example, lysyl oxidase is required for cross-linking collagen and elastin; therefore, ineffectual cross-linking leads to weakened vascular walls and laxity of the skin. Treatment of MND consists of restoring copper levels in the body to normal. Because copper cannot be absorbed via the gastrointestinal tract in MND patients, it must be administered by an alternative route such as subcutaneous injections. Patients treated this way have demonstrated some clinical improvement. However, none of the abnormalities is completely corrected or prevented. Based on studies in an animal model of MND, the brindled mouse, it has been proposed that treatment for MND would be most effective if it were started in mid-gestation. Consequently, prenatal therapy is being investigated.

In contrast to MND, in which a deficiency of copper causes disease, Wilson disease (WND) results from an excess of copper caused by defective excretion of copper into the biliary tract. This causes progressive liver disease and neurological abnormalities. WND, an autosomal recessive disorder, was first described by Kinnear Wilson in 1912 and was called hepatolenticular degeneration because of end-stage destruction of the liver and brain. Only in the 1940s did it become clear that these findings were due to the accumulation of copper. Further progress in understanding the defect underlying WND was not made until the 1990s.

Patients with WND usually present with acute or chronic liver disease in childhood. If WND is left untreated, the liver disease is progressive, resulting in liver insufficiency, cirrhosis, and failure. Adults commonly develop neurological symptoms such as dysarthria (the inability to correctly articulate words) and diminished coordination. Accumulation of copper can also cause arthropathy (inflammation of the joints), cardiomyopathy (abnormal function of the heart muscle), kidney damage, and hypoparathyroidism (diminished secretion of or response to parathormone). Deposition of copper in Descemet's membrane (at the limbus of the cornea) produces a characteristic finding in the eye (the Kayser–Fleischer ring), which is observed in 95% of all WND patients and 100% of WND patients with neurological symptoms.

Biochemical testing can be used to confirm the diagnosis of WND. Findings include decreased serum ceruloplasmin, increased serum nonceruloplasmin copper, increased urinary copper excretion, and increased deposition of copper in the liver. The most sensitive indicator of WND is the reduced incorporation of isotopes of copper into cells cultured in vitro. Treatment of WND consists of reducing the load of accumulated copper through the use of chelating agents such as penicillamine and ammonium tetrathiomolybdate.

In 1993 the gene responsible for MND, *ATP7A*, was cloned. It encodes an adenosine triphosphatase with six tandem copies of a heavy-metal-binding sequence homologous to previously identified bacterial proteins that confer resistance to toxic heavy metals. The high sequence conservation between human and bacterial binding sequences indicated that *ATP7A* has had an important role in regulating heavy-metal ion transport. *ATP7A* was expressed in a variety of tissues but not the liver, suggesting that a similar gene that is expressed in the liver might cause WND. Portions of the *ATP7A* gene were used as a probe to test for the presence of a similar gene on chromosome 13 (the known location of a *WND* gene, revealed by linkage analysis). This strategy led to the cloning in 1993 of *ATP7B*, the gene that, when mutated, causes WND. The protein product is highly homologous (76% amino acid homology) to that of *ATP7A*. In contrast to *ATP7A*, *ATP7B* is expressed predominantly in the liver and kidney, which are major sites of involvement in WND.

The ATP7A protein is usually localized to the Golgi network within a cell, where it supplies copper to various enzymes. When copper levels in an epithelial cell of the small intestine exceed a certain concentration, ATP7A redistributes to the plasma membrane and pumps copper into the blood stream. When copper levels drop, ATP7A returns to the Golgi network. Thus, it mediates the efflux of copper into the blood stream. ATP7A is also an important transporter of copper across the blood–brain barrier.

A variety of missense, nonsense, and splice site mutations in *ATP7A* have been found in MND patients. Approximately 15% to 20% of the mutations in *ATP7A* are large deletions. Several mutations of splice sites in *ATP7A* have been associated with another disorder called the occipital horn syndrome (also known as X-linked cutis laxa or Ehlers–Danlos syndrome type IX), which is characterized by mild mental retardation, bladder and ureteric diverticula (cul-de-sac-like herniations through the wall), skin and joint laxity, and ossified occipital (the most posterior bone of the calvarium) horns. These mutations permit the production of a small amount of normal protein and prevent development of severe neurological symptoms.

ATP7B plays a similar role as an effector of copper transport. However, it moves between the Golgi network and either endosomes or the cell membrane of hepatocytes, and it controls the excretion of copper into the biliary tree. ATP7B also aids in the incorporation of copper into ceruloplasmin. Several hundred different mutations have been described in patients with WND. A single missense mutation accounts for about 40% of disease-causing alleles in persons of northern European ancestry.

> Wilson disease (WND) is an autosomal recessive disorder characterized by progressive liver disease and neurological abnormalities. Menkes disease (MND) is an X-linked recessive disorder characterized by mental retardation, seizures, and death in early childhood. Accumulation of excess copper causes disease in WND, and MND results from a lack of copper and impaired enzyme function. WND and MND are caused by mutations in the highly homologous genes, *ATP7B* and *ATP7A*, respectively.

### Zinc

Acrodermatitis enteropathica (AE) is caused by a defect in the absorption of zinc from the intestinal tract. Persons with AE experience growth retardation, diarrhea, dysfunction of the immune system, and a severe dermatitis (inflammation of the skin) that typically affects the skin of the genitals and buttocks, around the mouth, and on the limbs (Fig. 7-10). Children usually present after weaning, and AE can be fatal if it is not treated with high doses of supplemental zinc, which is curative. AE is caused by mutations in *SLC39A4*, which encodes a putative zinc-transporter protein expressed on the apical membrane of the epithelial cell of the small intestine. It is unknown whether persons with AE can still absorb small amounts of zinc through a mutant form of this transporter or whether another transporter exists that can also transport zinc when it is given in high doses.

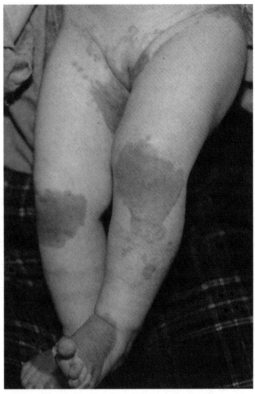

**FIGURE 7-10**
A child with acrodermatitis enteropathica caused by mutations in *SLC39A4*, encoding a protein necessary for intestinal absorption of zinc. The resulting deficiency of zinc produces a characteristic scaly, red rash around the mouth, genitals, buttocks, and limbs.
*(Courtesy of Dr. Virginia Sybert, University of Washington.)*

## Study Questions

1. Garrod found alkaptonuria to be more common in the offspring of consanguineous matings. Explain this finding. In general, what is the association between the coefficient of relationship and the prevalence of an inborn error of metabolism?

2. If many metabolic reactions can proceed in the absence of an enzyme, explain how diminished or absent enzyme activity can produce disease.

3. Despite the low prevalence of most metabolic disorders, why is it important to understand the pathogenesis of inborn errors of metabolism?

4. Describe three types of metabolic processes and give examples of disorders disturbing each of them.

5. Galactosemia is diagnosed in a 1-week-old neonate, yet enzymatic activity of GAL-1-P uridyl transferase is normal. Interpret these results and explain how mutations of different genes can produce a similar phenotype.

6. The prevalence of PKU varies from 1 in 10,000 to less than 1 in 100,000. Explain how prevalence rates of inborn errors of metabolism can vary so widely among different ethnic groups.

7. An 18-year-old woman comes to your office for prenatal counseling. She had a younger brother who died from a defect of mitochondrial fatty acid oxidation when he was a few months old. What is her risk of having an affected child? Explain your answer.

8. An 8-year-old girl develops hyperammonemia and is critically ill. Biochemical testing performed on a liver biopsy confirms that she has OTC deficiency. Which genetic test would you order next? Why?

9. Disorders of the OXPHOS system are commonly associated with elevated blood lactic acid levels. Explain this finding.

10. Polymorphisms that affect metabolism presumably have been maintained by selection because they offer heterozygotes a slight advantage. Provide an example of how these polymorphisms might have been advantageous to a group of hunter–gatherers 10,000 years ago.

## Suggested Readings

Adams PC, Barton JC. Haemochromatosis. Lancet 2007;370:1855–60.

Allen KJ, Gurrin LC, Constantine CC, et al. Iron-overload-related disease in HFE hereditary hemochromatosis. N Engl J Med 2008;358(3):221–30.

Andrews NC. Metal transporters and disease. Curr Opin Chem Biol 2002;6:181–6.

Beutler E. Hemochromatosis: Genetics and pathophysiology. Ann Rev Med 2006;57:331–47.

Bosch AM. Classical galactosemia revisited. J Inherit Metab Dis 2006;29:516–25.

Burrow TA, Hopkin RJ, Leslie ND, et al. Enzyme reconstitution/replacement therapy for lysosomal storage diseases. Curr Opin Pediatr 2007;19:628–35.

de Bie P, Muller P, Wijmenga C, Klomp LWJ. Molecular pathogenesis of Wilson and Menkes disease: Correlation of mutations with molecular defects and disease phenotypes. J Med Genet 2007;44:673–88.

James PM, Levy HL. The clinical aspects of newborn screening: Importance of newborn screening follow-up. Ment Retard Dev Disabil Res Rev 2006;12:246–54.

Merke DP, Bornstein SR. Congenital adrenal hyperplasia. Lancet 2005;365:2125–36.

Nassogne MC, Heron B, Touaita G, et al. Urea cycle defects: Management and outcome. J Inherit Metab Dis 2005;28:407–14.

Rinaldo P, Matern D. Fatty acid oxidation disorders. Annu Rev Physiol 2002;64:477–502.

Schulze A, Lindner M, Kohlmuller D, et al. Expanded newborn screening for inborn errors of metabolism by electrospray ionization-tandem mass spectrometry: Results, outcome, and implications. Pediatrics 2003;111:1399–406.

Scott CR. The genetic tyrosinemias. Am J Med Genet C Semin Med Genet 2006;142:121–26.

Scriver CR, Sly WS, Childs B, et al. The Metabolic and Molecular Bases of Inherited Disease. New York: McGraw–Hill, 2001.

Scriver CR. The *PAH* gene, phenylkentonuria, and a paradigm shift. Hum Mutation 2007;28:831–45.

Tishkoff SA, Reed FA, Ranciaro A, et al. Convergent adaptation of human lactase persistence in Africa and Europe. Nat Genet 2006;39:31–40.

Wappner R, Cho S, Kronmal RA, et al. Management of phenylketonuria for optimal outcome: A review of guidelines for phenylketonuria management and a report of surveys of parents, patients, and clinic directors. Pediatrics 1999;104:e68.

Yu H, Patel SB. Recent insights into Smith–Lemli–Opitz syndrome. Clin Genet 2005;68:383–91.

# Chapter 8

## GENE MAPPING AND IDENTIFICATION

The identification of mutations that cause disease is a central focus of medical genetics. This process often begins by mapping mutations that occur in affected persons to specific locations on chromosomes. With the completion of the Human Genome Project (see Box 8-2 later in this chapter), the locations of virtually every human gene in the genome are now known. The availability of these data, along with dramatic advances in molecular genetic technology and important developments in the statistical analysis of genetic data, has greatly expedited the mapping of disease-causing mutations. However, the specific genetic alterations responsible for most inherited disease phenotypes remain unknown (i.e., which gene or genes contribute to which disease). Much research is now devoted to discovering these genetic mutations and their consequences. As this work continues, our understanding of the biological basis of genetic disease will surely progress.

Mapping disease-causing genes is an important step in the understanding, diagnosis, and eventual treatment of a genetic disease. When a disease-causing gene's location has been pinpointed, it is often possible to provide a more accurate prognosis for persons at risk for a genetic disease. An important next step is often the cloning of the gene (as discussed in Chapter 3, *cloning* refers to inserting a gene into a vector so that copies, or clones, can be made). Once a gene has been cloned, its DNA sequence and protein product can be studied directly. This can contribute to our understanding of the actual cause of the disease. Furthermore, gene cloning can open the way to the manufacture of a normal gene product through recombinant DNA techniques, permitting more effective treatment of a genetic disease. For example, recombinant clotting factor VIII is used to treat hemophilia A, as discussed in Chapter 5, and recombinant insulin is used in the treatment of type 1 diabetes. Gene therapy—modifying genes of persons with a genetic disease—also becomes a possibility. Thus, gene mapping contributes directly to many of the primary goals of medical genetics.

This chapter discusses the approaches commonly used in gene mapping and identification. Two major types of gene mapping can be distinguished. In **gene mapping**, the frequency of meiotic crossovers between loci is used to estimate distances between loci. **Physical mapping** involves using cytogenetic, molecular, and computational methods to determine the actual physical locations of disease-causing mutations on chromosomes. Later sections in this chapter describe how these mapping techniques lead to the characterization of disease-causing genes and thus to more accurate prediction of disease risk in families.

## GENE MAPPING
### Linkage Analysis

One of Gregor Mendel's laws, the principle of independent assortment, states that an individual's genes will be transmitted to the next generation independently of one another (see Chapter 4). Mendel was not aware that genes are located on chromosomes and that genes located near one another on the same chromosome are transmitted together rather than independently. Thus, the principle of independent assortment holds true for most pairs of loci, but not for those that occupy the same region of a chromosome. Such loci are said to be **linked**.

Figure 8-1 depicts two loci, $A$ and $B$, that are located close together on the same chromosome. A third locus, $C$, is located on another chromosome. In the individual in our example, each of these loci has two alleles, designated *1* and *2*. $A$ and $B$ are linked, so $A_1$ and $B_1$ are inherited together. Because $A$ and $C$ are on different chromosomes and thus unlinked, their alleles do follow the principle of independent assortment. Hence, if the process of meiosis places $A_1$ in a gamete, the probability that $C_1$ will be found in the same gamete is 50%.

Recall from Chapter 2 that homologous chromosomes sometimes exchange portions of their DNA during prophase I (this is known as **crossing over** or **crossover**). The average chromosome experiences one to three crossover events during meiosis. As a result of crossover, new combinations of alleles can be formed on a chromosome. Consider again the linked loci, $A$ and $B$, in Figure 8-1. Alleles $A_1$ and $B_1$ are located close together on one chromosome, and alleles $A_2$ and $B_2$ are located on the homologous chromosome. The combination of alleles on each chromosome is a **haplotype** (from "haploid genotype"). The two haplotypes of this individual are denoted $A_1B_1/A_2B_2$. As Figure 8-2A shows, in the absence of crossover $A_1B_1$ will be found in one gamete and $A_2B_2$ in the other. But when there is a crossover, new

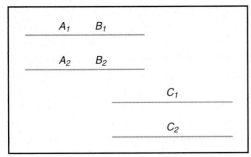

**FIGURE 8-1**
Loci *A* and *B* are linked on the same chromosome, so alleles *A₁* and *B₁* are usually inherited together. Locus *C* is on a different chromosome, so it is not linked to *A* and *B*, and its alleles are transmitted independently of the alleles of *A* and *B*.

allele combinations, $A_1B_2$ and $A_2B_1$, will be found in the two gametes (see Fig. 8-2B). The process of forming such new arrangements of alleles is called **recombination**. Crossover does not necessarily lead to recombination, however, because a **double crossover** can occur between two loci, resulting in no recombination (see Fig. 8-2C).

As Figure 8-3 shows, crossovers are more likely to occur between loci that are situated far apart on a chromosome than between loci that are situated close together. Thus, the distance between two loci can be inferred by estimating how frequently recombinations occur in families (this is called the **recombination frequency**). If, in a large series of meioses studied in families, the alleles of *A* and *B* undergo recombination 5% of the time, then the recombination frequency for *A* and *B* is 5%.

The genetic distance between two loci is measured in **centimorgans (cM)**, in honor of T. H. Morgan, who discovered the process of crossing over in 1910. One cM is approximately equal to a recombination frequency of 1%. The relationship between recombination frequency and genetic distance is approximate, because double crossovers produce no recombination. The recombination frequency thus underestimates map distance, especially as the recombination frequency increases above about 10%. Mathematical formulae have been devised to correct for this underestimate.

Loci that are on the same chromosome are said to be **syntenic** (meaning "same thread"). If two syntenic loci are 50 cM apart, they are considered to be unlinked. This is because their recombination frequency, 50%, is equivalent to independent transmission, as in the case of alleles of loci that are on different chromosomes. (To understand this, think of the chromosomes shown in Figure 8-1: If a person transmits allele $A_1$, the probability that he or she also transmits allele $C_1$, which is on another chromosome, is 50%, and the probability that he or she transmits allele $C_2$ is also 50%.)

> Crossovers between loci on the same chromosome can produce recombination. Loci on the same chromosome that experience recombination less than 50% of the time are said to be linked. The distance between loci can be expressed in centimorgans (cM); 1 cM represents a recombination frequency of approximately 1%.

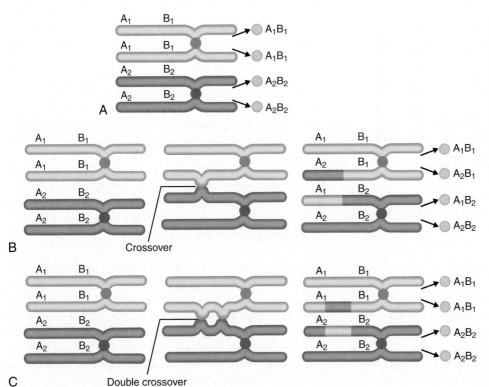

**FIGURE 8-2**
The genetic results of crossover. **A,** No crossover: $A_1$ and $B_1$ remain together after meiosis. **B,** A crossover between *A* and *B* results in a recombination: $A_1$ and $B_2$ are inherited together on one chromosome, and $A_2$ and $B_1$ are inherited together on another chromosome. **C,** A double crossover between *A* and *B* results in no recombination of alleles.

*(Modified from McCance KL, Huether SE: Pathophysiology: The Biologic Basis for Disease in Adults and Children, 5th ed. St Louis: Mosby, 2005.)*

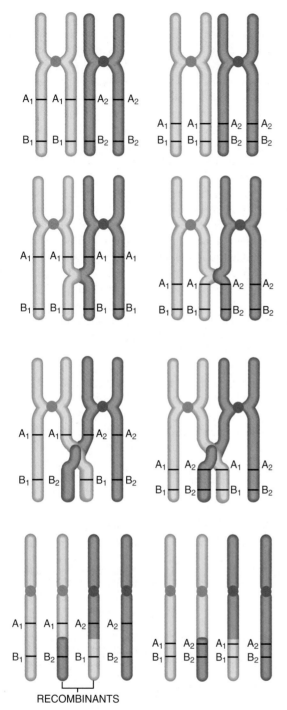

**RECOMBINANTS**

**FIGURE 8-3**
Crossover is more likely between loci that are far apart on chromosomes (*left*) than between those that are close together (*right*).

Recombination frequencies can be estimated by observing the transmission of genes in pedigrees. Figure 8-4 is an example of a pedigree in which neurofibromatosis type 1 (NF1) is being transmitted. The members of this pedigree have also been typed for a two-allele polymorphism, termed *1F10*, which, like the *NF1* gene, is located on chromosome 17.

The *1F10* genotypes are shown below each individual's number in the pedigree. Examination of generations I and II allows us to determine that, *under the hypothesis of linkage between NF1 and 1F10*, the disease-causing mutation in the *NF1* gene must be on the same copy of chromosome 17 as allele *1* of the *1F10* locus in this family, because individual I-2, who is homozygous for allele *2*, is unaffected with the disease. Only the affected father (I-1), who is a heterozygote for the *1F10* locus, could have transmitted a copy of chromosome 17 that contains both the *NF1* disease allele and *1F10* allele 1 to the daughter (II-2).

The arrangement of these alleles on each chromosome is referred to as **linkage phase**. With the linkage phase known, individual II-2's haplotypes would then be *N1/n2*, where *N* indicates the mutated allele causing *NF1*, *n* indicates the normal allele, and *1* and *2* are the two *1F10* alleles (in other words, individual II-2 has one copy of chromosome 17 that contains both the disease-causing mutation *N* and allele *1* of *1F10*, and her other copy of chromosome 17 contains the normal allele *n* and allele *2* of *1F10*). This woman's husband (individual II-1) is not affected with the disease and is a homozygote for allele *2* at *1F10*. He must have the haplotypes *n2/n2*. If the *NF1* and *1F10* loci are linked, the children of this union who are affected with *NF1* should usually have *1F10* allele *1*, and those who are unaffected should have allele *2*. In seven of eight children in generation III, we find this to be true. In one case, a recombination occurred (individual III-6). This gives a recombination frequency of 1/8, or 12.5%, supporting the hypothesis of linkage between the *NF1* and *1F10* loci. A recombination frequency of 50% would support the hypothesis that the two loci are not linked. Note that the pedigree allows us to determine linkage phase in individual II-2, but we cannot determine whether a recombination took place in the gamete transmitted to II-2 by her father. Thus, the recombination frequency is estimated only in the descendants of II-2.

In actual practice, a much larger sample of families would be used to ensure the statistical accuracy of this result. If this were done, it would show that *1F10* and *NF1* are in fact much more closely linked than indicated by this example, with a recombination frequency of less than 1%.

> Estimates of recombination frequencies are obtained by observing the transmission of alleles in families. Determination of the linkage phase (i.e., the chromosome on which each allele is located) is an important part of this procedure.

Polymorphisms such as *1F10*, which can be used to follow a disease-causing allele through a family, are termed markers (i.e., they can mark the chromosome on which a disease-causing allele is located). Because linked markers can be typed in an individual of any age (even in a fetus), they are useful for the early diagnosis of genetic disease (see Chapter 13). It is important to emphasize that a marker locus simply helps us to determine which member of a chromosome pair

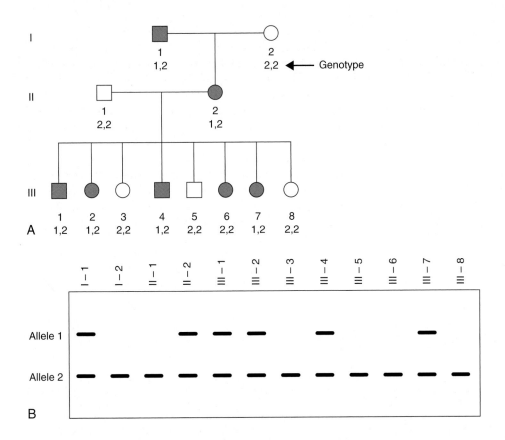

**FIGURE 8-4**
**A,** A neurofibromatosis type 1 pedigree in which each member has been typed for the *1F10* polymorphism. Genotypes for this two-allele marker locus are shown below each individual in the pedigree. Affected pedigree members are indicated by a *shaded* symbol. **B,** An autoradiogram for the *1F10* polymorphism in this family.

is being transmitted by a parent; it usually has nothing to do with the actual cause of the genetic disease.

In general, 1 cM corresponds to approximately 1 million base pairs (1 Mb) of DNA. However, this is only an approximate relationship, because several factors are known to influence crossover rates. First, crossovers are roughly 1.5 times more common during female meiosis (oogenesis) than during male meiosis (spermatogenesis). Crossovers tend to be especially common near the telomeres of chromosomes. Finally, some chromosome regions exhibit substantially elevated crossover rates. These regions, in which the recombination frequency is at least 10-fold higher than elsewhere, are termed **recombination hot spots**. Although the human genome contains approximately 50,000 recombination hot spots, it is not yet known what causes them.

Although there is a correlation between centimorgans and actual physical distances between loci, this relationship is complicated by sex differences in recombination, higher recombination frequencies near telomeres, and the existence of recombination hot spots.

## LOD Scores: Determining the Significance of Linkage Results

As in any statistical study, we must be careful to ensure that the results obtained in a linkage study are not due simply to chance. For example, consider a two-allele marker locus that has been typed in a pedigree. It is possible by chance for all affected offspring to inherit one allele and for all unaffected offspring to inherit the other allele even if the marker is not

linked to the disease-causing gene. This misleading result becomes less likely as we increase the number of subjects in our linkage study (just as the chance of a strong deviation from a 50/50 heads-to-tails ratio becomes smaller as we toss a coin many times).

How do we determine whether a linkage result is likely to be due to chance alone? In linkage analysis, a standard method is used. We begin by comparing the **likelihood** (a likelihood is similar in concept to a probability) that two loci are linked at a given recombination frequency (denoted $\theta$) versus the likelihood that the two loci are not linked (recombination frequency = 50%, or $\theta = 0.5$). Suppose we wish to test the hypothesis that two loci are linked at a recombination frequency of $\theta = 0.1$ versus the hypothesis that they are not linked. We use our pedigree data to form a **likelihood ratio**:

$$\frac{\text{likelihood of observing pedigree data if } \theta = 0.1}{\text{likelihood of observing pedigree data if } \theta = 0.5}$$

If our pedigree data indicate that $\theta$ is more likely to be 0.1 than 0.5, then the likelihood ratio (or odds) will be greater than 1. If, however, the pedigree data argue against linkage of the two loci, then the denominator will be greater than the numerator, and the ratio will be less than 1.0. For convenience, the common logarithm* of the ratio is usually taken;

---

*Recall that the common logarithm ($\log_{10}$) of a number is the power to which 10 is raised to obtain the number. The common logarithm of 100 is 2, the common logarithm of 1000 is 3, and so on.

this *logarithm of the odds* is termed a **LOD score**. Conventionally, a LOD score of 3.0 or more is accepted as evidence of linkage; a score of 3.0 indicates that the likelihood in favor of linkage is 1000 times greater than the likelihood against linkage. Conversely, a LOD score lower than –2.0 (odds of 100 to 1 against linkage) is considered to be evidence that two loci are not linked. Box 8-1 provides details on the calculation of LOD scores.

The statistical odds that two loci are a given number of centimorgans apart can be calculated by measuring the ratio of two likelihoods: the likelihood of linkage at a given recombination frequency divided by the likelihood of no linkage. The logarithm of this odds ratio is a LOD score. LOD scores of 3.0 or higher are taken as evidence of linkage, and LOD scores lower than –2.0 are taken as evidence that the two loci are not linked.

## Linkage Analysis and the Human Gene Map

Suppose that we are studying a disease-causing gene in a series of pedigrees, and we wish to map it to a specific chromosome location. Typically, we would type the members of our pedigree for marker loci whose locations on each chromosome have been established using a variety of molecular and statistical methods. Using the techniques just described, we test for linkage between the disease-causing gene and each marker. Most of these tests would yield negative LOD scores, indicating no linkage between the marker and the disease-causing gene. Eventually this exercise will reveal linkage between the disease-causing gene and a marker or group of markers. Because of the large size of the human genome, hundreds of markers are typically evaluated to find one or several that are linked to the disease-causing gene. Many important hereditary diseases have been localized using this approach, including cystic fibrosis, Huntington disease, Marfan syndrome, and neurofibromatosis type 1 (NF1).

---

**BOX 8-1**
## Estimating LOD Scores in Linkage Analysis

A simple example will help to illustrate the concepts of likelihood ratios and LOD scores. Consider the pedigree diagram in the figure below, which illustrates another family in which *NF1* is being transmitted. The family has been typed for the *1F10* marker, as in Figure 8-4. The male in generation II must have received the *1F10-1* allele from his mother, because she can transmit only this marker allele. Thus, his *1F10-2* allele had to come from his father, on the same chromosome copy as the *NF1* disease gene (under the hypothesis of linkage). This allows us to establish linkage phase in this pedigree: The affected male in generation II must have the haplotypes *N2/n1*. He marries an unaffected woman who is a homozygote for the *1F10-2* allele. Thus, the hypothesis of close linkage ($\theta$ = 0.0) predicts that each child in generation III who receives allele 2 from his or her father must also receive the *NF1* disease allele. Under the hypothesis of linkage, the father can transmit only two possible combinations: either the chromosome copy that carries both the disease-causing gene and the *1F10-2* allele (*N2* haplotype) or the other chromosome copy, which has the normal gene and the *1F10-1* allele (haplotype *n1*). The probability of each of these events is 1/2. Therefore, if $\theta$ = 0.0, the probability of observing five children with the genotypes shown in the figure below is $(1/2)^5$, or 1/32 (i.e., the multiplication rule is applied to obtain the probability that all five of these events will occur together). This is the numerator of the likelihood ratio.

Now consider the likelihood of observing these genotypes if *1F10* and *NF1* are not linked ($\theta$ = 0.5). Under this hypothesis, there is independent assortment of alleles at *1F10* and *NF1*. The father could transmit any of four combinations (*N1*, *N2*, *n1*, and *n2*) with equal probability (1/4). The probability of observing five children with the observed genotypes would then be $(1/4)^5$ = 1/1024. This likelihood is the denominator of the likelihood ratio. The likelihood ratio is then 1/32 divided by 1/1024, or 32. Thus, the data in this pedigree tell us that linkage, at $\theta$ = 0.0, is 32 times more likely than nonlinkage.

If we take the common logarithm of 32, we find that the LOD score is 1.5, which is still far short of the value of 3.0 usually accepted as evidence of linkage. To prove linkage, we would need to examine data from additional families. LOD scores obtained from individual families can be added together to obtain an overall score. (Note that, mathematically, adding LOD scores is the same as multiplying the odds of linkage in each family together and then taking the logarithm of the result. This is another example of the using the multiplication rule to assess the probability of co-occurrence.)

Suppose that a recombination had occurred in the meiosis producing III-5, the fifth child in generation III (i.e., she would retain the same marker genotype but would be affected with the disease rather than unaffected). This event is impossible under the hypothesis that $\theta$ = 0.0, so the numerator of the likelihood ratio becomes zero, and the LOD score for $\theta$ = 0.0 is $-\infty$. It is possible, however, that the marker and disease loci are still linked, but at a recombination frequency greater than zero. Let us test, for example, the hypothesis that $\theta$ = 0.1. This hypothesis predicts that the disease allele, *N*, will be transmitted along with marker allele 2 90% of the time and with marker allele 1 10% of the time (i.e., when a recombination has occurred). By the same reasoning, the normal allele, *n*, will be transmitted with marker allele 1 90% of the time and with marker allele 2 10% of the time. As in the previous example, the father can transmit either the normal allele or the disease allele with equal probability (0.5) to each child. Thus, the probability of inheriting the disease allele with marker allele 2 (haplotype *N2*) is 0.5 × 0.90 = 0.45, and the probability of inheriting the disease allele with marker allele 1 (haplotype *N1*) is 0.5 × 0.1 = 0.05. The probability of inheriting the normal allele with marker 1 (*n1*) is 0.45, and the probability of inheriting

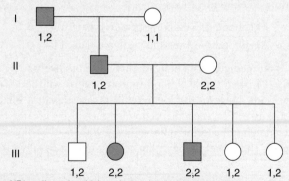

An *NF1* pedigree in which each member has been typed for the *1F10* polymorphism. The marker genotypes are shown below each individual in the pedigree.

## BOX 8-1
## Estimating LOD Scores in Linkage Analysis—cont'd

the normal allele with marker 2 (*n2*) is 0.05. In either case, then, the probability of receiving a nonrecombination (*N2* or *n1*) is 0.45, and the probability of receiving a recombination (*N1* or *n2*) is 0.05. We know that four of the children in generation III are nonrecombinants, and each of these events has a probability of 0.45. We know that one individual is a recombinant, and the probability of this event is 0.05. The probability of four nonrecombinations and one recombination occurring together in generation III is obtained by applying the multiplication rule: $0.45^4 \times 0.05$. This becomes the numerator for our LOD score calculation. As before, the denominator (the likelihood that $\theta = 0.5$) is $(1/4)^5$. The LOD score for $\theta = 0.1$, then, is given by $\log_{10}[(0.45^4 \times 0.05)/(1/4)^5] = 0.32$.

To test the hypothesis that $\theta = 0.2$, the approach just outlined is used again, with $\theta = 0.2$ instead of $\theta = 0.1$. This yields a LOD score of 0.42. It makes sense that the LOD score for $\theta = 0.2$ is higher than that for $\theta = 0.1$, because we know that one of five children (0.2) in generation III is a recombinant. Applying this formula to a series of possible $\theta$ values (0, 0.1, 0.2, 0.3, 0.4, and 0.5) shows that 0.2 yields the highest LOD score, as we would expect:

| θ | 0 | 0.1 | 0.2 | 0.3 | 0.4 | 0.5 |
|---|---|-----|-----|-----|-----|-----|
| LOD | $-\infty$ | 0.32 | 0.42 | 0.36 | 0.22 | 0.0 |

Sometimes the linkage phase in a pedigree is not known. For example, if the grandparents in the figure above had not been typed, we would not know the linkage phase of the father in generation II. It is equally likely that his haplotypes are *N2/n1* or *N1/n2* (i.e., each combination has a probability of 1/2). Thus, we need to take both possibilities into account. If he has the *N2/n1* haplotypes, then the first four children are nonrecombinants, each with a probability of $(1 - \theta)/2$, and the fifth child is a recombinant, with a probability of $\theta /2$ (using the reasoning outlined previously). Under the hypothesis that $\theta = 0.1$, the overall probability that the father has haplotypes *N2/n1* and that the five children have the observed genotypes is $1/2(0.45^4 \times 0.05) = 0.001$. We now need to take the alternative phase into account (i.e., that the father has haplotypes *N1/n2*). Here, the first four children would each be recombinants, with probability $\theta/2$, and only the fifth child would be a nonrecombinant, with probability $(1 - \theta)/2$. The probability that the father has the *N1/n2* haplotypes and that the children have the observed genotypes is $1/2(0.45 \times 0.05^4) = 0.000001$. This probability is considerably smaller than the probability of the previous phase, which makes sense when we consider that under the hypothesis of linkage at $\theta = 0.1$, four of five recombinants is an unlikely outcome. We can now consider the probability of either linkage phase in the father by adding the two probabilities together: $1/2(0.45^4 \times 0.05) + 1/2(0.45 \times 0.05^4)$. This becomes the numerator for the LOD score calculation. As before, the denominator (i.e., the probability that $\theta = 0.5$) is simply $(1/4)^5 = 1/1024$. Then,

the total LOD score for unknown linkage phase at $\theta = 0.1$ is $\log_{10}[(1/2[0.45^4 \times 0.05] + 1/2[0.45 \times 0.05^4])/(1/1024)] = 0.02$. As before, we can estimate LOD scores for each recombination frequency:

| θ | 0 | 0.1 | 0.2 | 0.3 | 0.4 | 0.5 |
|---|---|-----|-----|-----|-----|-----|
| LOD | $-\infty$ | 0.02 | 0.12 | 0.09 | 0.03 | 0.0 |

Notice that each LOD score is lower than the corresponding score when linkage phase is known. This follows from the fact that a known linkage phase contributes useful information to allow more accurate estimation of the actual genotypes of the offspring.

LOD scores are often graphed against the $\theta$ values, as shown in the figure below. The highest LOD score on the graph is the **maximum likelihood estimate** of $\theta$. That is, it is the most likely distance between the two loci being analyzed.

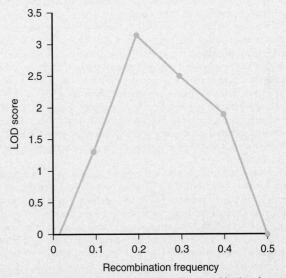

The LOD score *(y-axis)* is plotted against the recombination frequency *(x-axis)* to demonstrate the most likely recombination frequency for a pair of loci.

In practice, the analysis of human linkage data is not as simple as in these examples. Penetrance of the disease-causing gene may be incomplete, recombination frequencies differ between the sexes, and the mode of inheritance of the disease may be unclear. Consequently, linkage data are analyzed using one of several available computer software packages such as LIPED, MLINK, or MERLIN. Many of these packages also allow one to carry out **multipoint mapping**, an approach in which the map locations of several markers are estimated simultaneously.

---

Until the 1980s, linkage analyses had little chance of success because there were only a few dozen useful polymorphic markers in the entire human genome. Thus, it was unlikely that a disease-causing gene would be located near enough to a marker to yield a significant linkage result. This situation has changed dramatically as thousands of new polymorphic markers (restriction fragment length polymorphisms [RFLPs], variable number of tandem repeats [VNTRs], and short-tandem repeat polymorphisms [STRPs]; see Chapter 3) have been generated. With efficient genotyping techniques and large numbers of markers, it is now commonplace to map a disease-causing gene with only a few weeks or months of laboratory and statistical analysis.

To be useful for gene mapping, marker loci should have several properties. First, they should be codominant (i.e., homozygotes should be distinguishable from heterozygotes). This makes it easier to determine linkage phase. RFLPs, STRPs, and single nucleotide polymorphisms (SNPs) (Box 8-2) fulfill this criterion, but some of the older types of markers, such as the ABO and Rh blood groups (Chapter 3), do not. Second, marker loci should be numerous, so that close linkage to the disease-causing gene is likely. Many thousands of markers have now been identified throughout the genome, so this requirement has been fulfilled. Each chromosome is now saturated with markers (Fig. 8-5). Finally, marker loci are most useful when they are highly polymorphic (i.e., when the locus has many different alleles in the population). A high degree of polymorphism ensures that most parents will be heterozygous for the marker locus, making it easier to establish linkage phase in families. STRPs typically have many alleles and

---

## BOX 8-2
## The Human Genome Project

The Human Genome Project is one of the most widely publicized and ambitious undertakings in the history of biomedical research. Initiated in October 1990, this 15-year project had three major goals: a genetic marker map, a physical map, and the complete 3-billion bp sequence of the human genome.

The marker map was completed early in the course of the project and currently includes many thousands of polymorphisms distributed throughout the genome. These include RFLPs, VNTRs, and STRPs. On average, a useful polymorphism can be found at intervals of much less than 1 cM. Thus, a closely linked marker can be found for virtually any disease-causing gene. In addition to these polymorphisms, several million SNPs have been identified throughout the genome. SNPs are single-base variants that individually are less polymorphic than an STRP or VNTR polymorphism. However, they have a lower mutation rate than these polymorphisms, and they are especially amenable to computerized automated processing (e.g., DNA chips; see Chapter 3). They have thus added even further to the usefulness of the human genetic map.

The second goal, a physical map of known **sequence-tagged sites (STSs)** distributed at 100-kb intervals throughout the genome, is also complete. These physical signposts were invaluable in positional cloning experiments, where they were used to place a series of DNA sequences (e.g., those inserted into cloning vectors such as **yeast artificial chromosomes (YACs)**, **bacterial artificial chromosomes [BACs]**, or cosmids) in relative order.

The final goal, the complete genome sequence, has been the most challenging of all and has been pursued in both the public and private sectors. The publicly funded effort began by establishing a framework of overlapping cloned segments of human DNA. These DNA segments, which were cloned into vectors such as BACs and **bacteriophage P1 artificial chromosomes (PACs)**, were typically on the order of 100 to 200 kb in size. Establishing accurate overlaps and chromosome locations for these segments presented formidable technical challenges, and the effort was aided considerably by the physical STS map. Each DNA segment was then broken into small restriction fragments and sequenced, and the resulting data were placed into a publicly accessible database. In contrast, the privately funded effort began with much smaller DNA segments (several kb in size), cloned into plasmid vectors. Each of these small segments was sequenced and then searched for overlap in order to assemble the larger DNA sequence, making use of the publicly available data.

In February 2001, both groups announced that they had completed approximately 90% of the sequence of human euchromatic DNA (i.e., the portion of DNA that contains genes). A completed, highly accurate sequence, with an error rate lower than 1 in 10,000 bp, was unveiled in the spring of 2003, exactly 50 years after Watson and Crick first described the structure of DNA.

Completion of this project is providing many benefits. Already, gene mapping projects can often be completed in a matter of weeks because of dense and freely available marker maps. Positional cloning, once the bane of genetics laboratories, is now much more feasible because of the existing physical maps and DNA sequences. The amount of time required to identify genes via positional cloning and other approaches continues to decrease, and the number of disease-causing genes that have been pinpointed in this way is growing each year. The cloning of these genes yields many important benefits: improved genetic diagnosis, the potential for manufacture of gene products by recombinant DNA techniques, and improved treatment through more-specific drugs or gene therapy (see Chapter 13).

The complete genome sequence is likely to yield predictable—and unpredictable—results and benefits. Having a complete sequence in hand is greatly speeding the identification and characterization of all human genes. It provides the ultimate genetic blueprint of the human species. It is also quite possible that the vast expanses of noncoding DNA will surprise us with previously unknown insights about our biology and our origins.

The same technology used to sequence the human genome has been applied to dozens of other organisms: medically significant viruses and bacteria, agriculturally important crops such as rice and maize, and important experimental organisms such as yeast, fruit flies, mice, and rats. Similarities between the genes of these organisms and humans have helped us to understand the nature of many human genes.

It is a mistake to think of the completion of the human genome sequence as the end of an era of research. The genome sequence, while of immense value, is still nothing more than a long string of nucleotides. The challenge will be to use this vast pool of information to identify genes, understand their regulation and expression, and characterize the many complex interactions among genes and environment that ultimately give rise to phenotypes. In addition, the original public genome sequence represented, for each region of the genome, the sequence of only a single individual. (Because of the high degree of similarity of genes and their chromosomal locations in all humans, this single sequence is still highly useful for locating and identifying genes.) Full-scale DNA sequencing of numerous individuals, which has just begun in earnest, will help to inform us about differential susceptibility to genetic disease. The human genome sequence thus represents the beginning, rather than the end, of an era of fruitful and exciting biological research.

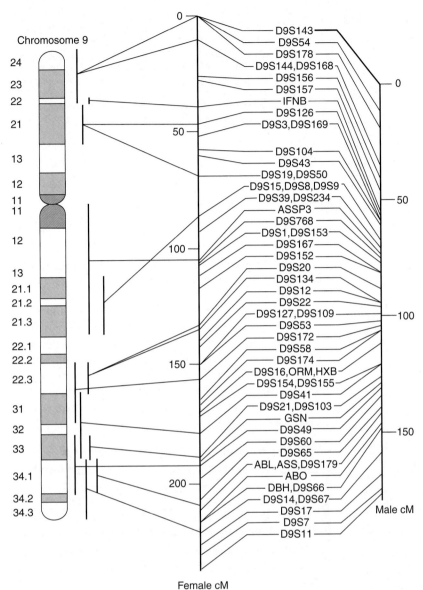

Chromosome 9

**FIGURE 8-5**
A genetic map of chromosome 9, showing the locations of a large number of polymorphic markers. Because recombination rates are usually higher in female meiosis, the distances between markers (in centimorgans) are larger for females than for males.
*(From Attwood J, Chiano M, Collins A, et al: CEPH consortium map of chromosome 9. Genomics 1994;19:203-214.)*

Female cM

are easy to assay; they are thus especially well suited to gene mapping.

An example illustrates this last point. Consider the pedigree in Figure 8-6*A*. The affected man is a homozygote for a two-allele marker locus that is closely linked to the disease locus. The man's wife is a heterozygote for the marker locus. Their affected daughter is homozygous for the marker locus. Based on these genotypes, it is impossible to determine linkage phase in this generation, so we cannot predict which children will be affected with the disorder and which will not. The mating in generation I is called an **uninformative mating**. In contrast, a marker locus with six alleles has been typed in the same family (see Fig. 8-6*B*). Because the mother in generation I has two alleles that differ from those of the affected father, it is now possible to determine that the affected daughter in generation II has inherited the disease-causing allele on the same copy of the chromosome that contains marker allele *1*. Because she married a man who has alleles *4* and *5*, we can predict that each offspring who receives allele *1* from her will be affected, and each one who receives allele *2* will be unaffected. Exceptions will be due to recombination. This example demonstrates the value of highly polymorphic markers for both linkage analysis and diagnosis of genetic disease (see Chapter 13).

> To be useful in gene mapping, linked markers should be codominant, numerous, and highly polymorphic. A high degree of polymorphism in the marker locus increases the probability that matings will be informative.

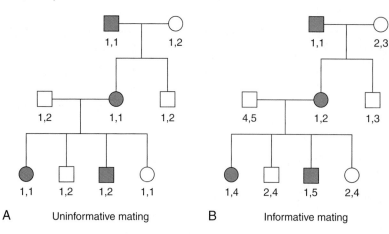

FIGURE 8-6
An autosomal dominant disease-causing gene is segregating in this family. **A,** A closely linked two-allele marker polymorphism has been typed for each member of the family, but linkage phase cannot be determined (uninformative mating). **B,** A closely linked six-allele short tandem repeat polymorphism (STRP) has been typed for each family member, and linkage phase can now be determined (informative mating).

**A**    Uninformative mating          **B**    Informative mating

The availability of many highly polymorphic markers throughout the genome helps researchers to narrow the location of a gene by direct observation of recombinations within families. Suppose that a series of marker polymorphisms, labeled $A$, $B$, $C$, $D$, and $E$, are all known to be closely linked to a disease-causing gene. The family shown in Figure 8-7 has been typed for each marker, and we observe that individual II-2 carries marker alleles $A_2$, $B_2$, $C_2$, $D_2$, and $E_2$ on the same copy of the chromosome that contains the disease-causing mutation (linkage phase). The other (normal) copy of this chromosome in individual II-2 carries marker alleles $A_1$, $B_1$, $C_1$, $D_1$, and $E_1$. Among the affected offspring in generation III, we see evidence of two recombinations. Individual III-2 clearly inherited marker allele $A_1$ from her affected mother (II-2), but she also inherited the disease-causing mutation from her mother. This tells us that there has been a recombination (crossover) between marker $A$ and the disease-causing gene. Thus, we now know that the region of the chromosome between marker $A$ and the telomere cannot contain the disease-causing gene.

We observe another recombination in the gamete transmitted to individual III-5. In this case, the individual inherited markers $D_1$ and $E_1$ but also inherited the disease-causing mutation from II-2. This indicates that a crossover occurred between marker locus $D$ and the disease-causing locus. We now know that the region between marker $D$ and the centromere (including marker E) cannot contain the disease-causing locus. These two key recombinations have thus allowed us to substantially narrow the region that contains the disease-causing locus. Additional analyses in other families could narrow the location even further, provided that additional recombinations can be observed. In this way, it is often possible to narrow the location of a disease-causing locus to a region that is several centimorgans or so in size.

Sometimes, a linkage analysis produces a total LOD score close to zero. This could mean simply that the pedigrees are uninformative (a LOD score of zero indicates that the likelihoods of linkage and nonlinkage are approximately equal, because $10^0 = 1$). However, a total LOD score of zero can also result when one subset of families has positive LOD scores (indicating linkage) and another subset has negative LOD scores (arguing against linkage). This result would provide evidence of locus heterogeneity for the disease in question (see Chapter 4). For example, osteogenesis imperfecta type I may be caused by mutations on either chromosome 7 or chromosome 17 (see Chapter 4). A study of families with this disease could show linkage to markers on chromosome 17 in some families and linkage to chromosome 7 in others. Linkage analysis has helped to define locus heterogeneity in a large number of diseases, including retinitis pigmentosa, a major cause of blindness (Clinical Commentary 8-1).

> The direct observation of recombinations between marker loci and the disease-causing locus can help to narrow the size of the region that contains the disease-causing locus. In addition, linkage analysis sometimes shows that some affected families demonstrate linkage to markers in a given chromosome region and others do not. This is an indication of locus heterogeneity.

## Linkage Disequilibrium: Nonrandom Association of Alleles at Linked Loci

Within families, one allele of a marker locus will usually be transmitted along with the disease-causing allele if the marker and disease loci are linked. For example, allele *1* of a linked two-allele marker could co-occur with the Huntington disease (HD) allele, located on chromosome 4, in a family. This association is part of the definition of linkage. However, if one examines a series of families for linkage between HD and the marker locus, allele *1* will co-occur with the disease in some families, and allele *2* of the marker will co-occur with the disease in others (Fig. 8-8). This reflects two things. First, disease-causing mutations might have occurred numerous times in the past, sometimes on a copy of chromosome 4 carrying marker allele *1* and other times on a copy of chromosome 4 carrying marker allele *2*. Second, even if the disease is the result of only one original mutation, crossovers occurring through time will eventually result in recombination of the marker and disease alleles. A disease-causing allele and a linked marker allele will thus be associated *within* families

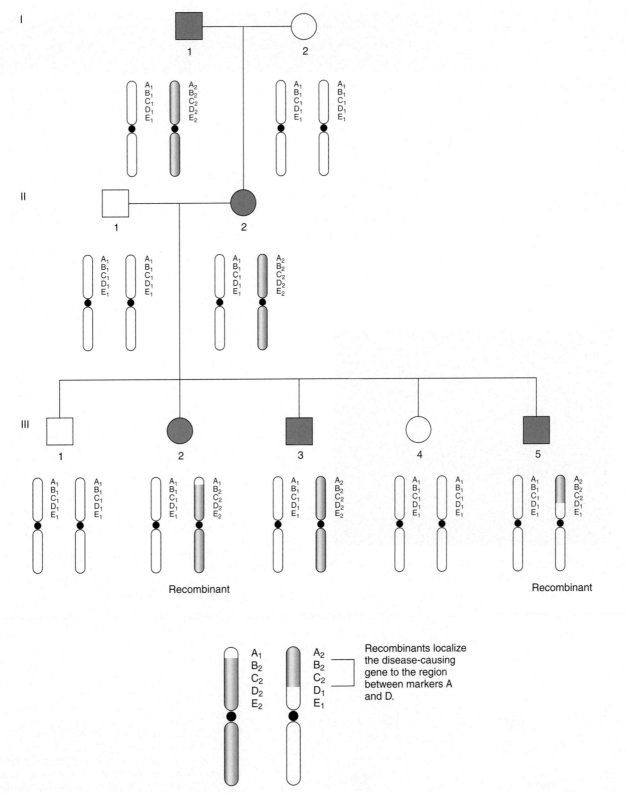

**FIGURE 8-7**
A family in which markers *A, B, C, D,* and *E* have been typed and assessed for linkage with an autosomal dominant disease–causing mutation.
As explained in the text, recombination is seen between the disease locus and marker *A* in individual III-2 and between the disease locus and marker *D* in individual III-5.

**CLINICAL COMMENTARY   8-1**
*Retinitis Pigmentosa: A Genetic Disorder Characterized by Locus Heterogeneity*

Retinitis pigmentosa (RP) describes a collection of inherited retinal defects that together are the most common inherited cause of human blindness, affecting 1 in 3000 to 1 in 4000 persons. The first clinical signs of RP are seen as the rod photoreceptor cells begin to die, causing night blindness. Rod electroretinogram (ERG) amplitudes are reduced or absent. With the death of rod cells, other tissue begins to degenerate as well. Cone cells die, and the vessels that supply blood to the retinal membranes begin to attenuate. This leads to a reduction in daytime vision. Patients develop tunnel vision, and most are legally blind by 40 years of age. The name *retinitis pigmentosa* comes from the pigments that are deposited on the retinal surface as pathological changes accumulate. RP is neither preventable nor curable, but there is evidence that its progress can be slowed somewhat by dietary supplementation with vitamin A.

RP is known to be inherited in different families in an autosomal dominant, autosomal recessive, or X-linked recessive fashion. These modes of inheritance

account for approximately 30% to 40%, 50% to 60%, and 5% to 15% of RP cases, respectively. In addition, a small number of cases are caused by mitochondrial mutations, and one form of RP is caused by the mutual occurrence of mutations at two different loci (*peripherin/RDS* and *ROM1*, both of which encode structural components of photoreceptor outer segment disc membranes). This mode of inheritance is termed **digenic**. Genetic studies have demonstrated that mutations in any of 45 different genes can cause RP, making this disease an example of extensive locus heterogeneity.

An early linkage analysis mapped an autosomal dominant form of RP to the long arm of chromosome 3. This was a significant finding, because the gene, *RHO*, that encodes rhodopsin had also been mapped to this region. Rhodopsin is the light-absorbing molecule that initiates the process of signal transduction in rod photoreceptor cells. Thus, *RHO* was a reasonable candidate gene (see text) for RP. Linkage analysis was performed using a polymorphism located within the *RHO*, and a LOD score of 20 was obtained for a recombination frequency of zero in a large Irish pedigree. Subsequently, more than 100 different mutations in *RHO* have been shown to cause RP, confirming the role of this locus in causing the disease. *RHO* mutations are estimated to account for 25% of autosomal dominant RP cases and about 10% of all RP cases.

Additional studies have identified mutations in genes involved in many different aspects of retinal degeneration. Some of these genes encode proteins involved, for example, in visual transduction (e.g., rhodopsin, the α subunit of the rod cyclic guanosine monophosphate [cGMP] cation-gated channel protein, and the α and β subunits of rod cGMP phosphodiesterase); photoreceptor structure (e.g., *peripherin/RDS* and *ROM1*); and retinal protein transport (e.g., *ABCR*). Additional genes have been implicated in syndromes that include RP as one feature. For example, RP is seen in Leber congenital amaurosis (LCA), the most common hereditary visual disorder of children. About 10% to 20% of persons with RP have Usher syndrome, which has a number of subtypes and typically also involves vestibular dysfunction and sensorineural deafness. Another 5% of RP cases occur as part of the Bardet–Biedl syndrome, in which mental retardation and obesity are also observed.

Collectively, the 45 genes identified thus far in causing RP account for about 60% of all disease cases. Additional studies of this genetically heterogeneous disorder will doubtless uncover additional genes, further enhancing our understanding of the causation of this disorder.

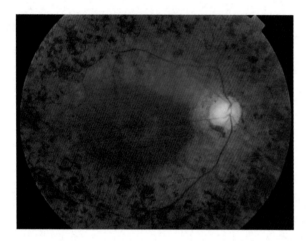

A fundus photograph illustrating clumps of pigment deposits and retinal blood vessel attenuation in retinitis pigmentosa.
*(Courtesy Dr. Donnell J. Creel, University of Utah Health Sciences Center.)*

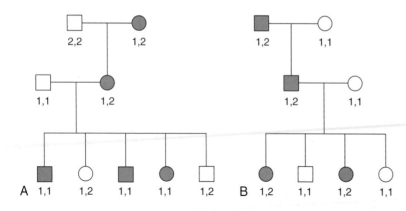

**FIGURE 8-8**
**A,** In this family, allele *1* of a marker polymorphism is in linkage phase with the disease-causing allele (i.e., both alleles are on the same copy of chromosome 4). **B,** In a second family, allele *2* of the same marker polymorphism is in linkage phase with the disease-causing allele. This difference in families can reflect an earlier recombination event between the marker locus and the disease-causing locus, or it can reflect the occurrence of two different mutation events in the ancestors of the two families.

but not necessarily *between* families. In other words, if we examine a marker locus and a disease-causing locus in a large series of families in a population, we do not necessarily expect that one specific marker allele will be associated with the disease-causing mutation in most or all families.

Sometimes, however, we do observe the preferential association of a specific marker allele and the disease-causing allele in a population. By this we mean that the chromosome haplotype consisting of one marker allele and the disease-causing allele is found more often than we would expect based on the frequencies of the two alleles in the population. Suppose, for example, that the disease-causing allele has a frequency of 0.1 in the population and the frequencies of the two alleles (labeled *1* and *2*) of the marker locus are 0.4 and 0.6, respectively.

Assuming statistical independence between the two loci (i.e., **linkage equilibrium**), the multiplication rule would predict that the population frequency of the haplotype containing both the disease-causing allele and marker allele *1* would be $0.1 \times 0.4 = 0.04$. By collecting family information, we can directly count the haplotypes in the population. If we find that the actual frequency of this haplotype is 0.09 instead of the predicted 0.04, then the assumption of independence has been violated, indicating preferential association of marker allele *1* with the disease allele. This association of alleles at linked loci is termed **linkage disequilibrium**.

Figure 8-9 illustrates how linkage disequilibrium can come about. Imagine two marker loci that are both linked to the myotonic dystrophy locus on chromosome 19. Marker *B* is

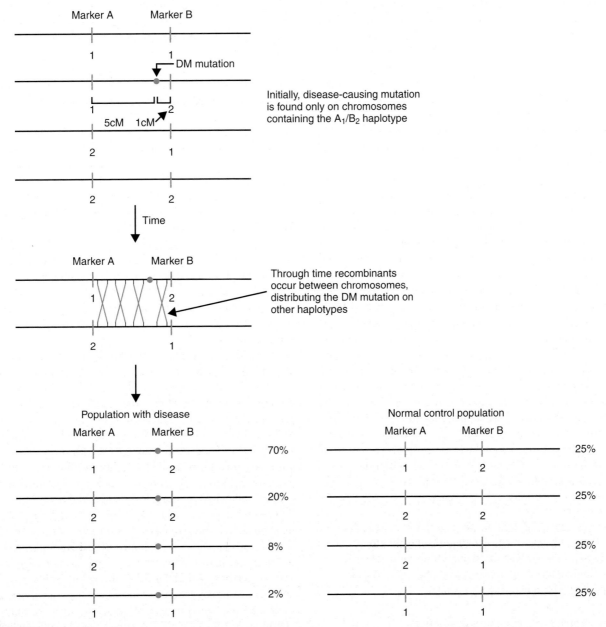

**FIGURE 8-9**
Linkage disequilibrium between the myotonic dystrophy (*DM*) locus and two linked loci, *A* and *B*. The *DM* mutation first arises on the chromosome with the $A_1B_2$ haplotype. After a number of generations have passed, most chromosomes carrying the *DM* mutation still have the $A_1B_2$ haplotype, but, as a result of recombination, the *DM* mutation is also found on other haplotypes. Because the $A_1B_2$ haplotype is seen in 70% of *DM* chromosomes but only 25% of normal chromosomes, there is linkage disequilibrium between *DM* and loci *A* and *B*. Because locus *B* is closer to *DM*, it has greater linkage disequilibrium with *DM* than does locus *A*.

closely linked, at less than 1 cM away. Marker A is less closely linked, at about 5 cM away. Because each of these marker loci has two alleles (denoted *1* and *2*), there are four possible combinations of marker alleles at the two loci, as shown in Figure 8-9. When a new myotonic dystrophy mutation first occurs in a population, it can be found on only one copy of a chromosome, in this case the one with the $A_1B_2$ marker combination. As the disease-causing mutation (allele) is passed through multiple generations, crossovers will occur between it and the two markers. Because the disease locus is more closely linked to marker B than marker A, fewer crossovers will occur between the disease-causing allele and marker B. As a result, the disease-causing allele is found on a $B_2$-containing chromosome 90% of the time, and it is found on an $A_1$-containing chromosome 72% of the time. The degree of linkage disequilibrium is stronger between marker B and the disease-causing allele than between marker A and the disease-causing allele. Notice also that both the $A_1$ and the $B_2$ alleles are still positively associated with the disease-causing allele, because each marker allele has a much lower frequency (50%) in the population of individuals who lack the disease-causing allele (see Fig. 8-9). If enough generations elapse, recombination would eventually eliminate the allelic associations completely, and the loci would be in linkage equilibrium.

Because linkage disequilibrium is a function of the distance between loci, it can be used to help infer the order of genes on chromosomes. Linkage disequilibrium provides an advantage over linkage analysis in that it reflects the action of recombinations that have occurred during dozens or hundreds of past generations (i.e., the number of generations that have elapsed since the disease-causing mutation first occurred in a population). Linkage analysis, in contrast, is limited to the recombinations that can be directly observed in only the past several generations. Consequently, there are seldom enough recombinants in a series of families to map a gene to a region smaller than several centimorgans using linkage analysis, whereas linkage disequilibrium analysis can sometimes map a gene to an interval of 0.1 cM or less. However, linkage disequilibrium can be influenced by evolutionary forces, such as natural selection or genetic drift, that have acted during the history of a population. For example, some loci in the major histocompatibility complex on chromosome 6 (see Chapter 9) are in disequilibrium, possibly because certain allelic combinations confer a selective advantage for immunity to some diseases.

> Linkage disequilibrium is the nonrandom association of alleles at linked loci. Linkage disequilibrium between loci diminishes through time as a result of recombination. It can be used to infer the order of genes on chromosomes.

## Linkage versus Association in Populations

The phenomena of linkage and association are sometimes confused. *Linkage* refers to the positions of loci on chromosomes.

When two loci are linked, specific combinations of alleles at these loci will be transmitted together within families because they are located together on the same chromosome. But, as the HD example cited previously showed, the specific combinations of alleles transmitted together can vary from one family to another. **Association**, on the other hand, refers to a statistical relationship between two traits in the *general population*. The two traits occur together in the same individual more often than would be expected by chance alone.

As just discussed, the alleles of two linked loci may be *associated* in a population (linkage disequilibrium, which is a form of association). In this case, a population association can lead to mapping of a disease-causing gene. An example is given by hereditary hemochromatosis, an autosomal recessive disorder discussed in Chapter 7. One association study showed that 78% of patients with hemochromatosis had the *A3* allele of the human leukocyte antigen (*HLA*) A locus (see Chapter 9 for further discussion of the HLA system), whereas only 27% of unaffected subjects (controls) had this allele. This strong statistical association prompted linkage analyses using HLA polymorphisms and led to the mapping of the major hemochromatosis-causing gene to a region of several centimorgans on chromosome 6. The large size of this region made it very difficult to pinpoint a specific gene. Linkage disequilibrium analysis was subsequently used to narrow the region to approximately 250 kb, prompting the ready identification of an *HLA*-like gene (*HFE*) in which a single mutation is responsible for the great majority of cases of hereditary hemochromatosis. The association between this hemochromatosis-causing gene and the linked *HLA-A* locus is probably the result of a recent hemochromatosis-causing mutation that occurred on a chromosome copy that contained the *HLA-A3* allele. Because the mutation occurred only 50 to 100 generations ago, linkage disequilibrium is still seen between the *HLA-A3* allele and the major hemochromatosis-causing mutation.

Population associations can also result from a causal relationship between an allele and a disease condition. An example of such an association involves ankylosing spondylitis, a disease that primarily affects the sacroiliac joint (Fig. 8-10). Inflammation of the ligaments leads to their ossification and eventually to fusion of the joints (ankylosis). The *HLA-B27* allele is found in about 90% of European Americans who have ankylosing spondylitis but in only 5% to 10% of the general European American population. Because the population incidence of ankylosing spondylitis is quite low (<1%), most people who have the *HLA-B27* allele do not develop ankylosing spondylitis. However, those who have the *HLA-B27* allele are about 90 times more likely to develop the disease than are those who do not have it (i.e., 9% of persons with *HLA-B27* shown in Table 8-1 have ankylosing spondylitis, and only about 0.1% of those without *HLA-B27* have the disease). Because of this strong association, a test for *HLA-B27* is sometimes included as part of the diagnosis of ankylosing spondylitis. Because ankylosing spondylitis is thought to be an autoimmune disorder, the

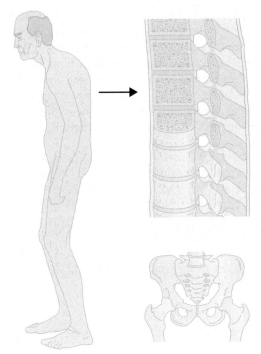

**FIGURE 8-10**
Ankylosing spondylitis, caused by ossification of discs, joints, and ligaments in the spinal column. Note the characteristic posture.
*(Modified from Mourad LA: Orthopedic Disorders. St Louis: Mosby, 1991.)*

**TABLE 8-1**

## Association of Ankylosing Spondylitis and the *HLA-B27* Allele in a Hypothetical Population*

| HLA-B27 | Ankylosing Spondylitis | |
|---------|------------|--------|
|         | **Present** | **Absent** |
| Present | 90 | 1000 |
| Absent | 10 | 9000 |

*This table shows that persons with ankylosing spondylitis are much more likely to have the HLA-B27 allele than are normal controls.*

association might reflect the fact that the HLA system is a key element in the body's immune response (see Chapter 9).

A population association is also seen between a nucleotide variant in the *HLA-DQβ* locus and type 1 diabetes (see Chapter 12). Because autoimmunity is a factor in the etiology of type 1 diabetes, there is likely a causative relationship between the *HLA-DQβ* locus and increased susceptibility to this form of diabetes.

> Population association refers to the nonrandom co-occurrence of factors at the population level. Associations are distinguished from linkage, which refers to the positions of loci on chromosomes. Linkage disequilibrium is a special case of association in which there is nonrandom association of specific alleles at linked loci.

## Gene Mapping by Association: Genome-Wide Association Studies

Recent technological developments, including microarrays (see Chapter 3) have enabled investigators to test for associations between disease phenotypes in populations and thousands to millions of marker loci distributed throughout the genome. These **genome-wide association studies (GWAS)** typically involve examining SNPs, using a microarray, in large numbers of affected cases. Microarrays are also used to assess copy number variants (CNVs, see Chapter 3), which can vary considerably among individuals. The allele frequency of each SNP among the cases is compared with the allele frequency of the same SNP in a sample of unaffected persons (controls). If a statistically significant difference is seen in the SNP frequencies in cases versus controls, then the SNP may be located in or very near a gene that contributes to disease susceptibility. The SNP itself might cause disease, or it might be in linkage disequilibrium with a nearby variant that causes the disease. When one million SNPs are typed, each SNP is located, on average, only 3 kb from the next SNP, making it very likely that the SNP will be located close to a disease-causing variant.

GWAS have been especially useful in discovering genes that contribute to common diseases, such as diabetes, cancers, and heart disease (Chapter 12). Because such diseases are the result of the action of multiple disease-causing loci, as well as nongenetic factors, traditional linkage analysis has sometimes been ineffective in detecting the loci. An advantage of GWAS is that no assumptions about the biology of the disease need to be made in order to choose which genes to study: Variants near every gene are tested. Indeed, the results of a GWAS often point to new biological pathways that had not been previously suspected to play a role in the disease being studied. In addition, it is not necessary to collect family data to detect associations in populations (although such data can be useful). Instead, unrelated cases and controls, which are easier to locate and sample than whole families, are typically used in GWAS.

Association studies should be interpreted cautiously, however, because many things can produce spurious associations between a disease and a potential risk factor. An example is ethnic stratification in a population: Certain diseases are more common in certain ethnic groups, and allele frequencies can also differ among these groups because of their evolutionary histories. Thus, if one compares disease cases and controls without proper matching for ethnicity, a false association, due simply to ethnic differences between the two groups, could be found. For example, type 2 diabetes (see Chapter 12) has been studied extensively among the Pima Native American population, where the disease is much more common than among European Americans. It was observed that the absence of haplotype Gm3,5,13,14 of human immunoglobulin G (abbreviated here as Gm3) was strongly associated with type 2 diabetes among the Pima. This suggested initially that the absence of Gm3 might be involved in causing type 2 diabetes. However, further analysis revealed that the proportion of European ancestry varied substantially among members of the Pima population and that the Gm3 frequency also varied with the degree of European ancestry: Gm3 is absent in Pimas with no European ancestry but has a frequency of 65% among Europeans. Because type 2 diabetes is much less common among Europeans, the apparent association between type 2 diabetes and the absence of Gm3 was most likely a consequence of the level of European mixture. Once the degree of European ancestry in the study subjects was taken into account, there was no evidence of an association.

Other factors that can produce false associations include imprecise definition of the disease state, inadequate sample sizes, and improper matching of cases and controls for variables such as age and sex. The inability to replicate an association in multiple study populations is an indication that the association may be invalid. An example is given by an association that was reported, but not generally replicated, between alcoholism and a polymorphism near the dopamine $D_2$ receptor locus. In addition, because a typical GWAS compares many thousands of markers in cases and controls, a small fraction of markers will appear to be associated with the disease just by chance. Statistical procedures are used to take this into account and correct for it.

Although GWAS typically incorporate 500,000 to 1 million SNPs distributed across the genome, human populations can contain several times this number of SNPs. How can we be sure that the subset of SNPs used in a GWAS adequately accounts for all of the SNPs in the genome? Investigators have applied the concept of linkage disequilibrium to address this issue. Suppose, for example, that SNP nucleotide C is known to be in very strong linkage disequilibrium with a nearby SNP nucleotide, T. This means that whenever a person has allele C, he or she almost always has allele T as well (i.e., the haplotype is C/T). Thus it is not necessary to type both SNPs in cases and controls in a study: Those who have C at the first SNP are also assumed to have T at the second SNP. By identifying sets of SNPs that are in strong disequilibrium, only one member of the set has to be typed (Fig. 8-11). This can

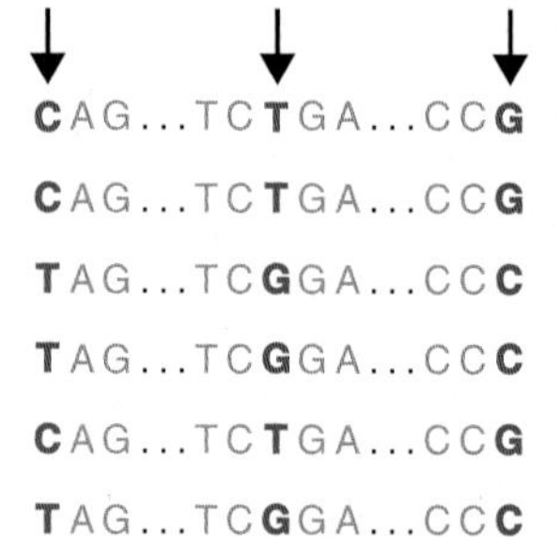

substantially reduce the cost of a GWAS. A large-scale effort to identify sets of SNPs in linkage disequilibrium with one another, the International Haplotype Map Project (HapMap), has established linkage disequilibrium patterns for millions of SNPs in the genomes of African, Asian, and European populations. This has in turn made it possible for researchers to focus gene-finding efforts on a reduced number of highly informative SNP markers in any region of the human genome.

> Genome-wide association studies (GWAS) test for association, or linkage disequilibrium, between a disease and a marker (or several markers) by testing many thousands of markers across the genome. Typically this is accomplished with microarray analysis of disease cases and unaffected controls. As in all case-control studies, considerable care must be taken to avoid spurious results by closely matching cases and controls.

## PHYSICAL MAPPING AND CLONING

Linkage analysis allows us to determine the relative distances between loci, but it does not assign specific chromosome locations to markers or disease-causing genes. Physical mapping, which has involved a variety of methods, accomplishes this goal, and considerable progress has been made in developing high-resolution physical mapping approaches.

### Chromosome Morphology

A simple and direct way of mapping disease-causing genes is to show that the disease is consistently associated with a cytogenetic abnormality, such as a duplication, deletion, or other variation in the appearance of a chromosome. Such abnormalities might have no clinical consequences themselves (thus serving as a marker), or they might cause the disease. Because these approaches are historically the oldest of the physical mapping approaches, they are discussed first.

### Heteromorphisms

A **heteromorphism** is a variation in the appearance of a chromosome. Conceptually, heteromorphisms are similar to polymorphisms: they are natural variations that occur among individuals in populations. The difference is that polymorphisms are not detectable microscopically, and heteromorphisms are.

A well-known example of a heteromorphism is an "uncoiled" (and therefore elongated) region of chromosome 1, a rare feature that is transmitted regularly in families. In the 1960s a researcher named R. P. Donahue was practicing cytogenetic analysis on his own chromosomes and found that he had this heteromorphism. He studied other members of his family and found that the heteromorphism was transmitted through his family as a mendelian trait. He then typed his family members for several blood groups. He found that the heteromorphism was perfectly associated in his family with allele A of the Duffy blood group. Linkage analysis of the uncoiler locus and the Duffy locus showed that they were closely linked, leading to the first assignment of a gene to a specific autosome.

SNP    SNP    SNP

CAG...TCTGA...CCG

CAG...TCTGA...CCG

TAG...TCGGA...CCC

TAG...TCGGA...CCC

CAG...TCTGA...CCG

TAG...TCGGA...CCC

**FIGURE 8-11**
DNA sequences from the same chromosome location have been examined in six individuals (one chromosome copy in each individual). Three polymorphic SNPs *(arrows)* in this sequence are shown in *red*. The other nucleotide bases do not vary among individuals. Because of linkage disequilibrium, alleles *C*, *T*, and *G* of the three SNPs occur together on some chromosome copies, and alleles *T*, *G*, and *C* occur together on other chromosome copies. Thus, it is necessary to type only one of the SNPs in order to know which allele an individual possesses at the other two SNPs.

It should be emphasized that heteromorphisms, like marker loci, do not cause a genetic variant or disease, but they may be associated with it within a family, thus indicating the gene's location. Although such heteromorphisms can be useful in mapping genes, they are not very common and therefore have been useful in only a few instances.

## Deletions

Karyotypes of patients with a genetic disease occasionally reveal deletions of a specific region of a chromosome. This provides a strong hint that the locus causing the disease might lie within the deleted region. The extent of a deletion can vary in several patients with the same disease. Deletions are compared in many patients to define the region that is deleted in all patients, thereby narrowing the location of the gene (Fig. 8-12). Deletion mapping has been used, for example, in locating the genes responsible for retinoblastoma (see Chapters 4 and 11), Prader–Willi and Angelman syndromes (see Chapter 4), and Wilms tumor, a childhood kidney tumor that can be caused by mutations on chromosome 11. In contrast to the heteromorphisms just discussed, deletions of genetic material are the direct cause of the genetic disease.

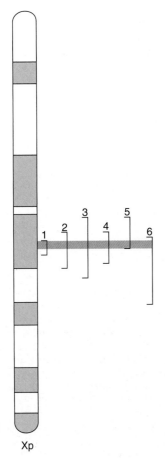

**FIGURE 8-12**
Localization of a disease-causing gene through deletion mapping. A series of overlapping deletions is studied in which each deletion produces the disease phenotype. The region of overlap of all deletions defines the approximate location of the gene.

Note that these deletions affect only one member of the homologous pair of chromosomes, making the patient heterozygous for the deletion. If a region large enough to be microscopically observable were missing from both chromosomes, it would usually produce a lethal condition.

## Translocations

As discussed in Chapter 6, balanced chromosome translocations often have no effect on a translocation carrier because the individual still has a complete copy of his or her genetic material. However, when a translocation happens to interrupt a gene, it can produce genetic disease. For example, after linkage analysis had mapped the *NF1* gene approximately to the long arm of chromosome 17, a more refined location was obtained when two patients were identified—one with a balanced translocation between chromosomes 17 and 22, and the other with a balanced translocation between chromosomes 17 and 1. The breakpoints of these translocations on chromosome 17 were located very close to each other, in the same region implicated by linkage analysis. They provided a physical starting point for experiments that subsequently led to cloning of the *NF1* gene.

A similar example is provided by translocations observed between the X chromosome and autosomes in females with Duchenne muscular dystrophy (DMD). Because this is a lethal X-linked recessive disorder, affected homozygous females are rare. The translocation breakpoint on the X chromosome was found to be in the same location (Xp21) in several affected females, suggesting that the translocation interrupted the DMD gene. This proved to be the case, and these translocations aided considerably in mapping and cloning of the DMD gene. (Although these females also carried a normal X chromosome, the normal X was preferentially inactivated, leaving only the interrupted X as the active chromosome.)

### Dosage Mapping Using Deletions and Duplications

When a deletion occurs on a chromosome, it stands to reason that the protein products encoded by genes in the deleted region will be present in only half the normal quantity. This is the basis of a simple approach known as **dosage mapping**. For example, it was observed that a 50% reduction in the level of the enzyme adenylate kinase was consistently associated with a deletion on chromosome 9, mapping the adenylate kinase gene to this chromosome region.

Similarly, a duplication of chromosome material should be associated with an increase in gene product levels. Because three genes are present instead of two, the increase should be approximately 50% above normal. This form of dosage mapping was used to assign the gene encoding superoxide dismutase-1 (SOD-1) to the long arm of chromosome 21.

> A gene can be physically mapped to a chromosome region by associating cytogenetically observable variations (heteromorphisms, translocations, deletions, duplications) with gene expression (including the presence of a genetic disease).

## Closing in on the Gene: Positional Cloning

Sometimes the gene product responsible for a genetic disease is known before the gene itself is identified. This was the case, for example, with the β-globin polypeptide and sickle cell disease. In such cases, one can deduce the DNA sequence from the amino acid sequence of the polypeptide; this DNA sequence can be used to make a probe in order to locate the disease-causing gene. This type of approach, in which the gene product and its function are used to pinpoint the gene, is an example of **functional cloning**.

More often, however, we have only a linkage result that has localized the disease-causing gene to a region near the linked marker polymorphism (the locations of these markers have been established previously [Box 8-3]). Because of the limited resolution of linkage analysis, the region that contains the disease-causing locus may be several megabases or larger and can easily contain dozens of genes interspersed with noncoding DNA (Fig. 8-13). A common approach is to begin with a linked marker and then canvas the region around the marker to locate and identify the disease-causing gene itself. Because this process begins with an approximate knowledge of the gene's position on a chromosome, it has traditionally been termed **positional cloning**.

Suppose that we know the approximate location of a disease-causing gene, determined through linkage analysis. The region that contains the disease-causing gene may be defined by the recombination fraction between the gene and a marker polymorphism (typically one to several centimorgans). Its boundaries are more commonly defined by markers that have not undergone observable recombination with the gene (see Fig. 8-7). The disease-causing gene could be located anywhere in this region, so the region's DNA sequence must be analyzed and evaluated to pinpoint the sequence corresponding to the gene. Until quite recently, this was a daunting task that could require years of work. Now that the human genome has been fully sequenced

---

### BOX 8-3
### Probes and Libraries: Their Construction and Use in Gene Mapping

Probes and libraries play central roles in gene mapping and cloning. A DNA library is much like an ordinary library, except that the library is composed of pieces of DNA rather than books.

There are several types of DNA libraries. The most general type, a **genomic library**, consists of fragments of DNA that are the result of a restriction digest of whole genomic DNA. The DNA is partially digested, so that some recognition sites are cleaved and others are not. This produces fragments that will overlap with one another. These fragments are then cloned into vectors such as phage, cosmids, or **yeast artificial chromosomes (YACs)**, using the recombinant DNA techniques discussed in Chapter 3. A genomic library contains all of the human genome: introns, exons, enhancers, promoters, and the vast stretches of noncoding DNA that separate genes.

A cDNA library is much more limited (and thus often easier to analyze); it contains only the DNA that corresponds to exons. It is obtained by purifying mRNA from a specific tissue, such as liver or skeletal muscle, and then exposing the mRNA to an enzyme called **reverse transcriptase**. This enzyme copies the mRNA into the appropriate complementary cDNA sequence. DNA polymerase can then be used to convert this single-stranded DNA to double-stranded DNA, after which it is cloned into phage or other vectors, as in the genomic library. The steps involved in making genomic and cDNA libraries are summarized in the figure below.

Yet another type of DNA library is the **chromosome-specific library**. Chromosomes are sorted by a method called **flow cytometry**, which separates chromosomes according to the fraction of AT base pairs in each one. The result is a library that consists mostly of DNA from only one chromosome. For example, after the gene for Huntington disease was mapped to a region on the short arm of chromosome 4, a library specific for that chromosome was used to refine the location of the gene.

DNA libraries are often used in making new polymorphic markers. For example, short tandem repeat polymorphisms can be obtained from DNA libraries by constructing a probe that contains multiple repeated DNA sequences (e.g., multiple CA repeats). Then the DNA library is screened with this probe to find fragments in the library that hybridize with the probe. These fragments can be tested in a series of individuals to see whether they are polymorphic. The polymorphism can then be mapped to a specific location by using physical mapping techniques such as **in situ hybridization** (see Chapter 6), in which a labeled probe is constructed that contains the PCR primer sequence that flanks the marker itself. The chromosome location at which the labeled probe undergoes hybridization (i.e., complementary DNA base pairing) defines the approximate physical location of the marker locus.

Probes are also highly useful in isolating specific disease-causing genes. In this context, they can be made in several different ways. If the defective protein (or part of it) has been identified, the protein's amino acid sequence can be used to deduce part of the DNA sequence of the gene. Generally, a short DNA sequence, only 20 to 30 bp long, is sufficiently distinctive so that it will hybridize only to the cDNA in the disease-causing gene. These sequences can be synthesized in a laboratory instrument to make oligonucleotide probes (see Chapter 3). Because of the degeneracy of the DNA code, more than one triplet codon may specify an amino acid. For this reason, different possible combinations of base pairs must be tried. These combinations of oligonucleotide probes are mixed together, and the mixture is then used to probe a DNA library (e.g., a cDNA library). When one of the oligonucleotide probes in the mixture hybridizes with a fragment in the library, a portion of the desired gene has been identified. This fragment can be mapped using the physical techniques mentioned in the text.

Often an investigator has no knowledge of the sequence of the gene product. In this case it is sometimes possible to isolate the gene by purifying mRNA produced by specialized cell types. For example, reticulocytes (immature red blood cells) produce mostly globin polypeptides. mRNA taken from these cells can be converted to cDNA using reverse transcriptase, as discussed previously, and then used as a probe to find additional fragments of the gene in a DNA library. It is sometimes easier to obtain mRNA from an experimental animal, such as a pig or rodent. Because of sequence similarity between these animals and humans, the animal-derived probe usually hybridizes appropriately with segments of a human DNA library.

BOX 8-3
## Probes and Libraries: Their Construction and Use in Gene Mapping—cont'd

Genomic DNA

Cells

mRNA purified

Cells

Restriction enzyme digestion

mRNA

Double-stranded DNA

*Reverse transcriptase*

Fragments are inserted into cloning vectors

Recombinant DNA inserted into *E.coli*

Fragments are inserted into cloning vectors

Recombinant DNA inserted into *E.coli*

Genomic DNA library

cDNA library

The creation of human DNA libraries. *Left,* A total genomic library is created using a partial restriction digest of human DNA and then cloning the fragments into vectors such as phages, cosmids, or yeast artificial chromosomes (YACs). *Right,* A cDNA library is created by purifying mRNA from a tissue and exposing it to reverse transcriptase to create cDNA sequences, which are then cloned into vectors. *(E. coli, Escherichia coli.)*

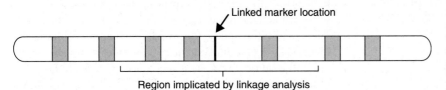

Linked marker location

Region implicated by linkage analysis

**FIGURE 8-13**
In a typical linkage analysis, a marker polymorphism is identified that is closely linked to the disease-causing gene. The chromosome location of the marker polymorphism has been established by previous linkage and physical mapping. The region surrounding the marker polymorphism can contain up to several megabases of DNA sequence, and each gene in this region is a potential candidate as the disease-causing gene.

(see Box 8-2), the process of DNA sequence evaluation can proceed much more rapidly. The finished human genome sequence is now available in computerized databases, so investigators often scan a region of interest simply by accessing the appropriate DNA sequence on a computer.

As we scan the DNA region containing a disease-causing gene, how do we know when we have actually reached the gene? Coding DNA (i.e., DNA that encodes proteins) must be distinguished from noncoding DNA, and the likely function of each gene in the region must be determined. Several approaches may be used to accomplish these tasks.

## Functional versus Nonfunctional DNA

Most of our DNA sequence has no known function and is unlikely to contribute to disease. Thus, in searching for disease-causing alterations, we typically wish to focus on DNA that encodes proteins or performs important regulatory functions (i.e., enhancer or promoter sequences). Because of their functional significance, coding DNA or regulatory DNA sequences generally cannot change very much through the course of evolution. This means that such DNA sequences will be **conserved**, or similar in base-pair sequence, in many different species. In contrast, nonfunctional DNA sequences are likely to change rapidly and to differ substantially among species. Published DNA sequences can be compared using computer algorithms (see later) to distinguish functional (conserved) DNA from nonfunctional (nonconserved) DNA. In some cases, a labeled probe containing the human DNA segment of interest may be constructed and then exposed to denatured DNA from several other species to determine whether the DNA sequences are similar enough to undergo complementary base pairing with the probe (this is whimsically termed a "zooblot"). If the human DNA sequence is nonconserved and thus likely nonfunctional, complementary base pairing between the probe DNA and that of other species is less likely to occur.

As discussed in Chapter 3, most CG dinucleotides are methylated. However, approximately 60% of human genes have unmethylated CG dinucleotides (**CG islands**) in their 5' region. (A lack of methylation in the 5' region of the gene probably makes it more accessible to transcription factors required for active expression.) The identification of a series of CG islands has often been used to pinpoint the locations of coding genes.

> The identification of DNA sequences that are highly conserved across multiple species, and the identification of unmethylated CG islands, are both ways to distinguish coding or regulatory (functional) DNA from nonfunctional DNA.

## Computer Analysis of DNA Sequence

Computer analysis, termed *in silico* research, has become an effective and popular approach for identifying genes. Sophisticated computer algorithms can test a DNA sequence for patterns that signal a coding gene (e.g., transcription initiation sites, stop codons, intron–exon boundaries). This approach was used, for example, to help identify and characterize one of the adult polycystic kidney disease genes (*PKD1*; see Chapter 4). In addition, these algorithms can often recognize patterns typical of genes that encode specific classes of proteins (e.g., transcription factors, transmembrane proteins).

Computerized databases of known DNA sequences also play an important role in gene identification. When studying a specific region of DNA to find a gene, it is common to search for similarity between DNA sequences from the region and other DNA sequences in the database. The sequences in the database might derive from genes with known function or tissue-specific expression patterns. Suppose, for example, that we have used linkage analysis to identify a region containing a gene that causes a developmental disorder such as a limb malformation. As we evaluate DNA sequences in the region, we would look for similarity between a DNA sequence from this region and a plausible sequence from the database (e.g., a sequence from a gene that encodes a protein involved in bone development, such as a fibroblast growth factor). Because genes that encode similar protein products usually have similar DNA sequences, a match between the sequence from our region and a sequence in the database could be a vital clue that this particular DNA sequence is actually part of the gene that causes the limb malformation.

A computerized database commonly used in these searches consists of small DNA sequences known as **expressed sequence tags (ESTs)**. ESTs are obtained by sequencing several hundred base pairs from both ends of cDNA clones taken from a cDNA library (see Box 8-3). Because these clones consist of DNA that is complementary to mRNA, the ESTs represent expressed portions of genes. Often, they are expressed only in certain tissues at certain points in time. The investigator can search for sequence similarity between DNA sequences in a region of interest and ESTs known to be located in the same region (and possibly known to be expressed in a tissue that would be consistent with the disease in question). This strategy was used, for example, to identify one of the genes known to cause Alzheimer disease (see Chapter 12).

The similarity search need not be confined to human genes. The complete DNA sequences of more than two dozen organisms, including chimpanzee, chicken, mouse, fruit fly, and yeast, are now available in computerized databases. Often, sequence similarity is observed with genes of known function in other organisms, such as mouse or even yeast or bacteria. Because genes with important protein products tend to be highly conserved throughout evolution, the identification of a similar gene in another can provide important information about the gene's function in the human. For example, many of the genes involved in cell cycle regulation are quite similar in yeast and human (e.g., portions of the *NF1* gene and the yeast *IRA2* gene). Indeed, approximately one third of human disease-causing genes identified to date have similar counterparts in yeast. These genes and their products can easily be manipulated in experimental organisms, so their well-understood functions in these organisms can provide useful inferences about their functions in humans. A number of important human disease-causing genes have been discovered because similar candidate genes had previously been identified in other organisms (e.g., the mouse "small eye" gene and aniridia, the mouse "splotch" gene and Waardenburg syndrome, the *Drosophila* and mouse "patched" genes and basal cell nevus syndrome, the mouse "pink eye" gene and oculocutaneous albinism type 2). In addition to providing insight about coding genes, interspecies sequence comparisons can also reveal highly conserved noncoding sequences that contain important regulatory elements. These, too, can contribute to disease risk.

> Many computer databases and algorithms are now used to infer whether DNA sequences are located within genes. The possible function of a gene can be inferred by comparing its sequence with that of other human or nonhuman genes whose functions are known.

### Screen for Mutations in the Sequence

Once a portion of coding DNA has been isolated, it can be examined for disease-causing mutations, usually by direct DNA sequencing (see Chapter 3). If a DNA sequence represents the disease-causing gene, then mutations should be found in individuals with the disease, but they should not be found in unaffected individuals. To help distinguish disease-causing mutations from polymorphisms that vary naturally among individuals, it is particularly useful to compare the DNA of patients whose disease is caused by a new mutation with the DNA of their unaffected parents. While a harmless polymorphism will be seen in both the affected offspring and the unaffected parents, a mutation that is responsible for the disease in the offspring will not be seen in the parents. This approach is especially useful for identifying mutations in genes for highly penetrant autosomal dominant diseases such as *NF1*.

Another type of mutation that can be tested is a submicroscopic deletion (i.e., a deletion too small to be observable under a microscope). Small deletions can be detected using Southern blotting techniques (see Chapter 3). A deletion will produce a smaller-than-normal restriction fragment, which migrates more quickly through a gel. Somewhat larger deletions can be detected through **pulsed-field gel electrophoresis**, a variation on the Southern blotting technique in which restriction digests are done with enzymes that, because their recognition sequences are rare, cleave the DNA infrequently. This produces restriction fragments that can be tens or hundreds of kilobases in length. These fragments are too large to be distinguished from one another using standard gel electrophoresis, but they will migrate differentially according to size when the electrical current is pulsed in alternating directions across the gel. Other techniques, including

fluorescent in situ hybridization (FISH; see Chapter 6), microarray analysis (which can assay single nucleotide polymorphisms or copy number variants), or multiplex ligation-dependent probe amplification (MLPA) (see Chapter 3) can also be used in some cases to detect small deletions.

### Test for Gene Expression

To help verify that a gene is responsible for a given disease, one can test various tissues to determine the ones in which the gene is expressed (i.e., transcribed into mRNA). This can be done by purifying mRNA from the tissue, placing it on a blot, and testing for hybridization with a probe made from the gene. This technique, known as **Northern blotting** (Fig. 8-14), is conceptually similar to Southern blotting except that mRNA, rather than DNA, is being probed. If the gene in question causes the disease, the mRNA may be expressed in tissues known to be affected by the disease

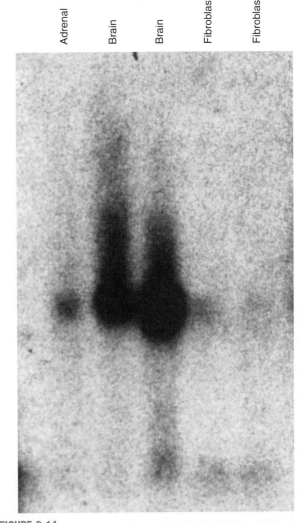

**FIGURE 8-14**
An example of a Northern blot, showing the hybridization of a cDNA probe from the *EVI2A* gene (a gene embedded within an intron of the *NF1* gene) with mRNA from adrenal gland, brain, and fibroblasts. This result indicates that *EVI2A* is expressed in the brain at a much higher level than in the other two tissues.

*(Courtesy Dr. Richard Cawthon, University of Utah Health Sciences Center.)*

(the same reasoning is applied to the analysis of ESTs obtained from tissue-specific cDNA libraries). For example, one would expect the phenylalanine hydroxylase gene (mutations in which can cause phenylketonuria [PKU]) to be expressed in the liver, where this enzyme is known to be synthesized.

Increasingly, microarrays (see Chapter 3) are being used to test for gene expression. Microarrays containing thousands of oligonucleotide sequences representing genes of interest have been manufactured, and these can be exposed to mRNA from relevant tissues to determine the tissues in which specific genes are expressed.

Another gene expression test involves inserting the normal version of the DNA sequence into a defective cell from an affected person (or animal model), using recombinant DNA techniques. If the normal sequence corrects the defect, then it is very likely to represent the gene of interest. This approach has been used, for example, to show that mutations in the *CFTR* gene can cause cystic fibrosis.

> In determining whether a gene contributes to the cause of a disease, a critical test is to screen the gene for mutations that are present in affected persons but absent in unaffected controls. Further evidence that a gene helps to cause a specific disease is to show that mRNA corresponding to that gene is expressed in tissues associated with or affected by the disease.

## Candidate Genes

The gene hunting process can be expedited considerably if a **candidate gene** is available. As the name implies, this is a gene whose known protein product makes it a likely candidate for the disease in question. For example, the various collagen genes were considered to be reasonable candidates for Marfan syndrome because collagen is an important component of connective tissue. However, linkage analysis using collagen gene markers in Marfan syndrome families consistently yielded negative results. Another candidate gene emerged when the gene that encodes Fibrillin-1 (FBN1) was identified on chromosome 15. Fibrillin, as discussed in Chapter 4, is also a connective tissue component. Linkage analysis had localized the Marfan syndrome gene to chromosome 15, so *FBN1* became an even stronger candidate. Analysis of *FBN1* mutations showed that they were consistently associated with Marfan syndrome, confirming these mutations as a cause of the disease. This combination of linkage analysis to identify the region containing a gene followed by a search of the region for plausible candidate genes is sometimes called the **positional candidate** approach.

> Candidate genes are those whose characteristics (e.g., protein product) suggest that they may be responsible for a genetic disease. The analysis of candidate genes in a region known to contain the disease-causing gene is termed the *positional candidate approach.*

Using the techniques described in this chapter, a large number of important disease-causing genes have now been mapped, and many of them have also been cloned. Some examples are given in Table 8-2.

Although the gene identification examples used in this chapter have focused on mendelian conditions, these same approaches are being used to identify genes that contribute to the cause of common, complex diseases such as diabetes, hypertension, and heart disease. Because these diseases are influenced by multiple genes (as discussed in Chapter 12), gene identification tends to be especially challenging. Nevertheless, the techniques discussed in this chapter are now being applied with considerable success in pinpointing genes that help to cause the common diseases that affect the majority of the human population.

**TABLE 8-2**

## Examples of Well-Known Mendelian Disease-Causing Genes that Have Been Mapped and Cloned*

| Disease | Chromosome Location | Gene Product |
|---|---|---|
| $\alpha_1$-Antitrypsin deficiency | 14q32.1 | Serine protease inhibitor |
| α-Thalassemia | 16p13.3 | α-Globin component of hemoglobin |
| β-Thalassemia | 11p15.5 | β-Globin component of hemoglobin |
| Achondroplasia | 4p16.3 | Fibroblast growth factor receptor 3 |
| Albinism, oculocutaneous (type 1) | 11q14-q21 | Tyrosinase |
| Albinism, oculocutaneous (type 2) | 15q11-q12 | Tyrosine transporter |
| Alzheimer disease* (familial) | 14q24.3 | Presenilin 1 |
| | 1q31-q42 | Presenilin 2 |
| | 21q21 | β-Amyloid precursor protein |

**TABLE 8-2**

Examples of Well-Known Mendelian Disease-Causing Genes that Have Been Mapped and Cloned*—cont'd

| Disease | Chromosome Location | Gene Product |
|---|---|---|
| Amyotrophic lateral sclerosis (familial) | 21q22.1 | Superoxide dismutase 1 |
| Angelman syndrome | 15q11-q13 | Ubiquitin-protein ligase E3A |
| Ataxia-telangiectasia | 11q22.3 | Cell cycle control protein |
| Bloom syndrome | 15q26.1 | RecQ helicase |
| Breast cancer (familial) | 17q21 | BRCA1 tumor suppressor/DNA repair protein |
| | 13q12.3 | BRCA2 tumor suppressor/DNA repair protein |
| | 22q12.1 | CHEK2 DNA repair protein |
| Li–Fraumeni syndrome | 17p13.1 | P53 tumor suppressor |
| | 22q12.1 | CHEK2 DNA repair protein |
| Cystic fibrosis | 7q31.2 | Cystic fibrosis transmembrane regulator (CFTR) |
| Duchenne/Becker muscular dystrophy | Xp21.2 | Dystrophin |
| Ehlers–Danlos syndrome* | 2q31 | Collagen (COL3A1); there are numerous types of this disorder, most of which are produced by mutations in collagen genes |
| Fragile X syndrome | Xq27.3 | FMR1 RNA-binding protein |
| Friedreich ataxia | 9q13 | Frataxin mitochondrial protein |
| Galactosemia | 9p13 | Galactose-1-phosphate-uridyltransferase |
| Hemochromatosis (adult)* | 6p21.3 | Transferrin receptor binding protein |
| Hemophilia A | Xq28 | Clotting factor VIII |
| Hemophilia B | Xq27 | Clotting factor IX |
| Hereditary nonpolyposis colorectal cancer | 3p21.3 | MLH1 DNA mismatch repair protein |
| | 2p22-p21 | MSH2 DNA mismatch repair protein |
| | 2q31-q33 | PMS1 DNA mismatch repair protein |
| | 7p22 | PMS2 DNA mismatch repair protein |
| | 2p16 | MSH6 DNA mismatch repair protein |
| | 14q24.3 | MLH3 DNA mismatch repair protein |
| Huntington disease | 4p16.3 | Huntingtin |
| Hypercholesterolemia (familial) | 19p13.2 | LDL receptor |
| LQT1* | 11p15.5 | KCNQ1 cardiac $K^+$ channel $\alpha$ subunit |
| LQT2 | 7q35-q36 | KCNH2 cardiac $K^+$ channel |
| LQT3 | 3p21 | SCN5A cardiac $Na^+$ channel |
| LQT4 | 4q25-q27 | Ankyrin-B |
| LQT5 | 21q22 | KCNE1 cardiac $K^+$ channel $\beta$ subunit |
| LQT6 | 21q22 | KCNE2 cardiac $K^+$ channel |

*Continued*

**TABLE 8-2**

Examples of Well-Known Mendelian Disease-Causing Genes that Have Been Mapped and Cloned*—cont'd

| Disease | Chromosome Location | Gene Product |
|---|---|---|
| Marfan syndrome type 1 | 15q21.1 | Fibrillin-1 |
| Marfan syndrome type 2 | 3p22 | TGF-β receptor type 2 |
| Melanoma (familial)* | 9p21 | Cyclin-dependent kinase inhibitor tumor suppressor |
| | 12q14 | Cyclin-dependent kinase 4 |
| Myotonic dystrophy type 1 | 19q13.2-q13.3 | Protein kinase |
| Myotonic dystrophy type 2 | 3q13.3-q24 | Zinc finger protein |
| Neurofibromatosis type 1 | 17q11.2 | Neurofibromin tumor suppressor |
| Neurofibromatosis type 2 | 22q12.2 | Merlin (schwannomin) tumor suppressor |
| Parkinson disease (familial) | 4q21 | α-Synuclein |
| Parkinson disease (autosomal recessive early-onset) | 6q25.2-q27 | Parkin |
| Phenylketonuria | 12q24.1 | Phenylalanine hydroxylase |
| Polycystic kidney disease | 16p13.3-p13.12 | Polycystin-1 membrane protein |
| | 4q21-q23 | Polycystin-2 membrane protein |
| | 6p21-p12 | Fibrocystin receptor-like protein |
| Polyposis coli (familial) | 5q21-22 | APC tumor suppressor |
| Retinitis pigmentosa* (more than 45 genes cloned to date; representative examples shown here) | 3q21-q24 | Rhodopsin |
| | 11q13 | Rod outer segment membrane protein-1 |
| | 6p21.1 | Peripherin/RDS |
| | 4p16.3 | Retinal rod photoreceptor cGMP phosphodiesterase, β subunit |
| | Xp21.1 | Retinitis pigmentosa GTPase regulator |
| Retinoblastoma | 13q14.1-q14.2 | pRB tumor suppressor |
| Rett syndrome | Xq28 | Methyl CpG-binding protein 2 |
| Sickle cell disease | 11p15.5 | β-Globin component of hemoglobin |
| Tay–Sachs disease | 15q23-q24 | Hexosaminidase A |
| Tuberous sclerosis type 1* | 9q34 | Hamartin tumor suppressor |
| Tuberous sclerosis type 2 | 16p13.3 | Tuberin tumor suppressor |
| Wilson disease | 13q14.3-q21.1 | Copper transporting ATPase |
| von Willebrand disease | 12p13.3 | von Willebrand clotting factor |

*Additional disease-causing loci have been mapped and/or cloned.

APC, adenomatosis polyposis coli; ATPase, adenosine triphosphatase; GMP, guanosine monophosphate; GTPase, guanosine triphosphatase; LQT, long QT [syndrome]; TGF, transforming growth factor.

## Study Questions

1. In Figure 8-15, a pedigree for an autosomal dominant disease, each family member has been typed for a four-allele STRP marker, as shown in the autoradiogram below the pedigree. Determine linkage phase for the disease and marker locus in the affected male in generation II. Based on the meioses that produced the offspring in generation III, what is the recombination frequency for the marker and the disease locus?

2. In the Huntington disease pedigree in Figure 8-16, the family has been typed for two two-allele markers, *A* and *B*. The genotypes for each marker are shown below the symbol for each family member, with the genotype for marker *A* above the genotype for marker *B*. Under the hypothesis that $\theta = 0.0$, what is the LOD score for linkage between each marker locus and the Huntington disease locus?

3. Interpret the following table of LOD scores and recombination frequencies ($\theta$):

| $\theta$ | 0.0 | 0.05 | 0.1 | 0.2 | 0.3 | 0.4 | 0.5 |
|---|---|---|---|---|---|---|---|
| LOD | $-\infty$ | 1.7 | 3.5 | 2.8 | 2.2 | 1.1 | 0.0 |

4. Two pedigrees for an autosomal dominant disease are shown in Figure 8-17. The families have been typed

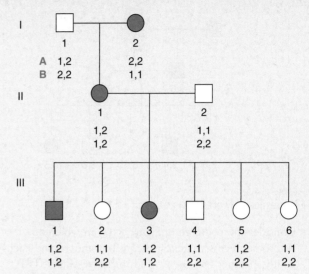

FIGURE 8-16
Pedigree for study question 2.

for a six-allele STRP. Based on the data in these families, what is the recombination frequency between the marker locus and the disease locus? What is the LOD score for linkage between the marker and disease loci under the hypothesis that $\theta = 0.0$?

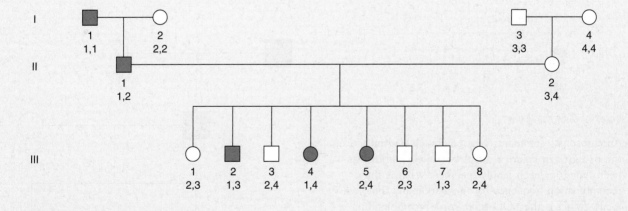

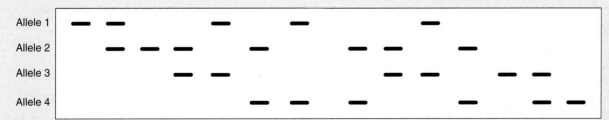

FIGURE 8-15
Pedigree for study question 1.

## Study Questions—cont'd

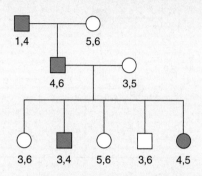

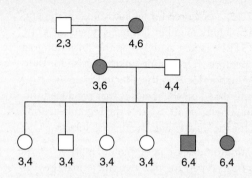

**FIGURE 8-17**
Pedigrees for study question 4.

**5.** Consider the pedigree in Figure 8-18, in which an autosomal recessive gene is being transmitted and each family member has been typed for a five-allele STRP. Carrier status in each individual has been established by an independent enzymatic assay. What is the recombination frequency between the STRP and the disease locus?

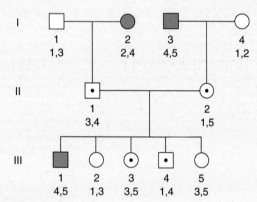

**FIGURE 8-18**
Pedigree for study question 5.

**6.** An autosomal dominant disease is being transmitted in the pedigree in Figure 8-19. A two-allele marker has been typed for each family member. What is the recombination frequency for the marker and the disease locus? What is the LOD score for a recombination frequency of 0.0? What is the LOD score for a recombination frequency of 0.1?

**7.** The family shown in the pedigree in Figure 8-20 has presented for genetic diagnosis. The disease in question is inherited in autosomal dominant fashion. The family members have been typed for a closely linked two-allele marker locus. What can you tell the family about the risks for their offspring?

**8.** Distinguish the differences among the concepts of synteny, linkage, linkage disequilibrium, and association.

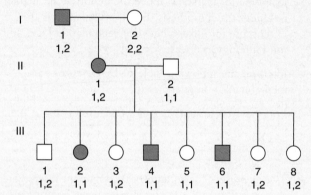

**FIGURE 8-19**
Pedigree for study question 6.

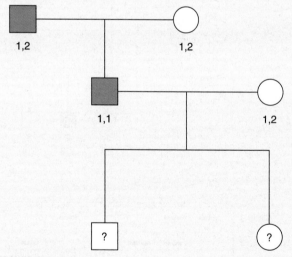

**FIGURE 8-20**
Pedigree for study question 7.

**9.** A published study showed no linkage disequilibrium between *NF1* and a very closely linked marker locus. Explain this, keeping in mind that the *NF1* gene has a high rate of new mutations.

## Suggested Readings

Altshuler D, Daly MJ, Lander ES. Genetic mapping in human disease. Science 2008;322(5903):881–8.

Christensen K, Murray JC. What genome-wide association studies can do for medicine. N Engl J Med 2007;356(11):1094–7.

Collins FS, Morgan M, Patrinos A. The Human Genome Project: Lessons from large-scale biology. Science 2003;300(5617):286–90.

Dawn Teare M, Barrett JH. Genetic linkage studies. Lancet 2005;366(9490):1036–44.

Frazer KA, Ballinger DG, Cox DR, et al. A second generation human haplotype map of over 3.1 million SNPs. Nature 2007;449(7164):851–61.

Hartong DT, Berson EL, Dryja TP. Retinitis pigmentosa. Lancet 2006;368(9549):1795–809.

Hirschhorn JN, Daly MJ. Genome-wide association studies for common diseases and complex traits. Nat Rev Genet 2005;6(2):95–108.

Hopper JL, Bishop DT, Easton DF. Population-based family studies in genetic epidemiology. Lancet 2005;366(9494):1397–406.

International Human Genome Sequencing Consortium. Finishing the euchromatic sequence of the human genome. Nature 2004;431(7011):931–45.

Kennan A, Aherne A, Humphries P. Light in retinitis pigmentosa. Trends Genet 2005;21(2):103–10.

Morton NE. LODs past and present. Genetics 1995;140:7–12.

Ott J. Analysis of Human Genetic Linkage, 3rd ed. Baltimore: Johns Hopkins University Press, 1999.

Palmer LJ, Cardon LR. Shaking the tree: Mapping complex disease genes with linkage disequilibrium. Lancet 2005;366(9492):1223–34.

Pearson TA, Manolio TA. How to interpret a genome-wide association study. JAMA 2008;299(11):1335–44.

Venter JC, Adams MD, Myers EW, et al. The sequence of the human genome. Science 2001;291(5507):1304–51.

## Internet Resources

Catalog of Published Genome-Wide Association Studies *http://www.genome.gov/26525384*

Computational Biology at the Oak Ridge National Laboratory (contains a number of useful links for the analysis of DNA sequences and protein structures) *http://compbio.ornl.gov/*

Ensembl (provides DNA and protein sequences for humans and other organisms, along with descriptive information; includes the BLAST sequence analysis algorithm) *http://www.ensembl.org/*

Genamics (contains hundreds of links for genetic analysis software, including programs for analyzing linkage, linkage disequilibrium, and DNA sequence variation) *http://genamics.com/*

GeneCards (database of human genes and their products, with information about the function of each gene product) *http://bioinfo.weizmann.ac.il/cards/*

International HapMap Project (public database of 1 million SNPs genotyped in 270 individuals from four human populations) *http://www.hapmap.org/*

National Center for Biotechnology Information (maps of chromosomes and disease loci, with links to other useful genomics sites such as SNP databases) *http://www.ncbi.nlm.nih.gov/genome/guide/human/*

UCSC Genome Browser (contains reference sequence for the genomes of many organisms, along with useful tools for sequence analysis) *http://genome.ucsc.edu/*

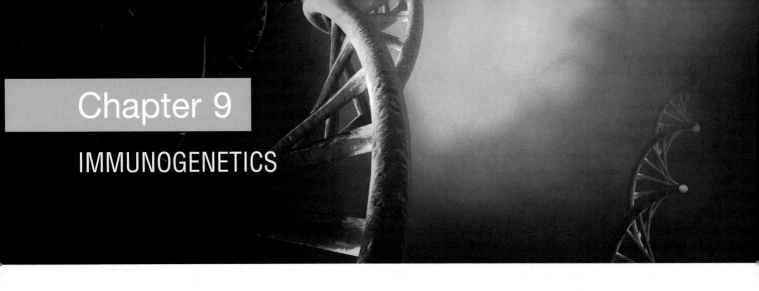

Each day, our bodies are confronted with a formidable series of invaders: viruses, bacteria, and many other disease-causing organisms whose goal is to overcome our natural defenses. These defenses, known collectively as the immune system, consist of a diverse collection of trillions of cells. The immune system must be able to cope with a multitude of invading microorganisms, and it must be able to distinguish "self" from "nonself" with a high degree of accuracy.

As one might expect, the genetic basis of the immune system is complex. The study of the genetics of the immune system, known as **immunogenetics**, has benefited enormously from new developments in gene mapping and cloning. Most of the techniques discussed in earlier chapters (e.g., linkage analysis, positional cloning, DNA sequencing) have been used to study genes responsible for the immune response. Many new genes have been discovered, and their functions and interactions have been studied intensely. This chapter provides a brief review of basic immunology and discusses the genes underlying the body's capacity to defend against a highly diverse array of pathogens. Aspects of autoimmune disease are examined, and some of the major immunodeficiency disorders are discussed.

## THE IMMUNE RESPONSE: BASIC CONCEPTS

### The Innate Immune System

When a pathogenic microorganism is encountered, the body's first line of defense includes **phagocytes** (a type of cell that engulfs and destroys the microorganism) and the **complement system**. The complement proteins can destroy microbes directly by perforating their cell membranes, and they can also attract phagocytes and other immune system agents to microbes by coating the microbial surface (it is because of this assisting role that the term *complement* originated). **Natural killer cells**, a specific type of lymphocyte, can respond to certain viral infections and some tumor cells. Phagocytes, complement, and natural killer cells are all part of the **innate immune system**, which is capable of responding very rapidly to pathogens.

The innate immune system is activated by general features that are detected in pathogens but are not found in the host. For example, gram-negative bacteria produce lipopolysaccharides, and gram-positive bacteria produce peptidoglycans.

Some bacteria have a high percentage of unmethylated CG sequences, and some viruses produce double-stranded RNA. These distinctive features of pathogenic organisms can be detected by receptor molecules located on the surfaces of innate immune system cells. An important example is the Toll-like family of receptors, named after a cell-surface receptor, Toll, that was first described in fruit flies. The genes that encode the human and fruit-fly versions of Toll-like receptors have a remarkable similarity to one another, attesting to their importance in maintaining an innate immune response in a wide variety of organisms. Indeed, all multicellular organisms are thought to possess innate immune systems.

> The innate immune system, which includes some phagocytes, natural killer cells, and the complement system, is an early part of the immune response and recognizes general features of invading microorganisms.

### The Adaptive Immune System

Although the innate immune system typically helps to hold an infection in check in its early phases, it is sometimes incapable of overcoming the infection. This becomes the task of a more specialized component of the immune response, the **adaptive immune system**. As its name suggests, this part of the immune system is capable of changing, or adapting to features of the invading microorganism in order to mount a more specific and more effective immune response. The adaptive immune system is a more recent evolutionary development than the innate immune system and is found only in vertebrates.

Key components of the adaptive immune response (Fig. 9-1) include **T lymphocytes** (or **T cells**) and **B lymphocytes** (or **B cells**). These cells develop in the body's primary lymphoid organs (bone marrow for B cells and the thymus for T cells). In the thymus, developing T cells are exposed to a wide variety of the body's peptides. Those that can recognize and tolerate the body's own peptides are selected, and those that would attack the body's peptides are eliminated. The B and T cells progress to secondary lymphoid tissues, such as the lymph nodes, spleen, and tonsils, where they encounter disease-causing microorganisms. Mature B

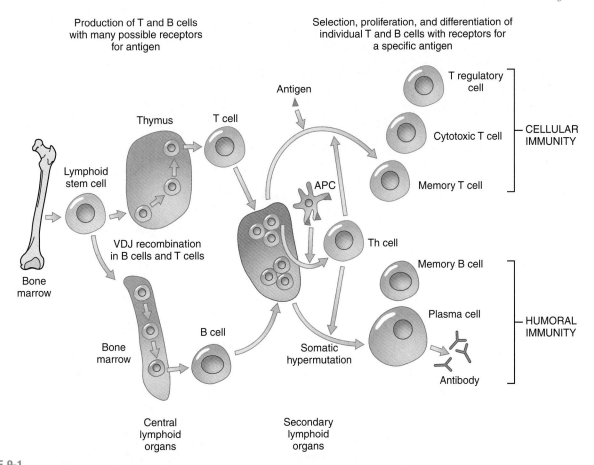

Production of T and B cells with many possible receptors for antigen

Selection, proliferation, and differentiation of individual T and B cells with receptors for a specific antigen

CELLULAR IMMUNITY

HUMORAL IMMUNITY

Bone marrow

Lymphoid stem cell

Thymus

T cell

Antigen

T regulatory cell

Cytotoxic T cell

Memory T cell

VDJ recombination in B cells and T cells

Bone marrow

B cell

APC

Th cell

Memory B cell

Plasma cell

Somatic hypermutation

Antibody

Central lymphoid organs

Secondary lymphoid organs

**FIGURE 9-1**

Overview of the adaptive immune response. Lymphoid stem cells from the bone marrow migrate to the central lymphoid organs, where they undergo cellular division and differentiation, producing either T cells (thymus) or B cells (bone marrow). At this stage, the B and T cells have not yet encountered antigens, but they have developed considerable diversity in their cell-surface receptors as a result of VDJ recombination and junctional diversity (see text). The cells enter the circulation and migrate to the secondary lymphoid organs (e.g., the spleen and lymph nodes), where they will encounter foreign antigens that are usually processed by antigen-presenting cells (APCs) for presentation to helper T (T$_H$) cells. Only a small subset of T and B cells have receptors capable of binding to a specific foreign antigen, and this subset is selected for further development and differentiation. Somatic hypermutation (see text) occurs in B cells at this stage and results in further receptor diversity and the ability to bind the foreign antigen with greater affinity. The end result of this process is a humoral (B cell) and/or cellular (T cell) response to a foreign antigen. The T cell response includes cytotoxic T cells that can kill infected cells, regulatory T cells that help to control the immune response, and memory T cells that enable a rapid response if the antigen is encountered later in life. Humoral immunity results in a population of mature B cells (plasma cells) that secrete antibodies into the circulation and a population of memory B cells.

lymphocytes secrete circulating **antibodies**, which combat infections. The B lymphocyte component of the immune system is sometimes called the **humoral immune system** because it produces antibodies that circulate in the blood stream. **Helper T lymphocytes** stimulate B lymphocytes and other types of T lymphocytes to respond to infections more effectively, and **cytotoxic T lymphocytes** can directly kill infected cells. Because of this direct interaction with infected cells, the T-cell component of the immune system is sometimes called the **cellular immune system**. It is estimated that the body contains several trillion B and T cells.

> B lymphocytes are a component of the adaptive immune system; they produce circulating antibodies in response to infection. T lymphocytes, another component of the adaptive immune system, interact directly with infected cells to kill these cells, and they aid in the B cell response.

## The B Cell Response: Humoral Immune System

A major element of the adaptive immune response begins when specialized types of phagocytes, which are part of the innate immune system, engulf invading microbes and then present peptides derived from these microbes on their cell surfaces. These cells, which include **macrophages** and **dendritic cells**, are termed **antigen-presenting cells (APCs)**. B cells are also capable of engulfing microbes and presenting foreign peptides on their cell surfaces.

The APCs alert the adaptive immune system to the presence of pathogens in two ways. First, the foreign peptide is transported to the surface of the APC by a **class II major histocompatibility complex (MHC)** molecule, which carries the foreign peptide in a specialized groove (Fig. 9-2). This complex, which projects into the extracellular environment, is recognized by T lymphocytes, which have receptors on their surfaces that are capable of binding to the MHC–peptide complex. In addition, the APCs, upon encountering

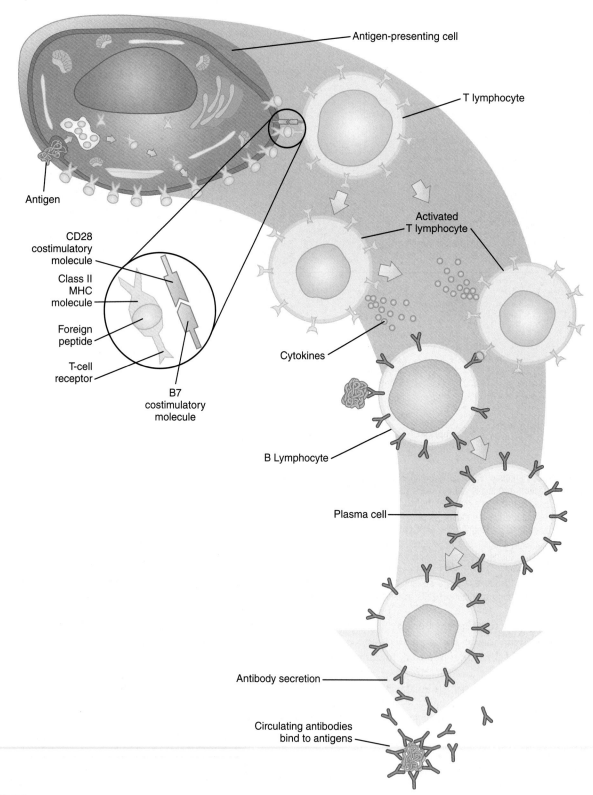

**FIGURE 9-2**
The humoral immune response. Class II MHC molecules in antigen-presenting cells carry foreign peptides to the surface of the cell, where the foreign peptide is recognized by a helper T cell. The T cell secretes cytokines, which stimulate B cells whose immunoglobulins will bind to the foreign peptide. These B cells become plasma cells, which secrete antibodies into the circulation to bind with the microbe, helping to combat the infection.
*(Modified from Nossal GJ: Life, death and the immune system. Sci Am 1993;269:53-62.)*

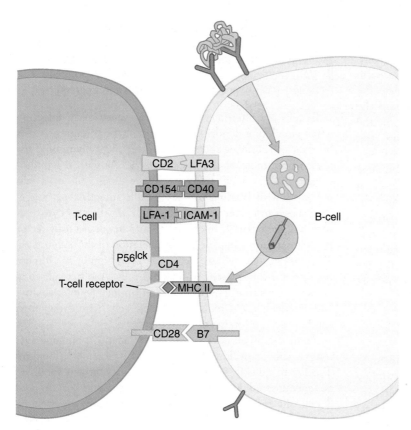

**FIGURE 9-3**
A detailed view of the binding between a helper T cell and a B cell. In addition to the binding of the T-cell receptor to the MHC–peptide complex, a number of other molecules interact with one another, such as the costimulatory B7–CD28 complex. MHC, major histocompatibility complex.
*(Modified from Roitt I, Brostoff J, Male D: Immunology, 6th ed. St. Louis: Mosby, 2001.)*

a pathogen, display **costimulatory molecules** on their cell surfaces as a signal that foreign pathogens have been encountered (Fig. 9-3). Binding to the MHC-peptide complex stimulates the helper T lymphocyte to secrete **cytokines**, which are signaling proteins that mediate communication between cells. In particular, these cytokines help to stimulate the subset of B lymphocytes whose cell surface receptors, termed **immunoglobulins**, can bind to the invading microorganism's peptides (see Fig. 9-3). The immunoglobulin's capacity to bind a specific foreign peptide (i.e., its affinity for the peptide) is determined by its shape and other characteristics.

> In the humoral immune response, foreign particles are displayed in conjunction with class II MHC molecules by antigen-presenting cells. These displayed molecules are recognized by helper T cells, which then stimulate proliferation of B cells whose receptors (immunoglobulins) can bind to the foreign pathogen.

It is estimated that upon initial exposure to a foreign microbe, as few as 1 in every 1 million B lymphocytes happens to produce cell-surface receptors capable of binding to the microbe. This number is too small to fight an infection effectively. Furthermore, the receptor's binding affinity is likely to be relatively poor. However, once this relatively small population of B lymphocytes is stimulated, they begin

an adaptive process in which additional DNA sequence variation is generated via the process of **somatic hypermutation** (see later discussion). These DNA mutations, which are confined to the genes that encode the cell-surface receptors, in turn produce alterations in the receptors' binding characteristics (e.g., the shape of the protein). Some of these variant receptors possess a higher level of binding affinity for the microorganism. The B cells that produce these receptors are favorably selected because they bind the pathogen for a longer period of time. They thus proliferate rapidly. These B cells then become **plasma cells**, which secrete their cell-surface receptors, or immunoglobulins, into the blood stream. The secreted molecules, which are structurally identical to the receptors on the B cell's surface, are antibodies. It can now be seen how the adaptive immune system gets its name: It involves the initial selection of B cells and T cells whose receptors can bind with the pathogen, followed by subsequent fine-tuning (adaptation) of these cells to achieve higher binding affinity.

> During the B-cell response to a foreign peptide, the binding affinity of immunoglobulins for the invading pathogen increases. When mature, the B cell becomes an antibody-secreting plasma cell.

After initial stimulation by the disease pathogen, the process of B-cell differentiation and maturation into antibody-producing plasma cells requires about 5 to 7 days for completion. Each plasma cell is capable of secreting approximately 10 million antibody molecules per hour. Antibodies bind to the pathogen's surface **antigens** (a term derived from "*antibody generating*") and may neutralize the microorganism directly. More often, the antibody tags the pathogen for destruction by other components of the immune system, such as complement proteins and phagocytes.

Another important activity of the humoral immune response is the creation of **memory B cells**, a subset of high-affinity-binding B cells that persist in the body after the infection has subsided. These cells, which have already been highly selected for response to the pathogen, provide a more rapid response should the pathogen be encountered again later in the individual's life. Vaccinations are effective because they induce the formation of memory cells that can respond to a specific pathogen.

## The Cellular Immune System

Some microorganisms, such as viruses, are very adept at quickly inserting themselves into the body's cells. Here, they are inaccessible to antibodies, which are water-soluble proteins that cannot pass through the cell's lipid membrane. A second component of the adaptive immune system, the cellular immune system, has evolved to combat such infections. A key member of the cellular immune response is the **class I MHC molecule**, which is found on the surfaces of nearly all of the body's cells. In a normal cell, the class I MHC molecule binds with small peptides (8 to 10 amino acids long) derived from the interior of the cell. It migrates to the cell's surface, carrying the peptide with it and displaying it outside the cell. Because this is one of the body's own peptides, no immune response is elicited. In an infected cell, however, the class I MHC molecule can bind to small peptides that are derived from the infecting organism. Cell-surface presentation of foreign peptides by the class I MHC molecule alerts the immune system, T cells in particular. Recall that T lymphocytes learn to recognize and tolerate self peptides (in conjunction with MHC molecules) while developing in the thymus, but they are highly intolerant of foreign peptides. The MHC-peptide complex binds to receptors on the appropriate T cell's surface, which prompts the T cell to emit a chemical that destroys the infected cell (Fig. 9-4). Because of their ability to destroy cells in this way, these T lymphocytes are termed **cytotoxic T lymphocytes** or **killer T lymphocytes**.* Each cytotoxic T lymphocyte can destroy one infected cell every 1 to 5 minutes.

> The immune system is capable of destroying the body's cells once they are infected. Peptides from the pathogen are displayed on cell surfaces by class I MHC molecules. These are recognized by cytotoxic (killer) T lymphocytes, which destroy the infected cell.

T cells are often alerted to the presence of an infection when circulating dendritic cells present the foreign peptides on their cell surfaces and migrate to secondary lymphoid tissues, where most of the T cells reside. As with B lymphocytes, only a small fraction of the body's T cells has receptors with binding affinity for the infecting pathogen. One subset of helper T cells, labeled $T_H1$, secretes cytokines such as interleukin-2, interferon-$\gamma$, and tumor necrosis factor $\beta$, which in turn stimulate the proliferation of those cytotoxic T lymphocytes whose receptors can bind to foreign peptides on the surfaces of infected cells. Another subset of helper T cells, the $T_H2$ cells, secrete primarily interleukin-4 and interleukin-5, which stimulate B cells whose receptors can bind to the foreign peptide.

As in the B-cell component of the adaptive immune system, a subset of long-lived T cells is retained (**memory T cells**) to quickly respond to a foreign pathogen should it be encountered again in the future. Yet another type of T cell, the **regulatory T cell**, helps to regulate the immune system so that self peptides are not inadvertently attacked.

> $T_H1$ cells are a subset of helper T cells that stimulate appropriate cytotoxic T cells to respond to an infection, and $T_H2$ cells are a subset of helper T cells that stimulate B cells to respond to infection. Memory T cells remain long after an infection and ensure rapid response to a subsequent infection by the same pathogen. Regulatory T cells help to prevent the immune system from attacking the body's own cells.

## The Innate, Humoral, and Cellular Immune Systems: A Comparison

Although the innate and adaptive immune systems are described separately, a great deal of interaction takes place between them, and the two systems fulfill complementary functions. The innate system, because it recognizes general features of pathogens, can react very quickly to foreign elements. While doing so, it signals the adaptive immune system to initiate a fine-tuned response to the pathogen. Without this signal, the adaptive immune system is incapable of responding to an infection. After several days during which the adaptive system "learns" the characteristics of the pathogen, it can launch a massive, specialized response. Through the creation of memory B and T cells, the adaptive immune system allows the organism to respond quickly and effectively to a pathogen should it be encountered again. No such memory cells exist for the innate immune system.

The humoral immune system is specialized to combat extracellular infections, such as circulating bacteria and viruses. The cellular immune system combats intracellular

---

*They are also known as CD8 T cells because of the presence of CD8 (cluster of differentiation antigen 8) coreceptors on their surfaces, which bind to the class I MHC molecule (see Fig. 9-4). Helper T cells have CD4 coreceptors on their cell surfaces, which bind to class II MHC molecules (see Fig. 9-3) and are thus known as CD4 T cells.

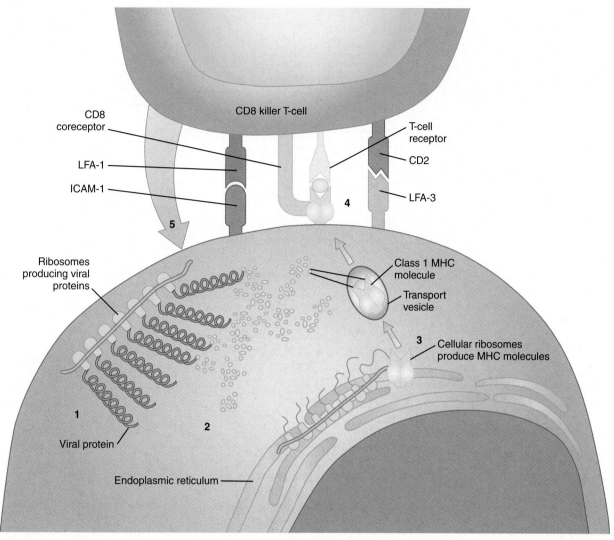

**FIGURE 9-4**

In a cell infected by a virus, the viral peptides *(1, 2)* are carried to the cell surface by class I MHC molecules *(3)*. The T-cell receptor of a CD8 cytotoxic T cell binds to the peptide–MHC complex *(4)*. Recognizing the peptide as foreign, the cytotoxic T cell secretes chemicals that either kill the infected cell directly or induce it to undergo programmed cell death (apoptosis; see Chapter 11) *(5)*. MHC, major histocompatibility complex.

*(Modified from Janeway CA Jr: How the immune system recognizes invaders. Sci Am 1996;269:73-79.)*

infections, such as parasites and viruses within cells. However, this division of labor is not strict, and there is again a great deal of interaction between the humoral and cellular components of the immune system.

## IMMUNE RESPONSE PROTEINS: GENETIC BASIS OF STRUCTURE AND DIVERSITY

### Immunoglobulin Molecules and Genes

As illustrated in Figure 9-5, each antibody (or immunoglobulin) molecule is composed of four chains: an identical pair of longer **heavy chains** and an identical pair of shorter **light chains**, which are linked together by disulfide bonds. There are five different types of heavy chains (termed $\gamma$, $\mu$, $\alpha$, $\delta$, and $\varepsilon$) and two types of light chains ($\kappa$ and $\lambda$). The five types of heavy chains determine the major **class** (or **isotype**) to which an immunoglobulin (Ig) molecule belongs: $\gamma$, $\mu$, $\alpha$,

$\delta$, and $\varepsilon$ correspond to the immunoglobulin isotypes IgG, IgM, IgA, IgD, and IgE, respectively. Immature B lymphocytes produce only IgM, but as they mature, a rearrangement of heavy chain genes called **class switching** occurs. This produces the other four major classes of immunoglobulins, each of which differs in amino acid composition, charge, size, and carbohydrate content. Each class tends to be localized in certain parts of the body, and each tends to respond to a different type of infection. The two types of light chains can be found in association with any of the five types of heavy chains.

The heavy and light chains both contain a **constant** and a **variable** region, which are located at the carboxyl (C)-terminal and amino (N)-terminal ends of the chains, respectively. The arrangement of genes encoding the constant region determines the major class of the Ig molecule (e.g., IgA, IgE). The variable region is responsible for antigen

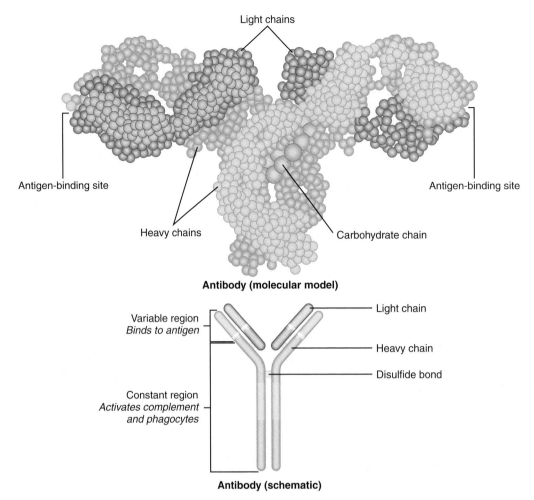

Light chains

Antigen-binding site

Antigen-binding site

Heavy chains

Carbohydrate chain

**Antibody (molecular model)**

Variable region
*Binds to antigen*

Light chain

Heavy chain

Disulfide bond

Constant region
*Activates complement
and phagocytes*

**Antibody (schematic)**

**FIGURE 9-5**
An antibody molecule consists of two identical light chains and two identical heavy chains. The light chain includes variable, joining, and constant regions; the heavy chain includes these regions as well as a diversity region located between its variable and joining regions. The *upper portion* of the figure depicts a molecular model of antibody structure.

recognition and binding and thus varies within immunoglobulin classes. Three distinct gene segments encode the light chains: C for the constant region, V for the variable region, and J for the region that joins the constant and variable regions. Four gene segments encode the heavy chains, with C, V, and J coding again for the constant, variable, and joining regions, respectively, and a "diversity" (D) region located between the joining and variable regions.

> Immunoglobulin molecules consist of two identical heavy chains and two identical light chains. The heavy chain constant region determines the major class to which an immunoglobulin belongs. The variable regions of the light and heavy chains recognize and bind antigens.

### The Genetic Basis of Antibody Diversity

Because the immune system cannot "know" in advance what types of microbes it will encounter, the system must contain a huge reservoir of structurally diverse immune cells so that

at least a few cells can respond (i.e., bind) to any invading microbe. Indeed, the humoral immune system is capable of generating at least 10 billion structurally distinct antibodies. At one time, it was thought that because each antibody has a unique amino acid sequence, each must be encoded by a different gene. However, this one gene–one antibody hypothesis could not possibly be correct, because the human genome has only 20,000 to 25,000 genes. Further study has shown that several mechanisms are responsible for generating antibody diversity in somatic cells:

### 1. Multiple Germline Immunoglobulin Genes

Molecular genetic studies (cloning and DNA sequencing) have shown that for each heavy and light chain, an individual has more than 80 different V segments located contiguously in his or her germline and six different J segments. There are at least 30 D segments in the heavy chain region.

### 2. Somatic Recombination (VDJ Recombination)

As immunoglobulin molecules are formed during B lymphocyte maturation, a specific combination of single V and J segments is selected for the light chain, and another combination of V, D, and J segments is selected for the heavy chain. This is accomplished by deleting the DNA sequences separating the single V, J, and D segments before they are transcribed into mRNA (Fig. 9-6). The deletion process is carried out in part by **recombinases** (encoded by the *RAG1* and *RAG2* genes), which initiate double-strand DNA breaks at specific DNA sequences that flank the V

and D gene segments. After the deletion of all but one V, D, and J segment, the nondeleted segments are joined by ligases. This cutting-and-pasting process is known as **somatic recombination** (in contrast to the germline recombination that takes place during meiosis). Somatic recombination produces a distinctive result: Unlike most other cells of the body, whose DNA sequences are identical to one another, mature B lymphocytes *vary* in terms of their rearranged immunoglobulin DNA sequences. Because there are many possible combinations of single V, J, and D segments, somatic recombination can generate approximately 100,000 to 1,000,000 different types of antibody molecules.

### 3. Junctional Diversity

As the V, D, and J regions are assembled, slight variations occur in the position at which they are joined, and small numbers of nucleotides may be deleted or inserted at the junctions joining the regions. This creates even more variation in antibody amino acid sequence.

### 4. Somatic Hypermutation

Typically, only a small subset of B cells has cell-surface receptors (immunoglobulins) that can bind to a specific foreign antigen, and their binding affinity is usually low. Once this subset of B cells is stimulated by a foreign antigen, they undergo an **affinity maturation** process characterized by somatic hypermutation of the V segments of immunoglobulin genes, as mentioned previously. An enzyme termed *activation-induced deaminase* causes cytosine bases to be replaced by uracil. Error-prone DNA polymerases are recruited, and DNA repair processes are modified so that mutations can persist in the DNA sequence. Consequently, the mutation rate of these gene segments is approximately $10^{-3}$ per base pair per generation (recall that the mutation rate in the human genome is normally only $10^{-9}$ per base pair per generation). This causes much additional variation in immunoglobulin-encoding DNA sequences and thus in the antigen-binding properties of the encoded immunoglobulins. Because mutation is a random process, most of the new receptors have poor binding affinity and are thus not selected. Eventually, however, somatic hypermutation produces a subset of immunoglobulins that have high-affinity binding to the foreign antigen, and the B cells that harbor these immunoglobulins are selected to proliferate extensively. The end result is a population of mature plasma cells that secrete antibodies that are highly specific to the invading pathogen.

### 5. Multiple Combinations of Heavy and Light Chains

Further diversity is created by the random combination of different heavy and light chains in assembling the immunoglobulin molecule.

Each of these mechanisms contributes to antibody diversity. Considering all of them together, it has been estimated that as many as $10^{10}$ to $10^{14}$ distinct antibodies can potentially be produced.

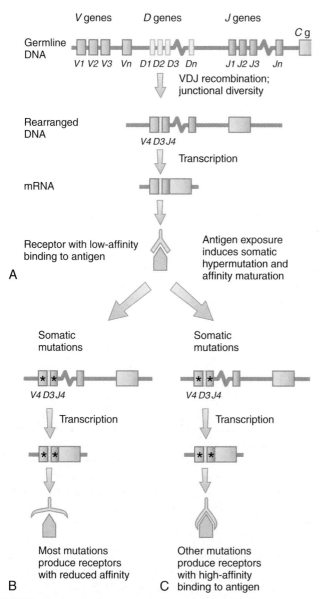

**FIGURE 9-6**

**A,** Somatic VDJ recombination in the formation of a heavy chain of an antibody molecule. The functional heavy chain is encoded by only one segment each from the multiple V, D, and J segments. This produces a subset of B cells whose receptors have low-affinity binding for a foreign antigen. Once the antigen is encountered in the secondary lymphoid tissue, somatic hypermutation (**B** and **C**) is initiated. Most of the mutated receptors have little binding affinity (**B**), but eventually somatic hypermutation produces a subset of receptors with high-affinity binding (**C**). The cells that contain these receptors become antibody-secreting plasma cells.

**FIGURE 9-7**
The T-cell receptor is a heterodimer that consists of either an α and a β chain or a γ and a δ chain. The complex of MHC molecule and antigen molecule is bound by the variable regions of the α and β chains.
*(Modified from Raven PH, Johnson GB: Biology, 3rd ed. St. Louis: Mosby, 1992.)*

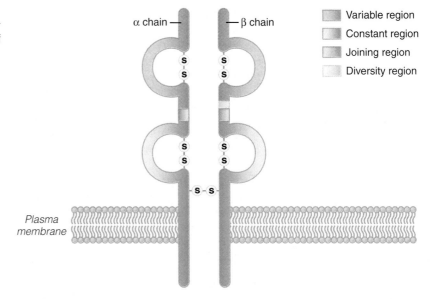

Mechanisms that produce antibody diversity include multiple germline immunoglobulin gene segments, somatic recombination of the immunoglobulin gene segments, junctional diversity, somatic hypermutation, and the potential for multiple combinations of heavy and light chains.

## T-Cell Receptors

The T-cell receptors are similar in many ways to the immunoglobulins, or B-cell receptors. Like the immunoglobulins, T-cell receptors must be able to bind to a large variety of peptides derived from invading organisms. Unlike immunoglobulins, however, T-cell receptors are never secreted from the cell, and T-cell activation requires the presentation of foreign peptide along with an MHC molecule. Approximately 90% of T-cell receptors are heterodimers composed of an α and a β chain, and approximately 10% are heterodimers composed of a γ and a δ chain (Fig. 9-7). A given T cell has a population of either α-β receptors or γ-δ receptors.

Most of the mechanisms involved in generating immunoglobulin diversity—multiple germline gene segments, VDJ somatic recombination, and junctional diversity—are also important in generating T-cell receptor diversity. However, somatic hypermutation does not occur in the genes that encode the T-cell receptors. This is thought to help avoid the generation of T cells that would react against the body's own cells (an autoimmune response, which is discussed later in the chapter).

T-cell receptors are similar in function to B-cell receptors (immunoglobulins). Unlike immunoglobulins, however, they can bind to foreign peptide only when it is presented by an MHC molecule. Their diversity is created by the same mechanisms that produce immunoglobulin diversity, with the exception of somatic hypermutation.

## THE MAJOR HISTOCOMPATIBILITY COMPLEX

### Class I, II, and III Genes

The MHC, which includes a series of more than 200 genes that lie in a 4-Mb region on the short arm of chromosome 6 (Fig. 9-8), is commonly classified into three groups: class I, class II, and class III. The class I MHC molecule forms a complex with foreign peptides that is recognized by receptors on the surfaces of cytotoxic T lymphocytes. Class I presentation is thus essential for the cytotoxic T cell response. Some viruses evade cytotoxic T cell detection by down-regulating the expression of MHC class I genes in the cells they infect.

Class I MHC molecules are composed of a single heavy glycoprotein chain and a single light chain called β₂-microglobulin (Fig. 9-9A). The most important of the class I loci are labeled **human leukocyte antigens** A, B, and C (**HLA-A,**

**FIGURE 9-8**
A map of the human major histocompatibility complex. The 4-Mb complex is divided into three regions: classes I, II, and III.

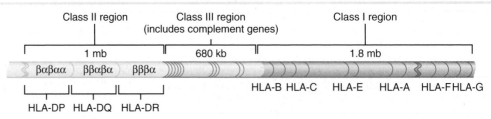

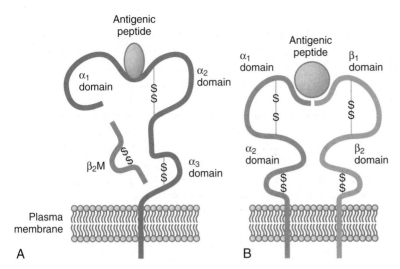

**FIGURE 9-9**
**A,** A class I major histocompatibility complex (MHC) molecule, showing the structure of the heavy chain, which consists of three extracellular domains ($\alpha_1$, $\alpha_2$, and $\alpha_3$), a membrane-spanning domain, and a cytoplasmic domain. A groove formed by the $\alpha_1$ and $\alpha_2$ domains carries peptide for presentation to T-cell receptors. The $\alpha_3$ domain associates closely with the $\beta_2$-microglobulin ($\beta_2$M) chain. **B,** A class II MHC molecule, showing the structure of the $\alpha$ and $\beta$ chains. Each has two globular extracellular domains, a membrane-spanning domain, and a cytoplasmic domain. The $\alpha_1$ and $\beta_1$ domains form a groove into which peptide nestles for presentation to T-cell receptors.
*(Modified from Huether SE, McCance KL: Pathophysiology: The Biological Basis for Disease in Adults and Children, 5th ed 5. St. Louis: Mosby, 2006, p 217.)*

-B, and -C).* Each of these loci has dozens or hundreds of alleles, resulting in a high degree of class I MHC variability among individuals. The class I region spans 1.8 Mb and includes a number of additional genes and **pseudogenes** (genes that are similar in DNA sequence to coding genes but that have been altered so that they cannot be transcribed or translated).

The class I molecules were first discovered in the 1940s by scientists who were experimenting with tissue grafts in mice. When the class I alleles in donor and recipient mice differed, the grafts were rejected. This is the historical basis for the term *major histocompatibility complex.* In humans, matching of the donor's and recipient's class I alleles increases the probability of graft or transplant tolerance. Because grafts and transplants are a relatively new phenomenon in human history, the MHC obviously did not evolve to effect transplant rejection. Instead, T cells, when confronted with foreign MHC molecules on donor cells, interpret these as foreign peptides and attack the cells.

> The class I MHC molecules are encoded by the highly polymorphic *HLA-A, -B,* and *-C* loci on chromosome 6. In addition to presenting foreign peptides on the surfaces of infected cells, they can also bring about transplant rejection when non-self MHC molecules stimulate cytotoxic T cells.

Whereas the class I MHC molecules are found on the surfaces of nearly all cells and can bind with cytotoxic T-cell receptors, the class II MHC molecules ordinarily are found only on the surfaces of the immune system's APCs (e.g., phagocytes and B lymphocytes). When associated with foreign peptides, they stimulate helper T cell activity after binding to the T cells' receptors, as described previously. The class II molecules are heterodimers consisting of an $\alpha$ and a $\beta$ chain, each of which is encoded by a different gene located

on chromosome 6 (see Fig. 9-9B). In addition to the genes in the major class II groups (*HLA-DP, -DQ,* and *-DR*), this region includes genes that encode peptide transporter proteins (*TAP1* and *TAP2*) that help to transport peptides into the endoplasmic reticulum, where they initially form complexes with class I molecules before migrating to the cell surface.

> Class II MHC molecules are heterodimers encoded by genes on chromosome 6. They present peptides on the surfaces of antigen-presenting cells. These peptides, in conjunction with class II MHC molecules, are bound by receptors on helper T cells.

Like the major class I MHC loci, the major class II loci are highly polymorphic, expressing hundreds of different alleles. Indeed, the MHC loci are, as a class, the most polymorphic loci known in humans. Each MHC allele encodes a molecule with slightly different binding properties: Some variants bind peptide from a given pathogen more effectively than others do.[†] Consequently, a person who expresses a greater variety of MHC molecules has a better chance of dealing effectively with a variety of infectious organisms. For example, someone who is homozygous for each of the major class I loci (*A, B,* and *C*) expresses only three different class I MHC molecules in each cell, whereas someone who is heterozygous for each of these loci expresses six different class I MHC molecules in each cell and can cope more successfully with pathogenic diversity (many thousands of MHC molecules are expressed on a typical cell's surface). A higher degree of polymorphism in the general population increases the chance that any individual in the population is heterozygous. For example, HIV-infected persons who are heterozygous for the *HLA-A, HLA-B,* and/or *HLA-C* loci have longer survival times than

---

*These molecules were named "human leukocyte antigens" because they were seen in early studies on leukocyte (white cell) surfaces. However, as mentioned previously, they are found on the surfaces of nearly all cells.

[†]Because of the relatively limited repertoire of different MHC molecules, the peptide-binding affinity of MHC molecules is usually much lower than the binding affinity of finely tuned T-cell receptors.

those who are homozygous at these loci. In addition, greater MHC polymorphism in a population decreases the chance that an infectious pathogen can spread easily through the population. Thus, the high degree of polymorphism in MHC genes is thought to be the result of natural selection for allelic variability.

In some cases, specific MHC alleles are known to produce proteins that are effective against specific pathogens. For example, the *HLA-B53* allele was shown to have a strong protective effect against severe malaria in the population of Gambia, and the *HLA-DRB1*1302* allele protects against hepatitis B infection in the same population. These alleles produce MHC molecules that have higher-affinity binding of the infectious agents.

> Like the class I MHC genes, class II genes are highly polymorphic. This increases the ability of individuals and populations to respond to a wide variety of pathogens.

Both class I and class II MHC molecules guide T-cell receptors (cytotoxic and helper, respectively) to specific cells. T-cell receptors recognize peptides only in combination with MHC molecules on cell surfaces, a phenomenon known as **MHC restriction**. Not all components of the immune system are MHC restricted. The complement system, for example, does not require direct interaction with MHC molecules.

Some virus-infected cells and tumor cells take advantage of MHC restriction: they suppress the expression of MHC molecules on their surfaces in an attempt to evade detection by T cells (Clinical Commentary 9-1). Fortunately, natural killer cells are activated by the absence, rather than the presence, of MHC molecules on cell surfaces. This activation is mediated by an important family of receptors found on the surfaces of natural killer cells, **killer cell immunoglobulin-like receptors (KIR)**. These receptors inhibit natural killer cells when they bind to MHC class I molecules on the surfaces of normal cells but activate them when MHC class I molecules are absent.

## CLINICAL COMMENTARY   9-1
### *The Immune Response as a Molecular Arms Race*

The vast majority of the pathogens that assault the human body are destroyed by our immune system. Consequently, there is strong natural selection for pathogens that can evade immune surveillance and destruction. These microorganisms often have high mutation rates, and their numbers are huge. Thus, despite their biological simplicity, viruses and other pathogens have evolved some very clever ways of overcoming the immune response. Our immune systems, in turn, are constantly creating new ways to deal with pathogenic ingenuity. Three examples of this molecular "arms race" are discussed here.

Cytomegalovirus (CMV) is a common infectious agent that can produce mononucleosis, hemolytic anemia, pneumonitis, congenital infections, and thrombocytopenia (a decrease in the number of platelets). Cells infected with CMV are the targets of destruction by cytotoxic T cells. However, some CMV strains (as well as other viruses and tumor cells) can evade T-cell detection by down-regulating the expression of class I molecules on the surfaces of infected cells. Without viral peptide presentation by class I molecules, the cytotoxic T cells are blind to the presence of CMV, and they do not destroy the infected cell. At this point, the cell would normally become the target of natural killer cells, which attack cells that lack MHC class I molecules on their surfaces. But the CMV has devised a way to outwit natural killer cells, too. The virus encodes a cell-surface protein that resembles class I molecules closely enough so that the natural killer cells mistake the viral protein for a true class I molecule. The viral protein is also sufficiently *different* from a true class I MHC molecule that it does not trigger destruction by the more finely tuned cytotoxic T cell. In this way, the CMV can avoid destruction by both T cells and natural killer cells.

Pregnancy presents an interesting immunological challenge in that placental cells express foreign class I MHC molecules derived from the father. Ordinarily, such cells would be quickly destroyed by the mother's cytotoxic T cells. To avoid this, class I MHC expression is down-regulated in these cells. As with viral down-regulation of class I molecules, this lack of MHC class I expression leaves the placental cells liable to destruction by the mother's natural killer cells. In this case, the cells are saved from

destruction by presenting HLA-G molecules on their surface. This relatively nonvariant MHC molecule does not stimulate a T cell response, which is limited to presentation of HLA-A, -B, and -C molecules. It does inhibit the natural killer cell, which has HLA-G receptors on its surface. The fetus, like CMV, has thus devised ways of avoiding destruction by both T cells and natural killer cells.

A third example of the molecular arms race is given by one of the most feared infectious agents of modern times, the human immunodeficiency virus (HIV). Some strains of this virus gain entry into macrophages and helper T cells via a cell-surface receptor, CC chemokine receptor 5 (*CCR5*). Once inside the cell, HIV inserts its genetic material into the nucleus and takes advantage of the cell's machinery to replicate itself (HIV is a retrovirus, a type of virus that is discussed further in Chapter 13). Helper T cells are a critical component of the body's immune system, and their destruction by HIV leads to severe secondary immunodeficiency. Persons who are homozygous for a 32-bp deletion of the *CCR5* gene lack the CCR5 receptor and are thus remarkably resistant to HIV infection. Among those who are heterozygous for this deletion, progression to AIDS symptoms after seroconversion is slowed by 2 to 4 years. This deletion is especially common in northeastern European populations, where the gene frequency reaches 0.20. It is virtually absent in Asian and African populations.

Analysis of linkage disequilibrium in the chromosomal region that contains CCR5 indicates that the deletion arose in European populations only 700 to 2000 years ago. Because HIV appeared in humans only a few decades ago, the high gene frequency in northeastern Europe cannot be due to HIV and must be due to another selective force or perhaps to genetic drift. Considering the age of the deletion, it is possible that it underwent positive selection because it provided resistance to pathogens, such as smallpox, that ravaged European populations in the past. Clearly, there would be strong selection in favor of this mutation in populations now heavily exposed to HIV, and the deletion would eventually rise in frequency in these populations. Even now, knowledge of the consequences of this deletion may help accelerate the medical arms race against HIV.

The class III MHC region spans 680 kb and contains at least 36 genes, only some of which are involved in the immune response. Among the most important of these are the genes encoding the complement proteins.

The genes encoding the immunoglobulins, the T-cell receptors, KIR, and the class I and class II MHC proteins all share similar DNA sequences and structural features. Thus, they are members of a gene family, like the globin genes, the color vision genes, and the collagen genes described in earlier chapters. Table 9-1 provides a summary of the major genes of the immune system and their chromosome locations.

It is important to emphasize that the class I and class II MHC molecules differ greatly *among* individuals, but each cell within an individual has the same class I and class II molecules (this uniformity is necessary for recognition by T cells). In contrast, after VDJ recombination the T-cell receptors and immunoglobulins differ from cell to cell *within* individuals, allowing the body to respond to a large variety of different infectious agents.

> The immunoglobulin, T-cell receptor, KIR, and MHC genes are members of a gene family. Whereas immunoglobulins and T-cell receptors vary between cells within an individual, MHC molecules vary between individuals.

## MHC and Disease Associations

A number of diseases show significant associations with specific MHC alleles: persons who have the allele are much more likely to develop the disease than are those who lack it. Some examples, mentioned in earlier chapters, include the association of *HLA-B27* (i.e., allele 27 of the *HLA-B* locus) with ankylosing spondylitis and of *HLA-DQB1* with type 1 diabetes. An especially strong association is seen between several *HLA-DR* and *-DQ* alleles and narcolepsy, a disorder characterized by sudden and uncontrollable episodes of sleep. As Table 9-2 shows, most of the HLA–disease associations involve the class II MHC genes.

In some cases the association between MHC alleles and a disease is caused by linkage disequilibrium. For example, the hemochromatosis locus is linked to *HLA-A*, and significant associations occur between *HLA-A3* and the hemochromatosis disease gene (see Chapter 8). There is no known causal link between *HLA-A3* and this disorder, however. More likely, the association represents a past event in which the primary hemochromatosis mutation arose on a copy of chromosome 6 that had the *HLA-A3* allele. Similarly, the association between *HLA-DQB1* and *HLA-DQA1* and narcolepsy is due to linkage disequilibrium between the *HLA-DQ* region and the nearby locus that causes narcolepsy (the hypocretin type 2 receptor gene).

In other cases a causal association may exist. Some MHC–disease associations involve **autoimmunity**, in which the body's immune system attacks its own normal cells. For example, type 1 diabetes is characterized by T-cell infiltration of the pancreas and subsequent T-cell destruction of the insulin-producing beta cells. In some cases, autoimmunity involves "molecular mimicry". Here, a peptide that stimulates an immune response is so similar to the body's own peptides that the immune system begins to attack the body's own cells. This phenomenon helps to explain the onset of ankylosing spondylitis, another autoimmune

**TABLE 9-1**

## Chromosome Location and Function of Major Immune Response Genes

| Gene System | Chromosome Location | Gene Product Function |
|---|---|---|
| Immunoglobulin heavy chain (*C*, *V*, *D*, and *J* genes) | 14q32 | Heavy chain, the first part of antibody molecule, which binds foreign antigens |
| Immunoglobulin κ light chain (*C*, *V*, and *J* genes) | 2p13 | Light chain, the second part of antibody molecule |
| Immunoglobulin λ light chain (*C*, *V*, and *J* genes) | 22q11 | Light chain, the second part of antibody molecule (either κ or λ may be used) |
| T-cell receptor α | 14q11 | One chain of the α-β T-cell receptor, which recognizes antigen with MHC molecule |
| T-cell receptor β | 7q35 | The other chain of the α-β T-cell receptor |
| T-cell receptor γ | 7p15 | One chain of the γ-δ T-cell receptor |
| T-cell receptor δ | 14q11 | The second chain of the γ-δ T-cell receptor |
| MHC (classes I, II, and III); includes *TAP1* and *TAP2* | 6p21 | Cell-surface molecules that present peptides to T-cell receptors. *TAP1* and *TAP2* are transporter molecules that process foreign peptides and carry them to the endoplasmic reticulum. |
| β2-microglobulin | 15q21-22 | Forms second chain of the class I MHC molecule |
| *RAG1, RAG2* | 11p13 | Recombinases that participate in VDJ somatic recombination |

**TABLE 9-2**
Examples of Major Histocompatibility Complex and Disease Associations

| Disease | MHC (HLA) Associated Allele | Approximate Relative Risk* |
|---|---|---|
| Type 1 diabetes | DQB1*0302 | 10 |
| Ankylosing spondylitis | B27 | 90 |
| Narcolepsy | DR2 and DQA1, DQB1 | >100 |
| Celiac disease | DR3, DR7 | 10 |
| Rheumatoid arthritis | DR1, DR4 | 5 |
| Myasthenia gravis | DR3, DR7 | 2.5 |
| Multiple sclerosis | DR2 | 4 |
| Pemphigus vulgaris | DR4 | 14 |
| Systemic lupus erythematosus | DR3 | 6 |
| Hemochromatosis | A3 | 20 |
| Malaria | B53 | 0.59 |
| Graves' disease | DR3 | 5 |
| Psoriasis vulgaris | Cw6 | 13 |
| Squamous cell cervical carcinoma | DQw3 | 7 |

*Relative risk can be interpreted loosely as the odds that a person who has a risk factor (in this case, an MHC antigen) will develop the disease, compared with a person who lacks the risk factor. Thus, a relative risk of 4 for DR2 and multiple sclerosis means that persons with DR2 are four times more likely to develop multiple sclerosis than are those without DR2. A relative risk <1 (as seen for malaria and B53) indicates that the factor is protective against the disease.

MHC, major histocompatibility complex.

Data from Bell JI, Todd JA, McDevitt HO: The molecular basis of HLA-disease association. Adv Hum Genet. 1989;18:1–41; Doherty DG, Nepom GT: The human major histocompatibility complex and disease susceptibility. In Rimoin DL, Connor JM, Pyeritz RE (eds): Emery and Rimoin's Principles and Practice of Medical Genetics. Vol 1. New York: Churchill Livingstone, 1997; Hill AV, et al: Common West African HLA antigens are associated with protection from severe malaria. Nature 1991;352:595-600; Klein J, Sato A: The HLA system. Second of two parts. N Engl J Med 2000;343:782-786; and Wank R, Thomssen C: High risk of squamous cell carcinoma of the cervix for women with HLA-DQW3. Nature 1991;352:723-725.

disease. Infections of *HLA-B27*–positive persons with specific microbes, such as *Klebsiella*, can lead to a **cross-reaction** in which the immune system mistakes peptides from some of the body's normal cells for microbial peptides. Another such example is given by rheumatic fever, in which a streptococcal infection initiates cross-reactivity between streptococcus and cardiac myosin. In each of these scenarios, the body already has a small population of self-reactive T cells, but they remain inactive and quite harmless until they are stimulated to proliferate by a foreign peptide that closely resembles a self peptide.

Autoimmunity can also be caused by specific defects in the regulation of immune system components. For example, regulatory T cells help to prevent the formation of self-reactive immune cells and require a transcription factor, encoded by *FOXP3*, for their normal development. Mutations in *FOXP3* result in a deficiency of regulatory T cells and an autoimmune disease called *IPEX* (immunodysregulation, polyendocrinopathy, enteropathy, X linked).

Other common diseases that involve autoimmunity include rheumatoid arthritis, systemic lupus erythematosus, psoriasis, and multiple sclerosis. It is estimated that approximately 5% of the population suffers from some type of autoimmune disease.

> A significant number of diseases are associated with specific MHC alleles. Some of these associations are the result of linkage disequilibrium, but most are likely to result from causal associations involving autoimmunity.

## The ABO and Rh Blood Groups

Another component of the immune system involves red blood cell surface molecules that can cause an immune reaction during blood transfusions. The ABO and Rh red-cell antigen systems were discussed in Chapter 3 as early examples of polymorphic marker loci. They are also the most important systems that determine blood transfusion compatibility.

## The ABO System

There are four major ABO blood types: A, B, AB, and O. The first three groups respectively represent persons who carry the A, B, or A and B antigens on their erythrocyte surfaces. Those with type O have neither the A nor the B antigen. Persons who have one of these antigens on their erythrocyte surfaces possess antibodies against all other ABO antigens in their blood stream. (These antibodies are formed early in life as a result of exposure to antigens that are identical to the A and B antigens but are present in various microorganisms.) Thus, if a type B person received type A or AB blood, his or her anti-A antibodies would produce a severe and possibly fatal reaction. Type O persons, who have neither the A nor the B antigen and thus both anti-A and anti-B antibodies, would react strongly to blood of the other three types (A, B, and AB). It was once thought that type O persons, because they lack both types of antigens, could be "universal donors" (anyone could accept their blood). Similarly, type AB persons were termed "universal recipients" because they lacked both anti-A and anti-B antibodies. However, when patients are given transfusions of whole blood containing large volumes of serum, the donor's antibodies can react against the recipient's erythrocyte antigens. Hence, complete ABO matching is nearly always done for blood transfusions.

> The ABO locus encodes red cell antigens that can cause a transfusion reaction if donors and recipients are not properly matched.

## The Rh System

The Rh blood group is encoded by two tightly linked loci, one of which is labeled *D*. The other locus produces Rh antigens labeled *C* and *E* through alternative splicing of the messenger RNA. The *D* locus is of primary interest because it is responsible for Rh maternal–fetal incompatibility and the resulting disease, hemolytic disease of the newborn (HDN). Persons with the *DD* or *Dd* genotype have the Rh antigen on their erythrocytes and are Rh-positive. The recessive homozygotes, with genotype *dd*, are Rh-negative and do not have the Rh antigen. About 85% of North Americans are Rh-positive and about 15% are Rh-negative.

Unlike the ABO system, in which antibodies normally are formed in response to antigens presented by other organisms, anti-Rh antibody production requires a stimulus by the human Rh antigen itself. An Rh-negative person does not begin to produce anti-Rh antibodies unless he or she is exposed to the Rh antigen, usually through a blood transfusion or during pregnancy. Maternal–fetal incompatibility results when an Rh-positive man and an Rh-negative woman produce children. If the man's genotype is *DD*, all of their offspring will be Rh-positive and will have Rh antigens on their erythrocytes. If the man is a heterozygote, with genotype *Dd*, half of their children will be Rh-positive, on average.

There are usually no difficulties with the first Rh-incompatible child, because very few of the fetus's red blood cells cross the placental barrier during gestation. When the placenta detaches at birth, a large number of fetal red blood cells typically enter the mother's blood stream. These cells, carrying the Rh antigens, stimulate production of anti-Rh antibodies by the mother. These antibodies persist in the blood stream for a long time, and if the next offspring is again Rh-positive, the mother's anti-Rh antibodies enter the fetus's bloodstream and destroy its red blood cells. As this destruction proceeds, the fetus becomes anemic and begins to release many erythroblasts (immature nucleated red cells) into its blood stream. This phenomenon is responsible for the descriptive term *erythroblastosis fetalis*. The anemia can lead to a spontaneous abortion or stillbirth. Because the maternal antibodies remain in the newborn's circulatory system, destruction of red cells can continue in the neonate. This causes a buildup of bilirubin and a jaundiced appearance shortly after birth. Without replacement transfusions, in which the child receives Rh-negative red cells, the bilirubin is deposited in the brain, producing cerebral damage and usually death. Infants who do not die can develop mental retardation, cerebral palsy, and/or high-frequency deafness.

Among North Americans of European descent, approximately 13% of all matings are Rh-incompatible. Fortunately, a simple therapy now exists to avoid Rh sensitization of the mother. During and after pregnancy, an Rh-negative mother is given injections of Rh immune globulin, which consists of anti-Rh antibodies. These antibodies destroy the fetal erythrocytes in the mother's blood stream before they stimulate production of maternal anti-Rh antibodies. Because the injected antibodies do not remain in the mother's blood stream for long, they do not affect subsequent offspring. To avoid sensitization, these injections must be administered with each pregnancy. The Rh-negative mother must also be careful not to receive a transfusion containing Rh-positive blood, because this would also stimulate production of anti-Rh antibodies.

> Maternal–fetal Rh incompatibility (Rh-negative mother and Rh-positive fetus) can produce hemolytic disease of the newborn if the mother's Rh antibodies attack the fetus. Administration of Rh immune globulin to the mother prevents this reaction.

A rarer form of maternal–fetal incompatibility can result when a mother with type O blood carries a fetus with type A or B blood. The HDN produced by this combination is usually so mild that it does not require treatment. Interestingly, if the mother is also Rh-negative and the child is Rh-positive, the ABO incompatibility *protects* against the more severe Rh incompatibility. This is because any fetal red blood cells entering the mother's circulatory system are quickly destroyed by her anti-A or anti-B antibodies before she can form anti-Rh antibodies.

## IMMUNODEFICIENCY DISEASES

**Immunodeficiency disease** results when one or more components of the immune system (e.g., T cells, B cells, MHC, complement proteins) are missing or fail to function normally. **Primary immunodeficiency diseases** are caused by abnormalities in cells of the immune system and are usually produced by genetic alterations. To date, more than 100 different primary immunodeficiency syndromes have been described, and it is estimated that these diseases affect at least 1 in 10,000 persons. **Secondary immunodeficiency** occurs when components of the immune system are altered or destroyed by other factors, such as radiation, infection, or drugs. For example, the human immunodeficiency virus (HIV), which causes acquired immunodeficiency syndrome (AIDS), attacks macrophages and helper T lymphocytes, central components of the immune system. The result is increased susceptibility to a multitude of opportunistic infections.

B-cell immunodeficiency diseases render the patient especially susceptible to recurrent bacterial infections, such as *Streptococcus pneumoniae*. An important example of a B-cell immunodeficiency is X-linked agammaglobulinemia (XLA). Patients with this disorder, the overwhelming majority of whom are male, lack B cells completely and have no IgA, IgE, IgM, or IgD in their serum. Because IgG crosses the placenta during pregnancy, infants with XLA have some degree of B cell immune response for the first several months of life. However, the IgG supply is soon depleted, and the infants develop recurrent bacterial infections. They are treated with large amounts of γ-globulin. XLA is caused by mutations in a gene (*BTK*) that encodes a B-cell tyrosine kinase necessary for normal B cell maturation. Mutations in the genes that encode the immunoglobulin heavy and light chains can cause autosomal recessive B-cell immunodeficiency.

T-cell immunodeficiency diseases directly affect T cells, but they also affect the humoral immune response, because B cell proliferation largely depends on helper T cells. Thus, affected patients develop severe combined immune deficiency (SCID) and are susceptible to many opportunistic infections, including *Pneumocystis jiroveci* (a protozoan that commonly infects AIDS patients). Without bone marrow transplants, these patients usually die within the first several years of life. About half of SCID cases are caused by X-linked recessive mutations in a gene encoding the γ chain that is found in six different cytokine receptors (those of interleukins 2, 4, 7, 9, 15, and 21). Lacking these receptors, T cells and natural killer cells cannot receive the signals they need for normal maturation. These receptors all interact with an intracellular signaling molecule called Jak3. As might be expected, persons who lack Jak3 as a result of autosomal recessive mutations in the *JAK3* gene experience a form of SCID that is very similar to the X-linked form just described.

About 15% of SCID cases are caused by adenosine deaminase (ADA) deficiency, an autosomal recessive disorder of purine metabolism that results in a buildup of metabolites that are toxic to B and T cells. This type of SCID, as well as the X-linked form, can be treated by bone marrow transplantation, and some cases are being treated experimentally with gene therapy (see Chapter 13).

SCID can also arise from mutations in *RAG1* or *RAG2*, two of the genes involved in VDJ recombination and the proper formation of T-cell and B-cell receptors. These mutations produce a combined B-cell and T-cell immunodeficiency, although normal natural killer cells are produced. Other examples of SCID are given in Table 9-3.

Several immune system defects result in lymphocytes that lack MHC molecules on their surfaces. These are collectively termed *bare lymphocyte syndrome*. One form of this syndrome is caused by mutations in the *TAP2* gene. The *TAP2* protein helps to transport peptides to the endoplasmic reticulum, where they are bound by class I MHC molecules. A defect in the TAP2 protein destabilizes the class I MHC molecules so that they are not expressed on the cell surface. Because exposure to MHC molecules is necessary for normal T-cell development in the thymus, bare lymphocyte syndrome

**TABLE 9-3**

## Examples of Primary Immunodeficiency Diseases

| Condition | Mode of Inheritance | Brief Description |
|---|---|---|
| X-linked agammaglobulinemia | XR | Absence of B cells leads to recurrent bacterial infections |
| SCID (γ chain cytokine receptor defect or ADA deficiency) | XR, AR | T-cell deficiency leading also to impairment of humoral immune response; fatal unless treated by bone marrow transplantation or gene therapy |
| SCID due to Jak3 deficiency | AR | Protein kinase deficiency leading to T-cell and B-cell deficiency |
| SCID due to *RAG1* or *RAG2* deficiency; Omenn syndrome | AR | Lack of recombinase activity impairs VDJ recombination, leading to B-cell and T-cell deficiency |
| SCID due to interleukin-7 α chain deficiency | AR | T-cell deficiency leading to impaired B cell response |
| Zap70 kinase deficiency | AR | Lack of cytotoxic T cells; defective helper T cells; impaired antibody response |
| Purine nucleoside phosphorylase deficiency | AR | Purine metabolism disorder leading to T-cell deficiency |

**TABLE 9-3**

Examples of Primary Immunodeficiency Diseases—cont'd

| Condition | Mode of Inheritance | Brief Description |
|---|---|---|
| Bare lymphocyte syndrome (BLS) | AR | Deficient MHC class I expression (*TAP2* mutation) leads to T-cell and B-cell deficiency in type 1 BLS; mutations in transcription factors for MHC class II genes lead to a relative lack of helper T cells in type 2 BLS |
| Complement system defects | Mostly AR | Increased susceptibility to bacterial and other infections |
| DiGeorge anomaly | AD, sporadic | Congenital malformations include abnormal facial features, congenital heart disease, and thymus abnormality leading to T-cell deficiency |
| Ataxia telangiectasia | AR | DNA repair defect characterized by unsteady gait (ataxia), telangiectasia (dilated capillaries), and thymus abnormality producing T-cell deficiency |
| Wiskott–Aldrich syndrome | XR | Abnormal, small platelets, eczema, and abnormal T cells causing susceptibility to opportunistic infections |
| Chediak–Higashi syndrome | AR | Partial albinism, defective lysosomal assembly, giant cytoplasmic granules, abnormal natural killer cells and neutrophils leading to recurrent bacterial infections |
| Leukocyte adhesion deficiency | AR | Mutations in integrin receptor gene produce phagocytes that cannot recognize and ingest microorganisms, resulting in severe bacterial infections |
| Chronic granulomatous disease | XR, AR | Phagocytes ingest microbes but cannot kill them; leads to formation of granulomas and recurrent infections |
| HyperIgE syndrome | AD, AR | Recurrent staphylococcal infections, markedly elevated serum IgE levels, coarse facial features |
| IRAK-4 deficiency | | Toll-like receptor/interleukin-1 defect caused by deficiency of interleukin-1 receptor associated kinase-4 (IRAK-4), resulting in extracellular bacterial (especially *Streptococcus pneumoniae*) and fungal infections. |

*AD, Autosomal dominant; AR, autosomal recessive; SCID, severe combined immune deficiency; XR, X-linked recessive.*

results in a severe reduction in the number of functional T and B cells. Bare lymphocyte syndrome can also be caused by defects in several different transcription factors that bind to promoters in the class II MHC region. The result is a lack of class II MHC molecules on APCs, a deficiency of helper T cells, and a consequent lack of antibody production.

Chronic granulomatous disease (CGD) is a primary immunodeficiency disorder in which phagocytes can ingest bacteria and fungi but are then unable to kill them. This brings about a persistent cellular immune response to the ingested microbes, resulting in the formation of the granulomas (nodular inflammatory lesions containing macrophages) for which the disease is named. These patients develop pneumonia, lymph node infections, and abscesses of the skin, liver, and other sites. The most common cause of CGD is an X-linked mutation, but there are also at least three autosomal recessive genes that can cause CGD. The gene that causes X-linked CGD was the first disease-causing gene to be isolated through positional cloning. It encodes a subunit of cytochrome b, a protein that the phagocyte requires for a burst of microbe-killing oxygen metabolism.

Multiple defects in the various proteins that make up the complement system have been identified. Most of these are inherited as autosomal recessive disorders, and most result in increased susceptibility to bacterial infections.

Finally, a number of syndromes include immunodeficiency as one of their features. One example is the DiGeorge sequence (see Chapter 6), in which a lack of thymic development leads to T-cell deficiency. Wiskott–Aldrich syndrome is an X-linked recessive disorder that involves deficiencies of platelets and B and T cells. It is caused by mutations in a gene (*WAS*) whose protein product is needed for normal formation of the cellular cytoskeleton. Several syndromes that involve DNA instability are characterized by immunodeficiency (e.g., ataxia telangiectasia, Bloom syndrome, Fanconi anemia; see Chapter 3).

> Primary immunodeficiency diseases involve intrinsic defects of immune response cells (B cells, T cells, MHC, complement system, or phagocytes) and are usually caused by genetic alterations. Secondary immunodeficiency disorders, of which AIDS is an example, are caused by external factors. Immunodeficiency is also seen in a number of genetic syndromes, including several DNA instability disorders.

## Study Questions

1. Compare the functions of class I and class II MHC molecules.

2. MHC molecules and immunoglobulins both display a great deal of diversity but in different ways. How and why do these types of diversity differ?

3. In what ways are T-cell receptors and immunoglobulins similar? In what ways are they different?

4. If there are 80 V segments, 6 J segments, and 30 D segments that can encode an immunoglobulin heavy chain of one particular class, how many different immunoglobulins can be formed on the basis of somatic recombination alone?

5. When matching donors and recipients for organ transplants, siblings are often desirable donors because they are more likely to be HLA-compatible with the recipient. If we assume no crossing over within the HLA loci and assume four distinct HLA haplotypes among the parents, what is the probability that a transplant recipient will be HLA-identical with a sibling donor?

6. What types of matings will produce Rh maternal–fetal incompatibility?

## Suggested Readings

Charron D. Immunogenetics today: HLA, MHC and much more. Curr Opin Immunol 2005;17(5): 493–497.

Cunningham-Rundles C, Ponda PP. Molecular defects in T- and B-cell primary immunodeficiency diseases. Nat Rev Immunol 2005;5(11):880–892.

Horton R, Wilming L, Rand V, et al. Gene map of the extended human MHC. Nature Rev Genet 2004; 5(12): 889–899.

Krogsgaard M, Davis MM. How T cells "see" antigen. Nat Immunol 2005;6(3):239–245.

Nossal GJ. The double helix and immunology. Nature 2003;421(6921):440–444.

O'Brien SJ, Nelson GW. Human genes that limit AIDS. Nat Genet 2004;36(6):565–574.

O'Neill LA. Immunity's early-warning system. Sci Am 2005;292(1):24–31.

Peled JU, Kuang FL, Iglesias-Ussel MD, et al. The biochemistry of somatic hypermutation. Annu Rev Immunol 2008; 26:481–511.

Rioux JD, Abbas AK. Paths to understanding the genetic basis of autoimmune disease. Nature 2005;435 (7042):584–589.

Roitt I, Brostoff J, Male DRoth DA. Immunology. 7th ed. St. Louis: Mosby; 2006.

Schwartz RS. Diversity of the immune repertoire and immuno-regulation. N Engl J Med 2003;348(11):1017–1026.

Trowsdale J. HLA genomics in the third millennium. Curr Opin Immunol 2005;17(5):498–504.

### Internet Resources

Immunogenetics Database *http://www.ebi.ac.uk/imgt/*

Molecular Immunology Tutorial *http://www.mi.interhealth.info/*

Developmental genetics is the study of how the instructions encoded in genes control and coordinate development of an organism, beginning with fertilization and ending with death. The field is important to health care providers because mutations in genes can perturb developmental processes, resulting in increased risk for birth defects, mental retardation, and cancer. In the United States, 2% to 3% of all live-born children have a major birth defect (i.e., one that has a substantial impact on health) so about 100,000 children are born each year with a birth defect. Birth defects are the leading cause of infant* death in the United States and are associated with substantial morbidity.

Birth defects can be isolated abnormalities, or they can be features of one of several thousand known genetic syndromes. The etiology of most birth defects is unknown; however, a substantial fraction is caused by mutations in genes, either alone or in combination, that control normal development. The characterization of genes and developmental processes that coordinate animal development has revolutionized our understanding of the molecular basis of human birth defects. This chapter provides a brief review of the genes and proteins that control development and then discusses some of the cardinal developmental processes that, when disturbed, cause birth defects.

## DEVELOPMENT

### Basic Concepts

Animal development can be defined as the process by which a fertilized ovum becomes a mature organism capable of reproduction. A single fertilized egg divides and grows to form different cell types, tissues, and organs, all of which are arranged in a species-specific **body plan** (i.e., the arrangement and pattern of body parts). Many of the instructions necessary for normal development are encoded by an animal's genes. Because the genes in each cell of an organism are identical, several questions arise: How do cells with identical genetic constitutions form a complex adult organism composed of many different cells and tissues? What controls the fate of each cell, instructing a cell to become, for example, a brain cell or a liver cell? How do cells organize into discrete tissues? How is the body plan of an organism determined? Answering such fundamental questions has been a major focus of developmental

biology for more than a century. The pace of discovery has accelerated dramatically in the past several decades, and these discoveries are helping us to understand the causes of human malformations and genetic syndromes.

For ethical and technical reasons, it is difficult to study early developmental events in human embryos. Consequently, a variety of nonhuman model organisms are used to facilitate the study of development (Table 10-1). This approach is feasible because the major elements (genes and pathways) that control animal development are conserved across a wide range of species and body plans. In addition, many regulatory switches and signaling pathways are used repeatedly during development to control various patterning and differentiation events. This underscores the point that the evolution of species proceeds, in part, by continual tinkering with similar developmental programs to effect changes in an organism's phenotype.

For example, **ectopic expression** (i.e., expression of the gene product in an abnormal location) of the *Drosophila* gene *eyeless* results in the formation of a well-formed but misplaced eye (Fig. 10-1). Mice have a homologous gene, *Pax6*,[†] in which mutations can produce abnormally small eyes. When inserted ectopically into *Drosophila*, *Pax6* again produces a misplaced fly eye. Mutations in the human homolog, *PAX6*, cause defects of the eye such as cataracts and aniridia (absence of the iris). *PAX6*, *Pax6*, and *eyeless* are homologous genes that encode DNA transcription factors (see Chapter 3). Although the ancestors of *Drosophila* and mouse diverged from the lineage leading to humans 500 million and 60 million years ago, respectively, the genes and pathways involved in development of the eye have been conserved.

> Approximately 2% to 3% of babies are born with a recognizable birth defect. Many birth defects are caused by mutations in genes encoding elements in pathways that control development. These pathways are highly conserved among animal species. Thus, studies of nonhuman animal models are invaluable for understanding human development and the causes of birth defects.

---

*An infant is a person who is younger than 1 year.

[†]It is conventional to capitalize all letters of the names of human genes but only the first letter of the names of mouse genes, except for recessive mutations, which begin with a lower-case letter.

**TABLE 10-1**
## Animal Models of Human Development

| Organism | Generation Time* | Advantages | Disadvantages |
|---|---|---|---|
| *Caenorhabditis elegans* (roundworm) | 9 days | Fate of every cell known<br>Genome well characterized | Alternative body plan compared to vertebrates |
| | | Easy to breed and maintain | Tissues cannot be cultured |
| *Drosophila melanogaster* (fruit fly) | 10 days | Easy to breed<br>Large populations<br>Vast database of mutants<br>Feasible and affordable to do large screens | Alternative body plan compared to vertebrates<br>Must be stored live; cannot be frozen<br>Pathology often different compared to humans |
| *Danio rerio* (zebrafish) | 3 months | Transparent embryo<br>Easy to maintain<br>Large populations<br>Feasible and affordable to do large screens | Targeted gene modification difficult |
| *Xenopus laevis* (frog) | 12 months | Transparent embryo is large and easy to manipulate | Tetraploid genome makes genetic experiments difficult |
| *Gallus gallus* (chicken) | 5 months | Easy to observe and manipulate embryo | Genetic experiments difficult |
| *Mus musculus* (mouse) | 2 months | Pathology similar to humans<br>Excellent tools for phenotypic characterization<br>Targeted gene modification straightforward<br>Fully annotated genome available | Relatively expensive to maintain<br>Manipulation of embryo is challenging |
| *Papio hamadryas* (baboon) | 60 months | Pathology and physiology similar to that of humans | Very expensive to maintain<br>Small populations<br>Long generation time<br>Strong ethical concerns with use of primates |

*Generation time is defined as the age at which the organism is first capable of reproduction.

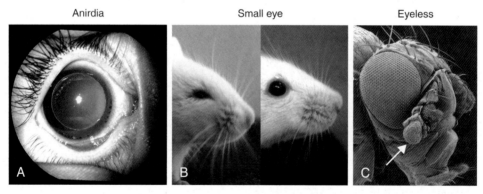

Anirdia        Small eye        Eyeless

**FIGURE 10-1**
Evolutionary relationships among homologs of *PAX6* and associated phenotypes illustrating functional conservation of *Pax* genes among humans, mice, and *Drosophila*. **A**, Mutations in human *PAX6* cause aplasia of the irides, or aniridia. **B**, Loss of function of murine *Pax6* causes small eyes (*left*) compared to a wild-type mouse (*right*). **C**, Misexpression of *eyeless* in tissue fated to become an antenna causes formation of a normal but misplaced eye *(tip of arrow)* instead. *(A from Jones KL: Smith's Recognizable Patterns of Human Malformation, 6th ed. Philadelphia: Mosby, 2006, p 53. B Courtesy of James Lauderdale, University of Georgia. C from Science. 1995 Mar 24; 267[5205]. Reprinted with permission From AAAS.)*

## A Brief Overview of Major Processes in Embryonic Development

Several major processes are involved in the development of the embryo. These include **axis specification**, **pattern formation**, and **organogenesis**. As the name suggests, pattern formation describes a series of steps in which differentiated cells are arranged spatially to form tissues and organs. The interactions of these cells are mediated by processes such as **induction**, which occurs when the cells of one embryonic region influence the organization and differentiation of cells in a second region.

Axis specification involves the definition of the major axes of the body: ventral/dorsal, anterior/posterior, medial/lateral, and left/right. Specification of **polarity** (direction) is an important part of this process. As the axes are specified, the formation of organs and limbs (organogenesis) begins. Each of these major processes involves many different proteins that form structures and provide signals to coordinate the development of the embryo. The major types of these proteins, and the genes that encode them, are described next.

> Embryonic development involves the processes of pattern formation, axis specification, and organogenesis. Each of these processes is controlled by a series of proteins that provide signals and form structures necessary for normal development of the embryo.

## GENETIC MEDIATORS OF DEVELOPMENT: THE MOLECULAR TOOLBOX

The genes required for normal development encode many different products, including signaling molecules and their receptors, DNA transcription factors, components of the extracellular matrix, enzymes, transport systems, and other proteins. Each of these genetic mediators is expressed in combinations of spatially and temporally overlapping patterns that control different developmental processes. As detailed in this chapter, mutations in the genes mediating development are a common cause of human birth defects.

### Paracrine Signaling Molecules

Interactions between neighboring cells are usually mediated by proteins that can diffuse across small distances to induce a response. These molecules are often called paracrine factors because they are secreted into the space surrounding a cell (unlike hormones, which are secreted into the blood stream). Closely related paracrine factors have been isolated from a variety of organisms, making it clear that homologous molecules are used throughout the animal kingdom. To date, four major families of paracrine signaling molecules have been described: the Fibroblast Growth Factor (FGF) family, the Hedgehog family, the Wingless (Wnt) family, and the Transforming Growth Factor β (TGF-β) family. Each of

these signaling molecules binds to one or more receptors to effect a response, and mutations in genes encoding these molecules may lead to abnormal communication between cells. Clinical Commentary 10-1 discusses the FGF family and associated receptors; the other three families are described in this section.

The first member of the Hedgehog family was originally isolated in a *Drosophila* mutant that has bristles in an area that is naked in the normal fly (hence the name "hedgehog"). Vertebrates have several homologs of *hedgehog*, the most widely used of which is termed *Sonic hedgehog* (*Shh*). Among its many roles, *Shh* participates in axis specification, induction of motor neurons within the neural plate, and patterning of the limbs. The primary receptor of *Shh* is a transmembrane protein encoded by a gene called *patched*. The normal action of *patched* is to inhibit the function of another transmembrane protein called *smoothened*, encoded by the gene *Smo*. Binding of *Shh* to the patched receptor results in disinhibition of *smoothened* and activation of an intracellular signaling cascade that targets members of the GLI family of transcription factors (Fig. 10-2).

Mutations in the human homolog, *PATCHED* (*PTC*), cause Gorlin syndrome (see Fig. 10-2), a disorder characterized by rib anomalies, cysts of the jaw, and basal cell carcinomas (a form of skin cancer). Somatic mutations in *PTC* have also been found in sporadic basal cell carcinomas. Thus, germline mutations in *PTC* alter the regulation of developing cells to cause birth defects, and somatic mutations in *PTC* can alter the regulation of terminally differentiated cells to cause cancer.

The Wnt family of genes is named after the *Drosophila* gene *wingless* and one of its vertebrate homologs, *integrated*. The *wingless* gene establishes polarity during *Drosophila* limb formation, and members of the Wnt family play similar roles in vertebrates. Wnt genes encode secreted glycoproteins that bind to members of the *frizzled* and low-density lipoprotein (LDL) receptor–related protein families. In humans, 19 different Wnt genes have been identified, and they participate in a wide variety of developmental processes, including specification of the dorsal/ventral axis and formation of the brain, muscle, gonads, and kidneys. Homozygosity for mutations in *WNT3* causes tetra-amelia (absence of all four limbs) in

## CLINICAL COMMENTARY 10-1
### *Disorders of Fibroblast Growth Factor Receptors*

Fibroblast growth factor receptors (FGFRs) are highly homologous glycoproteins, with a common structure that consists of a signal peptide (an amino acid sequence that helps to direct the protein to its proper cellular location), three immunoglobulin-like (Ig-like) domains, a membrane-spanning segment, and an intracellular tyrosine kinase domain. FGFRs are receptors for at least 22 fibroblast growth factors (FGFs) that participate in a wide variety of biological processes, including cell migration, growth, and differentiation. With varying affinities, FGFs bind FGFRs, leading to phosphorylation, and hence activation, of the tyrosine kinase domain.

FGFRs are widely expressed in developing bone, and many common human autosomal dominant disorders of generalized bone growth (i.e., skeletal dysplasias) are caused by mutations in FGFR genes. The most prevalent of these disorders, affecting more than 250,000 people worldwide, is achondroplasia (ACH), which is characterized by disproportionate short stature (i.e., the limbs are disproportionately shorter than the trunk) and macrocephaly (see Chapter 4). Nearly all persons with ACH have a glycine → arginine substitution in the transmembrane domain of *FGFR3*, resulting in constitutive *FGFR3* activation.

*Continued*

**CLINICAL COMMENTARY   10-1**
*Disorders of Fibroblast Growth Factor Receptors—cont'd*

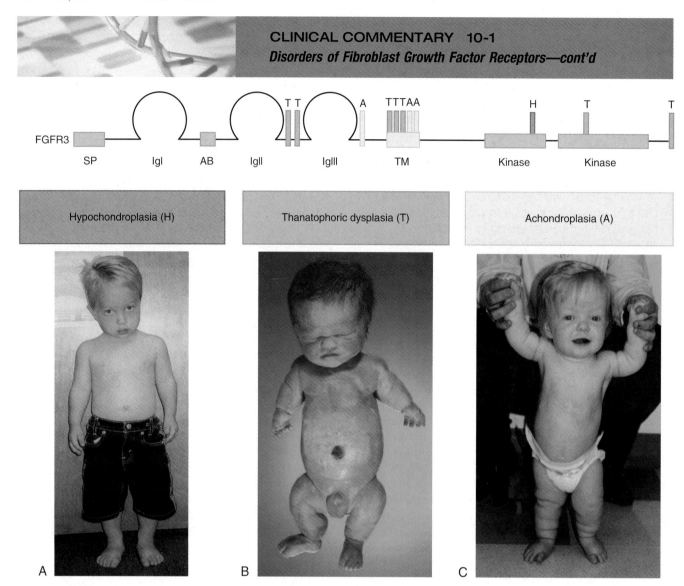

| Hypochondroplasia (H) | Thanatophoric dysplasia (T) | Achondroplasia (A) |

A                    B                    C

*Top*, Schematic drawing of fibroblast growth factor 3 receptor (*FGFR3*) protein. Important functional domains of *FGFR3* include a signal peptide (*SP*), three immunoglobulin-like (*Ig*) domains, an acid box (*AB*), a transmembrane (*TM*) domain, and a split tyrosine kinase (*Kinase*) domain. The locations of point mutations causing achondroplasia (*A; gray*), hypochondroplasia (*H; dark blue*), and thanatophoric dysplasia (*T; light blue*) are indicated. *Bottom*, Photographs of children with mutations in *FGFR3*. **A**, A boy with hypochondroplasia. He has mildly short limbs relative to his trunk. **B**, An infant with thanatophoric dysplasia, the most common of the lethal skeletal dysplasias. He has markedly shortened limbs and a very narrow thoracic cage. **C**, A girl with achondroplasia. She has short limbs relative to the length of her trunk (resulting in redundant skin folds in the arms and legs), a prominent forehead, and a depressed nasal root.

*(Modified from Webster MK, Donoghue DJ: FGFR activation in skeletal disorders: Too much of a good thing. Trends Genet 1997;13:178-182.)*

FGFR3 is normally expressed in resting chondrocytes, where it restrains chondrocyte proliferation and differentiation. Overactivation causes further inhibition of chondrocyte growth, producing skeletal defects. The degree of *FGFR3* activation, which varies depending on the domain that is altered, corresponds to the severity of long bone shortening. Lesser degrees of activation result in the milder skeletal abnormalities observed in hypochondroplasia, whereas markedly increased activation causes a virtually lethal short stature syndrome called *thanatophoric dysplasia*.

These findings suggest that a potential therapeutic strategy might be to reduce or block the activity of *FGFR3*. However, keeping some residual *FGFR3* function is necessary for normal development. In mice, *FGFR3*-inactivating mutations cause expansion of the zones of proliferating cartilage and increased long bone growth, resulting in mice that are longer than average. A similar average increase in height has been observed in humans with a rare syndrome that is caused by a *FGFR3* mutation predicted to reduce *FGFR3* function.

Abnormal bone growth is also a feature of a group of autosomal dominant disorders characterized by premature fusion (synostosis) of the cranial sutures, misshapen skulls, and various types of limb defects. Collectively, these disorders are called *craniosynostosis syndromes*. Mutations in *FGFR1*, *FGFR2*, and *FGFR3* cause at least six distinct craniosynostosis disorders, the best-known of which is Apert syndrome. The same mutation can sometimes cause two or more different craniosynostosis syndromes. For example, a cysteine → tyrosine substitution in *FGFR2* can cause either Pfeiffer or Crouzon syndrome. This suggests that additional factors, such as modifying genes, are partly responsible for creating different phenotypes.

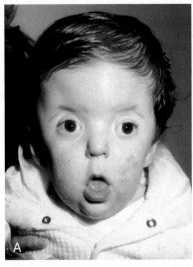
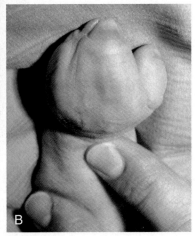

A, Face of a child with Apert syndrome. Note that the skull is tall and narrow. The eyes are protuberant because the bony orbits are shallow. B, Hand of a child with Apert syndrome showing a broad thumb and fusion of all of the digits (syndactyly).
*(From Jones KL: Smith's Recognizable Patterns of Human Malformation, 6th ed. Philadelphia: Mosby, 2006.)*

### Craniosynostosis Syndromes Caused by Mutations in Fibroblast Growth Factor Receptors

| Gene | Syndrome | Characteristic |
| --- | --- | --- |
| FGFR1 | Pfeiffer | Broad first digits, hypertelorism |
| FGFR2 | Apert | Mid-face hypoplasia, fusion of digits |
| | Pfeiffer | Broad first digits, hypertelorism |
| | Crouzon | Mid-face hypoplasia, ocular proptosis |
| | Beare–Stevenson | Mid-face hypoplasia, corrugated skin |
| | Jackson–Weiss | Mid-face hypoplasia, foot anomalies |
| FGFR3 | Crouzon | Mid-face hypoplasia, ocular proptosis, acanthosis nigricans* |
| | Muenke (nonsyndromic craniosynostosis) | Mid-face hypoplasia, brachydactyly, hearing loss |

*Acanthosis nigricans is characterized by hyperplastic and hypertrophic skin of varying pigmentation, most often covering the axillae, neck, genitalia, and flexural surfaces.*

humans, and abnormal Wnt signaling has been associated with the formation of tumors.

The TGF-β supergene family* is composed of a large group of structurally related genes that encode proteins that form homodimers or heterodimers. Members of the TGF-β supergene family include the TGF-β family itself, the bone morphogenetic protein (BMP) family, the activin family, and the Vg1 family. Although the role of BMPs is not limited to bone development, members of the BMP family were originally isolated because of their ability to induce bone formation.

Mutations in a member of the BMP family, *cartilage-derived morphogenetic protein* 1 (*CDMP*1), cause various skeletal abnormalities. Different mutations can produce distinct phenotypes (allelic heterogeneity; see Chapter 4). For example, a nonsense mutation in *CDMP*1 causes dominantly inherited brachydactyly (short digits). Persons homozygous for a 22-bp duplication in *CDMP*1 have brachydactyly as well as shortening of the long bones of the limbs in an autosomal recessive disorder called acromesomelic dysplasia. A homozygous missense mutation produces autosomal recessive Grebe chondrodysplasia, also characterized by severe shortening of the long bones and digits. The mutant protein is not secreted, and it is thought to inactivate other BMPs

---

*A supergene family is a group of related gene families.

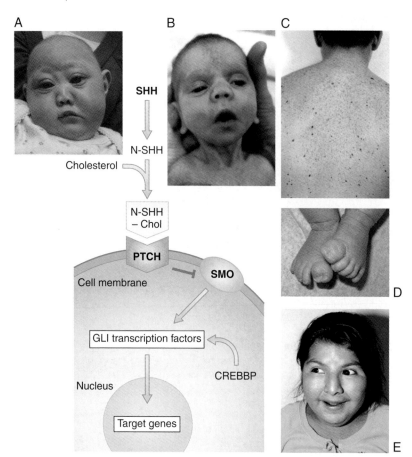

A    B    C

SHH

N-SHH

Cholesterol

N-SHH
– Chol

PTCH

Cell membrane    SMO

GLI transcription factors

CREBBP

Nucleus

Target genes

D

E

**FIGURE 10-2**

The Sonic hedgehog (*Shh*)–Patched (*Ptch*)–Gli pathway signaling pathway and associated disorders. Shh protein to which a cholesterol moiety has been attached binds to Patched. Smo is normally inhibited by Patched, but when bound by Shh protein, this inhibition is released, and Smo is available to activate downstream targets such as Gli transcription factors. cAMP response element-binding protein (CREBBP) is a cofactor of Gli transcription factors. Disorders caused by perturbation of proteins in this pathway include (clockwise from *top left*) holoprosencephaly (Shh; abnormal midline brain development), Smith–Lemli–Opitz (cholesterol biosynthesis; see Chapter 7), Gorlin or nevoid basal cell carcinoma syndrome (PTCH; rib anomalies, cysts of the jaw, and basal cell skin cancer), Greig cephalopolysyndactyly syndrome (Gli; craniosynostosis, polydactyly), and Rubenstein–Taybi syndrome (CREBBP; mental retardation, distinctive facial features and broad thumbs). cAMP, cyclic adenosine monophosphate.

*(Photos from Jones K: Smith's Recognizable Patterns of Human Malformation, 6th ed, pp 116. Philadelphia: Saunders, 2006; diagram modified from Turnpenny P, Ellard S: Emery's Elements of Medical Genetics, 13th ed. London: Churchill Livingstone, 2007.)*

by forming heterodimers with them and preventing their secretion. Thus, mutations causing Grebe chondrodysplasia produce a novel type of dominant negative effect by inactivating the products of other genes.

To effect a response, extracellular signals must be transduced by a cell. One of the best known of these signal-transduction systems is the RTK/Ras GTPase/MAPK* (RTK-MAPK) signaling pathway. The RTK-MAPK pathway regulates various cellular functions such as gene expression, division, differentiation, and death. Accordingly, the RTK-MAPK pathway is used widely during development. Recently, mutations in genes that encode several components of the RTK-MAPK pathway have been found to cause human malformation syndromes (Fig. 10-3). The best-known of these, Noonan syndrome, is characterized by short stature, characteristic facial features, webbing of the neck, and congenital heart disease—most commonly stenosis of the pulmonary outflow tract. Most cases of Noonan syndrome are caused by gain-of-function mutations in the *protein tyrosine phosphatase, non-receptor-type, 11 (PTPN11)* gene, which encodes a protein that interacts with the RTK-MAPK pathway. The clinical characteristics of Noonan syndrome

overlap with those of two rarer conditions, Costello syndrome and cardiofaciocutaneous syndrome (CFC) syndrome. These conditions are caused by mutations in other components of the RTK-MAPK pathway (see Fig. 10-3). Thus, disruption of different components of the same signaling or developmental pathway cause different malformation syndromes with overlapping clinical features.

Other secreted proteins inhibit the function of BMPs. In humans, mutations in the gene *Noggin*, which encodes one of these inhibitors, cause fusion of the bones in various joints. In some affected persons, the joints initially appear to be normal. However, they are progressively obliterated by formation of excess cartilage that ultimately fuses the bones of the joint together (i.e., a synostosis) as the person ages. The primary joints affected are those of the spine, the middle ear bones, and the limbs, particularly the hands and feet. Affected persons experience progressively limited movement of these joints and develop hearing loss.

> Paracrine signaling molecules are secreted, diffuse a short distance, and bind to a receptor that effects a response. There are four major families of paracrine signaling molecules: the fibroblast growth factor (FGF) family, the Hedgehog family, the Wingless (Wnt) family, and the Transforming Growth Factor β (TGF-β) family.

---

*GTPase, guanosine triphosphatase; MAPK, mitogen-activated protein kinase; RTK, receptor tyrosine kinase.

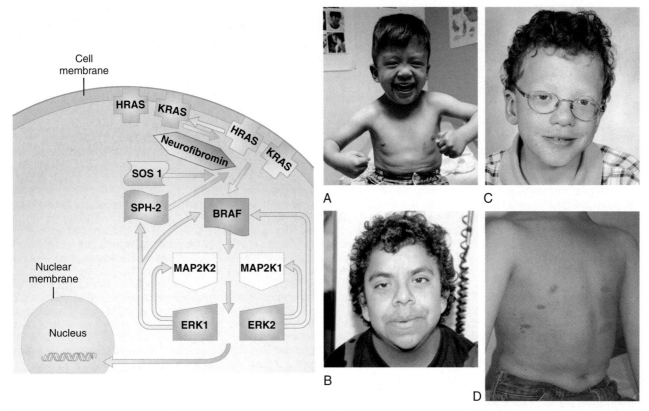

**FIGURE 10-3**
Mutations in genes in the RAS-MAPK signaling pathway cause Noonan syndrome **(A)** (short stature, webbing of the neck, and congenital heart disease), Costello syndrome **(B)** (mental retardation, thick lips, depressed nasal bridge, curly hair, cardiomyopathy), cardio-facial-cutaneous syndrome **(C)** (mental retardation, prominent forehead, hypertelorism, skin abnormalities), and neurofibromatosis type 1 (D; see chapter 4). RAS-MAPK (RTK/Ras GTPase/MAPK signaling system), GTPase, guanosine triphosphatase; MAPK, mitogen-activated protein kinase; RTK, receptor tyrosine kinase.

*(Photos from Jones KL: Smith's Recognizable Patterns of Human Malformation, 6th ed, pp 127. Philadelphia: Saunders, 2006; diagram modified from Turnpenny P, Ellard S: Emery's Elements of Medical Genetics, 13th ed. London: Churchill Livingstone, 2007.)*

## DNA Transcription Factors

There are many different ways to regulate the expression of a gene. For example, a gene might not be transcribed, the rate of transcription may be altered, or the transcribed mRNA might not be translated into protein. As discussed in Chapter 2, genes encoding proteins that turn on (activate) or turn off (repress) other genes are called *transcription factors*. Transcription factors commonly do not activate or repress only a single target. Often they regulate the transcription of many genes that, in turn, regulate other genes in a cascading effect. Consequently, mutations in transcription factor genes typically have pleiotropic effects.

There are many different families of transcription factors, members of which commonly share specific properties such as a common DNA-binding domain. Members of these different families have pivotal roles in controlling development, and alterations can cause birth defects. Examples include homeobox-containing genes such as the HOX, PAX, EMX, and MSX families; high-mobility group (HMG)-box–containing genes such as the SOX family; and the T-box family.

The HMG domain of SOX proteins appears to activate transcription indirectly by bending DNA so that other factors can make contact with promoter regions of genes. Several SOX genes act in different developmental pathways. The prototypic SOX gene is the *SRY* (sex-determining region of the Y chromosome) gene, which encodes the mammalian testis-determining factor (see Clinical Commentary 6-2 in Chapter 6). *Sox9* is expressed in the genital ridges of both sexes, but it is up-regulated in males and down-regulated in females before differentiation of the gonads. *Sox9* also regulates chondrogenesis and the expression of *Col2A1*, a collagen gene (see Chapter 2). As might be predicted from these expression and interaction patterns, mutations in *SOX9* cause a disorder characterized by skeletal defects (campomelic dysplasia) and sex reversal that produces XY females. Mutations in a related gene, *SOX10*, result in a syndrome characterized by Hirschsprung disease (hypomotility of the bowel caused by a reduced number of enteric nerve cells), pigmentary disturbances, and deafness (Clinical Commentary 10-2).

## CLINICAL COMMENTARY 10-2
### Defects of Neural Crest Development

During neurulation, neural crest cells migrate from the neuroepithelium along defined routes to tissues, where they differentiate into various cell types. One fate of neural crest cells is to populate the small and large bowel (i.e., the enteric tract) with nerve cells to create the enteric nervous system. These cells partly control and coordinate the normal movements of the enteric tract that facilitate digestion and transport of bowel contents. Reduced or absent nerve cells in the enteric tract causes a disorder called Hirschsprung disease (HSCR).

HSCR occurs in approximately 1 in 5000 live births, although its incidence varies among ethnic groups. Additionally, there is a sex bias, with males affected four times as often as females. The major characteristic of HSCR is hypomotility of the bowel, which leads to severe constipation. The disease often manifests in the newborn period, although it is also found in children and sometimes adults. If untreated, bowel hypomotility can lead to obstruction and gross distention of the bowel. Consequently, HSCR was formerly termed congenital megacolon.

In approximately 70% of cases, HSCR occurs as an isolated trait, and affected persons have no additional problems. However, HSCR is also a well-recognized feature of many multiple birth defect syndromes, such as trisomy 21 and Waardenburg syndrome. For the last few decades, HSCR has been considered an example of a disorder that fits a multifactorial model of inheritance (i.e., caused by a combination of genes plus environmental factors; see Chapter 12). However, it has become clear that half of all cases of familial HSCR and 15% to 20% of sporadic cases are caused by mutations in one of at least eight different genes. Study of these genes provides us with a window through which we can observe the development of neural crest cells.

Most commonly, HSCR is caused by mutations that inactivate the *RET* (*r*earranged during *t*ransfection) gene, which encodes a receptor tyrosine kinase (other mutations in *RET* have been associated with cancer; see Chapter 11). Dozens of different mutations have been found, including missense and nonsense mutations as well as deletions encompassing the *RET* gene. Thus, haploinsufficiency is the most likely mechanism by which mutations in *RET* cause HSCR. The penetrance of *RET* mutations is higher in males than in females, suggesting that sex-specific modifiers of the phenotype might exist.

Normal signaling through *RET* appears to be required for neural crest migration into the distal portions of the bowel and for differentiation into nerve cells. One ligand for *RET* is glial cell line derived neurotrophic factor (GDNF).

Mutations of another cell membrane receptor, endothelin-B (*EDNRB*), or its ligand, endothelin-3 (*EDN3*), also cause HSCR. Penetrance appears to vary by sex and genotype. In a large Mennonite community, persons homozygous for an *EDNRB* mutation were four times more likely to develop HSCR than were heterozygotes. In addition to HSCR, some persons with mutations in either *EDNRB* or *EDN3* have melanocyte abnormalities that produce hypopigmented patches of skin and sensorineural hearing loss (normal melanocytes are required for auditory development). This disorder is the Waardenburg–Shah syndrome. Thus, normal EDNRB and EDN3 signaling is required for the development of neural crest cells into enteric nerve cells and melanocytes.

More recently, mutations in the *SOX10* transcription factor gene have been found in persons with Waardenburg–Shah syndrome. Disruption of the homologous mouse gene, *Sox10*, causes coat spotting and aganglionic megacolon. Although SOX genes are involved in many different biological processes, the role of *SOX10* in neural crest cell development remains to be determined.

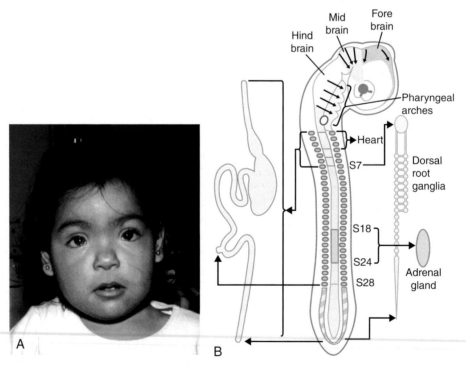

Fates of selected populations of neural crest cells migrating from different levels along the anterior/posterior axis of the developing embryo. Appropriate fate of neural crest derivatives depends on normal cell migration and terminal differentiation. Defects of neural crest cells can cause Hirschsprung disease or Waardenburg–Shah syndrome (see text).

*(Photo from Jones KL: Smith's Recognizable Patterns of Human Malformation, 6th ed. Philadelphia: Mosby, 2006; diagram modified from Gilbert S: Developmental Biology, 7th ed. Sunderland, MA: Sinauer, 2003, p 429.)*

> There are many different families of transcription factors, each of which regulates the transcription of specific genes. The same transcription factor is often used in different developmental pathways. Thus, disorders caused by mutations in genes encoding transcription factors are often pleiotropic.

### Extracellular Matrix Proteins

Extracellular matrix proteins (EMPs) are secreted macromolecules that serve as scaffolding for all tissues and organs. These molecules include collagens, fibrillins, proteoglycans, and large glycoproteins such as fibronectin, laminin, and tenascin. EMPs are not simply passive structural elements. By separating adjacent groups of cells and forming matrices on which cells can migrate, they are active mediators of development. For example, the proteins encoded by **fibrillin-1** and elastin coordinate microfibril assembly in the extracellular matrix. Mutations in these two genes result in Marfan syndrome (see Chapter 4) and supravalvular aortic stenosis (see Chapter 6), respectively. Both of these conditions are characterized by abnormalities of the heart and/or large blood vessels.

To facilitate cell migration, EMPs must transiently adhere to a cell's surface. This is commonly accomplished by two families of cell surface receptors: integrins and glycosyltransferases. Integrins are so named because they integrate the extracellular matrix and the cytoskeleton, allowing them to function in tandem. Attachment between cells and the extracellular matrix can be more permanent as well. One group of molecules that make such attachments are called laminins. Mutations in *LAMC2*, a gene that encodes a subunit of laminin, cause autosomal recessive junctional epidermolysis bullosa (JEB). Because of an inability of epithelial cells to properly anchor themselves to the basement membrane, the skin of persons with JEB spontaneously forms large blisters.

> EMPs are secreted macromolecules that serve as a dynamic scaffolding for tissues and organs. They are also active mediators of development.

## PATTERN FORMATION

The process by which ordered spatial arrangements of differentiated cells create tissues and organs is called **pattern formation**. The general pattern of the animal body plan is laid down during embryogenesis. This leads to the formation of semiautonomous regions of the embryo, in which the process of pattern formation is repeated to form organs and appendages. Such regional specification takes place in several steps: definition of the cells of a region, establishment of signaling centers that provide positional information, and differentiation of cells within a region in response to additional cues. For example, cells in the developing vertebrate upper limb differentiate into many cell types, including muscle (myocytes), cartilage (chondrocytes), and bone cells (osteocytes). However, these cells must also be arranged in a temporal–spatial pattern that creates functional muscle and bone. Additional information is required to determine whether a bone becomes an ulna or a humerus. How do

particular structures develop in specific places? How do cells acquire information about their relative positions? Answering such questions is an area of intense investigation.

For pattern formation to occur, cells and tissues communicate with each other through many different signaling pathways. It has become clear that these pathways are used repeatedly and are integrated to control specific cell fates (i.e., the eventual location and function of the cell). For example, the Shh protein is involved in patterning of the vertebrate neural tube, somites, and limbs, as well as the way the left side is distinguished from the right. Point mutations in the human *Shh* gene, *SHH*, can cause abnormal midline brain development (holoprosencephaly; see Fig. 10-2), severe mental retardation, and early death. (Not all affected persons have holoprosencephaly, however; some have only minor birth defects, such as a single upper central incisor.) Attachment of the SHH protein to cholesterol appears to be necessary for proper patterning of hedgehog signaling. This might partly explain how midline brain defects could be caused by some environmental substances that inhibit embryonic cholesterol biosynthesis and by genetic disorders of cholesterol metabolism such as Smith–Lemli–Opitz syndrome (see Fig. 10-2).

> The process by which ordered spatial arrangements of differentiated cells create tissues and organs is called pattern formation. Regional specification takes place in several steps: definition of the cells of a region, establishment of signaling centers that provide positional information, and differentiation of cells within a region in response to additional cues.

### Gastrulation

**Gastrulation** is the process of cell and tissue movements whereby the cells of the blastula are rearranged so that they have new positions and neighbors. In the human embryo, gastrulation occurs between days 14 and 28 of gestation. In this process, the embryo is transformed into a three-layer (trilaminar) structure composed of three germ layers: ectoderm (outer layer), endoderm (inner layer), and mesoderm (middle layer) (Fig. 10-4). The formation of these layers is a prerequisite for the next phase of development, organogenesis. The major structural feature of mammalian gastrulation is the **primitive streak**, which appears as a thickening of epiblast tissue extending along the anterior/posterior axis (see Fig. 10-4). In placental animals such as humans, gastrulation includes formation of the extraembryonic tissues. As might be predicted, the process of gastrulation is dominated by cell migration. Thus, many of the genes expressed during gastrulation encode proteins that facilitate cell movement.

> Human gastrulation is characterized by cell and tissue movements that result in the formation of three germ layers: ectoderm, endoderm, and mesoderm. The major structural feature of mammalian gastrulation is the primitive streak.

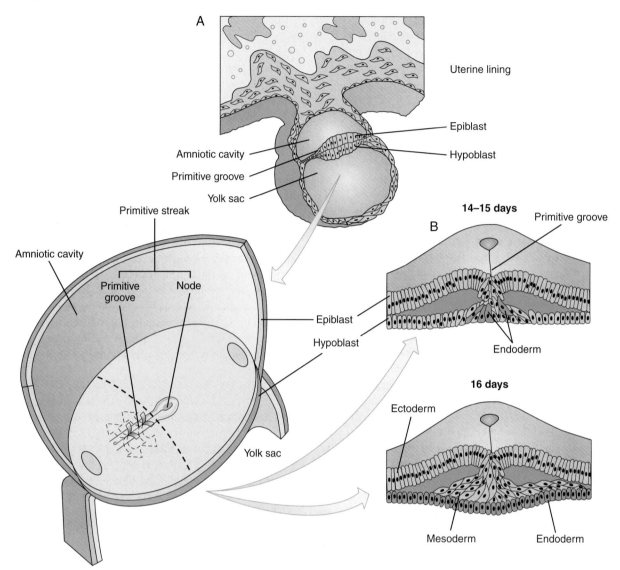

**FIGURE 10-4**
Human gastrulation. **A**, Sagittal section through the midline of an embryo embedded in the uterine lining. **B**, Dorsal surface of an embryo exposed by removing part of the embryonic mesoderm surrounding the amniotic cavity and yolk sac. *Arrows* denote ingressing epiblast cells. On days 14 to 15, epiblast cells replace hypoblast cells to form endoderm. A day later, migrating epiblast cells are creating a layer of mesoderm.

### Neurulation and Ectoderm

Once a trilaminar embryo is formed, the dorsal mesoderm and the overlying ectoderm interact to form the hollow neural tube. This event, called **neurulation**, is mediated by induction. In amphibians, induction of the neural tube and transformation of the flanking mesoderm to create an embryo with clear anterior/posterior and dorsal/ventral axes is controlled by a group of cells known as the **Spemann–Mangold organizer**. A number of proteins are expressed almost exclusively in the organizer. Chordin is a secreted protein that prevents dorsalized mesoderm from being ventralized. Another secreted protein, Noggin, induces neural tissue from dorsal ectoderm and dorsalizes the mesoderm. Understanding the major functions of the organizer and the molecules that mediate these functions is an area of active research.

Neurulation is a critical event in development because it initiates organogenesis and divides the ectoderm into three different cell populations: the neural tube, which will eventually form the brain and spinal cord; the epidermis of the skin; and the neural crest cells. In humans, neural tube closure begins at five separate sites, which correspond to the locations of common neural tube defects such as anencephaly (absence of the brain), occipital encephalocele, and lumbar spina bifida (see Chapter 12). Neural crest cells migrate from the neuroepithelium along defined routes to tissues, where they differentiate into cell types such as sensory neurons, melanocytes, neurons of the small bowel, and smooth muscle cells (see Clinical Commentary 10-2).

Induction is the process by which cells of one embryonic region influence the organization and differentiation of cells in a second embryonic region. Neurulation initiates organogenesis and induces the ectoderm to divide into the neural tube and neural crest cells. Defects of neural tube closure and neural crest migration or differentiation cause some types of birth defects.

## Mesoderm and Endoderm

The formation of a layer of mesoderm between the endoderm and ectoderm is one of the major events in gastrulation. Mesoderm can be divided into five components: the notochord; the dorsal, intermediate, and lateral mesoderms; and head **mesenchyme**.* The notochord is a transient midline structure that induces the formation of the neural tube and body axis. The adjacent dorsal mesoderm on either side of the notochord differentiates into elements that form the axial skeleton, skeletal muscles, and connective tissue of the skin. Intermediate mesoderm forms the kidneys and genitourinary system. Lateral plate mesoderm differentiates into the heart, the appendicular skeleton, the connective tissue of viscera and the body wall, and the connective tissue elements of the amnion and chorion. Finally, the muscles of the eyes and head arise from head mesenchyme.

The primary function of embryonic endoderm is to form the linings of the digestive tract and the respiratory tree. Outgrowths of the intestinal tract form the pancreas, gallbladder, and liver. A bifurcation of the respiratory tree produces the left and right lungs. The endoderm also produces the pharyngeal pouches, which, in conjunction with cells derived from the neural crest, give rise to endodermal-lined structures such as the middle ear, thymus, parathyroids, and thyroid.

A process common to endoderm-derived structures is budding and branching. This process appears to be controlled, in part, by FGFs, BMPs, and their respective receptors. Mutations in *fibroblast growth factor receptor 3* (*FGFR3*), one of four FGFRs, cause three different skeletal dysplasias (see Clinical Commentary 10-1). The most severe of these, thanatophoric dysplasia, is caused by mutations that activate *FGFR3*, resulting in shortened long bones, a poorly developed vertebral column, a small thoracic cage, and a relatively large skull. Children with thanatophoric dysplasia may also have pulmonary hypoplasia and brain anomalies, suggesting that *FGFR3* plays a role in formation of the lung and brain.

Formation of a layer of mesoderm between endoderm and ectoderm is one of the major events of gastrulation. Mesoderm contributes to the formation of the skeleton, urogenital system, and limbs. Endoderm lines the digestive and respiratory tracts and forms visceral organs and the lungs.

---

*Mesenchyme is tissue that forms the connective tissues, blood vessels, and lymphatic vessels.

## Axis Specification

Animal body plans have evolved into a wide variety of symmetries. Some animals, such as the sea anemones, are completely symmetrical. Other animals (e.g., starfish) exhibit only a dorsal/ventral symmetry. Many animals, such as worms, add an anterior/posterior axis. All chordates (animals that develop a notochord) have a third axis that is perpendicular to the first two, the left/right axis. Specification and formation of these axes are critical events in development because they determine the orientation of the body plan. The proteins mediating these processes are rapidly being discovered. Many of these mediators have additional roles in patterning of the body plan and tissues.

### Formation of the Anterior/Posterior Axis

The anterior/posterior axis of a developing mammalian embryo is defined by the primitive streak. At the anterior end of the primitive streak is a structure called the *node*. Expression of a gene called *nodal* is required for initiating and maintaining the primitive streak; later in gastrulation, its expression is important in forming the left/right axis (Clinical Commentary 10-3).

Patterning along the anterior/posterior axis is controlled by a cluster of genes that encode transcription factors that contain a DNA-binding region, the **homeodomain**, of approximately 60 amino acids. These genes compose the homeotic gene complex (HOM-C) in *Drosophila* (Fig. 10-5), the organism in which they were first isolated through mutation identification. (A classic example of such a mutation, termed antennapedia, disturbs axis patterning so that antennae are replaced by legs.) In contrast to *Drosophila*, four copies of *HOM-C* (termed *HoxA* through *HoxD*) are found in humans and mice. Each of these four 100-kb gene clusters is located on a different chromosome, and there are 39 Hox genes divided among the clusters. Mammalian Hox genes are numbered from 1 to 13, although not every cluster contains 13 genes. Equivalent genes in each complex (e.g., *Hoxa13*, *Hoxc13*, *Hoxd13*) are called **paralogs**.

Hox genes are expressed along the dorsal axis from the anterior boundary of the hindbrain to the tail. Within each cluster, 3′ Hox genes are expressed earlier than 5′ Hox genes (**temporal colinearity**). Also, the 3′ Hox genes are expressed anterior to the 5′ Hox genes (**spatial colinearity**). Thus, *Hoxa1* expression occurs earlier and in a more anterior location than does the expression of *Hoxa2* (see Fig. 10-5). These overlapping domains of Hox gene expression produce combinations of codes that specify the positions of cells and tissues. Collectively these codes identify various regions along the anterior/posterior axis of the trunk and limbs. To study the role of different Hox genes in mammalian development, it has been common to produce a knockout mouse—a mouse that lacks a functional copy of the gene of interest (Box 10-1).

The anterior/posterior axis of a developing mammalian embryo is defined by the primitive streak and patterned by combinations of Hox genes. Collectively, these combinations identify various regions along the anterior/posterior axis of the body and limbs. Disruption of Hox genes produces defects in body, limb, and organ patterning.

## CLINICAL COMMENTARY   10-3
### Laterality Defects: Disorders of the Left/Right Axis

Left/right (L/R) asymmetry is common in nature. For example, all animals use only L-amino acids and D-sugars. Likewise, all vertebrates have asymmetrical structures that are consistently oriented L/R of the midline. For example, looping of the heart tube toward the right, the first observable sign of L/R asymmetry in the embryo, is seen in all chordates. How did L/R asymmetry evolve? How and when is L/R asymmetry established? Substantial progress is being made toward understanding the molecular basis of L/R asymmetry so that answers may be obtained for these questions. This is important because disorders of L/R asymmetry (i.e., laterality defects) are found in approximately 1 in 10,000 live births.

The final position of asymmetrically placed vertebrate structures is determined by at least three different mechanisms. Unpaired organs in the chest and abdomen (e.g., heart, liver) begin their development in the midline and then lateralize to their adult positions. The mirror image of a paired structure can regress, leaving a lateralized, unpaired structure (e.g., some blood vessels). Some organs (e.g., lungs) begin as asymmetrical

outgrowths from a midline structure. It is unknown whether the molecular basis of laterality determination differs among these mechanisms. Disorders of L/R asymmetry can cause randomization (situs ambiguus) or L/R reversal of organ position (situs inversus). These defects may be limited to a single organ (as with a right-sided heart, or dextrocardia), or they may include many organs with L/R asymmetry (e.g., stomach and spleen). Midline defects (e.g., cleft palate, neural tube defects) are observed in nearly half of all persons with disorders of L/R asymmetry.

Initially, establishment of L/R asymmetry requires a mechanism that generates asymmetry. Molecular asymmetry around the node, the anterior end of the primitive streak, appears to play a critical role in this process. A number of genes and proteins that exhibit asymmetrical activity before node formation have been found in chicks, zebra fish, and frogs, but not in mice. In mice, certain nodal cells have motile cilia on their surface. The beating action of these cilia creates a leftward flow of perinodal fluid into which it appears that morphogens involved in symmetry breaking are released.

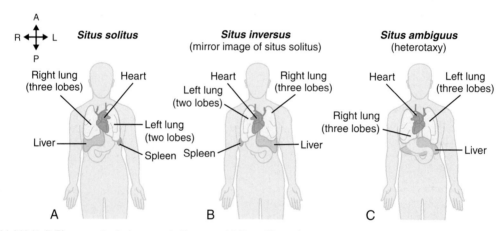

Abnormalities of left/right (L/R) asymmetry in humans. **A,** The normal L/R positions of organs arranged along midline (situs solitus). The apex of the heart points toward the left. The right lung is trilobed and the left lung is bilobed. In the abdominal cavity, the spleen and stomach are positioned on the left side and the liver is on the right. The small bowel is looped in a counterclockwise direction. **B,** A complete mirror image of organ arrangement along the midline is called *situs inversus*. Persons with situs inversus might have no symptoms. **C,** Randomization of the arrangement of the heart, lungs, liver, spleen, and stomach along the midline (situs ambiguus, or heterotaxy). Situs ambiguus is often associated with congenital heart defects.

The function of these cilia partly depends on the expression of two proteins, left-right dynein (lrd) and polycystin-2. Dynein abnormalities in humans cause a group of disorders called primary ciliary dyskinesias, in which most individuals have situs inversus. Abnormal ciliary function is also associated with recurrent sinusitis, infertility, and hydrocephalus. Mutations in *PKD1*, the gene encoding polycystin-2 produce laterality defects in mice and autosomal dominant polycystic kidney disease in humans (see Chapter 4).

Although more than 75 genes are known to be required for normal L/R development in model organisms, mutations in only a few of these have been found to cause laterality defects in humans. Mutations in *zinc-finger protein of the cerebellum* (*ZIC3*), a member of the Gli transcription factor family located on the X chromosome, are the most common known genetic cause of human laterality defects. Affected males exhibit randomization defects, and some carrier females have L/R reversal. In *Drosophila*, some members of the Gli family are known to be regulated by forming a complex with costal2, a motor molecule similar to dynein. This could explain how

mutations in genes encoding dissimilar proteins could cause both human laterality disorders. Other genes that cause laterality defects include *LEFTYA*, *CRYPTIC*, and *ACVR2B*.

Once L/R asymmetry has been established in the early embryo, the left and right sides of individual organs must also be patterned. For example, two related transcription factors, dHAND and eHAND, play roles in patterning the right and left ventricles of the heart. In mice, homozygous mutation of *dHAND* produces animals that fail to form a right ventricle, indicating that *dHAND* participates in cardiac differentiation.

Abnormalities of L/R asymmetry are found more often in human conjoined twins than in singletons or dizygotic twins. Most commonly, it is the twin arising on the right side that exhibits randomization of the L/R information. It has been suggested that randomization of the right-sided twin is caused by inadequate signaling from the left-sided embryo. A candidate for this signaling molecule, discovered in frogs, is Vg1. This suggests a possible molecular pathway for the formation of birth defects in human conjoined twins.

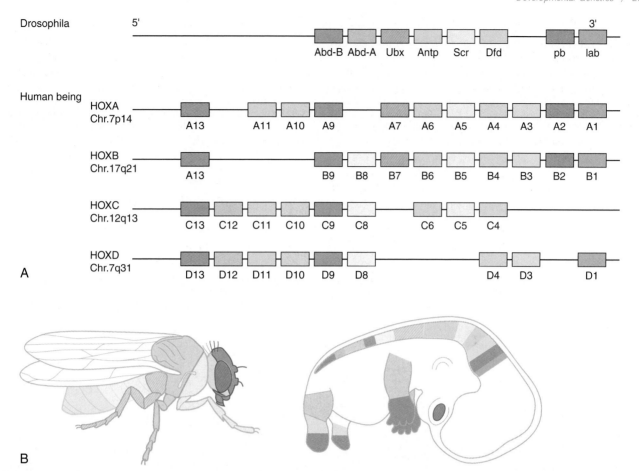

**FIGURE 10-5**

**A,** Distribution of 8 Hox genes in a single cluster in *Drosophila* and 39 Hox genes among clusters on 4 chromosomes in human (*HOX*). Individual Hox genes are labeled from 1 (3') to 13 (5') within each cluster. Hox genes that share the same number but are located in different clusters are called *paralogs* (e.g., *HOXA13* and *HOXD13* are paralogs). Paralogs often exhibit more sequence homology than do different Hox genes in the same cluster. Hox genes are expressed from 3' to 5' along the anterior/posterior axis of the embryo, and Hox genes located 3' are expressed earlier than Hox genes located 5'. **B,** Schematic diagram of combinatorial codes of overlapping Hox gene expression domains along the anterior/posterior body axis. Hox codes determine the identity of each segment. Thus, if expression of *Hoxb4* is eliminated (e.g., in a knockout), the combinatorial code in the third segment is altered from 1 + 2 + 3 to 1 + 2. This results in transformation of the third segment into another second segment. The transformation of one structure into another is called a homeotic transformation.

*(Modified from Verakas A, Del Campo M, McGinnis W: Developmental patterning genes and their conserved functions: From model organisms to humans. Mol Genet Metab 2000;69:85-100, with permission.)*

## Formation of the Dorsal/Ventral Axis

Dorsal/ventral patterning of the vertebrate depends on the interaction between dorsalizing and ventralizing signals. As previously mentioned, *noggin* and *chordin* encode secreted proteins that are capable of dorsalizing ventral mesoderm and restoring dorsal structures that have been ventralized. In contrast, *Bmp*-4 is expressed ventrally and induces ventral fates, patterning the dorsal/ventral axis. Noggin and chordin bind directly to Bmp-4 to prevent it from activating its receptor. Thus, the organizer promotes dorsalization by repressing a ventralizing signal encoded by *Bmp*-4. This mechanism, in which a signal promotes one process by repressing another competing process, is a common feature of embryonic development.

▶ Dorsal/ventral patterning of the embryo is an active process that is coordinated by signaling molecules and their antagonists.

## Formation of Organs and Appendages

The formation of organs and limbs (organogenesis) occurs after gastrulation. Many of the proteins used during this process are the same ones used earlier in embryonic development. As might be expected, a number of genes that were transcriptionally silent now become active. To date, most of the developmental genes known to cause human birth defects have prominent roles in this phase of development. This could represent an ascertainment bias, because mutations in genes that disrupt earlier developmental events may be lethal.

## Craniofacial Development

Development of the craniofacial region is directly related to the formation of the underlying central nervous system. In mammalian embryos, neural crest cells from the forebrain and midbrain contribute to the nasal processes, palate, and mesenchyme of the first pharyngeal pouch. This mesenchyme

## BOX 10-1
## Animal Models in the Study of Human Development

There are significant obstacles to the study of genes that affect human development. Many of these genes are expressed embryonically, and it is difficult (or in some cases ethically undesirable) to analyze human embryos directly. Humans have relatively small family sizes and a long generation time. Human mating patterns are often not conducive to genetic study. For these and other reasons, animal models of human diseases are a useful alternative to the direct study of the disease in humans.

The mouse is commonly used as an animal model of human disease because it is a well-understood and easily manipulated experimental system and because many developmental genes are conserved in most mammalian species. In some cases, a naturally

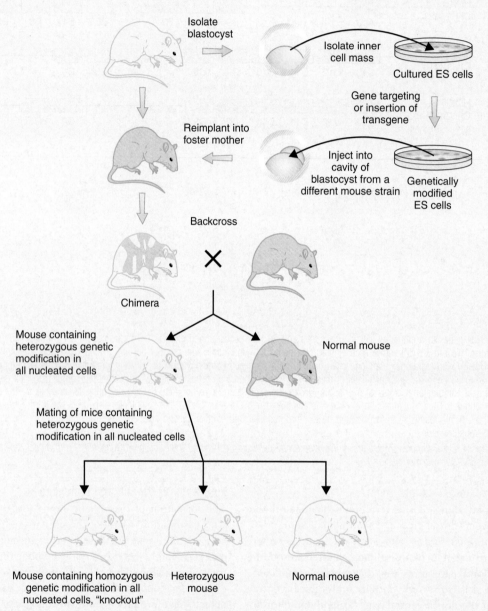

Construction of an animal model. Blastocysts are collected from a pregnant mouse that has a marker that identifies its strain (e.g., light coat color). The inner cell mass is isolated, and embryonic stem (ES) cells are cultured. ES cells can be modified to introduce foreign genes (creating a transgenic animal) or to disrupt the normal function of an endogenous gene (creating a knockout animal). Genetically modified ES cells are implanted into blastocysts from a different mouse strain that has a marker recessive to the marker in the modified strain (e.g., dark coat color is recessive to light coat color). Modified blastocysts are injected into a pseudopregnant surrogate mouse. Development of introduced blastocysts results in chimeric animals with two populations of cells (i.e., some cells have the genetic modification and other cells do not). Chimeras can be detected by the presence of two markers in the same mouse (e.g., two different coat colors in the same mouse). Backcrossing of chimeras and mating of heterozygotes can produce mice that are homozygous for a genetic modification (e.g., a knockout), heterozygous for a genetic modification, or normal.

*(Modified from Strachan T, Read AP: Human Molecular Genetics. Oxford: Bios Scientific, 1996.)*

## BOX 10-1
## Animal Models in the Study of Human Development—cont'd

occurring animal model of a human genetic disease exists (e.g., dog and mouse models for muscular dystrophy), but natural mouse models are relatively uncommon. To overcome this difficulty, human genes can be inserted directly into mouse embryonic stem cells, which are then placed into mouse embryos to create a **transgenic** mouse. The expression of the human gene can then be studied directly in mouse embryos. It is also possible to use **targeted disruption** to alter a specific mouse gene so that it is not expressed. This is a **knockout** model. Mice that are heterozygous for the disrupted gene can be bred to produce homozygotes. Many human genetic diseases have been studied using mouse knockouts, including neurofibromatosis type 1, Gaucher disease, Huntington disease, myotonic dystrophy, fragile X syndrome, cystic fibrosis, and subtypes of Alzheimer disease.

In addition to their roles in morphogenesis and organogenesis, some genes are critical for early embryogenesis. Consequently, knocking out their function results in embryonic lethality. This makes it difficult to study the role of these genes using targeted disruption. One way to overcome this problem is to condition the disruption of a gene so that it takes place only in a certain type of cell (e.g., neural crest), in a specific tissue (e.g., the limb), or at a specific time during development. This is a **conditional knockout**. For example, constitutional disruption of Fgf8 is lethal in early embryogenesis. To study the effect of inactivating *Fgf8* in the limb, a mouse can be engineered so that the function of *Fgf8* is disrupted only in the apical ectodermal ridge (AER) of the forelimb bud. The result is a live-born mouse whose forelimbs are severely truncated but whose other organs and body areas are all normal.

Animal models do not always mimic their human counterparts accurately. Sometimes this reflects differences in the interactions of gene products in the model system and the human. Such differences might account for the fact that a heterozygous mouse knockout of the retinoblastoma homolog (*RB*) develops pituitary tumors instead of retinoblastomas. In some cases the knockout has little detectable effect, possibly reflecting genetic redundancy: Even though the expression of one gene product is blocked, a backup system might compensate for its loss. Thus, the mouse knockout of either *Hoxa*11 or *Hoxd*11 alone has little phenotypic effect, but the simultaneous knockout of both genes produces a severe reduction in length of the radius and ulna. Despite these potential shortcomings, the introduction or disruption of genes in mice and other model systems can be a powerful approach for analyzing human genetic disease.

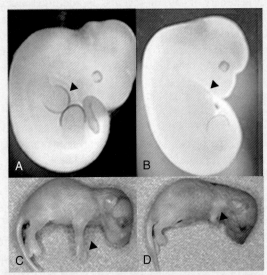

Conditional mutants that lack *Fgf8* expression in the apical ectodermal ridge (AER) of the forelimb. **A**, In situ hybridization demonstrating *Fgf8* expression (*dark band at tip of arrow*) in the AER of the developing limbs of a wild-type mouse. **B**, In the conditional mutant, no *Fgf8* is expressed in the forelimb (no dark band at *tip of arrow*), although it is expressed in the hindlimb bud. **C**, Normal forelimb in the wild-type mouse (*tip of arrow*). **D**, Severely hypoplastic limb in the conditional mutant (*tip of arrow*).
*(Courtesy of Dr. Anne Moon, University of Utah.)*

---

forms the maxilla, mandible, incus, and malleus. The neural crest cells of the anterior hindbrain migrate and differentiate to become the mesenchyme of the second pharyngeal pouch and the stapes and facial cartilage. Cervical neural crest cells produce the mesenchyme of the third, fourth, and sixth pharyngeal arches (in humans the fifth pharyngeal arch degenerates). This mesenchyme becomes the muscles and bones of the neck. The fate of each group of neural crest cells is specified by *Hox* genes. For example, functional inactivation of *Hoxa*3 results in mice with small or absent thymuses and thyroid and parathyroid glands, as well as malformations of the heart and major blood vessels. Although the number of neural crest cells in these mice is normal, they lack fate information and thus fail to proliferate and differentiate. These defects are similar to those found in children with deletions of chromosome 22q11 (see Chapter 6).

The bones of the skull develop directly from mesenchyme produced by neural crest cells. Complete fusion of these bones usually does not occur until adulthood. Premature fusion (synostosis) of the skull bones (craniosynostosis) causes the head to be misshapen and can impair brain growth. Often, craniosynostosis is associated with additional birth defects (e.g., hearing loss). Many of the craniosynostosis syndromes are caused by mutations in FGFR genes (see Clinical Commentary 10-1). Craniosynostosis also can be caused by mutations in *MSX*2, a transcription factor that may play a role in controlling the programmed death of neural crest cells in the skull.

Craniosynostosis is also a feature of Greig cephalopolysyndactyly, a disorder caused by mutations of the gene encoding GLI3, a zinc-finger transcription factor. *GLI*3 encodes at least seven conserved domains, including DNA-binding, zinc-finger, and microtubular anchor domains. Studies of the *Drosophila* homolog of *GLI*3 suggest that this gene may be regulated so that it can have either activator or repressor functions. Mutations that cause Greig cephalopolysyndactyly occur in the carboxy-terminal portion of *GLI*3, eliminating both its activator and repressor functions. Mutations in the region between the zinc-finger and the microtubular anchor domains produce a protein in which the amino terminal is cleaved so that it can migrate to the nucleus and repress transcription. Such mutations in *GLI*3 cause a disorder called Pallister–Hall syndrome, characterized by hypothalamic hamartomas, visceral anomalies, and posterior polydactyly. Mutations 3′ of the microtubular anchor domain produce a protein that retains both repressor and activator functions. These mutations have been described in persons with isolated posterior polydactyly, a relatively minor birth defect. Therefore, mutations in *GLI*3 alter the balance between its activator and repressor functions and cause three distinct disorders of varying severity. In addition, mutations that cause loss of function of a protein that acts as a cofactor of Gli proteins, CREBBP, cause Rubenstein–Taybi syndrome, a disorder characterized by mental retardation, distinctive facial features, and broad thumbs (see Fig. 10-2).

> The majority of craniofacial structures are derived from neural crest cells. The fate of each group of neural crest cells is specified by homeobox-containing genes. Some of the genes controlling craniofacial development have been isolated by analysis of craniosynostosis syndromes.

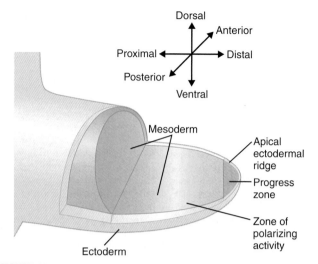

**FIGURE 10-6**

Schematic illustration of a limb bud. The apical ectodermal ridge (AER) extends from anterior to posterior along the dorsal/ventral boundary of the limb bud. Proximal to the AER is a region of rapidly proliferating mesodermal cells called the *progress zone* (PZ). Located in the posterior mesoderm is an important signaling center called the *zone of polarizing activity* (ZPA). The signaling pathways of the AER, PZ, and ZPA are interconnected so that pattern formation and growth partly depend on their coordinated function.

## Development of the Limb

The limb is our best-understood classic model of development. Surgical manipulation, ectopic gene expression, and targeted disruption of genes in animal models (see Box 10-1) have led to the isolation and characterization of many of the genes controlling growth and patterning of the limb. Many of the signaling pathways and transcriptional control elements that coordinate limb development in model organisms such as *Drosophila* and chick appear to be conserved in mammals. Because the newborn prevalence of limb defects is second only to that of congenital heart defects, the limb defect phenotypes are well documented. As a result, our knowledge of the molecular basis of human limb defects has expanded rapidly.

The vertebrate limb is composed of elements derived from lateral plate mesoderm (bone, cartilage, and tendons) and somitic mesoderm (muscle, nerve, and vasculature). The first step in the formation of a limb is its induction. The exact mediators and mechanism of limb induction remain controversial. The signal that initiates induction of forelimbs and hindlimbs appears to arise in the intermediate mesoderm, although that is not the only tissue involved in limb induction.

Once initiated, proximal/distal growth of the limb bud is dependent on a region of ectoderm called the *apical ectodermal ridge* (AER), which extends from anterior to posterior along the dorsal/ventral boundary of the limb bud (Fig. 10-6). Before AER differentiation, two genes, Radical fringe (*r-Fng*) and *Wnt7a*, are expressed in the dorsal ectoderm. In the ventral ectoderm, expression of *r-Fng* and *Wnt7a* is blocked by Engrailed-1 (*En-1*), a homeobox-containing transcription factor. Expression of *Wnt7a* instructs the mesoderm to adopt dorsal characteristics. Mesoderm in which *Wnt7a* expression is blocked becomes ventralized. Thus, the processes of AER formation and dorsal/ventral patterning are

interconnected and coordinated by *En-1*. In the mouse, functional inactivation of *Wnt7a* results in ventralization of the dorsal surface (i.e., pads on both sides of the foot). Ventralization of the dorsal surface of the limb has been described in humans, but the etiology remains unknown.

Mediation of proximal/distal growth by the AER is controlled in part by FGFs (e.g., *FGF2*, *FGF4*, *FGF8*) that stimulate proliferation of an underlying population of mesodermal cells in the progress zone (PZ). Maintenance of the AER is dependent on a signal from a posterior portion of the limb bud known as the *zone of polarizing activity* (ZPA). The signaling molecule of the ZPA is *Shh*, which is also responsible for dorsal/ventral patterning of the central nervous system and for establishing the embryonic left/right axis. The ZPA also specifies positional information along the anterior/posterior axis of the limb bud.

Defects of the anterior and posterior elements of the upper limb occur in the Holt–Oram syndrome and ulnar–mammary syndrome, respectively (Fig. 10-7). Holt–Oram syndrome is caused by mutations in the gene *TBX5*, and ulnar–mammary syndrome is caused by mutations in the tightly linked gene *TBX3*. *TBX3* and *TBX5* are members of a highly conserved family of DNA transcription factors containing a DNA-binding domain called a **T-box**. It appears that *TBX3* and *TBX5* have evolved from a common ancestral gene, and each has acquired specific yet complementary roles in patterning the anterior/posterior axis of the mammalian upper limb. *TBX3* and *TBX5* also play roles in the development of many other organs. For example, persons with Holt–Oram syndrome also have congenital heart defects, most commonly an atrial septal defect that allows blood in the left and right atria to mix. *TBX5* interacts with another transcription factor,

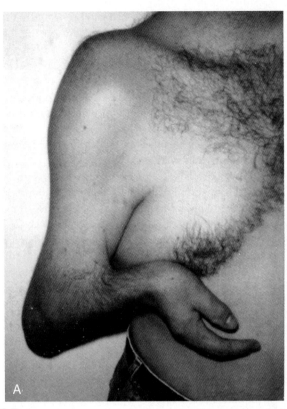
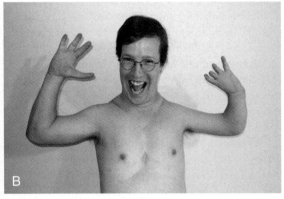

**FIGURE 10-7**
**A**, Absence of the right third, fourth, and fifth digits (i.e., posterior digits) accompanied by aplasia of the ulna and hypoplasia of the radius in a patient with ulnar–mammary syndrome. The fifth digit of the right hand is also missing. **B**, Bilateral absence of the thumb (i.e., an anterior digit) and radius with marked hypoplasia of the humerus in an adult with Holt–Oram syndrome.
*(From Jones KL: Smith's Recognizable Patterns of Human Malformation, 6th ed. Philadelphia: Mosby, 2006.)*

Nkx2-5, during heart development. Mutations in the gene encoding Nkx2-5 also cause atrial septal defects. Thus, perturbation of two different mediators in the same developmental program can produce the same type of birth defect.

If the early events of limb bud signaling provide positional information to developing cells, what controls the growth and differentiation of these cells? One important component is the transcription factors encoded by Hox genes. The expression patterns of *Hoxa*9 through *Hoxa*13 define overlapping domains along the proximal/distal axis of the developing limb bud. Combinations of Hox paralogs preferentially promote growth within different segments of the limb according to their 5′ position within a Hox cluster. For example, mice with *Hoxa*11 or *Hoxd*11 mutations have only minor abnormalities, but *Hoxa*11/*Hoxd*11 double mutants exhibit a marked reduction in the size of the radius and ulna. Similarly, deletion of increasing numbers of *Hox*-13 paralogs has a cumulative effect on phenotypic abnormalities of the hands or feet, presumably because the functions of Hox paralogs are partially redundant.

Because of this redundancy, it was suspected that mutations in Hox genes would be unlikely causes of human birth defects. However, Hox mutations have been described in persons with synpolydactyly and with hand–foot–genital syndrome. Synpolydactyly is characterized by duplication and fusion of the middle digits of the hands and feet. It is

caused by mutations in *HOXD*13 that produce an expansion of a polyalanine tract in the amino-terminal end of the *HOXD*13 protein. A similar defect can be reproduced by simultaneous disruption of *Hoxd*13, *Hoxd*12, and *Hoxd*11, suggesting that expansion of the polyalanine tract in *HOXD*13 results in the functional inactivation of its 3′ neighbors.

> The vertebrate limb is composed of elements derived from lateral plate mesoderm and somitic mesoderm. Growth and patterning are controlled by proteins secreted from specialized collections of cells called the apical ectodermal ridge, the progress zone, and the zone of polarizing activity.

## Organ Formation

Many developmental processes must be coordinated simultaneously to construct the specific arrangement of cells and tissues that compose an organ. As in limb development, the formation of organs involves numerous interactions. These interactions are mediated by secreted signaling molecules that bind to receptors, conduct signals through various interconnected pathways, and stimulate or repress DNA transcription. Use of the same elaborate networks to form different organs allows genomic economy while maintaining developmental flexibility.

Once a specialized cell within an organ is terminally differentiated, various proteins turn on its molecular machinery so that it may perform its intended function. Often, development of the organ and function of the differentiated cell are interrelated. For example, the endocrine pancreas is largely composed of three different cell types: alpha, beta, and gamma. Transcription of insulin in beta cells is stimulated by the binding of insulin promoter factor 1 (IPF1) to the insulin promoter region. Mutations in the gene encoding IPF1 prevent pancreatic development, indicating that IPF1 is necessary for the maturation and differentiation of pancreatic precursor cells.

Interactions between mesenchymal and epithelial cells are prominent in the development of cutaneous structures (e.g., hair, sweat glands, breasts), parenchymal organs (e.g., liver, pancreas), lungs, thyroid, kidneys, and teeth. These interactions are dynamic in that expression patterns in the epithelia and mesenchyme change over time and continue to influence one another. For example, during tooth development, the epithelium secretes *Bmp*-4, which signals the underlying mesenchyme to express a set of transcription factors, including Msx1. The mutual exchange of signals between epithelium and mesenchyme leads to formation of a dental papilla and cusp and finally to terminal differentiation of mesenchyme into tooth-forming odontoblasts. In humans, mutations in *MSX1* disrupt tooth formation and cause loss of the second premolars. Similarly, mutations in the human homolog of the mouse gene hairless cause loss of all body hair, including that of the scalp, eyebrows, axillae, and pubis.

The integrity of signals exchanged between the epithelium and the mesenchyme is dependent on the integrity of these tissues. Several proteins produced within the epithelium are known to promote the growth and differentiation of the epithelium. One of these proteins is p63, a homolog of the prototypic tumor suppressor gene product, p53. Mutations that perturb the function of p63 decrease the availability of epithelial progenitor cells. This results in abnormalities of the limbs, skin, teeth, hair, and nails in at least six different malformation syndromes.

One of the largest organs in the body is the skeleton. Skeletal formation is dependent on bone-forming cells called osteoblasts. The differentiation of osteoblasts is regulated by an osteoblast-specific transcription factor, Runx2. Targeted disruption of Runx2 produces mice with a complete lack of ossification of the skeleton. Heterozygous mice have widened cranial sutures, shortened digits, and abnormalities of the shoulder girdle. Similar defects are found in persons with cleidocranial dysplasia, which is caused by mutations in *Runx*2, the human homolog of mouse *Runx*2.

> Formation of organs involves reciprocal interactions between epithelium and mesenchyme. This interaction is mediated by secreted signaling molecules that bind to receptors, conduct signals through various interconnected pathways, and stimulate or repress DNA transcription.

## Study Questions

1. Explain how nonhuman animal models are useful for studying human development and birth defects. Give at least one example.

2. Mutations in fibroblast growth factor receptors (FGFRs) cause at least different craniosynostosis syndromes. Moreover, the same mutation in *FGFR*2 causes Pfeiffer syndrome in some families but Crouzon syndrome in other families. How can the same mutation produce two distinct disorders?

3. Disorders caused by mutations in genes encoding transcription factors are often pleiotropic. Explain this finding.

4. Define pattern formation, and give an example of a birth defect caused by disruption of this process.

5. If the control of developmental processes is tightly regulated, how can you explain that mutations in some developmental genes (e.g., Hox genes) produce subtle phenotypes?

6. Loss-of-function mutants in mammals are commonly studied by creating a knockout mouse model. Explain why it is impractical to use some organisms (e.g., baboons) to generate knockouts. Can you think of a way to circumvent some of these obstacles?

7. Give an example of a birth defect that can be caused by a defect in a ligand or its receptor.

8. Explain what may be some of the obstacles to using gene therapy to treat birth defects.

## Suggested Readings

Capecchi MR. Gene targeting in mice: Functional analysis of the mammalian genome for the twenty-first century. Nat Rev Genet 2005;6:507–12.

Chen D, Zhao M, Mundy GR. Bone morphogenetic proteins. Growth Factors 2004;22:233–41.

Epstein CP, Erickson RP, Wynshaw-Boris A, (eds): Inborn Errors of Development, 2nd ed. New York: Oxford University Press, 2008.

Farlie PG, McKeown SJ, Newgreen DF. The neural crest: Basic biology and clinical relationships in the craniofacial and enteric nervous systems. Birth Defects Res C Embryo Today 2004;72:173–89.

Gilbert SF. Developmental Biology, 8th ed. Sunderland, Mass: Sinauer, 2006.

Hoppler S, Kavanagh CL. Wnt signaling: Variety at the core. J Cell Sci 2007;120:385–93.

Iimura T, Pourquie O. Hox genes in time and space during vertebrate body formation. Develop Growth Differ 2007;49:265–75.

Isphording D, Leylek AM, Yeung J, et al. T-box genes and congenital heart/limb malformations. Clin Genet 2004; 66:253–64.

Itoh N, Ornitz DM. Evolution of the Fgf and Fgfr gene families. Trends Genet 2004;20:563–69.

Jorgensen EM, Mango SE. The art and design of genetic screens: *Caenorhabditis elegans*. Nat Rev Genet 2002;3:356–69.

Lang D, Powell SK, Plummer RS, et al. PAX genes: Roles in development, pathophysiology, and cancer. Biochem Pharm 2007;73:1–14.

Levin M. Left–right asymmetry in embryonic development: A comprehensive review. Mech Dev 2005; 122:3–25.

Lieschke G, Currie PD. Animal models of human disease: Zebra fish swim into view. Nat Rev Genet 2007;8: 353–66.

Webster MK, Donoghue DJ. FGFR activation in skeletal disorders: Too much of a good thing. Trends Genet 1997;13:178–82.

Wilkie AO. Bad bones, absent smell, selfish testes: The pleiotropic consequences of human FGF receptor mutations. Cytokine Growth Factor Rev 2005;16:187–203.

Zhu L, Belmont JW, Ware SM. Genetics of human heterotaxias. Eur J Hum Genet 2006;14:17–25.

# Chapter 11

# CANCER GENETICS

Present evidence indicates that approximately one in four deaths is now due to cancer and that invasive cancer will be diagnosed in more than half of the population at some point in their lives. Many cancers are increasing in incidence, largely as a result of the increased average age of our population.

As shown in this chapter, the causes of cancer are a mixture of environmental and genetic alterations that occur in our tissues. Inherited predisposition plays a role in some families. Dramatic advances in molecular biology and genetics have now clarified the basic molecular elements of cancer and provide a schematic outline of the cellular events leading to cancer. This understanding will be crucially important in the control of cancer, providing the beginnings of a base of knowledge that should lead to significantly improved therapies and possibly prevention.

"Cancer" is a collection of disorders that share the common feature of uncontrolled cell growth. This leads to a mass of cells termed a **neoplasm** (Greek, "new formation"), or **tumor**. The formation of tumors is called **tumorigenesis**. Several key events must occur if cells are to escape the usual constraints that prevent uncontrolled proliferation. Additional growth signals must be produced and processed, and cells must become resistant to signals that normally inhibit growth. Because these abnormal characteristics would typically trigger the process of programmed cell death (**apoptosis**), cells must somehow disable this process. The growing cell mass (tumor) requires nourishment, so a new blood supply must be obtained through **angiogenesis** (the formation of new blood vessels). Additional inhibitory signals must be overcome for the tumor to achieve a **malignant** state, in which neoplasms invade nearby tissues and **metastasize** (spread) to more distant sites in the body. The capacity to invade and metastasize distinguishes malignant from **benign** neoplasms.

Tumors are classified according to the tissue type in which they arise. Major types of tumors include those of epithelial tissue (**carcinomas**, the most common tumors), connective tissue (**sarcomas**), lymphatic tissue (**lymphomas**), glial cells of the central nervous system (**gliomas**), and hematopoietic organs (**leukemias**). The cells that compose a tumor are usually derived from a single ancestral cell, making them a single clone (**monoclonal**).

Many of the basic biological features of **carcinogenesis** (cancer development) are now understood. Throughout our lives, many of our cells continue to grow and differentiate. These cells form, for example, the epithelial layers of our lungs and colons and the precursor cells of our immune systems. Relatively undifferentiated stem cells produce large numbers of progeny cells to repopulate and renew our worn defensive layers. Through integration of information provided by a complex array of biochemical signals, the new cells eventually stop dividing and terminally differentiate into a cell type appropriate for their role in the body (Fig. 11-1). Alternatively, if the cell is abnormal or damaged, it may undergo apoptosis.

Occasionally one of these cells fails to differentiate and begins to divide without restraint. The descendants of such cells can become the founders of neoplasms, capable of further transformation into invasive, metastatic cancer. We wish to understand in detail what has gone wrong in these cells, to detect them early, and ultimately to intervene in their development so as to eliminate them.

> Cells in the body are programmed to develop, grow, differentiate, and die in response to a complex system of biochemical signals. Cancer results from the emergence of a clone of cells freed of these developmental programming constraints and capable of inappropriate proliferation.

## CAUSES OF CANCER
### Genetic Considerations

Genetic alterations of cell regulatory systems are the primary basis of carcinogenesis. We can create cancer in animal models by damaging specific genes. In cell culture systems, we can reverse a cancer phenotype by introducing normal copies of the damaged genes into the cell. Most of the genetic events that cause cancer occur in somatic cells. The frequency of these events can be altered by exposure to mutagens, thus establishing a link to environmental **carcinogens** (cancer-causing agents). However, these genetic events are not transmitted to future generations because they occur in somatic, rather than germline, cells. Even though they are *genetic* events, they are not *inherited*.

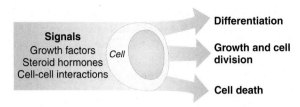

**FIGURE 11-1**
In response to environmental signals, a cell might continue to divide or it might differentiate or die (apoptosis).

It is also possible for cancer-predisposing mutations to occur in germline cells. This results in the transmission of cancer-causing genes from one generation to the next, producing families that have a high incidence of specific cancers (Fig. 11-2). Such "cancer families," although rare, demonstrate that inheritance of a damaged gene can cause cancer. In these families, inheritance of one mutant allele seems sufficient to cause a specific form of cancer: Almost all individuals who inherit the mutant allele will develop a tumor. This is because each of their cells now carries the altered gene and thus has already taken the first step down the cancer pathway. The childhood cancer of the eye, retinoblastoma, is a good example. As discussed in Chapter 4, those who inherit a mutant version of the retinoblastoma gene have approximately a 90% chance of developing one or more retinoblastoma tumors.

Although the transmission of cancer as a single-gene disorder is relatively uncommon, there is good evidence for more frequent clustering of some cancer types in families. For many kinds of cancer, such as those of the breast and colon, the diagnosis of the cancer in a first-degree relative implies at least a twofold increase in one's risk of developing the cancer. It is very likely that the inheritance of altered forms of specific genes is at least partly responsible for this increased risk.

The extent to which each of these mechanisms—inherited germline mutations versus mutations occurring in somatic cells—contributes to human cancer is an important question.

If inherited predispositions are significant determinants of a person's risk of acquiring a specific form of cancer, it should be possible to identify those whose risk is elevated. More intensive screening of defined high-risk groups could result in early detection and intervention, leading to better prognoses for patients and lowered morbidity and mortality.

> The basic cause of cancer is damage to specific genes. Usually, mutations in these genes accumulate in somatic cells over the years, until a cell accumulates a sufficient number of errors to initiate a tumor. If damage occurs in cells of the germline, however, an altered form of one of these genes can be transmitted to progeny and predispose them to cancer. The increased risk of cancer in such persons is due to the fact that each of their cells now carries the first step in a multistep cancer pathway.

## Environmental Considerations

What is the role of the nongenetic environment in carcinogenesis? At the level of the cell, cancer seems intrinsically genetic. Tumor cells arise when certain changes, or mutations, occur in genes that are responsible for regulating the cell's growth. However, the frequency and consequences of these mutations can be altered by a large number of environmental factors. It is well documented, for example, that many chemicals that cause mutation in experimental animals also cause cancer and are thus carcinogens. Furthermore, other environmental agents can enhance the growth of genetically altered cells without directly causing new mutations. Thus, it is often the interaction of genes with the environment that determines carcinogenesis; both play key roles in this process.

Two additional lines of argument support the idea that exposure to environmental agents can significantly alter a person's risk of cancer. The first is that a number of environmental agents with carcinogenic properties have been identified. For example, epidemiological studies and laboratory experiments have shown that cigarette smoke causes lung

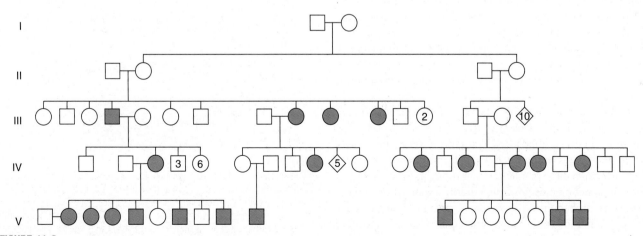

**FIGURE 11-2**
A familial colon cancer pedigree. Darkened symbols represent individuals with diagnosed colon cancer.

and other types of cancer. Roles for other environmental agents in specific cancers are also well documented (e.g., uranium dust in lung cancer among miners, asbestos exposure in lung cancer and mesothelioma).

The second line of argument is based on epidemiological comparisons among populations with differing lifestyles. Many kinds of cancer have quite different frequencies in different populations. Breast cancer, for example, is prevalent among northern Europeans and European Americans but relatively rare among women in developing countries. It is usually difficult to determine whether such dissimilarities reflect differences in lifestyle or in gene frequencies.

Examination of genetically similar populations under differing lifestyles, however, provides an opportunity to evaluate the genetic and environmental components of cancer. Epidemiological studies among migrant Japanese populations have yielded important findings with respect to colon cancer. This type of cancer was, until recently, relatively rare in the Japanese population living in Japan, with a lifetime risk of 0.5%, but it is 10 times more common in the United States. Stomach cancer, on the other hand, is common in Japan but relatively rare in the United States. These statistics in themselves cannot distinguish environmental from genetic influences in the two populations. However, because large numbers of Japanese have immigrated, first to Hawaii and then to the U.S. mainland, we can observe what happens to the rates of stomach and colon cancer among the immigrants. It is important to note that many of the Japanese immigrants maintained a genetic identity, marrying largely among themselves. Among the first-generation Japanese in Hawaii, the incidence of colon cancer rose several-fold—not yet as high as in the U.S. mainland, but higher than in Japan. Among second-generation Japanese Americans on the U.S. mainland, the colon cancer rate rose to 5%, equal to the U.S. average. At the same time, stomach cancer has become relatively rare among Japanese Americans.

These observations strongly suggest an important role for environment or lifestyle in the etiology of colon cancer. In each case, diet is a likely culprit: A high-fat, low-fiber diet in the United States is thought to increase the risk of colon cancer, whereas techniques used to preserve and season the fish commonly eaten in Japan are thought to increase the risk of stomach cancer. It is also interesting that the incidence of colon cancer in Japan has increased dramatically during the past several decades, as the Japanese population has adopted a diet more similar to that of North America and Europe.

Are we then to assume that genetic factors play no role in colon cancer? The fact remains that in the North American environment some people will get colon cancer and others will not. This distinction can result from differences within this environment (e.g., dietary variation) as well as differences in genetic predisposition: inherited genes that increase a person's probability of developing cancer. To account for the difference in colon cancer incidence between Japanese living in the U.S. and those living in Japan, it is argued that environmental features in Japan render the predisposing

genes less penetrant. Furthermore, a genetic component is strongly suggested by the several-fold increase in risk to a person when a first-degree relative has colon cancer. It is likely, then, that cancer risk is a composite of both genetics and environment, with interaction between the two components.

> Environmental factors are known to play important roles in carcinogenesis. However, a person's overall cancer risk depends on a combination of inherited factors and environmental components.

## CANCER GENES
### Genetic Control of Cell Growth and Differentiation

Cancers form when a clone of cells loses the normal controls over growth and differentiation. More than 100 cancer-causing genes that encode proteins participating in this regulation have now been identified. Characterization of the biochemical activities and interactions of these gene products has revealed an increasingly detailed picture of the normal regulation of cell growth and differentiation and the ways these processes become deregulated by the events of carcinogenesis.

Many features of this fundamental process are now well understood (Fig. 11-3). One component of cell regulation is mediated by external signals coming to the cell through

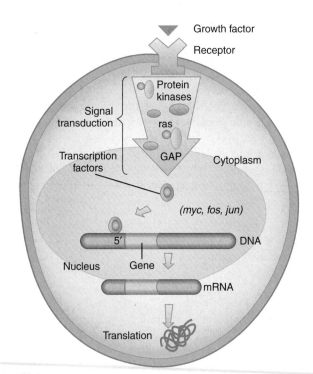

**FIGURE 11-3**
The major features of cellular regulation. External growth factors (proteins and steroid hormones such as epidermal growth factor) bind to membrane-spanning growth factor receptors on the cell surface, activating signal-transduction pathways in which genes such as *RAS* participate. Components of the signal-transduction pathway in turn interact with nuclear transcription factors, such as *MYC* and *FOS*, which can bind to regulatory regions in DNA. mRNA, mitochondrial RNA.

polypeptide **growth factors** (e.g., platelet-derived growth factor, epidermal growth factor, steroid hormones) produced in other cells. Each growth factor interacts with specific **growth factor receptors** located on the cell surface. Binding of a growth factor activates the receptor, triggering molecules that send messages to the cell's nucleus in the process of **signal transduction**. These signal transduction molecules include **protein kinases**, such as src tyrosine kinase, mitogen-activated protein kinase (MAPK), and jun kinase (JunK), which can alter the activity of target proteins by tagging them at a specific site with a phosphate molecule (phosphorylation). The ultimate stage of the signal transduction pathway is regulation of DNA transcription in the nucleus. Components of the signal transduction cascade interact with nuclear transcription factors that regulate the activity of specific genes whose protein products influence cellular growth and proliferation. Genes that encode these transcription factors include *MYC*, *FOS*, and *JUN*.

After several rounds of cell division, cells normally receive signals that tell them to stop growing and to differentiate into specialized cells. The signals may come from polypeptides, from steroid hormones, from direct contact with adjacent cells, or from internal programs that define the number of cell divisions that are allowed. The signals are transduced to the nucleus of the recipient cell. Here, by altering the transcription patterns of genes that govern the steps of the cell cycle, they repress genes that promote cell division and induce genes that inhibit entry into the cell division cycle.

> The regulation of cell growth is accomplished by substances that include: (1) growth factors that transmit signals from one cell to another; (2) specific receptors for the growth factors; (3) signal transduction molecules that activate a cascade of phosphorylating reactions within the cell; and (4) nuclear transcription factors. The cell integrates and interprets the host of signals it receives from its environment. Decisions to grow and divide, or to stop growing and differentiate, result from processing of these signals.

A cancer cell can emerge from within a population of growing cells through the accumulation of mutations in these genes. Although such mutations occur only rarely, these cells can fail to respond to differentiation signals and continue to divide instead of undergoing their normal differentiation program. Furthermore, cancers seem usually to result from a progressive series of events that incrementally increase the extent of deregulation within a cell lineage. Eventually, a cell emerges whose descendants multiply without appropriate restraints. Further changes give these cells the capacity to invade adjacent tissues and form metastases. Each of these changes involves mutations, and the requirement for more than one mutation has been characterized as the **multi-hit concept of carcinogenesis**. An example of this

concept is given by colorectal cancer, in which several genetic events are required to complete the progression from a benign growth to a malignant neoplasm (see later discussion).

> Mutations can occur in any of the steps involved in regulation of cell growth and differentiation. Accumulation of such mutations within a cell lineage can result in a progressive deregulation of growth, eventually producing a tumor cell.

### The Inherited Cancer Gene versus the Somatically Altered Gene

Although "cancer families" have long been recognized, it was not until the early 1970s that we began to understand the relationship between the inherited genetic aberrations and the carcinogenic events that occur in somatic tissue. In 1971, A. G. Knudson's analysis of retinoblastoma, a disease already mentioned as a model of inherited cancer, led him to a hypothesis that opened a new window into the mechanism of carcinogenesis. In the inherited form of retinoblastoma (see Chapter 4), an affected individual usually has an affected parent, and there is a 50% chance of genetic transmission to each of the offspring. In the sporadic (noninherited) form, neither parent is affected, nor is there additional risk to other progeny. A key feature distinguishing the two forms is that inherited retinoblastoma is usually bilateral (affecting both eyes), whereas sporadic retinoblastoma usually involves only a single tumor and therefore affects only one eye (unilateral).

Knudson reasoned that at least two mutations may be required to create a retinoblastoma. One of the mutations would alter the retinoblastoma gene; if this happened in the germline, it would be present in all cells of a child who received the mutant allele. The second mutation would be an additional, unspecified genetic event occurring in an already-altered cell. The hypothesis of a second event was required to explain why only a tiny fraction of the retinoblasts of a person who has inherited a mutant retinoblastoma gene actually give rise to tumors. Knudson's hypothesis is known as the **two-hit model of carcinogenesis**.

Familial retinoblastoma would thus be caused by the inheritance of one of the genetic "hits" as a **constitutional** mutation (i.e., a mutation present in all cells of the body). Persons who inherited one hit would require only one additional mutational event in a single retinoblast for that cell to seed a tumor clone. In sporadic cases, on the other hand, both mutations would have to occur somatically in the developing fetus (Fig. 11-4). This is a highly improbable combination of rare events, even considering the several million cells of the target tissue. The child who developed a retinoblastoma by this two-hit somatic route would be unlikely to develop more than one tumor. The child inheriting a mutant retinoblastoma gene, however, would need only a single, additional genetic hit in a retinoblast for a tumor clone to develop. Knudson argued that such an event was

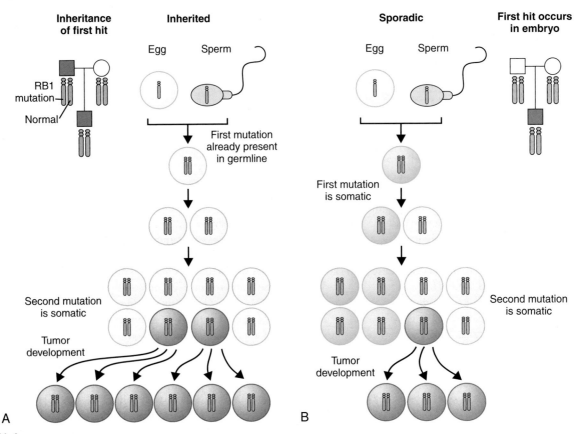

**FIGURE 11-4**

**A,** Persons inheriting an *RB1* mutation are heterozygous for the mutation in all cells of their body. The second hit occurs during embryonic development and typically affects more than one retinoblast, resulting in multiple tumors. **B,** In somatic retinoblastoma, both copies of the *RB1* gene must be disabled in the same retinoblast to cause tumor formation. Each process leads to homozygosity for the mutant *RB1* allele and thus to tumor development.

*(Data from Huether S, McCance K: Understanding Pathophysiology, 4th ed. St. Louis: Mosby, 2008, p 420.)*

likely to occur in several of the retinoblasts of each carrier of the inherited mutant gene, thus explaining the bilaterality of inherited retinoblastoma.

If the first event in the two-hit model is an inherited mutation, what is the nature of the second hit? Extensive molecular analysis of the region of chromosome 13 that contains the retinoblastoma-causing gene, *RB1*, showed that the second hit, like the first one, is a loss-of-function mutation. Several mechanisms, including point mutation, deletion, and hypermethylation of the *RB1* promoter region (associated with decreased transcription; see Chapters 3 and 5), can produce this effect. The second hit, which occurs in the fetus during the period in which retinoblasts are rapidly dividing and proliferating, has removed the remaining normal allele of this gene. This implies that a cell with one mutant *RB1* allele and one normal *RB1* allele cannot form a tumor. Thus, the product of the normal gene, even when present only in a single copy, prevents tumor formation.

An important corollary of this two-hit hypothesis is that the genes in which inherited mutations cause familial cancer syndromes may be the same as those that generate common cancers by somatic mutation. *By understanding the nature of the mutant alleles inherited in rare cancer families,*

*therefore, we will come to understand more about the somatic pathway to common cancer as well.* Indeed, somatic loss-of-function mutations of both copies of the *RB1* gene are seen frequently in many types of tumors, including small-cell lung carcinoma, breast carcinoma, glioblastoma (a brain tumor), and osteosarcoma.

Alfred Knudson's two-hit theory of carcinogenesis in retinoblastoma became the paradigm for a model to describe how inheritance of an altered gene predisposes the gene carrier to cancer. The theory states that a cell can initiate a tumor only when it contains two damaged alleles; therefore, a person who inherits one copy of a mutant retinoblastoma gene must experience a second, somatic mutation in one or more retinoblasts in order to develop one or more retinoblastomas. Two somatic mutations can also occur in a single retinoblast of a nonpredisposed fetus, producing sporadic retinoblastoma. Understanding mutated genes that are inherited in families can increase our understanding of the somatic pathway to common cancers.

## MAJOR CLASSES OF CANCER GENES

Cancer-causing genes can be classified into three major categories: those that normally inhibit cellular proliferation (tumor suppressors), those that activate proliferation (oncogenes), and those that participate in DNA repair.

### Tumor Suppressor Genes

The *RB1* gene was the first identified example of a **tumor suppressor gene**, a class of genes that control cell division and thus help to prevent tumors (Table 11-1). Characteristic of tumor suppressors is the somewhat perplexing feature that

**TABLE 11-1**

## Examples of Tumor Suppressor Genes and DNA Repair Genes and Their Roles in Inherited Cancer

| Gene (Related Genes in Parentheses) | Function of Gene Product | Disease Caused by Germline Mutations |
|---|---|---|
| **Tumor Suppressor Genes** | | |
| *RB1(p107, p130)* | Cell cycle brake; binds to E2F transcription factor complex | Retinoblastoma; osteosarcoma |
| *APC* | Interacts with β-catenin in Wnt signaling pathway | Familial adenomatous polyposis |
| *SMAD4* | Transmits signals from TGFβ | Juvenile polyposis |
| *NF1* | Down-regulates ras protein | Neurofibromatosis type 1 |
| *NF2* | Cytoskeletal protein regulation | Neurofibromatosis type 2 |
| *TP53* | Transcription factor; induces cell cycle arrest or apoptosis | Li–Fraumeni syndrome |
| *VHL* | Regulates multiple proteins, including p53 and NFκB | Von Hippel–Lindau disease (renal cysts and cancer) |
| *WT1* | Zinc finger transcription factor; binds to epidermal growth factor gene | Wilms' tumor |
| *CDKN2A (p14, p16)* | CDK4 inhibitor | Familial melanoma |
| *PTEN* | Phosphatase that regulates PI3K signaling pathway | Cowden syndrome (breast and thyroid cancer) |
| *CHEK2* | Phosphorylates p53 and BRCA1 | Li–Fraumeni syndrome |
| *PTCH* | Sonic hedgehog receptor | Gorlin syndrome (basal cell carcinoma, medulloblastoma) |
| *CDH1* | E-cadherin; regulates cell–cell adhesion | Gastric carcinoma |
| *DPC4* | Transduces transforming growth factor-β signals | Juvenile polyposis |
| *TSC2* | Downregulates mTOR (mammalian target of rapamycin) | Tuberous sclerosis |
| **DNA Repair Genes** | | |
| *MLH1* | DNA mismatch repair | HNPCC |
| *MSH2* | DNA mismatch repair | HNPCC |
| *BRCA1* | Interacts with BRCA2/RAD51 DNA repair protein complex | Familial breast and ovarian cancer |
| *BRCA2* | Interacts with RAD51 DNA repair protein | Familial breast and ovarian cancer |
| *ATM* | Protein kinase; phosphorylates BRCA1 in response to DNA damage | Ataxia telangiectasia; conflicting evidence for direct involvement in breast cancer |
| *XPA* | Nucleotide excision repair | Xeroderma pigmentosum |

inherited mutations are dominant alleles at the level of the individual (i.e., heterozygotes usually develop the disease), but they are recessive alleles at the level of the cell (heterozygous cells do not form tumors). This apparent contradiction is resolved by realizing that in individuals who have inherited the first hit, a second hit that occurs in any one cell will cause a tumor. Because there are several million target retinoblasts in the developing fetus, heterozygous persons form, on the average, several retinoblasts homozygous for an *RB1* mutation. Each of these can lead to a retinoblastoma. Thus, it is the strong predisposition to tumor formation (i.e., the first hit) that is inherited as an autosomal dominant trait. The incomplete penetrance of the retinoblastoma mutation (90%) is explained by the fact that some people who inherit the disease-causing mutation do not experience a second hit in any of their surviving retinoblasts.

A general property of tumor suppressors is that they normally block the uncontrolled cellular proliferation that can lead to cancer. Often, this is done by participating in pathways that regulate the cell cycle. For example, the protein encoded by *RB1* (pRb) is active when it is relatively unphosphorylated but is down-regulated when it is phosphorylated by cyclin-dependent kinases (CDKs) just before the S phase of the cell cycle (see Chapter 2). In its active, hypophosphorylated state, pRb binds to members of the E2F transcription complex, inactivating them (Fig. 11-5). E2F

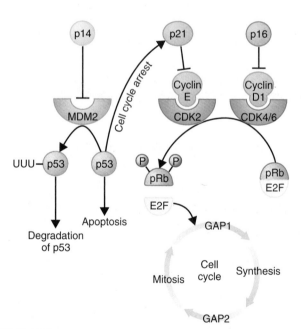

**FIGURE 11-5**
Regulation of the cell cycle is accomplished by a complex series of interactions among activators and repressors of the cycle. pRb acts as a master brake on the cell cycle by binding the E2F transcription complex, halting the cycle before S phase begins. The cyclin D–CDK4 complex inactivates pRb by phosphorylating it, thereby releasing the E2F complex and allowing the cell to progress through S phase. CDK inhibitors such as p16 and p21 inactivate CDKs, thus acting as another brake on the cycle. p53, acting through p21, can either halt the cell cycle or induce apoptosis in response to DNA damage. CDK, cyclin-dependent kinase.

activity is required for progression into S phase, so its inactivation by pRb halts the cell cycle. pRb thus serves as a cell-cycle brake that is normally released only when pRb is inactivated through phosphorylation by CDKs. This allows the cell to proceed through its mitotic cycle until pRb is activated again through removal of the phosphate groups. A loss-of-function mutation in *RB1*, a deletion of the gene, or hypermethylation of its 5′ region can lead to its permanent inactivation. Without this brake on the cell cycle, the cell can proceed through numerous uncontrolled divisions.

Loss-of-function mutations of other inhibitory factors can also lead to an unregulated cell cycle. A number of tumor suppressor genes encode **CDK inhibitors** (see Fig. 11-5), which inactivate CDKs and thus prevent them from phosphorylating target proteins such as pRb. Tumor suppressors can also control cell proliferation through their effects on transcription or on cell–cell interactions (some examples are discussed later). Again, mutations in these genes can lead to unrestricted cell division and ultimately to cancer.

> The discovery that retinoblastoma results when both alleles of the same locus on chromosome 13 are inactivated in the same retinoblast led to the concept of tumor suppressor genes. The products of such genes suppress tumor formation by controlling cell growth and can do so even if a cell contains only one normal version of the gene. Loss-of-function mutations that inactivate both copies of a cell's tumor suppressor gene can lead to uncontrolled cellular proliferation.

Because of the pivotal role of tumor suppressors in preventing tumor formation, their study is of considerable medical significance. By understanding how cancer is naturally suppressed by the body, we can ultimately develop more effective medical therapies for tumor prevention and treatment.

### Oncogenes

**Oncogenes** (i.e., "cancer genes") are a second category of genes that can cause cancer. Most oncogenes originate from **proto-oncogenes**, which are genes involved in the four basic regulators of normal cell growth mentioned previously (growth factors, growth factor receptors, signal transduction molecules, and nuclear transcription factors). When a mutation occurs in a proto-oncogene, it can become an oncogene, a gene whose excessively active product can lead to unregulated cell growth and differentiation. When a cell proceeds from regulated to unregulated growth, the cell is said to have been **transformed**.

Unlike tumor suppressor genes, oncogenes are usually dominant at the cellular level: Only a single copy of a mutated oncogene is required to contribute to the multistep process of tumor progression. Whereas tumor suppressors are typically

**TABLE 11-2**

## Comparison of Key Features of Tumor Suppressor Genes and Oncogenes

| Feature | Tumor Suppressor Genes | Oncogenes |
|---|---|---|
| Function of normal version | Regulates cell growth and proliferation; some can induce apoptosis | Promotes cell growth and proliferation |
| Mutation (at cell level) | Recessive (both copies of gene inactivated) | Dominant (only one copy of gene mutated) |
| Effect of mutation | Loss of function | Gain of function |
| Germline mutations resulting in inherited cancer syndromes | Seen in most tumor suppressor genes | Seen in only a few oncogenes |

disabled by deletions or loss-of-function mutations, oncogenes are typically activated by gain-of-function mutations, gene amplification (i.e., increased numbers of the gene through trisomy or other mechanisms), hypomethylation of the oncogene's 5′ region (which increases transcription), or chromosome rearrangements that upregulate the oncogene (e.g., the Philadelphia chromosome translocation, Chapter 6). Most tumor suppressor genes are known to exhibit germline mutations that can cause inherited cancer syndromes (e.g., retinoblastoma, Li–Fraumeni syndrome). In contrast, although oncogenes are commonly found in sporadic tumors, germline oncogene mutations that cause inherited cancer syndromes are uncommon (a few exceptions are noted later in the discussion). These and other differences are summarized in Table 11-2.

In this section, we review three approaches that have been used to identify specific oncogenes: retroviral definition, transfection experiments, and mapping in tumors.

> Proto-oncogenes encode products that control cell growth and differentiation. When mutated or amplified, they can become oncogenes, which can cause cancer. Most oncogenes act as dominant gain-of-function mutations that lead to the deregulation of cell cycle control. In contrast to tumor suppressor genes, most oncogenes do not exhibit germline mutations that cause inherited cancer syndromes. Instead, somatic mutations are seen that lead to sporadic cancers.

### IDENTIFICATION OF ONCOGENES

It has long been known that certain types of viruses can cause cancer. Especially significant are **retroviruses**, a type of RNA virus that is capable of using reverse transcriptase to transcribe its RNA into DNA. In this way, the RNA genome of the retrovirus is converted to DNA, which can be inserted into a chromosome of a host cell. Some retroviruses carry altered versions of growth-promoting genes into cells. These growth-promoting genes are oncogenes, which were first identified through the study of retroviruses

that cause cancer in chickens. When the retrovirus invades a new cell, it can transfer the oncogene into the genome of the new host, thus transforming the cell and initiating cancer.

A number of gene products that affect cell growth or differentiation have been identified through the study of oncogenes carried by transforming retroviruses. For example, retrovirus studies identified the gene encoding the receptor molecule for epidermal growth factor (EGF), through the ERBB oncogene. These studies also identified the *RAS* (*rat* sarcoma) oncogenes, which are altered in at least 25% of human cancers. Transforming retroviruses have also identified the nuclear transcription factor genes, *MYC*, *JUN*, and *FOS*, as other molecular components capable of initiating cell transformation. Table 11-3 provides some examples of proto-oncogenes.

Oncogenes have also been identified in experiments in which material from human tumor cells was transferred to nontumor cells (**transfection**), causing transformation of the recipients. A classic experiment began with the transfer of DNA from a human bladder-cancer cell line into mouse cells. A few recipient cells became fully transformed. Cloning and examination of the human-specific DNA sequences present in the transformed mouse cells revealed that the transforming gene was a mutant allele of the same *RAS* oncogene previously identified by retroviral studies. Thus, the same oncogene that could be transferred by retroviruses also occurs naturally, as a proto-oncogene, in the human genome.

Characterization of the protein product of mutant forms of *RAS* has revealed an important mechanism for the regulation of signal transduction. The RAS protein normally cycles between an *active* form bound to **guanosine triphosphate (GTP)** and an *inactive* form bound to **guanosine diphosphate (GDP)**. The biochemical consequence of *RAS* mutations is a RAS protein that is unable to shift from the active GTP form, which stimulates growth, to the inactive GDP form. The mutant RAS protein cannot extinguish its growth signal, contributing to excessive cell division.

A third approach for identifying oncogenes derives from the common observation of chromosomal rearrangements, such as translocations, in some types of tumor cells (Chapter 6). A well-known example is the Philadelphia chromosome, in which a translocation between chromosomes 9 and 22 places

**TABLE 11-3**
## Examples of Oncogenes and Their Roles in Cancer*

| Oncogene | Function | Associated Tumor |
|---|---|---|
| **Growth Factor Genes** | | |
| HST | Fibroblast growth factor | Stomach carcinoma |
| SIS | β subunit of platelet-derived growth factor | Glioma (brain tumor) |
| KS3 | Fibroblast growth factor | Kaposi sarcoma |
| **Growth Factor Receptor Genes** | | |
| RET[†] | Receptor tyrosine kinase | Multiple endocrine neoplasia; thyroid carcinoma |
| ERBB | Epidermal growth factor receptor | Glioblastoma (brain tumor); breast cancer |
| ERBA | Thyroid hormone receptor | Acute promyelocytic leukemia |
| NEU (ERBB2) | Receptor protein kinase | Neuroblastoma; breast carcinoma |
| MET[†] | Receptor tyrosine kinase | Hereditary papillary renal carcinoma; hepatocellular carcinoma |
| KIT[†] | Receptor tyrosine kinase | Gastrointestinal stromal tumor syndrome |
| **Signal Transduction Genes** | | |
| HRAS | GTPase | Carcinoma of colon, lung, pancreas |
| KRAS | GTPase | Melanoma, thyroid carcinoma, acute monocytic leukemia, colon carcinoma |
| NRAS | GTPase | Melanoma |
| BRAF | Serine/threonine kinase | Malignant melanoma; colon cancer |
| ABL | Protein kinase | Chronic myelogenous leukemia; acute lymphocytic leukemia |
| CDK4[†] | Cyclin-dependent kinase | Malignant melanoma |
| **Transcription Factor Genes** | | |
| NMYC | DNA-binding protein | Neuroblastoma; lung carcinoma |
| MYB | DNA-binding protein | Malignant melanoma; lymphoma; leukemia |
| FOS | Interacts with JUN oncogene to regulate transcription | Osteosarcoma |

*For additional examples, see Croce CM: Oncogenes and cancer. N Engl J Med 2008; 358(5):502-511.
[†]CDK4, KIT, MET, and RET are proto-oncogenes in which germline mutations can give rise to inherited cancer syndromes.

the *ABL* proto-oncogene next to the *BCR* gene, which enhances tyrosine kinase activity and produces chronic myelogenous leukemia. Another translocation, t(15;17) (q22;q11.2-12), is seen in acute promyelocytic leukemia (APL) and fuses two genes together: the retinoic acid receptor alpha (*RARα*) gene on chromosome 17 and the promyelocytic leukemia (*PML*) gene on chromosome 15. The fusion product (PML-RARα) interferes with the ability of the normal RARα protein to induce terminal differentiation of myeloid cells. (Interestingly, retinoic acid was already in use as a therapeutic agent for APL.) The fusion product also impairs the function of the PML protein, which acts as a tumor suppressor by helping to initiate apoptosis in damaged cells.

Retroviruses are capable of inserting oncogenes into the DNA of a host cell, thus transforming the host into a tumor-producing cell. The study of such retroviral transmission has identified a number of specific oncogenes. The transfection of oncogenes from tumor cells to normal cells can cause transformation of the normal cells. Some oncogenes were identified when specific rearrangements of chromosomal material were found to be associated with certain cancers. Because these translocations alter genes vital to cellular growth control, the sites of such rearrangements can be investigated to identify new oncogenes.

The identification of oncogenes has greatly increased our understanding of some of the underlying causes of cancer. In addition, oncogenes provide important targets for cancer therapy because of their key role in carcinogenesis. For example, the *ERBB2* oncogene, mentioned earlier and also known as *HER2/NEU*, is amplified in approximately 20% to 30% of invasive breast carcinomas. Its amplification, which can be identified by fluorescent in situ hybridization (FISH) or array comparative genomic hybridization (CGH) (see Chapter 6), is associated with aggressive cancer. The protein product of *HER2/NEU* is a growth factor receptor located on the surfaces of breast cancer cells. Identification of the oncogene and its product contributed to the development of a drug, trastuzumab, which binds to the amplified gene product, effectively downregulating it and helping to treat this form of breast cancer. Similar drugs have been developed to counter the effects of the upregulated *ABL* oncogene in chronic myelogenous leukemia, an upregulated epidermal growth factor receptor gene in non–small-cell lung cancer, and several others.

### DNA Repair Genes, Chromosome Integrity, and Tumorigenesis

Tumor cells typically are characterized by widespread mutations, chromosome breaks, and aneuploidy. This condition, termed **genomic instability**, contributes to tumorigenesis because mutations and chromosome defects can activate oncogenes or deactivate tumor suppressor genes. Genomic instability can occur because of defects in the proteins required for accurate cell division or in proteins responsible for DNA repair. It is also associated with hypomethylation of DNA, a common feature of many tumors. These defects are in turn the result of mutations. Sometimes, these mutations are inherited, resulting in relatively rare inherited cancer syndromes (see Table 11-1). More often, they arise in somatic cells and contribute to common, noninherited cancers.

There are a number of ways various types of genomic instability can give rise to cancer. Some breast cancers are caused by defective repair of double-stranded breaks that occur in DNA (e.g., from radiation exposure). This can result from mutations in genes such as *BRCA1*, *BRCA2*, or *ATM*. An inherited form of colon cancer, discussed later, can result from faulty DNA **mismatch repair** (so named because single-base mutations can lead to a DNA molecule in which base pairs are not complementary to each other: a mismatch). Xeroderma pigmentosum, an inherited condition that is characterized in part by multiple skin tumors (see Chapter 3), is the result of impaired nucleotide excision repair. Defects in proteins responsible for chromosome separation during mitosis (e.g., spindle fibers) can give rise to the multiple aneuploidies typically seen in tumor cells. Aneuploidy can contribute to tumorigenesis by creating extra copies of oncogenes or by deleting tumor suppressor genes.

> Genomic instability, which can result from defects in DNA repair, is often observed in tumor cells and is characterized by widespread mutations, chromosome breaks, and aneuploidy. These alterations can cause cancer when they affect pathways that regulate cellular proliferation.

### Genetic Alterations and Cancer Cell Immortality

Even after a tumor cell has escaped regulation by tumor suppressors or DNA repair proteins, it must overcome one more hurdle to unlimited proliferation: the intrinsic limitation on the number of cell divisions allowed to each cell. Ordinarily, a cell is restricted to about 50 to 70 mitotic divisions. After reaching this number, the cell typically becomes **senescent** and cannot continue to divide. Recent research has provided new insights on the mechanisms that count the number of cell divisions and has illustrated new ways tumor cells can circumvent the counting system.

Each time a cell divides, the telomeres of chromosomes shorten slightly because DNA polymerase cannot replicate the tips of chromosomes. Once the telomere is reduced to a critical length, a signal is transmitted that causes the cell to become senescent. This process would place severe limitations on the proliferating cells in a tumor, preventing further clonal expansion. Tumor cells overcome the process by activating a gene that encodes **telomerase**, a reverse transcriptase that replaces the telomeric segments that are normally lost during cell division. Activation of this enzyme, which is rarely present in normal cells but is found in 85% to 90% of tumor cells, is part of a process that allows a tumor cell to continue to divide without the restraint ordinarily imposed by telomere shortening. This uninhibited division allows the tumor to become large, and, by allowing continued DNA replication, it permits the accumulation of additional mutations that can further contribute to the aggressiveness of the tumor cell.

> Normally, progressive shortening of telomeres limits the number of divisions of a cell to about 50 to 70. Tumor cells overcome this limitation by activating telomerase, which replaces the telomere segments that are lost during each cell division. This appears to help tumor cells escape the restraint of cellular senescence.

### IDENTIFICATION OF INHERITED CANCER-CAUSING GENES

Although the methods described in the previous section have been successful in identifying many oncogenes, they are not well suited to the identification of tumor suppressor genes. These methods require dominant expression of the mutant phenotype, which is characteristic of oncogenes, whereas mutant tumor suppressor alleles seem to have a primarily recessive phenotype at the level of the cell. Alternative approaches to the identification of these tumor suppressor genes were necessary. The first, and most common, of these approaches is linkage mapping (Chapter 8) in families with hereditary cancer, where the chromosomal segment bearing a cancer-causing mutation can be identified by linkage with polymorphic markers. This approach has been used to identify mutations that cause inherited forms of breast and colon cancer (discussed later).

The second approach takes advantage of the frequent chromosomal losses associated with revealed tumor suppressor genes. As already described, inherited tumor suppressor mutations typically result in persons who are heterozygous for the mutation in all of their cells (a first hit). However, the tumor suppressor mutation is a recessive allele at the cellular level: There are two hits, resulting in the loss of both normal copies of the tumor suppressor gene. Often an inherited mutant allele is unmasked in tumor cells by a deletion of part or all of the other copy of the homologous chromosome that carried the normal allele. Therefore, the observation that a specific chromosome segment is deleted in a tumor suggests a map location for the inherited mutation.

The deleted chromosome regions in tumors can be pinpointed by examining a series of closely linked marker polymorphisms, such as STRs, in the region and determining which of the markers that are heterozygous in the patient's constitutional DNA have become homozygous in tumor DNA (i.e., which markers have lost an allele in the process of tumorigenesis). This **loss of heterozygosity** in tumor DNA indicates that the normal tumor suppressor gene, as well as polymorphic markers surrounding it, has been lost, leaving only the abnormal copy of the tumor suppressor gene (Fig. 11-6; see Fig. 11-4). This approach was used, for example, to help narrow the locations of the retinoblastoma gene on the long arm of chromosome 13 and of a gene for Wilms' tumor (nephroblastoma) on 11p. Another approach for identifying deleted regions in tumor DNA involves array CGH, described in Chapter 6.

Although gene mapping studies can often define a region in which an inherited cancer gene is located, they cannot alone identify the disease gene. As discussed in Chapter 8, detection of mutations carried in DNA from patients with the disease is crucial for identifying the specific disease-causing gene.

> Map locations of tumor-associated genes can be detected through linkage analysis or by showing that one homolog of a chromosome (or a part of it) is missing in DNA from a tumor. Confirmation of the etiological role of a possible cancer-causing gene is obtained by showing the consistent presence of mutations in the gene in DNA from patients.

### Neurofibromatosis Type 1

The initial evidence for mapping the neurofibromatosis type 1 gene (*NF1*) to chromosome 17 came from linkage studies in families. Subsequently, chromosomal translocations were discovered in the karyotypes of two unrelated patients with neurofibromatosis, each of whom had a breakpoint on chromosome 17q at a location indistinguishable from the map location of the *NF1* gene. These translocations were assumed to have caused neurofibromatosis in these persons by disrupting the *NF1* gene. Located only 50 kb apart, the breakpoints provided the physical clues necessary to define several candidate genes that were screened for mutations in patients with *NF1* (Fig. 11-7).

The nucleotide sequence of the *NF1* gene provided an early clue to function when its predicted amino acid sequence was compared with amino acid sequences of known gene products found in computerized databases. Extended similarities were observed with the mammalian GTPase-activating protein (GAP). This was an important finding, because at least one function of GAP is to decrease the amount of active GTP-bound RAS. The RAS protein is a key component of the signal transduction pathway, transmitting positive growth signals in its active form. The *NF1* gene product, neurofibromin, also plays a role in signal transduction by down-regulating RAS.

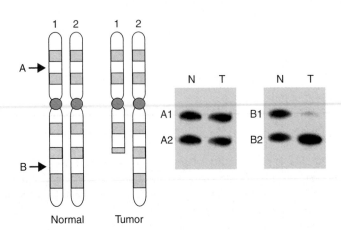

**FIGURE 11-6**

*A* and *B* represent two microsatellite polymorphisms that have been assayed using DNA from a cancer patient's normal cells *(N)* and tumor cells *(T)*. In normal cells, the patient is heterozygous for both marker loci. Deletion of the long arm from one of the paired chromosomes in tumor cells results in loss of heterozygosity (LOH) for locus *B* (i.e., the band corresponding to allele *B1* is missing, with only a faint signal due to residual traces of normal cells in the tumor specimen). LOH is a signpost for a tumor-suppressor gene near the deleted locus.

*(Courtesy of Dr. Dan Fults, University of Utah Health Sciences Center.)*

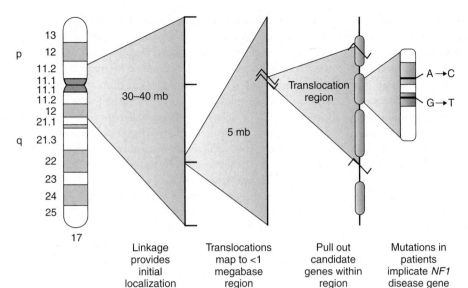

**FIGURE 11-7**
Localization of the *NF1* gene on chromosome 17q involved linkage analysis and identification of two translocation breakpoints that interrupted the disease gene. Candidate genes were isolated from this region and tested for mutations in patients with NF1 and in normal controls.

Linkage provides initial localization

Translocations map to <1 megabase region

Pull out candidate genes within region

Mutations in patients implicate *NF1* disease gene

With the identification of the *NF1* gene product as a component in signal transduction, a picture began to form of how the inheritance of a mutation in one *NF1* allele might contribute to the development of neurofibromas and *café-au-lait* spots. Reduced expression of the *NF1* gene permits increased RAS activity and allows the cell to escape from differentiation and continue its growth. Loss of the remaining normal allele in some cells (e.g., Schwann cells) further encourages unchecked growth. Discovery of the *NF1* gene has led to the identification of a key tumor suppressor that helps to regulate the fundamental process of signal transduction.

> The gene responsible for *NF1* was mapped to chromosome 17q by linkage in families and was identified through translocations and point mutations in patients. DNA sequencing of the gene predicted a protein product with a domain related to GAP, and a similar role in down-regulating the RAS signal transduction protein was confirmed by biochemical experiments.

### The *TP53* Gene

Somatic mutations in the *TP53* gene are found in more than half of all human tumors, making this the most commonly altered cancer-causing gene. Somatic *TP53* mutations are seen, for example, in approximately 70% of colorectal tumors, as well as 40% of breast tumors and 60% of lung tumors. Approximately 80% to 90% of *TP53* mutations are concentrated in the portion of the gene that encodes a DNA-binding domain, typically preventing the p53 protein from binding to DNA of other genes.

Like *RB1* and *NF1*, *TP53* functions as a tumor suppressor gene. Its protein product, p53, increases in quantity in response to cell damage (e.g., double-stranded DNA breaks caused by ionizing radiation). Acting as a transcription factor, p53 helps to regulate dozens of genes that affect cell growth, proliferation, and survival. For example, p53 binds to the promoter of *CDKN1A*, whose protein product, p21, is a CDK inhibitor that blocks CDK4's inactivation of pRb (see Fig. 11-5). This halts the cell cycle in the G1 phase, before DNA replication occurs in S phase. Arrest of the cell cycle before S phase provides time for repair of the damaged DNA. If the cell's DNA is severely damaged, p53 may instead induce programmed cell death (apoptosis). This response is more likely if the pRb pathway for cell cycle arrest is not intact. Lacking the possibility of cell cycle arrest to repair damage, the p53 protein "chooses" cell death by interacting with genes involved in apoptotic pathways (e.g., *PTEN, BAX*). In this way, p53 prevents the proliferation of an abnormal, potentially carcinogenic cell.

When *TP53* is mutated, cells with damaged DNA can evade both repair and destruction, and continued replication of the damaged DNA can lead to tumor formation. For this to happen, other components of cell cycle control must also be compromised. For example, several DNA tumor viruses, such as the human papilloma virus that is responsible for most cases of cervical cancer, inactivate both pRb and p53. This produces cells that can neither repair their DNA nor undergo apoptosis in response to damage, leading in some cases to cancer.

Carcinogenic substances can induce specific *TP53* mutations. Dietary ingestion of aflatoxin $B_1$, which can produce liver cancer, is associated with a mutation that produces an arginine-to-serine substitution at position 249 of the p53 protein. Exposure to benzopyrene, a powerful mutagen and carcinogen found in cigarette smoke, leads to alterations of specific *TP53* base pairs in lung tumors. This demonstrates a direct molecular link between cigarette smoking and lung cancer. Thus, examination of the type of *TP53* mutation seen in a tumor can provide clues to the identity of the causative carcinogenic agent.

Although tumor-causing *TP53* mutations have been observed mostly in somatic cells, germline mutations of *TP53* are responsible for an inherited cancer condition known as the Li–Fraumeni syndrome (LFS). This rare syndrome is transmitted in autosomal dominant fashion and involves breast and colon carcinomas, soft-tissue sarcomas, osteosarcomas, brain tumors, leukemia, and adrenocortical carcinomas. These tumors usually develop at early ages in members of LFS families, and multiple primary tumors are commonly seen in an affected person. The demonstration of consistent *TP53* mutations in the constitutional DNA of patients with LFS confirmed the causative role of this gene. As in retinoblastoma, the inheritance of a mutated *TP53* gene greatly increases the person's susceptibility to subsequent cell transformation and tumor development when a cell loses the other, normal copy of *TP53* (two-hit model). Among LFS family members who inherit an abnormal *TP53* gene, approximately 50% develop invasive cancer by 30 years of age, and more than 90% develop invasive cancer by age 70 years.

*TP53* mutations account for only about 75% of LFS cases. Some of the remaining cases are the result of mutations in another tumor suppressor gene, *CHEK2*. This gene encodes a kinase that normally phosphorylates p53 in response to ionizing radiation, resulting in the accumulation and activation of p53. Loss-of-function mutations in *CHEK2* result in a lack of p53 activation, causing LFS via the p53 pathway.

*TP53* is medically important in at least two ways. First, the presence of *TP53* mutations in tumors, particularly those of the breast and colon, often signals a more-aggressive cancer with relatively poor survival prospects. It is thus a useful prognostic indicator. Second, *TP53* might ultimately prove important in tumor prevention. Laboratory experiments show that the insertion of a normal *TP53* gene into tumor cells can induce tumor regression by inducing abnormal cancer cells to undergo apoptosis. This has led to gene therapy protocols (see Chapter 13) in which normal *TP53* copies are inserted into tumors in an effort to eliminate cancerous cells.

> Somatic mutations of the *TP53* gene are found in the majority of tumors. This gene encodes a protein that can induce either cell cycle arrest or apoptosis in response to damaged DNA. Inherited *TP53* mutations can cause Li–Fraumeni syndrome.

## The Familial Adenomatous Polyposis Gene, *APC*

Colon cancer affects approximately 1 in 20 Americans, and, like most common cancers, it is more likely to occur in persons with a positive family history. One's risk of developing colon cancer approximately doubles if a first-degree relative is affected. In addition, a small fraction of colon cancer cases are inherited as autosomal dominant syndromes. The two most important of these syndromes are discussed next.

Familial adenomatous polyposis (FAP), also called adenomatous polyposis coli (APC), is characterized by the appearance of large numbers adenomas, or polyps, of the colon in the second or third decade of life. Colonic adenomas are now understood to be the immediate precursors to colorectal cancer. The multiple adenomas of the patient with FAP therefore present a grave risk of early malignancy. Because early detection and removal of adenomatous polyps can significantly reduce the occurrence of cancer, it is important to understand the causative gene and its role in the development of polyps (Clinical Commentary 11-1).

The gene responsible for FAP was localized to the long arm of chromosome 5 by linkage analysis in families. Discovery of small, overlapping deletions in two unrelated patients provided the key to isolation of the disease-causing gene, termed *APC*. Among the genes that lay within the 100-kb region that was deleted in both patients, one showed apparent mutations in other patients. This mutation was seen in one patient but not in his unaffected parents, confirming the identification of the *APC* gene.

Like *RB1* and *TP53*, *APC* is a tumor suppressor gene, and both copies of *APC* must be inactivated in a cell for tumor progression to begin. Persons who inherit an *APC* mutation typically experience somatic loss-of-function mutations in hundreds of their colonic epithelial cells, giving rise to multiple adenomas. In some cases, loss of function of *APC* occurs because of hypermethylation of *APC*'s promoter region, which results in reduced transcription (see Chapter 5). (Hypermethylation has been observed in the inactivation of a number of tumor suppressor and DNA repair genes, including those associated with retinoblastoma [*RB1*], breast cancer [*BRCA1*], hereditary nonpolyposis colorectal cancer [*MLH1*], malignant melanoma [*CDKN2A*], and von Hippel–Lindau disease [*VHL*]).

Identification of the *APC* gene has been important in diagnosing and managing colon cancer in families with FAP (see Clinical Commentary 11-1). In addition, and perhaps even more importantly, *APC* mutations are found in 85% of *all* sporadic, noninherited cases of colon cancer. Somatic *APC* mutations (i.e., those that disable both copies of the gene in a colonic cell) are among the earliest alterations that give rise to colon cancer. But *APC* mutations are not sufficient by themselves to complete the progression to metastatic disease. As shown in Figure 11-8, other genes may also be altered. For example, gain-of-function mutations are seen in the *KRAS* gene in approximately 50% of colon tumors. This gene encodes a signal transduction molecule, and a gain-of-function mutation increases signaling and thus cellular proliferation. Loss-of-function mutations in the *TP53* gene are also seen in more than 50% of colon tumors and usually occur relatively late in the pathway to cancer. Ordinarily, p53 would be activated in response to mutations like those of *APC* and *KRAS*, leading to DNA repair or apoptosis. Cells that lack p53 activity are free to continue along the path to malignancy in spite of their damaged DNA. Still another tumor suppressor gene, *SMAD4*, appears also to be

## CLINICAL COMMENTARY 11-1
### *The APC Gene and Colorectal Cancer*

Colorectal cancer will be diagnosed in about 1 in 20 Americans. Currently, the mortality rate for this cancer is approximately one third. Genetic and environmental factors, such as dietary fat and fiber, are known to influence the probability of occurrence of colorectal cancer.

As indicated in the text, familial adenomatous polyposis (FAP), also called adenomatous polyposis coli (APC), is an autosomal dominant subtype of colon cancer that is characterized by a large number of adenomatous polyps. These polyps typically develop during the second decade of life and number in the hundreds or more (polyposis itself is defined as the presence of >100 polyps). Penetrance of FAP is virtually 100%. Germline mutations in the *APC* gene are consistently identified in family members affected with FAP, and approximately one third of cases are the result of new mutations in *APC*. More than 700 different mutations of the *APC* gene have been reported, most of which are nonsense or frameshift mutations. Because these mutations produce a truncated protein product, a protein truncation test can be used to help determine whether the at-risk person has inherited an *APC* mutation. (As discussed in Chapter 3, this test involves the in vitro manufacture and analysis of a protein product from the gene of interest.) It is now more common to rely on DNA-based tests, including direct sequencing, to identify *APC* mutations. These diagnostic tests, which identify mutations in about 80% of FAP cases, are important

for family members because a positive result alerts them to the need for frequent surveillance and possible colectomy.

FAP is relatively rare, affecting only about 1 in 10,000 to 1 in 20,000 persons and accounting for less than 1% of all cases of colon cancer. The broader significance of the APC gene derives from the fact that somatic mutations in this gene are seen in approximately 85% of all colon cancers. Furthermore, APC mutations typically occur early in the development of colorectal malignancies. A better understanding of the APC gene product, how it interacts with other proteins, and how it interacts with environmental factors such as diet, can provide important clues for the prevention and treatment of common colon cancer. In this way, the mapping and cloning of a gene responsible for a relatively rare cancer syndrome can have widespread clinical implications.

Treatment for colorectal cancer usually involves surgical resection of the colon and chemotherapy. However, because colorectal carcinoma is usually preceded by the appearance of benign polyps, it is one of the most preventable of cancers. The National Polyp Study Workgroup estimates that colonoscopic removal of polyps could reduce the nationwide incidence of colon cancer by as much as 90%. The importance of early intervention and treatment further underscores the need to understand early events in colorectal cancer, such as somatic mutations in the *APC* gene.

Because polyps usually begin to appear in the second decade of life in those who inherit an *APC* mutation, an annual colonoscopy is recommended in these persons starting at 12 years of age. Upper gastrointestinal polyps are seen in 90% of FAP patients by age 70. Thus, endoscopy of the upper gastrointestinal tract is recommended every 1 to 2 years, starting at ages 20 to 25 years. Because classic FAP results in hundreds to thousands of polyps, colectomy is often necessary by age 20 years. There is evidence that the use of nonsteroidal anti-inflammatory drugs causes some degree of polyp regression. Persons with FAP have increased risks of other cancers, including gastric cancer (<1% lifetime risk), duodenal adenocarcinoma (5%-10% lifetime risk), hepatoblastoma (1% risk), and thyroid cancer.

Mutations in the *APC* gene can also produce a related syndrome, termed *attenuated familial adenomatous polyposis*. This syndrome differs from FAP in that patients have fewer than 100 polyps (typically 10-20). Most of the mutations that produce this form of FAP are located in the 5′ or 3′ regions of *APC*.

FAP can also result from recessive mutations in *MUTYH*, a gene that encodes a DNA repair protein. These mutations are estimated to account for about 30% of cases of attenuated FAP and approximately 10% to 20% of classic FAP cases in which an *APC* mutation is not detected.

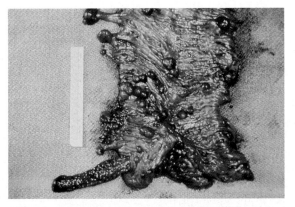

A portion of a colon removed from a patient with familial adenomatous polyposis (FAP), illustrating a large number of adenomatous polyps covering the colon. Each of these benign neoplasms has the potential to become a malignant tumor.

mutated in the colon cancer pathway. Thus, at least seven mutations are required to produce a colon tumor (two in each of the three tumor suppressor genes, as well as one dominant gain-of-function mutation in *KRAS* or another signal transduction gene).

Extensive studies have revealed at least three ways the APC protein acts as a tumor suppressor. Perhaps most importantly, it participates in the phosphorylation and degradation of β-catenin, a key molecule in the Wnt signal transduction pathway. Among other things, this pathway is involved in

activation of the MYC transcription factor. By reducing β-catenin levels, APC dampens signals that lead to cellular proliferation. Examination of colon carcinomas that do not carry mutations in the *APC* gene revealed that some of them have gain-of-function mutations in the β-catenin gene, thus confirming the potential etiological role of this gene in colon cancer. APC mutations are also thought to affect cell-to-cell and cell-to-matrix adhesion properties (this is important because alteration of cell adhesion control permits cells to invade other tissues and to metastasize to other

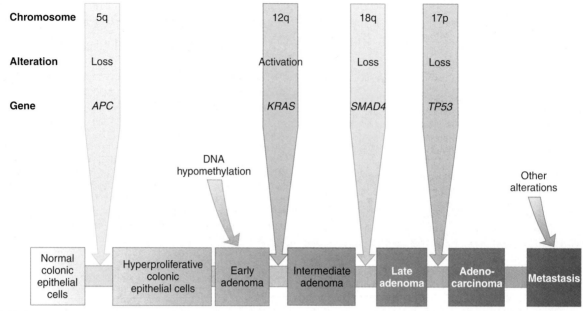

**FIGURE 11-8**

The pathway to colon cancer. Loss of the *APC* gene transforms normal epithelial tissue lining the gut to hyperproliferating tissue. Hypomethlyation of DNA (which can cause genomic instability and up-regulate proto-oncogenes), activation of the *KRAS* proto-oncogene, and loss of the *SMAD4* gene are involved in the progression to a benign adenoma. Loss of the *TP53* gene and other alterations are involved in the progression to malignant carcinoma and metastasis. Note that these alterations are present in varying frequencies in colon tumor cells.

*(Modified from Vogelstein B, Kinzler KW: The multistep nature of cancer. Trends Genet 1993;9:138-141.)*

sites). Again, this activity is mediated through β-catenin, which interacts with a cell-surface molecule (E-cadherin) whose loss of function leads to abnormal cell adhesion properties. Finally, APC is expressed in the microtubules that pull chromosomes apart during meiosis (Chapter 2). Alterations in APC result in altered microtubule activity, such that aneuploidies and chromosome breaks arise during mitosis. Thus, *APC* mutations also promote cancer by increasing genomic instability.

> The adenomatous polyposis coli gene (*APC*), which strikingly predisposes to colon cancer, was ultimately identified from mutations in patients. *APC* is also involved in the great majority of sporadic cases of colon cancer and is in fact one of the earliest alterations leading to colon tumorigenesis. This tumor suppressor gene has been shown to function as a major regulator of the *Wnt* signal transduction pathway via its interaction with β-catenin. It is also involved in cell adhesion control and in maintaining chromosome stability during mitosis.

## The Hereditary Nonpolyposis Colon Cancer Genes

Hereditary nonpolyposis colon cancer (HNPCC, or Lynch syndrome), a second form of inherited colon cancer, accounts for approximately 1% to 5% of all colorectal cancer cases. Like FAP, HNPCC is an autosomal dominant, high-penetrance cancer syndrome, with a lifetime colorectal cancer risk of 70% to 90% in heterozygotes. In addition, the risk of endometrial cancer among females with HNPCC is approximately 50%, and the risk of ovarian cancer is 5% to 10%. Cancers of the small bowel, stomach, brain, pancreas, renal pelvis, and ureter are seen in a smaller percentage of mutation carriers. In contrast to FAP, HNPCC patients do not have polyposis; they typically have a relatively small number of polyps. Also, polyps in HNPCC patients are more likely to occur in the proximal colon, whereas those of FAP patients are more likely to be concentrated in the distal colon.

Approximately 40% to 60% of HNPCC cases are caused by mutations in a gene called *MSH2*, and another 25% to 30% of cases are caused by mutations in the *MLH1* gene. Mutations in two other genes, *PMS2* and *MSH6*, help to account for a small percentage of additional cases. Each of these genes is known to play an important role in DNA mismatch repair in many different organisms (in fact, a critical clue in their identification was the existence of highly similar DNA repair genes in yeast and bacteria). Inactivation of both alleles of any one of these genes increases the genome-wide mutation rate in an affected cell by as much as 1000-fold. This increased rate of mutation results in the alteration of a number of cellular regulatory genes, thus leading to an increased incidence of cancer. A characteristic feature of tumors from HNPCC patients is a high degree of instability of microsatellite loci (see Chapter 3), which generates many new microsatellite alleles. Such microsatellite instability is also seen in about 15% of sporadic colorectal carcinomas, but somatic loss-of-function mutations in the HNPCC genes seem to occur only infrequently in these tumors. Instead, the most common alteration seen in these sporadic tumors is hypermethylation of the *MLH1* gene, resulting in its inactivation.

A comparison of FAP and HNPCC reveals interesting differences in the way each syndrome leads to colon cancer. In FAP, an inherited *APC* mutation results in hundreds of polyps, each of which has a relatively low probability of incurring all of the other genetic alterations required for progression to metastatic cancer. But, because the number of polyps is large, there is a high probability (almost 100%) that at least one of them will produce a cancerous tumor by approximately the age of 45 years. In HNPCC, the number of polyps is much smaller (hence the term *nonpolyposis*), but, because of a relative lack of DNA repair, each polyp has a high probability of experiencing the multiple alterations necessary for tumor development. Consequently, the average age of onset of colon cancer in HNPCC is similar to that of FAP.

> HNPCC is an inherited form of colorectal cancer that is caused by mutations in any of six genes involved in DNA mismatch repair. It represents an example of a cancer syndrome associated with microsatellite instability.

### Inherited Breast Cancer

The lifetime prevalence of breast cancer in women is 1 in 8, and a woman's risk of developing breast cancer doubles if a first-degree relative is affected. Two genes, *BRCA1* and *BRCA2*, have been identified as major contributors to inherited breast cancer. This section addresses three critical questions regarding these genes: What percentage of breast cancer cases are the result of *BRCA1* and *BRCA2* mutations? Among those who inherit a mutation, what is the risk of developing cancer? How do mutations in these genes contribute to cancer susceptibility?

Population-based studies show that only a small percentage of all breast cancers—approximately 1% to 3%—can be attributed to inherited mutations in *BRCA1* or *BRCA2*. Among women with breast cancer who also have a positive family history of the disease, the percentage with inherited mutations in either of these genes increases to approximately 20%. Among affected women who have a positive family history of breast and ovarian cancer, 60% to 80% have inherited a *BRCA1* or *BRCA2* mutation. Inherited mutations in these genes are also more common among women with early-onset breast cancer and among those with bilateral breast cancer.

Women who inherit a mutation in *BRCA1* experience a 50% to 80% lifetime risk of developing breast cancer; the lifetime risk for those who inherit a *BRCA2* mutation is slightly slower, averaging approximately 50%. *BRCA1* mutations also increase the risk of ovarian cancer among women (20%-50% lifetime risk), and they confer a modestly increased risk of prostate and colon cancers. *BRCA2*

mutations confer an increased risk of ovarian cancer (10%-20% lifetime prevalence). The lifetime risk of ovarian cancer in the general female population is about 1/70. Approximately 6% of males who inherit a *BRCA2* mutation develop breast cancer, which represents a 70-fold increase over the risk in the general male population (see Chapter 12 for further discussion of breast cancer risk factors).

*BRCA1* and *BRCA2* were both identified by linkage analysis in families, followed by positional cloning. Most *BRCA1* and *BRCA2* mutations result in truncated protein products and a consequent loss of function. As for the *RB1* and *APC* genes, affected persons inherit one copy of a *BRCA1* or *BRCA2* mutation and then experience a somatic loss of the remaining normal allele in one or more cells (following the two-hit model for tumor suppressor genes). In contrast to *RB1* and *APC*, somatic mutations affecting these genes are seldom seen in sporadic (noninherited) breast tumors. The large size of these genes, together with extensive allelic heterogeneity, poses challenges for genetic diagnosis (see Chapter 13), which is done primarily by direct DNA sequencing of the coding and regulatory regions of both genes.

Although *BRCA1* and *BRCA2* share no significant DNA sequence similarity, they both participate in the DNA repair process. The protein product of *BRCA1* is phosphorylated (and thus activated) by the ATM and CHEK2 kinases in response to DNA damage (Fig. 11-9). The *BRCA1* protein product binds to the *BRCA2* product, which in turn binds to RAD51, a protein involved in the repair of double-stranded DNA breaks (as with the *HNPCC* genes, yeast and bacteria both have DNA-repair genes similar to RAD51). *BRCA1* and *BRCA2* thus participate in an important DNA repair pathway, and their inactivation results in incorrect DNA repair and genomic instability. In addition to their roles in the RAD51 pathway, *BRCA1* and *BRCA2* help to suppress tumor formation through their interactions with previously discussed proteins such as p53, pRb, and Myc.

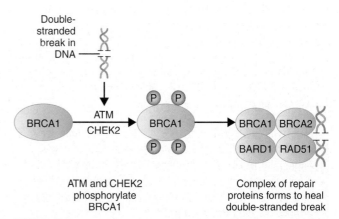

**FIGURE 11-9**
The roles of BRCA1 and BRCA2 in DNA repair. BRCA1 is phosphorylated by ATM and CHEK2 in response to double-stranded DNA breaks (produced, for example, by ionizing radiation). BRCA1 binds to BRCA2, which interacts with RAD51 to form a complex involved in DNA repair.

Because all of the genes illustrated in Figure 11-9 are involved in a DNA repair pathway, it might be anticipated that mutations in genes other than BRCA1 or BRCA2 could lead to DNA repair defects and possibly cancer. This is indeed the case. As already discussed, mutations in CHEK2 can cause LFS. Mutations in the ATM gene can cause ataxia telangiectasia (see Chapter 3), an autosomal recessive disease that involves extensive genomic instability, cerebellar ataxia, dilated vessels in the eyes and skin (telangiectasia), and cancers of mainly lymphoid origin. Another autosomal recessive chromosome instability syndrome, Fanconi anemia, can be caused by the inheritance of two copies of a BRCA2 mutation.

Although BRCA1 and BRCA2 mutations are the most common known causes of familial breast cancer, this disease can also be caused by inherited mutations in several other tumor suppressor genes (e.g., the previously discussed CHEK2 and TP53 genes). Germline mutations in a tumor suppressor gene called PTEN are responsible for Cowden disease, which is characterized by multiple benign tumors and an increased susceptibility to breast cancer. The risk of breast cancer among heterozygous carriers of mutations in the ATM gene is approximately double that of the general population.

It is estimated that the major breast cancer genes, such as BRCA1, BRCA2, PTEN, and CHEK2, account for less than 25% of the overall inherited predisposition to breast cancer. Other breast cancer–causing genes are likely to exist, but their individual effects on cancer risk are thought to be relatively small. Large-scale genome-wide association studies (Chapter 8) have now identified multiple additional inherited genetic variants that contribute small increases in breast cancer risk. For example, a variant in the gene that encodes fibroblast growth factor receptor 2 (FGFR2) increases breast cancer risk by about 25%.

Most studies indicate that the clinical course of breast cancer among patients with BRCA1 or BRCA2 mutations is not substantially different from that of other breast cancer patients. However, the substantial increase in risk of ovarian cancer (which has a high mortality rate and is difficult to detect early), has led to the recommendation that women who have completed child bearing should undergo a prophylactic oophorectomy. This reduces ovarian cancer risk by about 90%, and it reduces breast cancer risk by about 50%. Bilateral prophylactic mastectomy, an option chosen by some BRCA1 and BRCA2 carriers, reduces the risk of breast cancer by approximately 90%.

> Mutations in *BRCA1* and *BRCA2* are responsible for a significant fraction of inherited breast cancer cases, especially those of early onset. These mutations usually result in a truncated protein product and loss of function. The protein products of these genes play important roles in DNA repair.

## Familial Melanoma

Largely as a result of increased exposure to ultraviolet radiation, the incidence of melanoma has increased about 20-fold in the United States during the past 70 years. It is now one of the most common cancers, with 54,000 new cases per year. The risk of developing melanoma increases by a factor of 2 when a first-degree relative is affected. The risk increases further, to approximately 6.5, when the first-degree relative is affected before 50 years of age. It is estimated that approximately 5% to 10% of melanoma cases occur in inherited, familial forms.

Linkage analysis in families, studies of loss of heterozygosity in melanoma tumor cells, and positional cloning led to the identification of the CDKN2A gene as a cause of familial melanoma. Mutations in this gene are estimated to be involved in 10% to 40% of familial melanoma cases. CDKN2A encodes two different proteins, both of which are important components of the cell cycle. The first protein, p16, is cyclin-dependent kinase inhibitor which interacts negatively with a cyclin-dependent kinase (CDK4) that phosphorylates and down-regulates the pRb protein (see Fig. 11-5). Because pRb acts as a brake on the cell cycle, its down-regulation by CDK4 promotes progression through the cell cycle and can lead to cell proliferation. By negatively regulating CDK4, p16 acts as an upstream brake on the cell cycle. When p16 activity is lost through loss-of-function mutations in CDKN2A, CDK4 activity is increased and cell proliferation can ensue. This can lead to melanomas. The second protein encoded by CDKN2A, p14, binds to and inhibits MDM2. The MDM2 protein, as shown in Figure 11-5, binds to and degrades p53. Thus, by inhibiting MDM2, p14 increases p53 expression. As discussed earlier, p53 expression is needed to arrest the cell cycle for DNA repair in response to cell damage, and it can cause damaged cells to undergo apoptosis. A loss of p14 activity leads to a loss of p53 activity, and this can also produce melanomas.

Inherited mutations in the gene that encodes CDK4 can also result in familial melanoma. These gain-of-function mutations convert the cyclin-dependent kinase from a proto-oncogene to an activated oncogene. The activated CDK4 down-regulates pRb, resulting again in a lack of cell cycle control and tumor formation. Melanoma provides an example in which the same tumor type can result from either the activation of a proto-oncogene (CDK4) or the loss of a tumor suppressor gene (CDKN2A).

CDKN2A plays a role not only in familial melanoma but also in most sporadic melanomas, in which somatic loss-of-function mutations of this gene lead to inactivation of the p16 tumor suppressor protein. About 50% of sporadic melanomas contain somatic deletions of CDKN2A, and loss-of-function point mutations are seen in another 9% of tumors. Hypermethylation of the promoter region, which down-regulates the gene, is seen in ¼ to ¾ of all melanomas. As might be expected, somatic mutations in other genes are also seen in sporadic melanomas. About 9% of these melanomas have somatic TP53 mutations, and about 6% have somatic RB1 mutations (persons who inherit an RB1 mutation also have an increased risk of melanoma). Approximately two thirds of these tumors contain somatic gain-of-function mutations in BRAF, a gene that encodes a kinase involved in the RAS signal-transduction pathway. In addition, one of the RAS genes, NRAS, is mutated in 15% to 30% of sporadic melanomas.

> Familial melanoma can be caused by loss-of-function mutations in the *CDKN2A* tumor suppressor gene or by gain-of-function mutations in the *CDK4* proto-oncogene. Both mutations result in a loss of cell cycle control via the pRb and p53 pathways. Somatic mutations of *CDKN2A* and of *BRAF* are seen in most sporadic melanomas.

## The *RET* Proto-Oncogene and Multiple Endocrine Neoplasia

The *RET* proto-oncogene, initially identified by a transfection assay (rearranged during transfection; see previous discussion), encodes a receptor tyrosine kinase that includes an extracellular receptor domain, a transmembrane domain, and an intracellular tyrosine kinase domain. RET is involved in embryonic neural crest cell migration (see Chapter 10), and it is normally activated by a complex consisting of glial-derived neurotrophic factor (GDNF) and a coreceptor termed GFRα. The RET protein interacts with several signal transduction pathways, including the well-known RAS pathway.

Inherited loss-of-function mutations in *RET* can produce Hirschsprung disease (a lack of enteric nerve cells, resulting in severe, chronic constipation and bowel distention). Gain-of-function mutations in the same gene result in excess tyrosine kinase activity and increased signal transduction, leading ultimately to cellular proliferation and, depending on the type and location of the mutation, any of three forms of autosomal dominant multiple endocrine neoplasia type 2 (MEN2): (1) MEN2A, which accounts for 80% of MEN2 cases, is characterized by medullary thyroid carcinomas (MTC) in nearly 100% of patients, parathyroid hyperplasia in 30% of patients, and pheochromocytoma (an adrenal tumor) in 50% of patients. More than 98% of MEN2A cases are caused by missense mutations that affect cysteine residues in RET's extracellular domain. (2) MEN2B is similar to MEN2A but lacks parathyroid hyperplasia and includes multiple mucosal neuromas and a marfanoid appearance. Nearly all MEN2B alterations are missense mutations that affect RET's tyrosine kinase domain. MEN2B accounts for about 5% of MEN2 cases and is the most aggressive form of MEN2. (3) A syndrome consisting only of familial MTC can be caused by mutations in both the extracellular and tyrosine kinase domains of RET.

*RET* is one of only a few proto-oncogenes in which mutations can cause inherited cancer syndromes (see Table 11-3 for other examples).* Identification of the mutations responsible for each of these inherited cancer syndromes has provided an accurate means of early diagnosis. Patients with apparently sporadic MTC are advised to undergo genetic testing for *RET* mutations because 1% to 7% of these patients have germline *RET* mutations and thus familial, rather than sporadic, MTC. Prophylactic thyroidectomy

---

*Another inherited form of multiple endocrine neoplasia, MEN1, is characterized by tumors of the parathyroids, anterior pituitary, and pancreas. MEN1 is caused by germline mutations in a gene that encodes the menin tumor.

before 6 years of age is recommended for children who inherit a disease-causing mutation (thyroidectomy before age 3 years may be indicated for the more aggressive MEN2B tumors).

Somatic alterations of RET can produce papillary thyroid carcinomas, the most common type of thyroid tumor. The prevalence of this tumor has increased substantially among persons who were exposed to radioactive fallout from the Chernobyl nuclear reactor accident (see Chapter 3). Sixty percent of the papillary thyroid carcinomas in these persons have contained somatic *RET* alterations.

The *RET* gene provides an example of extraordinary allelic heterogeneity. Loss-of-function mutations in this gene produce defects in the embryonic development of the bowel, and gain-of-function mutations result in increased signal transduction and various forms of endocrine neoplasia. This example illustrates the critical connection between normal development and cancer: Both involve a finely tuned genetic regulation of cell growth and differentiation.

> Loss-of-function mutations in the *RET* proto-oncogene can produce Hirschsprung disease, a disorder of embryonic development. Germline gain-of-function mutations in the same gene can lead to any of three different types of inherited multiple endocrine neoplasia. Somatic alterations in *RET* can produce noninherited papillary thyroid carcinoma.

Many additional genes responsible for various inherited cancer syndromes have been identified. These include, for example, the genes that cause neurofibromatosis type 2, von Hippel–Lindau syndrome, and Beckwith–Wiedemann syndrome (see Table 11-1). With the current generation of genomic resources, including the complete human DNA sequence, it is reasonable to expect the identification of more such genes.

## IS GENETIC INHERITANCE IMPORTANT IN COMMON CANCERS?

The term "common cancers" is often used to designate those cancers, such as carcinomas of the breast, colon, or prostate, that are not usually part of an inherited cancer syndrome (e. g., Li–Fraumeni syndrome or familial adenomatous polyposis). Recall that germline mutations in genes such as *BRCA1* or *APC*, although very important in aiding our understanding of the basis of carcinogenesis, are responsible for only a small proportion of breast or colon cancer cases. Yet these cancers do cluster in families. Typically, the presence of one affected first-degree relative increases one's risk of developing a common cancer by a factor of two or more. It is likely that additional genes, as well as nongenetic factors that are shared in families, contribute to this increased familial risk (see Chapter 12 for further discussion of these genetic and nongenetic factors in common cancers).

It is likely that many common cancer-predisposing alleles exist, each with a relatively small effect on overall cancer risk (an example was given by an allele of the *FGFR2* gene in breast cancer). How will we identify these relatively minor risk alleles in the population? One way, perhaps, could exploit new approaches of gene mapping that identify genes involved in cellular pathways to cancer. Each of these becomes a candidate gene that might confer a predisposition to cancer. The use of animal models, as well as the increasing application of genome-wide association studies, will continue to uncover new cancer-predisposing alleles. The most critical test will be to determine whether there are functional mutations in a suspected cancer-causing gene among cancer patients. The identification of such mutations, and characterization of the genes in which they occur, can be expected to create new diagnostic and therapeutic tools that ultimately will reduce the cancer burden.

## Study Questions

1. The *G6PD* locus is located on the X chromosome. Studies of *G6PD* alleles in tumor cells from women show that all tumor cells usually express the same single *G6PD* allele, even though the women are heterozygous at the *G6PD* locus. What does this finding imply about the origin of the tumor cells?

2. If we assume that the somatic mutation rate at the *RB1* locus is three mutations per million cells and that there is a population of 2 million retinoblasts per individual, what is the expected frequency of *sporadic* retinoblastoma in the population? What is the expected number of tumors per individual among those who *inherit* a mutated copy of the *RB1* gene?

3. Compare and contrast oncogenes and tumor suppressor genes. How have the characteristics of these classes of cancer-causing genes affected our ability to detect them?

4. Members of Li–Fraumeni syndrome families nearly always develop tumors by 65 to 70 years of age, but mutation carriers do not necessarily develop the same type of tumor (e.g., one has breast cancer, and another has colon cancer). Explain this.

## Suggested Readings

Campeau PM, Foulkes WD, Tischkowitz MD. Hereditary breast cancer: New genetic developments, new therapeutic avenues. Hum Genet 2008;124(1):31–42.

Croce CM. Oncogenes and cancer. N Engl J Med 2008; 358(5):502–11.

de la Chapelle A. Genetic predisposition to colorectal cancer. Nat Rev Cancer 2004;4(10):769–80.

Damber JE, Aus G. Prostate cancer. Lancet 2008; 371(9625):1710–21.

Deng Y, Chan SS, Chang S. Telomere dysfunction and tumour suppression: The senescence connection. Nat Rev Cancer 2008;8(6):450–8.

Desai TK, Barkel D. Syndromic colon cancer: Lynch syndrome and familial adenomatous polyposis. Gastroenterol Clin North Am 2008;37(1):47–72.

Esteller M. Epigenetics in cancer. Engl J Med 2008; 358(11):1148–59.

Finkel T, Serrano M, Blasco MA. The common biology of cancer and ageing. Nature 2007;448(7155):767–74.

Foulkes WD. Inherited susceptibility to common cancers. N Engl J Med 2008;359(20):2143–53.

Galiatsatos P, Foulkes WD. Familial adenomatous polyposis. Am J Gastroenterol 2006;101(2):385–98.

Knudson AG. Two genetic hits (more or less) to cancer. Nat Rev Cancer 2001;1(2):157–62.

Lakhani VT, You YN, Wells SA. The Multiple endocrine neoplasia syndromes. Ann Rev Med 2007; 58(1):253–65.

Leiderman YI, Kiss S, Mukai S. Molecular genetics of *RB1*—the retinoblastoma gene. Semin Ophthalmol 2007;22 (4):247–54.

Lu X. *P53*: A heavily dictated dictator of life and death. Curr Opin Genet Dev 2005;15(1):27–33.

Robson M, Offit K. Management of an inherited predisposition to breast cancer. N Engl J Med 2007; 357(2):154–162.

Sekulic A, Haluska P Jr, Miller AJ, et al. Malignant melanoma in the 21st century: The emerging molecular landscape. Mayo Clin Proc 2008;83(7):825–46.

Van Dyke T. P53 and tumor suppression. N Engl J Med 2007;356(1):79–81.

Vogelstein B, Kinzler KW, editors. The Genetic Basis of Human Cancer. 2nd ed. New York: McGraw–Hill; 2002.

## Internet Resources

National Cancer Institute Information site (general information on cancer and cancer genetics, as well as links to other useful Web sites) *http://cancer.gov/cancerinformation*

## Chapter 12

# MULTIFACTORIAL INHERITANCE AND COMMON DISEASES

The focus of previous chapters has been on diseases that are caused primarily by mutations in single genes or by abnormalities of single chromosomes. Much progress has been made in identifying specific mutations that cause these diseases, leading to better risk estimates and in some cases more effective treatment. However, these conditions form only a small portion of the total burden of human genetic disease. A much larger component of our disease burden is composed of congenital malformations and common adult diseases, such as cancer, heart disease, and diabetes. Although they are not the result of single-gene mutations or chromosome abnormalities, these diseases have significant genetic components. They are the result of a complex interplay of multiple genetic and environmental factors. Because of the significance of common diseases in health care, it is vital to understand the ways in which genes contribute to causing them.

## PRINCIPLES OF MULTIFACTORIAL INHERITANCE

### The Basic Model

Traits in which variation is thought to be caused by the combined effects of multiple genes are called **polygenic** ("many genes"). When environmental factors are also believed to cause variation in the trait, which is usually the case, the term **multifactorial** is used. Many quantitative traits (those, such as blood pressure, that are measured on a continuous numerical scale) are multifactorial. Because they are caused by the additive effects of many genetic and environmental factors, these traits tend to follow a normal, or bell-shaped, distribution in populations.

Let us use an example to illustrate this concept. To begin with the simplest case, suppose (unrealistically) that height is determined by a single gene with two alleles, *A* and *a*. Allele *A* tends to make people tall, and allele *a* tends to make them short. If there is no dominance at this locus, then the three possible genotypes, *AA*, *Aa*, and *aa*, will produce three phenotypes: tall, intermediate, and short. Assume that the gene frequencies of *A* and *a* are each 0.50. If we assemble a population of individuals, we will observe the height distribution depicted in Figure 12-1A.

Now suppose, a bit more realistically, that height is determined by two loci instead of one. The second locus also has two alleles, *B* (tall) and *b* (short), and they affect height in exactly the same way as alleles *A* and *a* do. There are now nine possible genotypes in our population: *aabb*, *aaBb*, *aaBB*, *Aabb*, *AaBb*, *AaBB*, *AAbb*, *AABb*, and *AABB*. Because an individual might have zero, one, two, three, or four "tall" alleles, there are now five distinct phenotypes (see Fig. 12-1B). Although the height distribution in our population is not yet normal, it approaches a normal distribution more closely than in the single-gene case.

We now extend our example so that many genes and environmental factors influence height, each having a small effect. Then there are many possible phenotypes, each differing slightly, and the height distribution approaches the bell-shaped curve shown in Figure 12-1C.

It should be emphasized that the individual genes underlying a multifactorial trait such as height follow the mendelian principles of segregation and independent assortment, just like any other genes. The only difference is that many of them act together to influence the trait.

Blood pressure is another example of a multifactorial trait. There is a correlation between parents' blood pressures (systolic and diastolic) and those of their children, and there is good evidence that this correlation is due in part to genes. But blood pressure is also influenced by environmental factors, such as diet and stress. One of the goals of genetic research is identification of the genes responsible for multifactorial traits such as blood pressure and of the interactions of those genes with environmental factors.

> Many traits are thought to be influenced by multiple genes as well as by environmental factors. These traits are said to be multifactorial. When they can be measured on a continuous scale, they often follow a normal distribution.

### The Threshold Model

A number of diseases do not follow the bell-shaped distribution. Instead, they appear to be either present or absent in individuals. Yet they do not follow the patterns expected of single-gene diseases. A commonly used explanation is that there is an underlying **liability distribution** for these diseases in a population (Fig. 12-2). Persons who are on the low end of the distribution have little chance of developing the disease in question (i.e., they have few of the alleles or environmental factors that would cause the disease). Those who are closer to the high end of the distribution have more of the

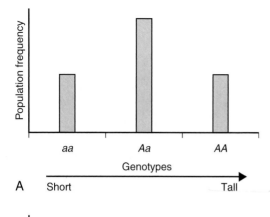

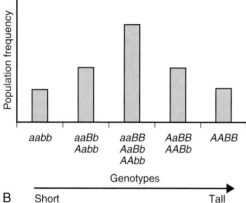

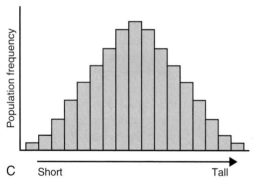

**FIGURE 12-1**
**A,** The distribution of height in a population, assuming that height is controlled by a single locus with genotypes *AA, Aa,* and *aa.* **B,** The distribution of height, assuming that height is controlled by two loci. There are now five distinct phenotypes instead of three, and the distribution begins to look more like the normal distribution. **C,** Distribution of height, assuming that multiple factors, each with a small effect, contribute to the trait (the multifactorial model).

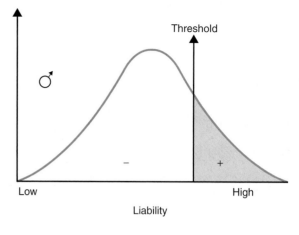

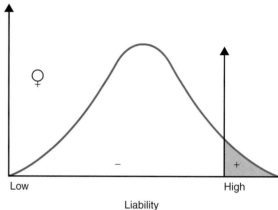

**FIGURE 12-2**
A liability distribution for a multifactorial disease in a population. To be affected with the disease, a person must exceed the threshold on the liability distribution. This figure shows two thresholds, a lower one for males and a higher one for females (as in pyloric stenosis).

disease-causing genes and environmental factors and are more likely to develop the disease. For multifactorial diseases that are either present or absent, it is thought that a **threshold of liability** must be crossed before the disease is expressed. Below the threshold, the person appears normal; above it, he or she is affected by the disease.

A disease that is thought to correspond to this threshold model is pyloric stenosis, a disorder that manifests shortly after birth and is caused by a narrowing or obstruction of the pylorus, the area between the stomach and intestine. Chronic vomiting, constipation, weight loss, and electrolyte imbalance result from the condition, which sometimes resolves spontaneously or can be corrected by surgery. The prevalence of pyloric stenosis among whites is about 3/1000 live births. It is much more common in males than in females, affecting 1/200 males and 1/1000 females. It is thought that this difference in prevalence reflects two thresholds in the liability distribution: a lower one in males and a higher one in females (see Fig. 12-2). A lower male threshold implies that fewer disease-causing factors are required to generate the disorder in males.

The liability threshold concept might explain the pattern of sibling recurrence risks for pyloric stenosis, shown in Table 12-1. Notice that males, having a lower threshold, always have a higher risk than females. However, the recurrence risk also depends on the sex of the proband. It is higher when the proband is female than when the proband is male. This reflects the concept that females, having a higher liability threshold, must be exposed to more disease-causing factors than males in order to develop the disease. Thus, a family with an affected female must have

**TABLE 12-1**

## Recurrence Risks (%) for Pyloric Stenosis, Subdivided by Gender of Affected Probands and Relatives*

| Relatives | Male Probands | | Female Probands | |
|---|---|---|---|---|
| | *London* | *Belfast* | *London* | *Belfast* |
| Brothers | 3.8 | 9.6 | 9.2 | 12.5 |
| Sisters | 2.7 | 3.0 | 3.8 | 3.8 |

*Note that the risks differ somewhat between the two populations.
(Adapted from Carter CO: Genetics of common single malformations. Br Med Bull 1976;32:21-26.)

more genetic and environmental risk factors, producing a higher recurrence risk for pyloric stenosis in future offspring. In such a situation, we would expect that the highest risk category would be *male* relatives of *female* probands; Table 12-1 shows that this is indeed the case.

A number of other congenital malformations are thought to correspond to this model. They include isolated* cleft lip and/or cleft palate, neural tube defects (anencephaly and spina bifida), club foot (talipes), and some forms of congenital heart disease.

---

*In this context, the term "isolated" means that this is the only observed disease feature (i.e., the feature is not part of a larger constellation of findings, as in cleft lip/palate secondary to trisomy 13).

> The threshold model applies to many multifactorial diseases. It assumes that there is an underlying liability distribution in a population and that a threshold on this distribution must be passed before a disease is expressed.

### Recurrence Risks and Transmission Patterns

Whereas recurrence risks can be given with confidence for single-gene diseases (50% for a completely penetrant autosomal dominant disease, 25% for autosomal recessive diseases, and so on), risk estimation is more complex for multifactorial diseases. This is because the number of genes contributing to the disease is usually not known, the precise allelic constitution of the parents is not known, and the extent of environmental effects can vary substantially. For most multifactorial diseases, **empirical risks** (i.e., risks based on direct observation of data) have been derived. To estimate empirical risks, a large series of families is examined in which one child (the proband) has developed the disease. The relatives of each proband are surveyed in order to calculate the percentage who have also developed the disease. For example, in North America neural tube defects are seen in about 2% to 3% of the siblings of probands with this condition (Clinical Commentary 12-1). Thus, the recurrence risk for parents who have had one child with a neural tube defect is 2% to 3%. For conditions that are not lethal or severely debilitating, such as cleft lip and cleft palate, recurrence risks can also be estimated for the offspring of affected parents. Because risk factors vary among diseases, empirical recurrence risks are specific for each multifactorial disease.

## CLINICAL COMMENTARY 12-1
### *Neural Tube Defects*

Neural tube defects (NTDs) include anencephaly, spina bifida, and encephalocele, as well as several other less-common forms. They are one of the most important classes of birth defects, with a newborn prevalence of 1 to 3 per 1000. There is considerable variation in the prevalence of NTDs among various populations, with an especially high rate among some northern Chinese populations (as high as 6 per 1000 births). For reasons that are not fully known, the prevalence of NTDs has been decreasing in many parts of the United States and Europe during the past two decades.

Normally the neural tube closes at about the fourth week of gestation. A defect in closure or a subsequent reopening of the neural tube results in an NTD. Spina bifida is the most commonly observed NTD and consists of a protrusion of spinal tissue through the vertebral column (the tissue usually includes meninges, spinal cord, and nerve roots). About 75% of spina bifida patients have secondary hydrocephalus, which sometimes in turn produces mental retardation. Paralysis or muscle weakness, lack of sphincter control, and club feet are often observed. A study conducted in British Columbia showed that survival rates for spina bifida patients have improved dramatically over the past several decades. Less than 30% of such children born between 1952 and 1969 survived to 10 years of age, but 65% of those born between 1970 and 1986 survived to this age.

Anencephaly is characterized by partial or complete absence of the cranial vault and calvarium and partial or complete absence of the cerebral hemispheres. At least two thirds of anencephalics are stillborn; term deliveries do not survive more than a few hours or days. Encephalocele consists of a protrusion of the brain into an enclosed sac. It is seldom compatible with survival.

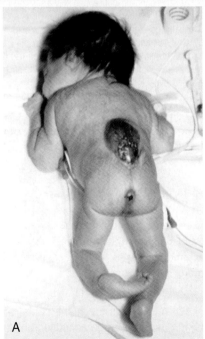

A

The major neural tube defects (NTDs). A, An infant with an open spina bifida (meningomyelocele).

*Continued*

## CLINICAL COMMENTARY  12-1
### Neural Tube Defects—cont'd

NTDs are thought to arise from a combination of genetic and environmental factors. In most populations surveyed thus far, empirical recurrence risks for siblings of affected individuals range from 2% to 5%. Consistent

children with NTDs. This result has been replicated in several different populations and is thus well confirmed. It has been estimated that approximately 50% to 70% of NTDs can be avoided simply by dietary folic acid

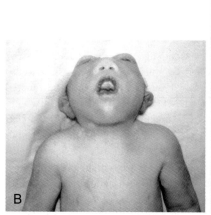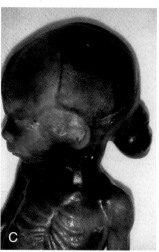

Major NTDs—cont'd. **B,** A fetus with anencephaly. Note the abnormalities of the orbits of the eye and the cranial defect. **C,** An occipital encephalocele.
*(A and B from Jones KL: Smith's Recognizable Patterns of Human Malformation, 6th ed. Philadelphia: Saunders, 2006, p 705.)*

with a multifactorial model, the recurrence risk increases with additional affected siblings. A Hungarian study showed that the overall prevalence of NTDs in that country was 1 in 300 births and that the sibling recurrence risks were 3%, 12%, and 25% after one, two, and three affected offspring, respectively. Recurrence risks tend to be slightly lower in populations with lower NTD prevalence rates, as predicted by the multifactorial model. Recurrence risk data support the idea that the major forms of NTDs are caused by similar factors. An anencephalic conception increases the recurrence risk for subsequent spina bifida conceptions, and vice versa.

NTDs can usually be diagnosed prenatally, sometimes by ultrasound and usually by an elevation in α-fetoprotein (AFP) in the maternal serum or amniotic fluid (see Chapter 13). A spina bifida lesion can be either open or closed (i.e., covered with a layer of skin). Open spina bifida is more likely to be detected by AFP assays.

A major epidemiological finding is that mothers who supplement their diet with folic acid at the time of conception are less likely to produce

supplementation. (Traditional prenatal vitamin supplements would not have an effect because administration does not usually begin until well after the time that the neural tube closes.) Because mothers would be likely to ingest similar amounts of folic acid from one pregnancy to the next, folic acid deficiency could well account for at least part of the elevated sibling recurrence risk for NTDs.

Dietary folic acid is an important example of a nongenetic factor that contributes to familial clustering of a disease. However, it is likely that there is genetic variation in response to folic acid, which helps to explain why most mothers with folic acid deficiency do not bear children with NTDs and why some who ingest adequate amounts of folic acid nonetheless bear children with NTDs. To address this issue, researchers are testing for associations between NTDs and variants in several genes whose products (e.g., methylene tetrahydrofolate reductase) are involved in folic acid metabolism (see Clinical Commentary 15-6 in Chapter 15 for further information on dietary folic acid supplementation and NTD prevention).

---

In contrast to most single-gene diseases, recurrence risks for multifactorial diseases can change substantially from one population to another (notice the differences between the London and Belfast populations in Table 12-1). This is because gene frequencies as well as environmental factors can differ among populations.

▶ Empirical recurrence risks for multifactorial diseases are based on studies of large collections of families. These risks are specific to a given population.

It is sometimes difficult to differentiate polygenic or multifactorial diseases from single-gene diseases that have reduced penetrance or variable expression. Large data sets and good family history data are necessary to make the distinction. Several criteria are usually used to define multifactorial inheritance.

• *The recurrence risk is higher if more than one family member is affected.* For example, the sibling recurrence risk for a ventricular septal defect (VSD, a type of congenital heart defect) is 3% if one sibling has had a VSD but increases to approximately 10% if two siblings have had VSDs. In contrast, the recurrence risk for single-gene diseases remains the same regardless of the number of affected siblings. This increase does not mean that the family's risk has actually changed. Rather, it means that we now have more information about the family's true risk: because they have had two affected children, they are probably located higher on the liability distribution than a family with only one affected child. In other words, they have more risk factors (genetic and/or environmental) and are more likely to produce an affected child.

- *If the expression of the disease in the proband is more severe, the recurrence risk is higher.* This is again consistent with the liability model, because a more-severe expression indicates that the affected person is at the extreme tail of the liability distribution (see Fig. 12-2). His or her relatives are thus at a higher risk to inherit disease-causing genes. For example, the occurrence of a bilateral (both sides) cleft lip/palate confers a higher recurrence risk on family members than does the occurrence of a unilateral (one side) cleft.

- *The recurrence risk is higher if the proband is of the less commonly affected sex* (see, for example, the previous discussion of pyloric stenosis). This is because an affected individual of the less susceptible sex is usually at a more extreme position on the liability distribution.

- *The recurrence risk for the disease usually decreases rapidly in more remotely related relatives* (Table 12-2). Although the recurrence risk for single-gene diseases decreases by 50% with each degree of relationship (e.g., an autosomal dominant disease has a 50% recurrence risk for offspring of affected persons, 25% for nieces or nephews, 12.5% for first cousins, and so on), it decreases much more quickly for multifactorial diseases. This reflects the fact that many genetic and environmental factors must combine to produce a trait. All of the necessary risk factors are unlikely to be present in less closely related family members.

- *If the prevalence of the disease in a population is f (which varies between zero and one), the risk for offspring and siblings of probands is approximately $\sqrt{f}$.* This does not hold true for single-gene traits, because their recurrence risks are independent of

population prevalence. It is not an absolute rule for multifactorial traits either, but many such diseases do tend to conform to this prediction. Examination of the risks given in Table 12-2 shows that these three diseases follow the prediction fairly well.

> Risks for multifactorial diseases usually increase if more family members are affected, if the disease has more severe expression, and if the affected proband is a member of the less commonly affected sex. Recurrence risks decrease rapidly with more-remote degrees of relationship. In general, the sibling recurrence risk is approximately equal to the square root of the prevalence of the disease in the population.

## Multifactorial versus Single-Gene Inheritance

It is important to clarify the difference between a multifactorial disease and a single-gene disease in which there is locus heterogeneity. In the former case, a disease is caused by the simultaneous influence of multiple genetic and environmental factors, each of which has a relatively small effect. In contrast, a disease with locus heterogeneity, such as osteogenesis imperfecta, requires only a single mutation to cause it. Because of locus heterogeneity, a single mutation at either of two loci can cause disease; some affected persons have one mutation while others have the other mutation.

In some cases, a trait may be influenced by the combination of both a single gene with large effects and a multifactorial background in which additional genes and environmental factors have small individual effects (Fig. 12-3). Imagine that variation in height, for example, is caused by a single locus (termed a **major gene**) and a multifactorial component. Individuals with the *AA* genotype tend to be taller, those with the *aa* genotype tend to be shorter, and those with *Aa* tend

**TABLE 12-2**

## Recurrence Risks for First-, Second-, and Third-Degree Relatives of Probands

| Disease | Prevalence in General Population | Degree of Relation | | |
| --- | --- | --- | --- | --- |
| | | *First Degree* | *Second Degree* | *Third Degree* |
| Cleft lip/palate | 0.001 | 0.04 | 0.007 | 0.003 |
| Club foot | 0.001 | 0.025 | 0.005 | 0.002 |
| Congenital hip dislocation | 0.002 | 0.005 | 0.006 | 0.004 |

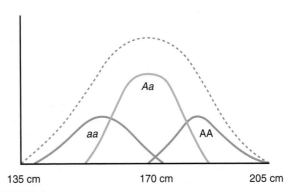

**FIGURE 12-3**
The distribution of height, assuming the presence of a major gene (genotypes *AA*, *Aa*, and *aa*) combined with a multifactorial background. The multifactorial background causes variation in height among individuals of each genotype. If the distributions of each of the three genotypes were superimposed, then the overall distribution of height would be approximately normal, as shown by the dotted line.

to be intermediate. But additional variation is caused by other factors (the multifactorial component). Thus, those with the *aa* genotype vary in height from 130 cm to about 170 cm, those with the *Aa* genotype vary from 150 cm to 190 cm, and those with the *AA* genotype vary from 170 to 210 cm. There is substantial overlap among the three major genotypes because of the influence of the multifactorial background. The total distribution of height, which is bell-shaped, is caused by the superposition of the three distributions about each genotype.

Many of the diseases to be discussed later can be caused by a major gene and/or multifactorial inheritance. That is, there are subsets of the population in which diseases such as colon cancer, breast cancer, or heart disease are inherited as single-gene disorders (with additional variation in disease susceptibility contributed by other genetic and environmental factors). These subsets usually account for only a small percentage of the total number of disease cases. It is nevertheless important to identify the responsible major genes, because their function can provide important clues to the pathophysiology and treatment of the disease.

> Multifactorial diseases can be distinguished from single-gene disorders caused by mutations at different loci (locus heterogeneity). Sometimes a disease has both single-gene and multifactorial components.

## NATURE AND NURTURE: DISENTANGLING THE EFFECTS OF GENES AND ENVIRONMENT

Family members share genes and a common environment. Family resemblance in traits such as blood pressure therefore reflects both genetic and environmental commonality ("nature" and "nurture," respectively). For centuries, people have debated the relative importance of these two types of factors. It is a mistake, of course, to view them as mutually exclusive. Few traits are influenced only by genes or only by environment. Most are influenced by both.

Determining the relative influence of genetic and environmental factors can lead to a better understanding of disease etiology. It can also help in the planning of public health strategies. A disease in which hereditary influence is relatively small, such as lung cancer, may be prevented most effectively through emphasis on lifestyle changes (avoidance of tobacco). When a disease has a relatively larger hereditary component, as in breast cancer, examination of family history should be emphasized in addition to lifestyle modification.

In the following sections, we review two research strategies that are often used to estimate the relative influence of genes and environment: twin studies and adoption studies. We then discuss methods that aim to delineate the individual genes responsible for multifactorial diseases.

### Twin Studies

Twins occur with a frequency of about 1/100 births in white populations. They are slightly more common among

**FIGURE 12-4**
Monozygotic twins, showing a striking similarity in physical appearance. Both twins developed myopia as teenagers.

Africans and a bit less common among Asians. **Monozygotic** (MZ, or identical) twins originate when the developing embryo divides to form two separate but identical embryos. Because they are genetically identical, MZ twins are an example of natural clones. Their physical appearances can be strikingly similar (Fig. 12-4). **Dizygotic** (DZ, or fraternal) twins are the result of a double ovulation followed by the fertilization of each egg by a different sperm.* Thus, DZ twins are genetically no more similar than other siblings. Because two different sperm cells are required to fertilize the two eggs, it is possible for each DZ twin to have a different father.

Because MZ twins are genetically identical, any differences between them should be due only to environmental effects. MZ twins should thus resemble each other very closely for traits that are strongly influenced by genes. DZ twins provide a convenient comparison: their environmental differences should be similar to those of MZ twins, but their genetic differences are as great as those between siblings. Twin studies thus usually consist of comparisons between MZ and DZ twins. If both members of a twin pair share a trait (e.g., cleft lip), they are said to be **concordant**. If they do not share the trait, they are **discordant**. For a trait determined completely by genes, MZ twins should always be concordant, and DZ twins should be concordant less often. Like siblings, DZ twins share only 50% of their DNA because each parent transmits half of his or her DNA to each offspring. Concordance rates can differ between

---

*While MZ twinning rates are quite constant across populations, DZ twinning rates vary somewhat. DZ twinning increases with maternal age until about age 40 years, after which the rate declines. The frequency of DZ twinning has increased dramatically in developed countries during the past two decades because of the use of ovulation-inducing drugs.

opposite-sex DZ twin pairs and same-sex DZ pairs for some traits, such as those that have different frequencies in males and females. For such traits, only same-sex DZ twin pairs should be used when comparing MZ and DZ concordance rates.

A concordance estimate would not be appropriate for quantitative traits, such as blood pressure or height. Here the **intraclass correlation coefficient** is used. This statistic varies between −1.0 and +1.0 and measures the degree of homogeneity of a trait in a sample of individuals. For example, we may wish to assess the degree of similarity between twins for a trait such as height. The measurements are made in a collection of twins, and correlation coefficients are estimated separately for the MZ sample and the DZ sample. If a trait were determined entirely by genes, we would expect the correlation coefficient for MZ pairs to be 1.0 (i.e., each pair of twins would have exactly the same height). A correlation coefficient of 0.0 would mean that the similarity between MZ twins for the trait in question is no greater than chance. Because DZ twins share half of their DNA, we would expect a DZ correlation coefficient of 0.50 for a trait determined entirely by genes.

> Monozygotic (identical) twins are the result of an early cleavage of the embryo, whereas dizygotic (fraternal) twins are caused by the fertilization of two eggs by two sperm cells. Comparisons of concordance rates and correlations in MZ and DZ twins help to estimate the extent to which a trait is influenced by genes.

Concordance rates and correlation coefficients for a number of traits are given in Table 12-3. The concordance rates for contagious diseases like measles are quite similar in MZ and DZ twins. This is expected, because most contagious diseases are unlikely to be influenced markedly by genes. On the other hand, the concordance rates for schizophrenia are quite dissimilar between MZ and DZ twins, indicating a sizable genetic component for this disease. The MZ correlation for dermatoglyphics (fingerprints), a series of traits determined almost entirely by genes, is close to 1.0.

Correlations and concordance rates in MZ and DZ twins can be used to measure the **heritability** of multifactorial traits. Essentially, heritability is the percentage of population variation in a trait that is due to genes (statistically, it is the proportion of the total **variance** of a trait that is caused by genes). A simple formula for estimating heritability (h) from twin correlations or concordance rates is as follows:

$$h = 2(c_{MZ} - c_{DZ})$$

where $c_{MZ}$ is the concordance rate (or intraclass correlation) for MZ twins and $c_{DZ}$ is the concordance rate (or intraclass correlation) for DZ twins.* As this formula illustrates, traits

---

*This formula represents one of the simplest ways of estimating heritability. A description of more complex and accurate approaches can be found in the books by Cavalli-Sforza and Bodmer (1971) and Neale and Cardon (1992) cited at the end of this chapter.

**TABLE 12-3**
## Concordance Rates in Twins for Selected Traits and Diseases*

| Trait or Disease | Concordance Rate MZ Twins | DZ Twins | Heritability |
|---|---|---|---|
| Affective disorder (bipolar) | 0.79 | 0.24 | >1.0[‡] |
| Affective disorder (unipolar) | 0.54 | 0.19 | 0.70 |
| Alcoholism | >0.60 | <0.30 | 0.60 |
| Autism | 0.92 | 0.0 | >1.0 |
| Blood pressure (diastolic)[†] | 0.58 | 0.27 | 0.62 |
| Blood pressure (systolic)[†] | 0.55 | 0.25 | 0.60 |
| Body fat percentage[†] | 0.73 | 0.22 | >1.0 |
| Body mass index[†] | 0.95 | 0.53 | 0.84 |
| Cleft lip/palate | 0.38 | 0.08 | 0.60 |
| Club foot | 0.32 | 0.03 | 0.58 |
| Dermatoglyphics (finger ridge count)[†] | 0.95 | 0.49 | 0.92 |
| Diabetes mellitus | 0.45–0.96 | 0.03–0.37 | >1.0 |
| Diabetes mellitus (type 1) | 0.35–0.50 | 0.05–0.10 | 0.60–0.80 |
| Diabetes mellitus (type 2) | 0.70–0.90 | 0.25–0.40 | 0.90–1.0 |
| Epilepsy (idiopathic) | 0.69 | 0.14 | >1.0 |
| Height[†] | 0.94 | 0.44 | 1.0 |
| IQ[†] | 0.76 | 0.51 | 0.50 |
| Measles | 0.95 | 0.87 | 0.16 |
| Multiple sclerosis | 0.28 | 0.03 | 0.50 |
| Myocardial infarction (males) | 0.39 | 0.26 | 0.26 |
| Myocardial infarction (females) | 0.44 | 0.14 | 0.60 |
| Schizophrenia | 0.47 | 0.12 | 0.70 |
| Spina bifida | 0.72 | 0.33 | 0.78 |

*These figures were compiled from a large variety of sources and represent primarily European and U.S. populations.

[†]Because these are quantitative traits, correlation coefficients are given rather than concordance rates.

[‡]Several heritability estimates exceed 1.0. Because it is impossible for >100% of the variance of a trait to be genetically determined, these values indicate that other factors, such as shared environmental factors, must be operating.

DZ, dizygotic; IQ, intelligence quotient; MZ, monozygotic.

that are largely determined by genes result in a heritability estimate that approaches 1.0 (i.e., $c_{MZ}$ approaches 1.0, and $c_{DZ}$ approaches 0.5). As the difference between MZ and DZ concordance rates becomes smaller, heritability approaches zero. Correlations and concordance rates in other types of relatives (e.g., between parents and offspring) can also be used to measure heritability.

Like recurrence risks, heritability values are specific for the population in which they are estimated. However, there is usually agreement from one population to another regarding the general range of heritability estimates of most traits (e.g., the heritability of height is almost always high, and the heritability of contagious diseases is almost always low). The same is true of empirical recurrence risks.

> Comparisons of correlations and concordance rates in MZ and DZ twins allow the estimation of heritability, a measure of the percentage of population variation in a disease that can be attributed to genes.

At one time, twins were thought to provide a perfect "natural laboratory" in which to determine the relative influences of genetics and environment. But several difficulties arise. One of the most important is the assumption that the environments of MZ and DZ twins are equally similar. MZ twins are often treated more similarly than DZ twins. The eminent geneticist L. S. Penrose once joked that, if one were to study the clothes of twins, it might be concluded that clothes are inherited biologically. A greater similarity in environment can make MZ twins more concordant for a trait, inflating the apparent influence of genes. In addition, MZ twins may be more likely to seek the same type of environment, further reinforcing environmental similarity. On the other hand, it has been suggested that some MZ twins tend to develop personality differences in an attempt to assert their individuality.

Another difficulty is that the uterine environments of different pairs of MZ twins can be more or less similar, depending on whether there are two amnions and two chorions, two amnions and one shared chorion, or one shared amnion and one shared chorion. In addition, somatic mutations can occur during mitotic divisions of the cells of MZ twin embryos after cleavage occurs. Thus, the MZ twins might not be quite "identical," especially if a mutation occurred early in the development of one of the twins. Finally, recent studies indicate that methylation patterns, which can influence the transcription of specific genes, become more dissimilar in MZ twin pairs as they age. This dissimilarity is greater when the twins adopt markedly different habits and lifestyles (e.g., when one twin smokes cigarettes and the other does not).

Of the various problems with the twin method, the greater degree of environmental sharing among MZ twins is perhaps the most serious. One way to circumvent this problem, at least in part, is to study MZ twins who were raised in separate environments. Concordance among these twin pairs should be caused by genetic, rather than environmental, similarities. As one might expect, it is not easy to find such twin pairs. A major effort to do so has been undertaken by researchers at the University of Minnesota, whose studies have shown a remarkable congruence among MZ twins reared apart, even for many behavioral traits. However, these studies must be viewed with caution, because the sample sizes are relatively small and because many of the twin pairs had at least some contact with each other before they were studied.

> Although twin studies provide valuable information, they are also affected by certain biases. The most serious is greater environmental similarity between MZ twins than between DZ twins. Other biases include somatic mutations that might affect only one MZ twin and differences in the uterine environments of twins.

### Adoption Studies

Studies of adopted children are also used to estimate the genetic contribution to a multifactorial trait. Offspring who were born to parents who have a disease but who were adopted by parents lacking the disease can be studied to find out whether the offspring develop the disease. In some cases, these adopted persons develop the disease more often than do children in a comparative control population (i.e., adopted children who were born to parents who do not have the disease). This provides evidence that genes may be involved in causing the disease, because the adopted children do not share an environment with their affected natural parents. For example, schizophrenia is seen in 8% to 10% of adopted children whose natural parent had schizophrenia, whereas it is seen in only 1% of adopted children of unaffected parents.

As with twin studies, several precautions must be exercised in interpreting the results of adoption studies. First, prenatal environmental influences could have long-lasting effects on an adopted child. Second, children are sometimes adopted after they are several years old, ensuring that some nongenetic influences have been imparted by the natural parents. Finally, adoption agencies sometimes try to match the adoptive parents with the natural parents in terms of attributes such as socioeconomic status. All of these factors could exaggerate the apparent influence of biological inheritance.

> Adoption studies provide a second means of estimating the influence of genes on multifactorial diseases. They consist of comparing disease rates among the adopted offspring of affected parents with the rates among adopted offspring of unaffected parents. As with the twin method, several biases can influence these studies.

These reservations, as well as those summarized for twin studies, underscore the need for caution in basing conclusions on twin and adoption studies. These approaches do not provide definitive measures of the role of genes in multifactorial disease, nor can they identify specific genes responsible for disease. Instead, they provide a preliminary indication of the extent to which a multifactorial disease may be influenced by genetic factors. Methods for the direct detection of genes underlying multifactorial traits are summarized in Box 12-1.

## THE GENETICS OF COMMON DISEASES

Having discussed the principles of multifactorial inheritance, we turn next to a discussion of the common multifactorial disorders themselves. Some of these disorders, the congenital malformations, are by definition present at birth. Others,

including heart disease, cancer, diabetes, and most psychiatric disorders, are seen primarily in adolescents and adults. Because of their complexity, unraveling the genetics of these disorders is a daunting task. Nonetheless, significant progress is now being made.

### Congenital Malformations

Approximately 2% of newborns present with a **congenital malformation** (i.e., one that is present at birth); most of these conditions are considered to be multifactorial in etiology. Some of the more common congenital malformations are listed in Table 12-4. In general, sibling recurrence risks for most of these disorders range from 1% to 5%.

Some congenital malformations, such as cleft lip/palate and pyloric stenosis, are relatively easy to repair and thus

---

**BOX 12-1**
## Finding Genes that Contribute to Multifactorial Disease

As mentioned in the text, twin and adoption studies are not designed to reveal specific genes that cause multifactorial diseases. The identification of specific causative genes is an important goal, because only then can we begin to understand the underlying biology of the disease and undertake to correct the defect. For complex multifactorial traits, this is a formidable task because of locus heterogeneity, the interactions of multiple genes, decreased penetrance, age-dependent onset, and phenocopies (persons who have a phenotype, such as breast cancer, but who do not carry a known disease-causing mutation, such as a *BRCA1* alteration). Fortunately, recent advances in gene mapping and molecular biology promise to make this goal more attainable. Here, we discuss several approaches that are used to identify the genes underlying multifactorial traits.

One way to search for these genes is to use conventional linkage analysis, as described in Chapter 8. Disease families are collected, a single-gene mode of inheritance is assumed, and linkage analysis is undertaken with a large series of marker polymorphisms that span the genome (this is termed a **genome scan**). If a sufficiently large LOD score (see Chapter 8) is obtained with a polymorphism, it is assumed that the region around this polymorphism might contain a disease-causing gene. This approach is sometimes successful, especially when there are subsets of families in which a single-gene mode of inheritance is seen (e.g., autosomal dominant, autosomal recessive). This was the case, for example, with familial breast cancer, where some families presented a clear autosomal dominant mode of inheritance.

With many multifactorial disorders, however, such subsets are not readily apparent. Because of obstacles such as heterogeneity and phenocopies, traditional linkage analysis may be impractical. One alternative to traditional linkage analysis is the **affected sib-pair method**. The logic of this approach is simple: if two siblings are both affected by a genetic disease, we would expect to see increased sharing of marker alleles in the genomic region that contains a susceptibility gene. To conduct an analysis using this approach, we begin by collecting DNA samples from a large number of sib pairs in which both members of the pair are affected by the disease. Then a genome scan is undertaken, and the proportion of affected sib pairs who share the same allele is estimated for each polymorphism. Because siblings share half their genes (see Chapter 4), we would expect this proportion to be 50% for marker polymorphisms that are not linked to a disease-susceptibility locus. However, if we find that siblings share the same allele

for a marker polymorphism more than half the time (say, 75% of the time), this would be evidence that the marker is linked to a susceptibility locus. This approach was used, for example, to show that the genes in the HLA region contribute to susceptibility for type 1 diabetes.

The affected sib-pair method has the advantage that one does not have to assume a specific mode of inheritance. In addition, the method is unaffected by reduced penetrance, because both members of the sib pair must be affected to be included in the analysis. It is especially useful for disorders with late age of onset (e.g., prostate cancer), for which it would be difficult to assemble multigenerational families from whom DNA samples could be taken. A weakness of this method is that it tends to require large sample sizes to yield significant results, and it tends to have low resolution (i.e., the genomic region implicated by the analysis tends to be quite large, often 10 cM or more).

Affected sib-pair analyses are sometimes made more powerful by selecting subjects with extreme values of a trait (e.g., sib pairs with very high blood pressure) to enrich the sample for genes likely to contribute to the trait. A variation on this approach is to sample sib pairs that are highly discordant for a trait (e.g., one with very high blood pressure and one with very low blood pressure) and then to look for markers in which there is less allele sharing than the expected 50%.

Association tests such as linkage disequilibrium (see Chapter 8) can also be used in the course of a genome scan (these are typically termed **genome-wide association studies**). These methods became more practical after the Human Genome Project developed dense sets of polymorphic markers (microsatellites, and, more recently, single nucleotide polymorphisms, or SNPs). It is now common to use microarrays that can assay one million SNPs in a collection of cases and controls. Because the gene frequency differences in disease-causing variants can be quite small, thousands of cases and thousands of controls are often tested in these studies. The likelihood of finding disease-causing genes using these approaches, as well as sib-pair and traditional linkage methods, may be enhanced by analyzing isolated populations (e.g., island populations, such as those mentioned in Chapter 3). Because these populations are typically derived from a small number of founders and have experienced little admixture with other populations, it is thought that the number of mutations contributing to a multifactorial disease may be reduced and thus easier to pinpoint.

*Continued*

**BOX 12-1**
## Finding Genes that Contribute to Multifactorial Disease—cont'd

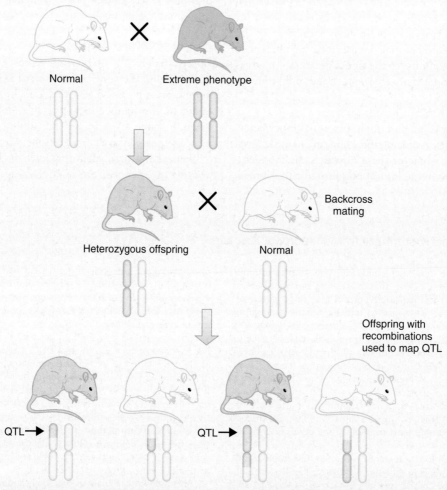

Basic steps involved in linkage analysis of a multifactorial (quantitative) trait using an animal model (see text for details). QTL, quantitative trait locus.

Another method combines genome scanning and the use of animal models. It consists basically of the following steps:

1. Breeding experiments are carried out with experimental animals, such as rats or mice, to select progeny that have extreme values of a trait (e.g., rats that have high blood pressure). These are then crossed with normal animals to produce offspring that, for each chromosome pair, have one normal chromosome and one "affected" chromosome that presumably contains genes that cause high blood pressure. These offspring are in turn mated with the normal animal (a backcross). This produces a third generation of animals in whom one chromosome has only the normal genes, while the homologous chromosome has experienced recombinations between the normal and the affected chromosomes (as a result of crossovers during meiosis in the parents). This series of matings produces progeny that are useful for linkage analysis.
2. High-resolution genetic maps of the experimental organism must be available. This means that polymorphic markers must be identified at regular intervals (ideally, at least every 10 cM) throughout the organism's genome.
3. Linkage analysis (see Chapter 8) is performed, comparing each polymorphic marker against the trait. Because animals with extreme values were selected, this procedure should uncover markers that are linked to loci that produce the extreme phenotype.

4. Once a linked marker (or markers) has been found, it may be possible to isolate the actual functional gene responsible for the trait using the gene-cloning techniques outlined in Chapter 8.
5. When a functional gene has been isolated and cloned in the experimental organism, it is used as a probe to search the human genome for a gene with high DNA sequence homology that may have the same function (a candidate gene). This approach is feasible because the DNA sequences of functionally important genes are often similar in humans and experimental animals such as rodents.

This approach has been applied in studies of type 1 diabetes and hypertension. It has the advantage that animals can easily be selected with extreme values of a trait, and any desired breeding scheme can be used to generate useful recombinants. Animals, of course, do not necessarily model humans accurately. Furthermore, this technique detects only individual genes that cause disease in the animal model; it cannot assess the pattern of interactions of these genes. There is evidence that the nature of these interactions may be critically important, and they may well differ in humans and experimental animals. Despite these reservations, this approach demonstrates effectively the way new developments in molecular genetics and gene mapping can increase our knowledge of the genes responsible for multifactorial disease.

**TABLE 12-4**
## Prevalence Rates of Common Congenital Malformations in Persons of European Descent

| Disorder | Approximate Prevalence per 1000 Births |
|---|---|
| Cleft lip/palate | 1.0 |
| Club foot | 1.0 |
| Congenital heart defects | 4.0–8.0 |
| Hydrocephaly | 0.5–2.5 |
| Isolated cleft palate | 0.4 |
| Neural tube defects | 1.0–3.0 |
| Pyloric stenosis | 3.0 |

are not considered to be serious problems. Others, such as the neural tube defects, usually have more severe consequences. Although some cases of congenital malformations can occur in the absence of any other problems, it is quite common for them to be associated with other disorders. For example, hydrocephaly and club foot are often seen secondary to spina bifida, cleft lip/palate is often seen in babies with trisomy 13, and congenital heart defects are seen in many syndromes, including trisomy of chromosomes 13, 18, and 21.

Considerable progress is now being made in isolating single genes that can cause congenital malformations. Many of these, including the HOX, PAX, and TBX families of genes, were discussed in Chapter 10. Another example is the *RET* proto-oncogene, which is responsible for some cases of Hirschsprung disease. However, the causes of most cases of this disorder remain undiscovered. Indeed, most of the genetic factors that contribute to important congenital malformations (e.g., neural tube defects, common congenital heart defects, cleft lip/palate) are as yet unidentified.

Environmental factors have also been shown to cause some congenital malformations. An example is thalidomide, a sedative used during pregnancy in the early 1960s (and recently reintroduced for the treatment of dermatological conditions such as leprosy). When ingested during early pregnancy, this drug often caused phocomelia (severely shortened limbs) in babies. Maternal exposure to retinoic acid, which is used to treat acne, can cause congenital defects of the heart, ear, and central nervous system. Maternal rubella infection can cause congenital heart defects. Other environmental factors that can cause congenital malformations are discussed in Chapter 15.

> Congenital malformations are seen in roughly 1 of every 50 live births. Most of them are considered to be multifactorial disorders. Specific genes and environmental causes have been detected for some congenital malformations, but the causes of most congenital malformations remain largely unknown.

## Multifactorial Disorders in the Adult Population

Until recently, very little was known about specific genes responsible for common adult diseases. With more powerful laboratory and analytical techniques, this situation is changing. We next review recent progress in understanding the genetics of the major common adult diseases. Table 12-5 gives

**TABLE 12-5**
## Prevalence Figures and Annual Costs for Common Adult Diseases

| Disease | Number of Affected Americans (approximate) | Annual Cost ($billion)* |
|---|---|---|
| Alcoholism | 14 million | 185 |
| Alzheimer disease | 4 million | 90 |
| Arthritis | 43 million | 65 |
| Asthma | 17 million | 13 |
| Cancer | 8 million | 157 |
| Cardiovascular disease (all forms) | | 300 |
| Coronary artery disease | 13 million | |
| Congestive heart failure | 5 million | |
| Congenital defects | 1 million | |
| Hypertension | 50 million | |
| Stroke | 5 million | |
| Depression and bipolar disorder | 17 million | 44 |
| Diabetes (type 1) | 1 million | |
| Diabetes (type 2) | 15 million | 100 (type 1 + type 2) |
| Epilepsy | 2.5 million | 3 |
| Multiple sclerosis | 350,000 | 5 |
| Obesity† | 60 million | 117 |
| Parkinson disease | 500,000 | 5.5 |
| Psoriasis | 3-5 million | 3 |
| Schizophrenia | 2 million | 30 |

*Cost estimates include direct medical costs as well as associated costs such as lost economic productivity.

†Body mass index >30.

Data from National Center for Chronic Disease Prevention and Health Promotion; American Heart Association (2002 Heart and Stroke Statistical Update); National Institute on Alcohol Abuse and Alcoholism; Office of the U.S. Surgeon General; American Academy of Allergy, Asthma and Immunology; Cown WM, Kandel ER: Prospects for neurology and psychiatry. JAMA 2001;285:594–600; Flegal KM, Carroll MD, Ogden CL, Johnson CL: Prevalence and trends in obesity among US adults, 1999-2000. JAMA 2002;288:1723-1727.

approximate prevalence figures for these disorders in the United States.

## Cardiovascular Disorders

### Heart Disease

Heart disease is the leading cause of death worldwide, and it accounts for approximately 25% of all deaths in the United States. The most common underlying cause of heart disease is coronary artery disease (CAD), which is caused by atherosclerosis (a narrowing of the coronary arteries resulting from the formation of lipid-laden lesions). This narrowing impedes blood flow to the heart and can eventually result in a myocardial infarction (death of heart tissue caused by an inadequate supply of oxygen). When atherosclerosis occurs in arteries that supply blood to the brain, a stroke can result. A number of risk factors for CAD have been identified, including obesity, cigarette smoking, hypertension, elevated cholesterol level, and positive family history (usually defined as having one or more affected first-degree relatives). Many studies have examined the role of family history in CAD, and they show that a person with a positive family history is at least twice as likely to suffer from CAD than is a person with no family history. Generally, these studies also show that the risk is higher if there are more affected relatives, if the affected relative is female (the less commonly affected sex) rather than male, and if the age of onset in the affected relative is early (before 55 years of age). For example, one study showed that men between the ages of 20 and 39 years had a three-fold increase in CAD risk if they had one affected first-degree relative. This risk increased to 13-fold if there were two first-degree relatives affected with CAD before 55 years of age.

What part do genes play in the familial clustering of CAD? Because of the key role of lipids in atherosclerosis, many studies have focused on the genetic determination of variation in circulating lipoprotein levels. An important advance was the isolation and cloning of the gene that encodes the low-density lipoprotein (LDL) receptor. Heterozygosity for a mutation in this gene roughly doubles LDL cholesterol levels and is seen in approximately 1 in 500 persons. (This condition, known as familial hypercholesterolemia, is described further in Clinical Commentary 12-2.) Mutations in the gene encoding apolipoprotein B, which are seen in about 1 in 1000 persons, are another common genetic cause of elevated LDL cholesterol. These mutations occur in the portion of the gene that is responsible for binding of apolipoprotein B to the LDL receptor, and they increase circulating LDL cholesterol levels by 50% to 100%. More than a dozen other genes involved in lipid metabolism and transport have been identified, including those genes that encode various apolipoproteins (these are the protein components of lipoproteins) (Table 12-6). In addition, several genes whose protein products contribute to inflammation have been associated with CAD, reflecting the critical role of inflammation in generating atherosclerotic plaques. Functional analysis of these genes is leading to increased understanding and more effective treatment of CAD.

Environmental factors, many of which are easily modified, are also important causes of CAD. There is abundant epidemiological evidence that cigarette smoking and obesity increase the risk of CAD, whereas exercise and a diet low in saturated fats decrease the risk. Indeed, the approximate 60% reduction in age-adjusted mortality due to CAD and stroke in the United States since 1950 is usually attributed to a decrease in the percentage of adults who smoke cigarettes, decreased consumption of saturated fats, improved medical care, and increased emphasis on healthy lifestyle factors such as exercise.

Another form of heart disease is cardiomyopathy, an abnormality of the heart muscle that leads to inadequate cardiac function. Cardiomyopathy is a common cause of heart failure, resulting in approximately 10,000 deaths annually in the United States. Hypertrophic cardiomyopathy, one major form of the disease, is characterized by thickening (hypertrophy) of portions of the left ventricle and is seen in as many as 1 in 500 adults. About half of hypertrophic cardiomyopathy cases are familial and are caused by autosomal dominant mutations in any of the multiple genes that encode various components of the cardiac sarcomere. The most commonly mutated genes are those that encode the β-myosin heavy chain (35% of familial cases), myosin-binding protein C (20% of cases), and troponin T (15% of cases).

In contrast to the hypertrophic form of cardiomyopathy, dilated cardiomyopathy, which is seen in about 1 in 2500 persons, consists of increased size and impaired contraction of the ventricles. The end result is impaired pumping of the heart. This disease is familial in about one third of affected persons; although autosomal dominant mutations are most common, mutations can also be X-linked or mitochondrial. The genes affected by these mutations encode various cytoskeletal proteins, including actin, cardiac troponin T, desmin, and components of the dystroglycan–sarcoglycan complex. (Recall from Chapter 5 that abnormalities of the latter proteins can also cause muscular dystrophies.)

Mutations have also been identified in several genes that cause the long QT (LQT) syndrome. LQT describes the characteristic elongated QT interval in the electrocardiogram of affected individuals, indicative of delayed cardiac repolarization. This disorder, which can be caused either by inherited mutations or by exposure to drugs that block potassium channels, predisposes affected person to potentially fatal cardiac arrhythmia. An autosomal dominant form, known as Romano–Ward syndrome, can be caused by loss-of-function mutations in genes that encode potassium channels (such as KCNQ1, KCNH2, KCNE1, KCNE2, or KCNJ2). These mutations delay cardiac repolarization. Gain-of-function mutations in several of these same genes have been shown to produce a shortened QT interval, as might be expected. Romano–Ward syndrome can also be caused by gain-of-function mutations in sodium or calcium channel genes (SCN5A and CACNA1C, respectively), which result in a prolonged depolarizing current. (Other examples of mutations that can cause LQT are given in Table 12-7.)

## CLINICAL COMMENTARY 12-2
### *Familial Hypercholesterolemia*

Autosomal dominant familial hypercholesterolemia (FH) is an important cause of heart disease, accounting for approximately 5% of myocardial infarctions (MIs) in persons younger than 60 years. FH is one of the most common autosomal dominant disorders: in most populations surveyed to date, about 1 in 500 persons is a heterozygote. Plasma cholesterol levels are approximately twice as high as normal (i.e., about 300-400 mg/dL), resulting in substantially accelerated atherosclerosis and the occurrence of distinctive cholesterol deposits in skin and tendons, called xanthomas. Data compiled from five studies showed that approximately 75% of men with FH developed coronary artery disease, and 50% had a fatal MI, by age 60 years. The corresponding percentages for women were lower (45% and 15%, respectively), because women generally develop heart disease at a later age than men.

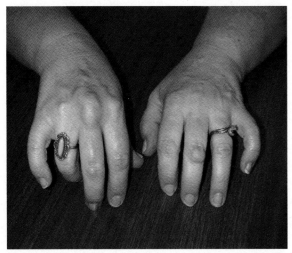

Xanthomas (fatty deposit), seen here on the knuckles, are often seen in patients with familial hypercholesterolemia.

Consistent with Hardy–Weinberg predictions (see Chapter 4), about 1/1,000,000 births is homozygous for the FH gene. Homozygotes are much more severely affected, with cholesterol levels ranging from 600 to 1200 mg/dL. Most homozygotes experience MIs before 20 years of age, and an MI at 18 months of age has been reported. Without treatment, most FH homozygotes die before the age of 30 years.

All cells require cholesterol as a component of their plasma membrane. They can either synthesize their own cholesterol, or preferentially they obtain it from the extracellular environment, where it is carried primarily by low-density lipoprotein (LDL). In a process known as **endocytosis**, LDL-bound cholesterol is taken into the cell via LDL receptors on the cell's surface. FH is caused by a reduction in the number of functional LDL receptors on cell surfaces. Because the person lacks the normal number of LDL receptors, cellular cholesterol uptake is reduced, and circulating cholesterol levels increase.

Much of what we know about endocytosis has been learned through the study of LDL receptors. The process of endocytosis and the processing of LDL in the cell is described in detail in the accompanying figure. These processes result in a fine-tuned regulation of cholesterol levels within cells, and they influence the level of circulating cholesterol as well.

The cloning of the LDL receptor gene *(LDLR)* in 1984 was a critical step in understanding exactly how LDL receptor defects cause FH. This gene, located on chromosome 19, is 45 kb in length and consists of 18 exons and 17 introns. More than 900 different mutations, two thirds of which are missense and nonsense substitutions, have been identified. Most of the remaining mutations are insertions and deletions, many of which arise from unequal crossovers (see Chapters 5 and 6) that occur between *Alu* repeat sequences (see Chapter 2) scattered throughout the gene. The *LDLR* mutations can be grouped into five broad classes, according to their effects on the activity of the receptor:

- Class I mutations in *LDLR* result in no detectable protein product. Thus, heterozygotes would produce only half the normal number of LDL receptors.
- Class II mutations result in production of the LDL receptor, but it is altered to the extent that it cannot leave the endoplasmic reticulum. It is eventually degraded.
- Class III mutations produce an LDL receptor that is capable of migrating to the cell surface but is incapable of normal binding to LDL.
- Class IV mutations, which are comparatively rare, produce receptors that are normal except that they do not migrate specifically to coated pits and thus cannot carry LDL into the cell.
- Class V mutations produce an LDL receptor that cannot disassociate from the LDL particle after entry into the cell. The receptor cannot return to the cell surface and is degraded.

Each class of mutations reduces the number of effective LDL receptors, resulting in decreased LDL uptake and hence elevated levels of circulating cholesterol. The number of effective receptors is reduced by about half in FH heterozygotes, and homozygotes have virtually no functional LDL receptors.

Understanding the defects that lead to FH has helped in the development of effective therapies for the disorder. Dietary reduction of cholesterol (primarily through the reduced intake of saturated fats) has only modest effects on cholesterol levels in FH heterozygotes. Because cholesterol is reabsorbed into the gut and then recycled through the liver (where most cholesterol synthesis takes place), serum cholesterol levels can be reduced by the administration of bile-acid absorbing resins, such as cholestyramine. The absorbed cholesterol is excreted. It is interesting that reduced recirculation from the gut causes the liver cells to form additional LDL receptors, lowering circulating cholesterol levels. However, the decrease in intracellular cholesterol also stimulates cholesterol synthesis by liver cells, so the overall reduction in plasma LDL is only about 15% to 20%. This treatment is much more effective when combined with one of the statin drugs (e.g., lovastatin, pravastatin), which reduce cholesterol synthesis by inhibiting 3-hydroxy-3-methylglutaryl coenzyme A (HMG-CoA) reductase. Decreased synthesis leads to further production of LDL receptors. When these therapies are used in combination, serum cholesterol levels in FH heterozygotes can often be reduced to approximately normal levels.

The picture is less encouraging for FH homozygotes. The therapies mentioned can enhance cholesterol elimination and reduce its synthesis, but they are largely ineffective in homozygotes because these persons have few or no LDL receptors. Liver transplants, which provide hepatocytes that have normal LDL receptors, have been successful in some cases, but this option is often limited by a lack of donors. Plasma exchange, carried out every 1 to 2 weeks, in combination with drug therapy, can reduce cholesterol levels by about 50%. However, this therapy is difficult to continue for long periods. Somatic cell gene therapy, in which hepatocytes carrying normal LDL receptor genes are introduced into the portal circulation, is now being tested (see Chapter 13). It might eventually prove to be an effective treatment for FH homozygotes.

The FH story illustrates how medical research has made important contributions to both the understanding of basic cell biology and advancements in clinical therapy. The process of receptor-mediated endocytosis, elucidated largely by research on the LDL receptor defects, is of fundamental significance for cellular processes throughout the body. Equally, this research, by clarifying how cholesterol synthesis and uptake can be modified, has led to significant improvements in therapy for this important cause of heart disease.

*Continued*

## CLINICAL COMMENTARY 12-2
### *Familial Hypercholesterolemia—cont'd*

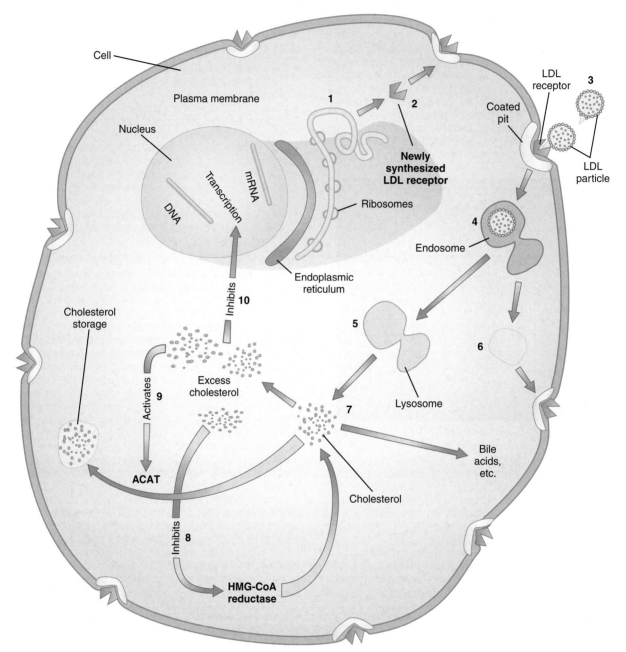

The process of receptor-mediated endocytosis. *1*, The low-density lipoprotein (LDL) receptors, which are glycoproteins, are synthesized in the endoplasmic reticulum of the cell. *2*, They pass through the Golgi apparatus to the cell surface, where part of the receptor protrudes outside the cell. *3*, The circulating LDL particle is bound by the LDL receptor and localized in cell-surface depressions called *coated pits* (so named because they are coated with a protein called clathrin). *4*, The coated pit invaginates, bringing the LDL particle inside the cell. *5*, Once inside the cell, the LDL particle is separated from the receptor, taken into a lysosome, and broken down into its constituents by lysosomal enzymes. *6*, The LDL receptor is recirculated to the cell surface to bind another LDL particle. Each LDL receptor goes through this cycle approximately once every 10 minutes, even if it is not occupied by an LDL particle. *7*, Free cholesterol is released from the lysosome for incorporation into cell membranes or metabolism into bile acids or steroids. Excess cholesterol can be stored in the cell as a cholesterol ester or removed from the cell by association with high-density lipoprotein (HDL). *8*, As cholesterol levels in the cell rise, cellular cholesterol synthesis is reduced by inhibition of the rate-limiting enzyme, HMG-CoA reductase. *9*, Rising cholesterol levels also increase the activity of acyl-coenzyme A:cholesterol acyltransferase (ACAT), an enzyme that modifies cholesterol for storage as cholesterol esters. *10*, In addition, the number of LDL receptors is decreased by lowering the transcription rate of the LDL receptor gene itself. This decreases cholesterol uptake.

**TABLE 12-6**
Lipoprotein Genes Known to Contribute to Coronary Heart Disease Risk

| Gene | Chromosome Location | Function of Protein Product |
|------|--------------------|-----------------------------|
| Apolipoprotein A-I | 11q | HDL component; LCAT cofactor |
| Apolipoprotein A-IV | 11q | Component of chylomicrons and HDL; may influence HDL metabolism |
| Apolipoprotein C-III | 11q | Allelic variation associated with hypertriglyceridemia |
| Apolipoprotein B | 2p | Ligand for LDL receptor; involved in formation of VLDL, LDL, IDL, and chylomicrons |
| Apolipoprotein D | 2p | HDL component |
| Apolipoprotein C-I | 19q | LCAT activation |
| Apolipoprotein C-II | 19q | Lipoprotein lipase activation |
| Apolipoprotein E | 19q | Ligand for LDL receptor |
| Apolipoprotein A-II | 1p | HDL component |
| LDL receptor | 19p | Uptake of circulating LDL particles |
| Lipoprotein(a) | 6q | Cholesterol transport |
| Lipoprotein lipase | 8p | Hydrolysis of lipoprotein lipids |
| Hepatic triglyceride lipase | 15q | Hydrolysis of lipoprotein lipids |
| LCAT | 16q | Cholesterol esterification |
| Cholesterol ester transfer protein | 16q | Facilitates transfer of cholesterol esters and phospholipids between lipoproteins |

*HDL, high-density lipoprotein; IDL, Intermediate-density lipoprotein; LCAT, lecithin cholesterol acyltransferase; LDL, low-density lipoprotein; VLDL, very-low-density lipoprotein.*
*Adapted in part from King RA, Rotter JI (eds): The Genetic Basis of Common Diseases, 2nd ed. New York: Oxford University Press, 2002.*

An autosomal recessive form of LQT syndrome, known as Jervell–Lange–Nielsen syndrome, is less common than the Romano–Ward syndrome but is associated with a longer QT interval, a higher incidence of sudden cardiac death, and sensorineural deafness. This syndrome is caused by mutations in either *KCNQ1* or *KCNE1*. Because LQT syndrome can be difficult to diagnose accurately, linked markers and mutation detection are used to enable more accurate diagnosis of affected family members. In addition, the identification of disease-causing genes and their protein products is now guiding the development of drug therapy to activate the encoded ion channels. Because cardiac arrhythmias account for most of the 300,000 sudden cardiac deaths that occur annually in the United States, a better understanding of the genetic defects underlying arrhythmia is of considerable public health significance.

> Heart disease aggregates in families. This aggregation is especially strong if there is early age of onset and if there are several affected relatives. Specific genes have been identified for some subsets of families with heart disease, and lifestyle changes (exercise, diet, avoidance of tobacco) can modify heart disease risks appreciably.

## Stroke

Stroke, which refers to brain damage caused by a sudden and sustained loss of blood flow to the brain, can result from arterial obstruction (ischemic stroke, which accounts for 80% of stroke cases) or breakage (hemorrhagic stroke). This disease is the third leading cause of mortality in the United States, accounting for approximately 150,000 deaths per year. As with heart disease, strokes cluster in families: One's risk of having a stroke increases by two- to three-fold if a parent has had a stroke. The largest twin study conducted to date showed that concordance rates for stroke death in MZ and DZ twins were 10% and 5%, respectively. These figures imply that genes might influence one's susceptibility to this disease.

Stroke is a well-known consequence of several single-gene disorders, including sickle cell disease (see Chapter 3), MELAS (*m*itochondrial myopathy, *e*ncephalopathy, *l*actic *a*cidosis, and *s*troke, a mitochondrial disorder discussed in Chapter 5), and cerebral autosomal dominant arteriopathy with subcortical infarcts and leukoencephalopathy (CADASIL, a condition characterized by recurrent strokes and dementia and caused by mutations in the *NOTCH3* gene). Because blood clots are a common cause of stroke, it is expected that mutations in genes that encode coagulation factors might affect stroke

**TABLE 12-7**
Examples of Mendelian Subtypes of Complex Disorders*

| Mendelian Subtype | Protein (Gene) | Consequence of Mutation |
|---|---|---|
| **Heart Disease** | | |
| Familial hypercholesterolemia | LDL receptor (*LDLR*) | Elevated LDL level |
| Tangier disease | ATP-binding cassette 1 (*ABC1*) | Reduced HDL level |
| Familial defective apoB-100 | Apolipoprotein B (*APOB*) | Elevated LDL level |
| Familial dilated cardiomyopathy | Cardiac troponin T (*TNNT2*) | Reduced force generation by sarcomere |
| | Cardiac β-myosin heavy chain (*MYH7*) | Reduced force generation by sarcomere |
| | β-Sarcoglycan (*SGCB*) | Destabilized sarcolemma and signal transduction |
| | δ-Sarcoglycan (*SGCD*) | Destabilized sarcolemma and signal transduction |
| | Dystrophin | Destabilized sarcolemma in cardiac myocytes |
| Familial hypertrophic cardiomyopathy | Cardiac β-myosin heavy chain (*MYH7*) | Reduced force generation by sarcomere |
| | Cardiac troponin T (*TNNT2*) | Reduced force generation by sarcomere |
| | Myosin-binding protein C (*MYBPC*) | Sarcomere damage |
| Long QT syndrome | Cardiac potassium channel α subunit (*LQT1, KCNQ1*) | Prolonged QT interval on electrocardiogram, arrhythmia |
| | Cardiac potassium channel α subunit (*LQT2, KCNH2*) | Prolonged QT interval on electrocardiogram, arrhythmia |
| | Cardiac sodium channel (*LQT3, SCN5A*) | Prolonged QT interval on electrocardiogram, arrhythmia |
| | Ankyrin B anchoring protein (*LQT4, ANK2*) | Prolonged QT interval on electrocardiogram, arrhythmia |
| | Cardiac potassium channel β subunit (*LQT5, KCNE1*) | Prolonged QT interval on electrocardiogram, arrhythmia |
| | Cardiac potassium channel subunit (*LQT6, KCNE2*) | Prolonged QT interval on electrocardiogram, arrhythmia |
| **Hypertension** | | |
| Liddle syndrome | Renal epithelial sodium channel subunits (*SCNN1B, SCNN1G*) | Severe hypertension, low renin and suppressed aldosterone |
| Gordon syndrome | *WNK1* or *WNK4* kinase genes | High serum potassium level and increased renal salt reabsorption |
| Glucocorticoid-remediable aldosteronism | Fusion of genes that encode aldosterone synthase and steroid 11β-hydroxylase | Early-onset hypertension with suppressed plasma renin and normal or elevated aldosterone levels |
| Syndrome of apparent mineralocorticoid excess | 11β-Hydroxysteroid dehydrogenase (*11β-HSD2*) | Early-onset hypertension, low potassium and renin levels, low aldosterone |
| **Diabetes** | | |
| MODY1 | Hepatocyte nuclear factor-4α (*HNF4A*) | Decreased insulin secretion |
| MODY2 | Glucokinase (*GCK*) | Impaired glucose metabolism, leading to mild nonprogressive hyperglycemia |
| MODY3 | Hepatocyte nuclear factor-1α (*HNF1A*) | Decreased insulin secretion |
| MODY4 | Insulin promoter factor-1 (*IPF1*) | Decreased transcription of insulin gene |
| MODY5 | Hepatocyte nuclear factor-1β (*HNF1B*) | β-cell dysfunction leads to decreased insulin secretion |
| MODY6 | NeuroD transcription factor (*NEUROD1*) | Decreased insulin secretion |

**TABLE 12-7**
## Examples of Mendelian Subtypes of Complex Disorders*—cont'd

| Mendelian Subtype | Protein (Gene) | Consequence of Mutation |
|---|---|---|
| **Alzheimer Disease** | | |
| Familial Alzheimer disease | Amyloid-β precursor protein (*APP*) | Alteration of cleavage sites in amyloid-β precursor protein, producing longer amyloid fragments |
| | Presenilin 1 (*PS1*) | Altered cleavage of amyloid-β precursor protein, producing larger proportion of long amyloid fragments |
| | Presenilin 2 (*PS2*) | Altered cleavage of amyloid-β precursor protein, producing larger proportion of long amyloid fragments |
| **Parkinson Disease** | | |
| Familial Parkinson disease (autosomal dominant) | α-Synuclein (*PARK1, SNCA*) | Formation of α-synuclein aggregates |
| Familial Parkinson disease (autosomal recessive) | Parkin: E3 ubiquityl ligase, thought to ubiquinate α-synuclein (*PARK2*) | Compromised degradation of α-synuclein |
| Familial Parkinson disease (autosomal dominant) | Ubiquitin C-hydrolase-L1 (*PARK5*) | Accumulation of α-synuclein |
| **Amyotrophic Lateral Sclerosis** | | |
| (Lou Gehrig's Disease) | | |
| Familial amyotrophic lateral sclerosis | Superoxide dismutase 1 (*SOD1*) | Neurotoxic gain of function |
| Juvenile amyotrophic lateral sclerosis (autosomal recessive) | Alsin (*ALS2*) | Presumed loss of function |
| **Epilepsy** | | |
| Benign neonatal epilepsy, types 1 and 2 | Voltage-gated potassium channels (*KCNQ2* and *KCNQ3*, respectively) | Reduced M current increases neuronal excitability |
| Generalized epilepsy with febrile seizures plus type 1 | Sodium channel β1 subunit (*SCN1B*) | Sodium current persistence leading to neuronal hyperexcitability |
| Autosomal dominant nocturnal frontal lobe epilepsy | Neuronal nicotinic acetylcholine receptor subunits (*CHRNA4* and *CHRNB2*) | Increased neuronal excitability in response to cholinergic stimulation |
| Generalized epilepsy with febrile seizures plus type 3 | GABA_A receptor (*GABRG2*) | Loss of synaptic inhibition leading to neuronal excitability |

*See Table 8-2 for genes involved in other diseases, including hearing loss and blindness. This table is not meant to provide an exhaustive list of genes; additional genes are discussed in the review papers cited at the end of Chapter 12.
*HDL, high-density lipoprotein; LDL, low-density lipoprotein; MODY, maturity-onset diabetes of the young.*

susceptibility. For example, inherited deficiencies of protein C and protein S, both of which are coagulation inhibitors, are associated with an increased risk of stroke, especially in children. A specific mutation in clotting factor V, the factor V Leiden allele, causes resistance to activated protein C and thus produces an increased susceptibility to clotting. Heterozygosity for this allele, which is seen in approximately 5% of whites, produces a seven-fold increase in the risk of venous thrombosis (clots). In homozygotes, the risk increases to 100-fold. However, the evidence for an association between the factor V Leiden allele and stroke is inconsistent.

In addition to family history and specific genes, several factors are known to increase the risk of stroke. These include hypertension, obesity, atherosclerosis, diabetes, and smoking.

▶ **Stroke, which clusters in families, is associated with several single-gene disorders and with some inherited coagulation disorders.**

### Hypertension

Systemic hypertension, which has a worldwide prevalence of approximately 27%, is a key risk factor for heart disease, stroke, and kidney disease. Studies of blood pressure correlations within families yield heritability estimates of approximately 20% to 40% for both systolic and diastolic blood pressure. Heritability estimates based on twin studies tend to be higher (about 60%) and may be inflated because of greater similarities in the environments of MZ compared with DZ twins. The fact that the heritability estimates are substantially less than 100% indicates that environmental factors must also be significant causes of blood pressure variation. The most important environmental risk factors for hypertension are increased sodium intake, decreased physical activity, psychosocial stress, and obesity (as discussed later, the latter factor is itself influenced both by genes and environment).

Blood pressure regulation is a highly complex process that is influenced by many physiological systems, including various aspects of kidney function, cellular ion transport, vascular tone, and heart function. Because of this complexity, much research is now focused on specific components that might influence blood pressure variation, such as the renin–angiotensin system (Fig. 12-5), (involved in sodium reabsorption and vasoconstriction), vasodilators such as nitric oxide and the kallikrein–kinin system, and ion-transport systems such as adducin and sodium–lithium countertransport. These individual factors are more likely to be under the control of smaller numbers of genes than is blood pressure itself, simplifying the task of identifying these genes

and their role in regulating blood pressure. For example, linkage and association studies have implicated several genes involved in the renin–angiotensin system (e.g., the genes that encode angiotensinogen, angiotensin-converting enzyme type 1, and angiotensin II type 1 receptor) in causing hypertension.

A small percentage of hypertension cases are the result of rare single-gene disorders, such as Liddle syndrome (low plasma aldosterone and hypertension caused by mutations that alter the ENaC epithelial sodium channel) and Gordon syndrome (hypertension, high serum potassium level, and increased renal salt reabsorption caused by mutations in the *WNK1* or *WNK4* kinase genes; see Table 12-7 for additional examples. More than 20 genes have been identified that can lead to rare inherited forms of hypertension, and all of them affect the reabsorption of water and salt by the kidney, which in turn affects blood volume and blood pressure. It is hoped that isolation and study of these genes will lead to the identification of genetic factors underlying essential hypertension.*

Large-scale genome scans, undertaken in humans and in experimental animals such as mice and rats, have sought to identify quantitative trait loci (see Box 12-1) that might underlie essential hypertension. These studies have identified

---

*The term "essential" refers to the 95% of hypertension cases that are not caused by a known mutation or syndrome.

**FIGURE 12-5**
The renin–angiotensin–aldosterone system.
↑, increased; ↓, decreased; AT₁, angiotensin type II receptor 1.
*(Modified from King RA, Rotter JI, Motulsky AG [eds]: The Genetic Basis of Common Diseases. New York: Oxford University Press, 1992.)*

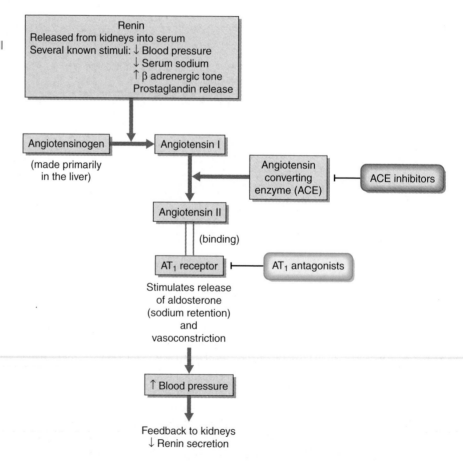

a number of regions in which LOD scores (see Chapter 8) offer statistical support for the presence of genes that influence susceptibility to hypertension, and in some cases several studies have implicated the same genomic region. Such results might help to pinpoint specific genes that underlie susceptibility to essential hypertension.

> Heritability estimates for systolic and diastolic blood pressure range from 20% to 40%. A number of genes responsible for rare hypertension syndromes have been identified, and genome scans have implicated regions that might contain genes that underlie susceptibility to essential hypertension. Other risk factors for hypertension include increased sodium intake, lack of exercise, psychosocial stress, and obesity.

## Cancer

Cancer is the second leading cause of death in the United States, although it is estimated that it might soon surpass heart disease as the leading cause of death. It is well established that many major types of cancer (e.g., breast, colon, prostate, ovarian) cluster strongly in families. This is due both to shared genes and shared environmental factors. Although numerous cancer genes have been isolated, environmental factors also play an important role in causing cancer by inducing somatic mutations. In particular, tobacco use is estimated to account for one third of all cancer cases in developed countries, making it the most important known cause of cancer. Diet (i.e., carcinogenic substances and the lack of "anticancer" components such as fiber, fruits, and vegetables) is another leading cause of cancer and may also account for as much as one third of cancer cases. It is estimated that approximately 15% of worldwide cancer cases are caused primarily by infectious agents (e.g., human papilloma virus for cervical cancer, hepatitis B and C for liver cancer). Because cancer genetics was the subject of Chapter 11, we confine our attention here to genetic and environmental factors that influence susceptibility to some of the most common cancers.

### Breast Cancer

Breast cancer is the second most commonly diagnosed cancer (after skin cancer) among women, affecting approximately 12% of American women who live to age 85 years or older. It was diagnosed in approximately 180,000 American women in 2008, and about 40,000 women die from this disease each year. Breast cancer was formerly the leading cause of cancer death among women, but it has been surpassed by lung cancer. Breast cancer can also occur in men, with a lifetime prevalence that is roughly 100 times lower than that of women. The familial aggregation of breast cancer has been recognized for centuries, having been described by physicians in ancient Rome. If a woman has one affected first-degree relative, her risk of developing breast cancer doubles. The risk increases further with additional affected

relatives, and it increases if those relatives developed cancer at a relatively early age (before 45 years of age).

Several genes are now known to predispose women to developing hereditary breast cancer. Most important among these are *BRCA1* and *BRCA2*, two genes involved in DNA repair (see Chapter 11). Germline mutations in the *TP53* and *CHK2* genes can cause Li–Fraumeni syndrome, which also predisposes to breast cancer. Cowden disease, a rare autosomal dominant condition that includes multiple hamartomas and breast cancer, is caused by mutations in the *PTEN* tumor suppressor gene (see Chapter 11). Ataxia-telangiectasia, an autosomal recessive disorder caused by defective DNA repair, includes breast cancer in its presentation. Mutations in the *MSH2* and *MLH1* DNA repair genes, which lead to hereditary nonpolyposis colorectal cancer (HNPCC), also confer an increased risk of breast cancer. Despite the significance of these genes, it should be emphasized that more than 90% of breast cancer cases are not inherited as mendelian diseases.

A number of environmental factors are known to increase the risk of developing breast cancer. These include nulliparity (never bearing children), bearing the first child after 30 years of age, a high-fat diet, alcohol use, and estrogen replacement therapy.

### Colorectal Cancer

It is estimated that 1 in 20 Americans will develop colorectal cancer, and roughly one third of those with this cancer will die from it. With approximately 150,000 new cases and 50,000 deaths in the United States in 2008, colorectal cancer is second only to lung cancer in the total number of annual cancer deaths. Like breast cancer, it clusters in families; familial clustering of this form of cancer was reported in the medical literature as early as 1881. The risk of colorectal cancer in people with one affected first-degree relative is two to three times higher than that of the general population.

As discussed in Chapter 11, familial colon cancer can be the result of mutations in the APC tumor suppressor gene or in one of several DNA mismatch-repair genes (HNPCC). Another, less common, inherited cause of colon cancer is the autosomal dominant Peutz–Jeghers syndrome. About half of Peutz–Jeghers cases are caused by mutations in the *STK11* tumor suppressor gene, which encodes a protein kinase. Juvenile intestinal polyposis, an autosomal dominant disease defined by the presence of 10 or more polyps before adulthood, can be caused by mutations in *SMAD4* (see Chapter 11), in *BMPRA1* (a receptor serine–threonine kinase gene), or, in rare cases, in *PTEN*. *PTEN* mutations can also cause Cowden disease, which, in addition to breast tumors, often includes polyps in the intestinal tract.

As with breast cancer, most colon cancer cases (>90%) are not inherited as mendelian conditions and are likely to be caused by a complex interaction of inherited and somatic genetic alterations and environmental factors. The latter risk factors include a lack of physical activity and a high-fat, low-fiber diet.

## Prostate Cancer

Prostate cancer is the second most commonly diagnosed cancer in men (after skin cancer), with approximately 185,000 new cases annually in the United States. Prostate cancer is second only to lung cancer as a cause of cancer death in men, causing more than 29,000 deaths in 2008. Having an affected first-degree relative increases the risk of developing prostate cancer by a factor of two to three. It is estimated that about 5% to 10% of prostate cancer cases are the result of inherited mutations.

The relatively late age of onset of most prostate cancer cases (median age, 72 years) makes genetic analysis especially difficult. However, loss of heterozygosity (see Chapter 11) has been observed in a number of genomic regions in prostate tumor cells, possibly indicating the presence of genetic alterations in these regions. In addition, genome scans have indicated that several chromosome regions might contain prostate cancer susceptibility genes. One of these regions, 8q24, has been associated with a significantly increased risk of prostate cancer in several populations. The *RNASEL* gene has also been associated with prostate cancer risk in several studies. The product of this gene, ribonuclease L, regulates cell proliferation and apoptosis. Mutations in *RNASEL* account for a small percentage of familial prostate cancer cases.

Nongenetic risk factors for prostate cancer may include a high-fat diet. Because prostate cancer usually progresses slowly and because it can be detected by digital examination and by the prostate-specific antigen (PSA) test, fatal metastasis can usually be prevented.

> Most common cancers have genetic components. Recurrence risks tend to be higher if there are several affected relatives and if those relatives developed cancer at an early age. Specific genes have been discovered that cause inherited colon, breast, and prostate cancer in some families.

## Diabetes Mellitus

Like the other disorders discussed in this chapter, the etiology of diabetes mellitus is complex and not fully understood. Nevertheless, progress is being made in understanding the genetic basis of this disorder, which is the leading cause of adult blindness, kidney failure, and lower-limb amputation and a major cause of heart disease and stroke. An important advance has been the recognition that diabetes mellitus is actually a heterogeneous group of disorders, all characterized by elevated blood sugar. We focus here on the three major types of diabetes, type 1 (formerly termed insulin-dependent diabetes mellitus, or IDDM), type 2 (formerly termed non–insulin-dependent diabetes mellitus, or NIDDM), and maturity-onset diabetes of the young (MODY).

### Type 1 Diabetes

Type 1 diabetes, which is characterized by T-cell infiltration of the pancreas and destruction of the insulin-producing beta cells, usually (though not always) manifests before 40 years of age. Patients with type 1 diabetes must receive exogenous insulin to survive. In addition to T-cell infiltration of the pancreas, autoantibodies are formed against pancreatic cells, insulin, and enzymes such as glutamic acid decarboxylase; these autoantibodies can be observed long before clinical symptoms occur. These findings, along with a strong association between type 1 diabetes and the presence of several human leukocyte antigen (HLA) class II alleles, indicate that this is an autoimmune disease.

Siblings of persons with type 1 diabetes face a substantial elevation in risk: approximately 6%, as opposed to a risk of about 0.3% to 0.5% in the general population. The recurrence risk is also elevated when there is a diabetic parent, although this risk varies with the sex of the affected parent. The risk for offspring of diabetic mothers is only 1% to 3%, but it is 4% to 6% for the offspring of diabetic fathers. (Because type 1 diabetes affects males and females in roughly equal proportions in the general population, this risk difference is inconsistent with the sex-specific threshold model for multifactorial traits.) Twin studies show that the empirical risk for MZ twins of type 1 diabetes patients ranges from 30% to 50%. In contrast, the concordance rate for DZ twins is 5% to 10%. The fact that type 1 diabetes is not 100% concordant among identical twins indicates that genetic factors are not solely responsible for the disorder. There is evidence that specific viral infections contribute to the cause of type 1 diabetes in at least some persons, possibly by activating an autoimmune response.

The association of specific HLA class II alleles and type 1 diabetes has been studied extensively, and it is estimated that the HLA loci account for about 40% to 50% of the genetic susceptibility to type 1 diabetes. Approximately 95% of whites with type 1 diabetes have the *HLA DR3* and/or *DR4* alleles, whereas only about 50% of the general white population has either of these alleles. If an affected proband and a sibling are both heterozygous for the *DR3* and *DR4* alleles, the sibling's risk of developing type 1 diabetes is nearly 20% (i.e., about 40 times higher than the risk in the general population). This association may in part reflect linkage disequilibrium between alleles of the *DR* locus and those of the *HLA-DQ* locus. The absence of aspartic acid at position 57 of the *DQ* polypeptide is strongly associated with susceptibility to type 1 diabetes; in fact, those who do not have this amino acid at position 57 (and instead are homozygous for a different amino acid) are 100 times more likely to develop type 1 diabetes. The aspartic acid substitution alters the shape of the HLA class II molecule and thus its ability to bind and present peptides to T cells (see Chapter 9). Altered T-cell recognition might help to protect persons with the aspartic acid substitution from an autoimmune episode.

The insulin gene, which is located on the short arm of chromosome 11, is another logical candidate for type 1 diabetes susceptibility. Polymorphisms within and near this gene have been tested for association with type 1 diabetes. Intriguingly, a strong risk association is seen with allelic variation in a VNTR polymorphism (see Chapter 3) located just 5' of the insulin gene. Differences in the number of VNTR repeat units might affect transcription of the insulin gene (possibly by altering chromatin structure), resulting in variation in susceptibility. It is estimated that inherited genetic variation in the insulin region accounts for approximately 10% of the familial clustering of type 1 diabetes.

Affected sib-pair and genome-wide association analyses have been used extensively to map additional genes that can cause type 1 diabetes. In addition, an animal model, the nonobese diabetic (NOD) mouse, has been used to identify diabetes susceptibility genes that could have similar roles in humans (see Box 12-1). These studies have identified at least 20 additional candidate regions that might contain type 1 diabetes susceptibility genes. One of these regions, 2q33, contains the *CTLA4* (cytotoxic lymphocyte associated-4) gene, which encodes an inhibitory T-cell receptor. Several studies have demonstrated that alleles of *CTLA4* are associated with an increased risk of type 1 diabetes. There is growing evidence that variation in *CTLA4* is also associated with other autoimmune diseases, such as rheumatoid arthritis and celiac disease. Another gene associated with type 1 diabetes susceptibility, *PTPN22*, is involved in T-cell regulation and is also associated with other autoimmune disorders, including rheumatoid arthritis and systemic lupus erythematosus.

### Type 2 Diabetes

Type 2 diabetes accounts for more than 90% of all diabetes cases, and its incidence is rising rapidly in populations with access to high-calorie diets. It currently affects approximately 10% to 20% of the adult populations of many developed countries. One study estimates that because of the rapid rate of increase of this disease, one third of Americans born in 2000 will eventually develop type 2 diabetes.

A number of features distinguish type 2 diabetes from type 1 diabetes. Persons with type 2 diabetes usually have some degree of endogenous insulin production, at least in earlier stages of the disease, and they can sometimes be treated successfully with dietary modification, oral drugs, or both. In contrast to those with type 1 diabetes, patients with type 2 diabetes have insulin resistance (i.e., their cells have difficulty using insulin) and are more likely to be obese. This form of diabetes has traditionally been seen primarily in patients older than 40 years, but because of increasing obesity among adolescents and young adults, it is now increasing rapidly in this segment of the population. Neither HLA associations nor autoantibodies are seen commonly in this form of diabetes. MZ twin concordance rates are substantially higher than in type 1 diabetes, often exceeding 90% (because of age dependence, the concordance rate increases if older subjects are studied). The empirical recurrence risks for first-degree relatives of patients with type 2 diabetes are higher than those for type 1 patients, generally ranging from 15% to 40%. The differences between type 1 and type 2 diabetes are summarized in Table 12-8.

The two most important risk factors for type 2 diabetes are a positive family history and obesity; the latter increases insulin resistance. The disease tends to rise in prevalence

**TABLE 12-8**

## Comparison of the Major Features of Type 1 and Type 2 Diabetes Mellitus

| Feature | Type 1 Diabetes | Type 2 Diabetes |
|---|---|---|
| Age of onset | Usually <40 yr | Usually >40 yr |
| Insulin production | None | Partial |
| Insulin resistance | No | Yes |
| Autoimmunity | Yes | No |
| Obesity | Not common | Common |
| MZ twin concordance | 0.35-0.50 | 0.90 |
| Sibling recurrence risk | 1%-6% | 15%-40% |

*MZ, monozygotic.*

when populations adopt a diet and exercise pattern typical of United States and European populations. Increases have been seen, for example, among Japanese immigrants to the United States and among some native populations of the South Pacific, Australia, and the Americas. Several studies, conducted on both male and female subjects, have shown that regular exercise can substantially lower one's risk of developing type 2 diabetes, even among persons with a family history of the disease. This is partly because exercise reduces obesity. However, even in the absence of weight loss, exercise increases insulin sensitivity and improves glucose tolerance.

Extensive linkage and genome-wide association analyses have been undertaken to identify genes that might contribute to type 2 diabetes susceptibility. The most significant gene identified thus far is *TCF7L2*, which encodes a transcription factor involved in secreting insulin. A variant of *TCF7L2* is associated with a 50% increased risk of developing type 2 diabetes. A significant association has also been observed between type 2 diabetes and a common allele of the gene that encodes peroxisome proliferator-activated receptor-γ (PPAR-γ), a transcription factor that is involved in adipocyte differentiation and glucose metabolism. Although this allele confers only a 25% increase in the risk of developing type 2 diabetes, it is found in more than 75% of persons of European descent and thus helps to account for a significant fraction of type 2 diabetes cases. Variation in *KCNJ11*, which encodes a potassium channel necessary for glucose-stimulated insulin secretion, confers an additional 20% increase in type 2 diabetes susceptibility. The associations between diabetes susceptibility and each of these genes have been widely replicated in numerous populations.

### Maturity-Onset Diabetes of the Young

MODY, which accounts for 1% to 5% of all diabetes cases, typically occurs before 25 years of age and follows an autosomal dominant mode of inheritance. In contrast to type 2 diabetes, it is not associated with obesity. Studies of MODY pedigrees have shown that about 50% of cases are caused by mutations in the gene that encodes glucokinase, a rate-limiting enzyme in the conversion of glucose to glucose-6-phosphate in the pancreas. Another 40% of MODY cases are caused by mutations in any of five genes that encode transcription factors involved in pancreatic development or insulin regulation: hepatocyte nuclear factor 1-α (*HNF1*α), hepatocyte nuclear factor 1-β (*HNF1*β), hepatocyte nuclear factor 4-α (*HNF4*α), insulin promoter factor 1 (*IPF1*), and neurogenic differentiation 1 (*NEUROD1*). Mutations in these genes, all of which are expressed in pancreatic beta cells, lead to beta cell abnormalities and thus to diabetes.

> Type 1 (insulin-dependent) and type 2 (non–insulin-dependent) diabetes both cluster in families, with stronger familial clustering observed for type 2 diabetes. Type 1 has an earlier average age of onset, is HLA-associated, and is an autoimmune disease. Type 2 is not an autoimmune disorder and is more likely to be seen in obese persons. Several genes have been identified that increase susceptibility to type 1 or type 2 diabetes. Most cases of autosomal dominant MODY are caused by mutations in any of six specific genes.

## Obesity

The worldwide prevalence of obesity is increasing rapidly among adults and children. Approximately 70% of American adults and 60% of British adults are overweight (body mass index [BMI] >25),* and about half of these overweight persons are obese (BMI >30). Although obesity itself is not a disease, it is an important risk factor for several common diseases, including heart disease, stroke, hypertension, and type 2 diabetes.

As one might expect, there is a strong correlation between obesity in parents and obesity in their children. This could easily be ascribed to common environmental effects: parents and children usually share similar diet and exercise habits. However, there is good evidence for genetic components as well. Four adoption studies each showed that the body weights of adopted persons correlated significantly with their natural parents' body weights but not with those of their adoptive parents. Twin studies also provide evidence for a genetic effect on body weight, with most studies yielding heritability estimates between 0.60 and 0.80. The heritability of "fatness" (measured, for example, by skinfold thickness) is approximately 0.40 to 0.50.

Research, aided substantially by mouse models, has shown that several genes each play a role in human obesity. Important among these are the genes that encode leptin (Greek, "thin") and its receptor. The leptin hormone is secreted by adipocytes (fat storage cells) and binds to receptors in the hypothalamus, the site of the body's appetite control center. Increased fat stores lead to an elevated leptin level, which produces satiety and a loss of appetite. Lower leptin levels lead to increased appetite. Mice with loss-of-function mutations in the leptin gene have uncontrolled appetites and become obese. When injected with leptin, these mice lose weight. Mice with mutations in the leptin receptor gene cannot respond to increased leptin levels and also develop obesity.

Identification of the leptin gene and its receptor in mice led to their identification in humans, which in turn prompted optimistic predictions that leptin could be a key to weight loss in humans (without the perceived unpleasantness of dieting and exercise). However, most obese humans have *high* levels of leptin, indicating that the leptin gene is functioning normally. Leptin receptor defects were then suspected, but these are also uncommon in humans. Although mutations in the human leptin gene and its receptor have now been identified in a few humans with severe obesity (BMI >40), they both appear to be extremely rare. Unfortunately, these genes will not solve the problem of human obesity. However, clinical trials using recombinant leptin have demonstrated moderate weight loss in a subset of obese individuals.

In addition, leptin participates in important interactions with other components of appetite control, such as neuropeptide Y, as well as α-melanocyte-stimulating hormone and its receptor, the melanocortin-4 receptor (MC4R). Mutations in the gene that encodes MC4R have been found in 3% to 5% of severely obese individuals. Several genome-wide association studies have demonstrated an association between a variant in the brain-expressed *FTO* gene and obesity in whites. Homozygosity for this variant, which is seen in about 16% of whites, confers increased risks of overweight and obesity of 40% and 70%, respectively. Identification of these and other obesity-presdisposing genes is leading to a better understanding of appetite control in humans and could eventually lead to effective treatments for some cases of obesity.

> Adoption and twin studies indicate that at least half of the population variation in obesity may be caused by genes. Specific genes and gene products involved in appetite control and susceptibility to obesity, including leptin and its receptor, *MC4R*, and *FTO*, are now being studied.

## Alzheimer Disease

Alzheimer disease (AD), which is responsible for 60% to 70% of cases of progressive cognitive impairment among the elderly, affects approximately 10% of the population older than 65 years and 40% of the population older than 85 years. Because of the aging of the population, the number of Americans with AD, which was about 5 million in 2007, continues to increase. Alzheimer disease is characterized by

---

*BMI is defined as $W/H^2$, where W is weight in kilograms and H is height in meters.

progressive dementia and memory loss and by the formation of amyloid plaques and neurofibrillary tangles in the brain, particularly in the cerebral cortex and hippocampus. The plaques and tangles lead to progressive neuronal loss, and death usually occurs within 7 to 10 years after the first appearance of symptoms.

The risk of developing AD doubles in persons who have an affected first-degree relative. Although most cases do not appear to be caused by single genes, approximately 10% follow an autosomal dominant mode of transmission. About 3% to 5% of AD cases occur before age 65 years and are considered early onset; these are much more likely to be inherited in autosomal dominant fashion.

Alzheimer disease is a genetically heterogeneous disorder. Approximately half of early-onset cases can be attributed to mutations in any of three genes, all of which affect amyloid-β deposition. Two of the genes, presenilin 1 (*PS1*) and presenilin 2 (*PS2*), are very similar to each other, and their protein products are involved in cleavage of the amyloid-β precursor protein (APP) by γ-secretase (posttranslational modification; see Chapter 2). Gain-of-function mutations in *PS1* or *PS2* affect the cleavage of APP such that amyloid-producing forms of it accumulate excessively and are deposited in the brain (Fig. 12-6). This is thought to be a primary cause of AD. Mutations in *PS1* typically result in especially early onset of AD, with the first occurrence of symptoms in the fifth decade of life.

A small number of cases of early-onset AD are caused by mutations of the gene (*APP*) that encodes APP itself, which is located on chromosome 21. These mutations disrupt normal secretase cleavage sites in APP (see Fig. 12-6), again leading to the accumulation of the longer protein product. It is interesting that this gene is present in three copies in persons with trisomy 21, where the extra gene copy leads to amyloid deposition and the occurrence of AD in Down syndrome patients (see Chapter 6).

An important risk factor for the more common late-onset form of AD is allelic variation in the apolipoprotein E (*APOE*) locus, which has three major alleles: *ε2*, *ε3*, and *ε4*. Studies conducted in diverse populations have shown that

persons who have one copy of the *ε4* allele are 2 to 5 times more likely to develop AD, and those with two copies of this allele are 5 to 10 times more likely to develop AD. The risk varies somewhat by population, with higher *ε4*-associated risks in whites and Japanese and relatively lower risks in Latin Americans and African Americans. Despite the strong association between *ε4* and AD, approximately half of persons who develop late-onset AD do not have a copy of the *ε4* allele, and many who are homozygous for *ε4* remain free of AD even at advanced age. The apolipoprotein E protein product is not involved in cleavage of APP but instead appears to be associated with clearance of amyloid from the brain.

Genome scans indicate that there are additional AD genes, with especially strong evidence for susceptibility loci in regions of chromosomes 10 and 12. A gene located within the chromosome 12 region encodes α₂-macroglobulin, a protease inhibitor that interacts with apolipoprotein E. Another gene in this region encodes the low-density lipoprotein receptor–related protein (LRP), which also interacts with apolipoprotein E. Some studies support an association between alleles of these genes and late-onset AD, and others fail to replicate the association. It remains to be seen whether these genes play significant roles in causing AD.

AD has several features that have made it refractory to genetic analysis. Its genetic heterogeneity has already been described. In addition, because a definitive diagnosis can be obtained only by a brain autopsy, it is often difficult to diagnose AD in living family members (although clinical features and brain imaging studies can provide strong evidence that a person is affected with AD). Finally, because onset of the disease can occur very late in life, persons carrying an AD-predisposing mutation could die from another cause before developing the disease. They would then be misidentified as noncarriers. These types of difficulties arise not only with AD but with many other common adult diseases as well. Despite these obstacles, several AD genes have now been identified, leading to a better understanding of the disorder and to the possibility of more effective AD treatment.

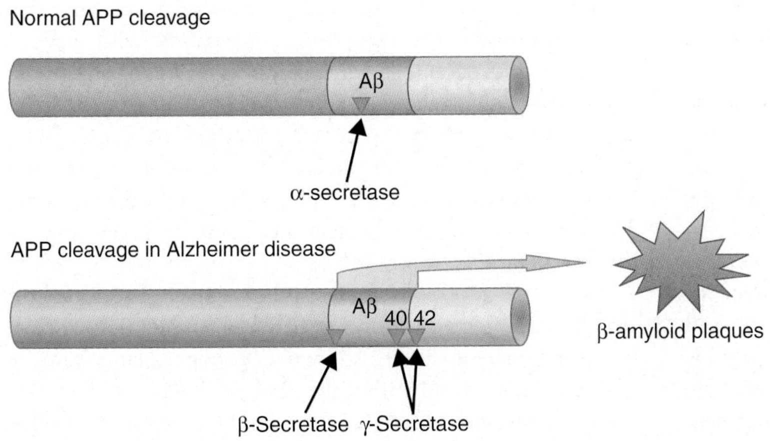

Normal APP cleavage

APP cleavage in Alzheimer disease

β-amyloid plaques

**FIGURE 12-6**
**A,** Cleavage of the amyloid-β precursor protein (APP) by α-secretase disrupts the amyloid-β protein and prevents the formation of amyloid plaques. **B,** An alternate cleavage pathway involves cleavage of APP by β-secretase at the N terminus and γ-secretase at the C terminus, producing a protein product of 40 to 42 amino acids. Gain-of-function mutations in the presenilin genes result in increased cleavage activity via this pathway. This yields an excessive amount of the 42–amino acid form of APP, which leads to formation of amyloid plaques. Mutations in the APP gene can also alter the α-secretase cleavage sites such that excess amounts of the longer form of APP are produced.

Approximately 10% of AD cases are caused by autosomal dominant genes. Early-onset cases cluster more strongly in families and are more likely to follow an autosomal dominant inheritance pattern. This disease is genetically heterogeneous: At least four AD susceptibility genes have been identified. Three of the genes (encoding presenilin 1, presenilin 2, and amyloid-β precursor protein) cause early-onset AD and affect the cleavage and processing of the amyloid precursor protein. A fourth encodes the apolipoprotein E protein and is strongly associated with late-age onset of AD.

## Alcoholism

At some point in their lives, alcoholism is diagnosed in approximately 10% of men and 3% to 5% of women in the United States (see Table 12-7). More than 100 studies have shown that this disease clusters in families: the risk of developing alcoholism among persons with one affected parent is three to five times higher than for those with unaffected parents. Most twin studies have yielded concordance rates for DZ twins of less than 30% and for MZ twins in excess of 60%. Adoption studies have shown that the offspring of an alcoholic parent, even when raised by nonalcoholic parents, have a four-fold increased risk of developing the disorder. To control for possible prenatal effects from an alcoholic mother, some studies have included the offspring of alcoholic fathers only. The results have remained the same. These data argue that there may be genes that predispose some people to alcoholism.

Some researchers distinguish two major subtypes of alcoholism. Type I is characterized by a later age of onset (after 25 years of age), occurrence in both males and females, and greater psychological dependency on alcohol. Type I alcoholics are more likely to be introverted, solitary drinkers. This form of alcoholism is less likely to cluster in families (one study yielded a heritability estimate of 0.21), has a less-severe course, and is more easily treated. Type II alcoholism is seen predominantly in males, typically occurs before 25 years of age, and tends to involve persons who are extroverted and thrill-seeking. This form is more difficult to treat successfully and tends to cluster more strongly in families, with heritability estimates ranging from 0.55 to more than 0.80.

It has long been known that an individual's physiological response to alcohol can be influenced by variation in the key enzymes responsible for alcohol metabolism: alcohol dehydrogenase (ADH), which converts ethanol to acetaldehyde, and aldehyde dehydrogenase (ALDH), which converts acetaldehyde to acetate. In particular, an allele of the *ALDH2* gene (*ALDH2*2*) results in excessive accumulation of acetaldehyde and thus in facial flushing, nausea, palpitations, and lightheadedness. Because of these unpleasant effects, persons who have the *ALDH2*2* allele are much less likely to become alcoholics. This protective allele is common in some Asian populations but is rare in other populations.

A number of genome scans have been undertaken in large cohorts of alcoholics and controls. One of the most consistent findings is that variants in genes that encode components of gamma-aminobutyric acid (GABA) receptors are associated with addiction to alcohol. This finding is biologically plausible, because the GABA neurotransmitter system inhibits excitatory signals in neurons, exerting a calming effect. Alcohol has been shown to increase GABA release, and allelic variation in GABA receptor genes might modulate this effect.

It should be underscored that we refer to genes that might increase one's *susceptibility* to alcoholism. This is obviously a disease that requires an environmental component, regardless of one's genetic constitution.

Twin and adoption studies show that alcoholism clusters quite strongly in families, reflecting a possible genetic contribution to this disease. Familial clustering is particularly strong for type II alcoholism (early-onset form primarily affecting males).

## Psychiatric Disorders

Two of the major psychiatric diseases, schizophrenia and bipolar disorder, have been the subjects of numerous genetic studies. Twin, adoption, and family studies have shown that both disorders aggregate in families.

### Schizophrenia

Schizophrenia is a severe emotional disorder characterized by delusions, hallucinations, retreat from reality, and bizarre, withdrawn, or inappropriate behavior (contrary to popular belief, schizophrenia is not a "split personality" disorder). The lifetime recurrence risk for schizophrenia among the offspring of one affected parent is approximately 8% to 10%, which is about 10 times higher than the risk in the general population. The empirical risks increase when more relatives are affected. For example, a person with an affected sibling and an affected parent has a risk of about 17%, and a person with two affected parents has a risk of 40% to 50%. The risks decrease when the affected family member is a second- or third-degree relative. Details are given in Table 12-9. On inspection of this table, it may seem puzzling that the proportion of schizophrenic probands who have a schizophrenic parent is only about 5%, which is substantially lower than the risk for other first-degree relatives (e.g., siblings, affected parents, and their offspring). This can be explained by the fact that schizophrenics are less likely to marry and produce children than are other persons. There is thus substantial selection against schizophrenia in the population.

Twin and adoption studies indicate that genetic factors are likely to be involved in schizophrenia. Data pooled from five different twin studies show a 47% concordance rate for MZ twins, compared with only 12% for DZ twins. The concordance rate for MZ twins reared apart, 46%, is about the same as the rate for MZ twins reared together. The risk of

**TABLE 12-9**

## Recurrence Risks for Relatives of Schizophrenic Probands, Based on Multiple Studies of Western European Populations

| Relationship to Proband | Recurrence Risk (%) |
| --- | --- |
| Monozygotic twin | 44.3 |
| Dizygotic twin | 12.1 |
| Offspring | 9.4 |
| Sibling | 7.3 |
| Niece or nephew | 2.7 |
| Grandchild | 2.8 |
| First cousin | 1.6 |
| Spouse | 1.0 |

*Adapted from McGue M, Gottesman II, Rao DC: The analysis of schizophrenia family data. Behav Genet 1986;16:75-87.*

developing the disease for offspring of a schizophrenic parent who are adopted by normal parents is about 10%, approximately the same as the risk when the offspring are raised by a schizophrenic biological parent.

Dozens of genome scans have been performed in an effort to locate schizophrenia genes. Linkage to several chromosome regions has been replicated in several populations, and specific genes in these regions are being analyzed. Some of the techniques discussed in Chapter 8 (linkage disequilibrium, candidate gene analysis) have identified promising associations between schizophrenia and several brain-expressed genes whose products interact with glutamate receptors. These include dysbindin (*DTNBP1*; chromosome 6p), neuregulin 1 (*NRG1*; chromosome 8p), and D-amino-acid oxidase activator (*G30*; chromosome 13q). Another susceptibility gene is *DISC1* (disrupted-in-schizophrenia-1), which was originally identified by its consistent translocation in affected members of a large schizophrenia pedigree. Each of these associations has been replicated in numerous populations. However, the mechanisms through which mutations in these genes contribute to schizophrenia susceptibility are not yet known.

### Bipolar Disorder

**Bipolar disorder**, also known as manic–depressive disorder, is a form of psychosis in which extreme mood swings and emotional instability are seen. The prevalence of the disorder in the general population is approximately 0.5% to 1%, but it rises to 5% to 10% among those with an affected first-degree relative. A study based on the Danish twin registry yielded concordance rates of 79% and 24% for MZ and DZ twins, respectively. The corresponding concordance rates for unipolar disorder (major depression) were 54% and 19%. Thus, it appears that bipolar disorder is more strongly influenced by genetic factors than is unipolar disorder.

As with schizophrenia, many linkage and genome-wide association studies have been carried out to identify genes contributing to bipolar disorder. These studies have implicated several chromosome regions in multiple population samples. In addition, some evidence has been found for modest associations between bipolar disorder and alleles in candidate loci. Some of these loci were identified because their products are involved in neurotransmitter systems that are targets of drugs used to treat the disease (e.g., the serotonin, dopamine, and noradrenaline systems). Examples of these genes include those that encode monoamine oxidase A (*MAOA*), the serotonin transporter (*5HTT*), and catechol-O-methyltransferase (*COMT*), a gene that has also been associated with schizophrenia susceptibility. In addition, the *DAOA*, *NRG1*, and *DISC1* genes, which were discussed previously because of their association with schizophrenia, have been shown in some studies to be associated with susceptibility to bipolar disorder. Although these associations are promising, they have often been difficult to replicate reliably in different populations, and the precise roles of mutations in causing disease susceptibility remain to be discovered.

These results reveal some of the difficulties encountered in genetic studies of complex disorders in general and psychiatric disorders in particular. These diseases are undoubtedly heterogeneous, reflecting the influence of numerous genetic and environmental factors. Also, definition of the phenotype is not always straightforward, and it can change through time. Several measures are being taken to improve the likelihood of finding genes underlying these conditions. Phenotypes are being defined in standardized and rigorous fashion. Larger sample sizes of affected persons, with more rigorous phenotype definition, are being collected in efforts to increase the power to detect linkage and association. Heterogeneity can be decreased by studying clinically defined subtypes of these diseases and by carrying out studies in genetically homogeneous populations.

> Marked familial aggregation has been observed for schizophrenia and for bipolar disorder. Genes that encode neurotransmitters, receptors, and neurotransmitter-related enzymes have been studied in families, and many genome scans have been carried out.

### Other Complex Disorders

The disorders discussed in this chapter represent some of the most common multifactorial disorders and those for which significant progress has been made in identifying genes. Many other multifactorial disorders are being studied as well, and in some cases specific susceptibility genes have been identified. These include, for example, Parkinson disease, hearing loss, multiple sclerosis, amyotrophic lateral sclerosis, epilepsy, asthma, inflammatory bowel disease, and some forms of blindness (see Table 12-7 and Table 8-2 in Chapter 8).

## SOME GENERAL PRINCIPLES AND CONCLUSIONS

Some general principles can be deduced from the results obtained thus far regarding the genetics of complex disorders. First, the more strongly inherited forms of complex disorders generally have an earlier age of onset (examples include breast cancer, Alzheimer disease, and heart disease). Often, these represent subsets of cases in which there is single-gene inheritance. Second, when there is laterality, bilateral forms sometimes cluster more strongly in families (e.g., cleft lip/palate). Third, while the sex-specific threshold model fits some of the complex disorders (e.g., pyloric stenosis, cleft lip/palate, autism, heart disease), it fails to fit others (e.g., type 1 diabetes).

There is a tendency, particularly among the lay public, to assume that the presence of a genetic component means that the course of a disease cannot be altered ("If it's genetic, you can't change it"). This is incorrect. Most of the diseases discussed in this chapter have both genetic and environmental components. Thus, environmental modification (e.g., diet, exercise, stress reduction) can often reduce risk significantly. Such modifications may be especially important for persons who have a family history of a disease, because they are likely to develop the disease earlier in life. Those with a family history of heart disease, for example, can often add many years of productive living with relatively minor lifestyle alterations. By targeting those who can benefit most from intervention, genetics helps to serve the goal of preventive medicine.

In addition, it should be stressed that the identification of a specific genetic alteration can lead to more effective prevention and treatment of the disease. Identification of mutations causing familial colon cancer can allow early screening and prevention of metastasis. Pinpointing a gene responsible for a neurotransmitter defect in a psychiatric disorder such as schizophrenia could lead to the development of more effective drug treatments. In some cases, such as familial hypercholesterolemia, gene therapy may be useful. It is important for health care practitioners to make their patients aware of these facts.

Although the genetics of common disorders is complex and often confusing, the public health impact of these diseases and the evidence for hereditary factors in their etiology demand that genetic studies be pursued. Substantial progress is already being made. The next decade will undoubtedly witness many advancements in our understanding and treatment of these disorders.

## Study Questions

1. Consider a multifactorial trait that is twice as common in females as in males. Indicate which type of mating is at higher risk for producing affected children (affected father and normal mother versus normal father and affected mother). Is the recurrence risk higher for their sons or their daughters?

2. Consider a disease that is known to have a 5% sibling recurrence risk. This recurrence risk could be the result of either multifactorial inheritance or a single autosomal dominant gene with 10% penetrance. How would you test which of these possibilities is correct?

3. One member of a pair of monozygotic twins is affected by an autosomal dominant disease, and the other is not. List two different ways this could happen.

4. Suppose that the heritability of body fat percentage is 0.80 when correlations between siblings are studied but only 0.50 when correlations between parents and offspring are studied. Suppose also that a significant positive correlation is observed in the body fat percentages of spouses. How would you interpret these results?

## Suggested Readings

Abrahams BS, Geschwind DH. Advances in autism genetics: On the threshold of a new neurobiology. Nat Rev Genet 2008;9(5):341–55.

Bell CG, Walley AJ, Froguel P. The genetics of human obesity. Nat Rev Genet 2005;6:221–34.

Bird TD. Genetic aspects of Alzheimer disease. Genet Med 2008;10(4):231–39.

Blennow K, de Leon MJ, Zetterberg H. Alzheimer's disease. Lancet 2006;368(9533):387–403.

Boomsma D, Busjahn A, Peltonen L. Classical twin studies and beyond. Nat Rev Genet 2002;3(11):872–82.

Burmeister M, McInnis MG, Zollner S. Psychiatric genetics: Progress amid controversy. Nat Rev Genet 2008;9(7):527–40.

Cowley AW Jr. The genetic dissection of essential hypertension. Nat Rev Genet 2006;7(11):829–40.

Edenberg HJ, Foroud T. The genetics of alcoholism: Identifying specific genes through family studies. Addict Biol 2006;11(3–4):386–96.

Hirschhorn JN, Daly MJ. Genome-wide association studies for common diseases and complex traits. Nat Rev Genet 2005;6:95–108.

Kibar Z, Capra V, Gros P. Toward understanding the genetic basis of neural tube defects. Clin Genet 2007;71(4):295–310.

King RA, Rotter JI, Motulsky AG (eds). The Genetic Basis of Common Diseases, 2nd ed. New York: Oxford University, 2002.

MacGregor AJ, Snieder H, Schork NJ, Spector TD. Twins: Novel uses to study complex traits and genetic diseases. Trends Genet 2000;16:131–4.

Manolio TA, Brooks LD, Collins FS. A hapmap harvest of insights into the genetics of common disease. J Clin Invest 2008;118(5):1590–605.

McCarthy MI, Abecasis GR, Cardon LR, et al. Genome-wide association studies for complex traits: Consensus, uncertainty and challenges. Nat Rev Genet 2008;9(5):356–69.

Morita H, Seidman J, Seidman CE. Genetic causes of human heart failure. J Clin Invest 2005;115:518–26.

Neale MC, Cardon LR. Methodology for Genetic Studies of Twins and Families. Dordrecht, Netherlands: Kluwer, 1992.

Newton-Cheh C, Shah R. Genetic determinants of QT interval variation and sudden cardiac death. Curr Opin Genet Dev 2007;17(3):213–21.

Owen KR, McCarthy MI. Genetics of type 2 diabetes. Curr Opin Genet Dev 2007;17(3):239–44.

Roden DM. Long-QT syndrome. N Engl J Med 2008;358(2): 169–76.

Shih PA, O'Connor DT. Hereditary determinants of human hypertension: Strategies in the setting of genetic complexity. Hypertension 2008;51(6):1456–64.

Stumvoll M, Goldstein BJ, van Haeften TW. Type 2 diabetes: Principles of pathogenesis and therapy. Lancet 2005;365:1333–1346.

Visscher PM, Hill WG, Wray NR. Heritability in the genomics era—concepts and misconceptions. Nat Rev Genet 2008;9(4):255–66.

Watkins H, Farrall M. Genetic susceptibility to coronary artery disease: From promise to progress. Nat Rev Genet 2006;7(3):163–73.

## Internet Resources

Human Genome Epidemiology Network Reviews (contains links to review articles on genetics of mendelian and common diseases) *http://www.cdc.gov/genomics/hugenet/reviews.htm*

International Clearinghouse for Birth Defects Web Guide *http://www.icbdsr.org/page.asp?n=WebGuide*

# Chapter 13

## GENETIC TESTING AND GENE THERAPY

As we have seen in previous chapters, significant advances have occurred in many areas of medical genetics, including DNA technology, gene mapping and cloning, and cytogenetics. These developments have paved the way for more accurate and efficient testing of genetic disorders. **Genetic testing** can be defined as the analysis of chromosomes, DNA, RNA, proteins, or other **analytes*** to detect abnormalities that can cause a genetic disease. Examples of genetic testing include prenatal diagnosis, heterozygote carrier detection, and presymptomatic diagnosis of genetic disease. The principles and applications of genetic testing in these contexts are one focus of this chapter.

The other focus is the treatment of genetic disease. Many aspects of disease management involve areas of medicine, such as surgery and drug treatment, that are beyond the scope of this book. However, gene therapy, in which patients' cells are genetically altered to combat specific diseases, is discussed here in some detail.

### POPULATION SCREENING FOR GENETIC DISEASE

Screening tests represent an important component of routine health care. These tests are usually designed to detect treatable human diseases in their presymptomatic stage. Papanicolaou (Pap) tests for detecting cervical dysplasia and population screening for hypercholesterolemia are well-known examples of this public health strategy. **Population screening** is large-scale testing of populations for disease in an effort to identify persons who probably have the disease and those who probably do not. Screening tests are not intended to provide definitive diagnoses; rather, they are aimed at identifying a subset of the population on whom further, more exact, diagnostic tests should be carried out. This important distinction is commonly misunderstood by the lay public and seldom clarified by the popular media.

**Genetic screening** is population screening for a gene that can cause the disease in the person carrying the gene or in the descendents of the carrier. Newborn screening for inherited metabolic diseases (see Chapter 7) is a good example of the first type of genetic screening, and heterozygote detection for Tay–Sachs disease (discussed later) exemplifies the

second. These two examples involve screening of populations, but genetic screening can also be applied to members of families with a positive history of a genetic condition. An example is testing for a balanced reciprocal translocation in families in which one or more members have had a chromosome disorder (see Chapter 6). Box 13-1 lists the various types of genetic screening, including several forms of prenatal diagnosis, that are discussed in this chapter.

> The goal of screening is early recognition of a disorder so that intervention will prevent or reverse the disease process (as in newborn screening for inborn errors of metabolism) or so that informed reproductive decisions can be made (as in screening for heterozygous carriers of an autosomal recessive mutation). A positive result from a genetic screening test is typically followed by a more precise diagnostic test.

### Principles of Screening

The basic principles of screening were developed in the 1960s and are still widely recognized. Characteristics of the disease, the test, and the health care system should be considered when deciding whether population screening is appropriate.

First, the condition should be serious and relatively common. This ensures that the benefits to be derived from the screening program will justify its costs. The natural history of the disease should be clearly understood. There should be an acceptable and effective treatment, or, in the case of some genetic conditions, prenatal diagnosis should be available. As for the screening test itself, it should be acceptable to the population, easy to perform, and relatively inexpensive. The screening test should be valid and reliable. Finally, the resources for diagnosis and treatment of the disorder must be accessible. A strategy for communicating results efficiently and effectively must be in place.

Screening programs typically use tests that are widely applicable and inexpensive to identify an at-risk population (e.g., the phenylketonuria (PKU) screening program, discussed in Clinical Commentary 13-1). Members of this population are then targeted for subsequent tests that are more accurate but also more expensive and time consuming. In this context, the screening test should be able to effectively

---

*An analyte is any substance that is subject to analysis.

## BOX 13-1 Types of Genetic Screening and Prenatal Diagnosis

### POPULATION SCREENING FOR GENETIC DISORDERS
#### Newborn Screening
Blood
- Phenylketonuria, all 50 states in the United States
- Galactosemia, all 50 states in the United States
- Hypothyroidism, all 50 states in the United States
- Hemoglobinopathies, nearly all states
- Other: maple syrup urine disease, homocystinuria, tyrosinemia, and several other diseases are screened in many states

Universal newborn hearing screening (>60% of congenital hearing loss is due to genetic factors)

#### Heterozygote Screening
Tay–Sachs disease, Ashkenazi Jewish population
Sickle cell disease, African-American population
Thalassemias in at-risk ethnic groups
Cystic fibrosis in some populations (persons of European descent, Ashkenazi Jews)

### PRENATAL DIAGNOSIS OF GENETIC DISORDERS
#### Diagnostic Testing (Invasive Prenatal Diagnosis)
Amniocentesis
Chorionic villus sampling

Percutaneous umbilical blood sampling (PUBS)
Preimplantation genetic diagnosis

#### Fetal Visualization Techniques
Ultrasonography
Radiography
Magnetic resonance imaging

#### Population Screening
Maternal age >35 years
Family history of condition diagnosable by prenatal techniques
Quadruple screen: maternal serum α-fetoprotein, estriol, human chorionic gonadotropin, inhibin-A
First trimester screening: ultrasonography, PAPP-A, and free β subunit of human chorionic gonadotropin

### FAMILY SCREENING FOR GENETIC DISORDERS
Family history of chromosomal rearrangement (e.g., translocation)
Screening female relatives in an X-linked pedigree (e.g., Duchenne muscular dystrophy, fragile X syndrome)
Heterozygote screening within at-risk families (e.g., cystic fibrosis)
Presymptomatic screening (e.g., Huntington disease, breast cancer, colon cancer)

---

## CLINICAL COMMENTARY 13-1
### *Neonatal Screening for Phenylketonuria*

### DISEASE CHARACTERISTICS
Population screening of newborns for PKU represents an excellent example of the application of the screening model to genetic disease. As discussed in Chapter 4, the prevalence of this autosomal recessive disorder of phenylalanine metabolism is about 1 in 10,000 to 15,000 white births. The natural history of PKU is well understood. More than 95% of untreated PKU patients become moderately to severely mentally retarded. The condition is not identified clinically in the first year of life, because the physical signs are subtle and PKU usually manifests only as developmental delay. Dietary restriction of phenylalanine, when begun before 4 weeks of age, is highly effective in altering the course of the disease. A low-phenylalanine diet, although not particularly palatable, largely eliminates the IQ loss that would otherwise occur (an important exception is those who have a defect in biopterin metabolism, in whom a different therapy is used).

### TEST CHARACTERISTICS
PKU is typically detected by the measurement of blood phenylalanine using a bacterial inhibition assay, the Guthrie test. Blood is collected in the newborn period, usually by heel stick, and placed on filter paper. The dried blood is placed on an agar plate and incubated with a strain of bacteria (*Bacillus subtilis*) that requires phenylalanine for growth. Measurement of bacterial growth permits quantification of the amount of phenylalanine in the blood sample. Increasingly, tandem mass spectrometry is used to screen for PKU. Positive test results are usually repeated and followed by a quantitative assay of plasma phenylalanine and tyrosine.

If the test is performed after 2 days of age and after regular feeding on a protein diet, the detection rate (sensitivity) is about 98%. If it is performed at less than 24 hours of age, the sensitivity is about 84% and a repeat test is recommended a few weeks after birth. Specificity is close to 100%.

### SYSTEM CHARACTERISTICS
Because of the requirement of normal protein in the diet, many states request rescreening at 2 to 4 weeks of age. At that point, sensitivity approaches 100%. A high sensitivity level is important because of the severe impact of a misdiagnosis.

Phenylalanine levels in children with classic PKU typically exceed 20 mg/dL. For every 20 positive PKU screening results, only one infant has classic PKU. The others represent either false-positive findings (usually due to a transient reversible tyrosinemia) or infants with a form of hyperphenylalaninemia (elevated phenylalanine) not caused by classic PKU.

The cost of a Guthrie test is typically less than a few dollars. Several studies have shown that the cost of nationwide PKU screening is significantly less than the savings it achieves by avoiding institutionalization costs and lost productivity.

---

separate persons who have the disease from those who do not. This attribute, which defines the test's **validity**, involves two components: **sensitivity** and **specificity**. Sensitivity reflects the ability of the test to correctly identify those *with* the disease. It is measured as the fraction of affected persons in whom the test is positive (i.e., **true positives**). Specificity is the ability of the test to correctly identify those *without* the disease. It is measured as the fraction of unaffected persons in whom the test is negative (i.e., **true negatives**). Sensitivity and specificity are determined by comparing the screening results with those of a definitive diagnostic test (Table 13-1).

**TABLE 13-1**

## Definitions of Sensitivity and Specificity*

| Result of Screening Test | Actual Disease State | |
|---|---|---|
| | **Affected** | **Unaffected** |
| Test positive (+) | a (true positives) | b (false positives) |
| Test negative (−) | c (false negatives) | d (true negatives) |

*a, b, c, and d represent the number of individuals in a population who were found to have the disease and test result combinations shown. The test sensitivity = a/(a + c); specificity = d/(b + d); positive predictive value = a/(a + b); and negative predictive value = d/(c + d).*

Screening tests are seldom, if ever, 100% sensitive and 100% specific. This is because the range of test values in the disease population overlaps that of the unaffected population (Fig. 13-1). Thus, results of a screening test (as opposed to the definitive follow-up diagnostic test) will be incorrect for some members of the population. Usually a cutoff value is designated to separate the diseased and nondiseased portions of the population. A tradeoff exists between the impact of nondetection or low sensitivity (i.e., an increased **false-negative rate**) and the impact of low specificity (an increased **false-positive rate**). If the penalty for missing affected persons is high (as in untreated PKU), then the cutoff level is lowered so that nearly all disease cases will be detected (higher sensitivity). This also lowers the specificity by increasing the number of unaffected persons with positive test results (false positives) who are targeted for subsequent diagnostic testing. If confirmation of a positive test is expensive or hazardous, then false-positive rates are minimized (i.e., the cutoff level is increased, producing high specificity at the expense of sensitivity).

> The basic elements of a test's validity include its sensitivity (proportion of true positives accurately detected) and specificity (proportion of true negatives accurately detected). When sensitivity is increased, specificity decreases, and vice versa.

A primary concern in the clinical setting is the accuracy of a positive screening test. One needs to know the fraction of persons with a positive test result who truly have the disease in question (i.e., a/(a + b) in Table 13-1). This quantity is defined as the **positive predictive value**. It is also useful to know the **negative predictive value**, which is the fraction of persons with a negative result who truly do not have the disease (d/(c + d)).

The concepts of sensitivity, specificity, and positive predictive value can be illustrated by an example. Congenital adrenal hyperplasia (CAH) due to a deficiency of 21 hydroxylase is an inborn error of steroid biosynthesis that can produce ambiguous genitalia in females and adrenal crises in both sexes. The screening test, a 17-hydroxyprogesterone assay, has a sensitivity of about 95% and a specificity of 99% (Table 13-2). The prevalence of CAH is about 1/10,000 in most white populations, but it rises to about 1/400 in the Yupik Native Alaskan population.

Let us assume that a screening program for CAH has been developed in both of these populations. In a population of 500,000 whites, the false-positive rate (1 − specificity) is 1%. Thus, about 5000 unaffected persons will have a positive test. With 95% sensitivity, 47 of the 50 white persons who have CAH will be detected through a positive test. Note that the great majority of people who have a positive test result would not have CAH: the positive predictive value is

**TABLE 13-2**

## Hypothetical Results of Screening for Congenital Adrenal Hyperplasia in a Low-Prevalence White Population and in a High-Prevalence Yupik Population*

| Result of Screening Test | CAH Present | CAH Absent |
|---|---|---|
| **Positive** | | |
| White | 47 | 5000 |
| Yupik | 24 | 100 |
| **Negative** | | |
| White | 3 | 494,950 |
| Yupik | 1 | 9875 |

*White positive predictive value = 47/(47 + 5000) ≈ 1%; Yupik positive predictive value = 24/(24 + 100) ≈ 19%.*
*CAH, congenital adrenal hyperplasia.*

**FIGURE 13-1**
The distribution of creatine kinase (CK) in normal women and in women who are heterozygous carriers for a mutation in the Duchenne muscular dystrophy gene. Note the overlap in distribution between the two groups: About two thirds of carriers have CK levels that exceed the 95th percentile in normal women. If the 95th percentile is used as a cutoff to identify carriers, then the sensitivity of the test is 67% (i.e., two thirds of carriers will be detected), and the specificity is 95% (i.e., 95% of normal women will be correctly identified).

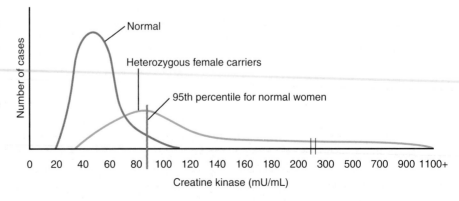

47/(47 + 5000), or less than 1%. Now suppose that 10,000 members of the Yupik population are screened for CAH. As Table 13-2 shows, 24 of 25 persons with CAH will test positive, and 100 persons without CAH will also test positive. Here, the positive predictive value is much higher than in the white population: 24/(24 + 100) = 19%. This example illustrates an important principle: *For a given level of sensitivity and specificity, the positive predictive value of a test increases as the prevalence of the disease increases.*

> The positive predictive value of a screening test is defined as the percentage of positive tests that are true positives. It increases as the prevalence of the target disorder increases.

### Newborn Screening for Inborn Errors of Metabolism

Newborn screening programs represent an ideal opportunity for presymptomatic detection and prevention of genetic disease. At present, all states in the United States screen newborns for PKU, galactosemia (see Chapter 7), and hypothyroidism. All of these conditions fulfill the previously stated criteria for population screening. Each is a disorder in which the person is at significant risk for mental retardation, which can be prevented by early detection and effective intervention.

In recent years, most states of the United States and many other nations have instituted screening programs to identify neonates with hemoglobin disorders (e.g., sickle-cell disease). These programs are justified by the fact that up to 15% of untreated children with sickle-cell disease die from infections before 5 years of age (see Chapter 3). Effective treatment, in the form of prophylactic antibiotics, is available.

Some communities have begun screening for Duchenne muscular dystrophy by measuring creatine kinase levels in newborns. The objective is not presymptomatic treatment; rather it is identification of families who should receive genetic counseling in order to make informed reproductive decisions. Conditions for which newborn screening is commonly performed are summarized in Table 13-3.

Many U.S. states and European countries have established **expanded newborn screening**. Tandem mass spectrometry

(Chapter 3) can detect abnormalities in the intermediary metabolism of sugars, fats, and proteins that characterize more than 30 metabolic disorders (see Chapter 7). Programs to deal with positive results and to provide rapid treatment of verified disease conditions have been developed

> Newborn screening is an effective public health strategy for treatable disorders such as PKU, hypothyroidism, galactosemia, and sickle cell disease. The use of tandem mass spectrometry has recently expanded the number of diseases that can be detected by newborn screening.

### Heterozygote Screening

The aforementioned principles of population screening can be applied to the detection of unaffected carriers of disease-causing mutations. The target population is a group known to be at risk. The intervention consists of the presentation of risk figures and options such as prenatal diagnosis. Genetic diseases amenable to heterozygote screening are typically autosomal recessive disorders for which prenatal diagnosis and genetic counseling are available, feasible, and accurate.

An example of a highly successful heterozygote screening effort is the Tay–Sachs screening program in North America. Infantile Tay–Sachs disease is an autosomal recessive lysosomal storage disorder in which the lysosomal enzyme β-hexosaminidase A (HEX A) is deficient (see Chapter 7), causing a buildup of the substrate, $G_{M2}$ ganglioside, in neuronal lysosomes. The accumulation of this substrate damages the neurons and leads to blindness, seizures, hypotonia, and death by about 5 years of age. Tay–Sachs disease is especially common among Ashkenazi Jews, with a heterozygote frequency of about 1 in 30. Thus, this population is a reasonable candidate for heterozygote screening. Accurate carrier testing is available (assays for HEX A or direct DNA testing for mutations). Because the disease is uniformly fatal, options such as pregnancy termination or artificial insemination by noncarrier donors are acceptable to most

**TABLE 13-3**

## Characteristics of Selected Newborn Screening Programs

| Disease | Inheritance | Prevalence | Screening Test | Treatment |
|---|---|---|---|---|
| Phenylketonuria | Autosomal recessive | 1/10,000–1/15,000 | Tandem mass spectrometry | Dietary restriction of phenylalanine |
| Galactosemia | Autosomal recessive | 1/50,000–1/100,000 | Transferase assay | Dietary restriction of galactose |
| Congenital hypothyroidism | Usually sporadic | 1/5000 | Measurement of thyroxine ($T_4$) or thyroid stimulating hormone (TSH) | Hormone replacement |
| Sickle cell disease | Autosomal recessive | 1/400–1/600 blacks | Isoelectric focusing or DNA diagnosis | Prophylactic penicillin |
| Cystic fibrosis | Autosomal recessive | 1/2500 | Immunoreactive trypsinogen confirmed by DNA diagnosis | Antibiotics, chest physical therapy, pancreatic enzyme replacement if needed |

*Data from guidelines of the American College of Medical Genetics, http://www.acmg.net/resources/policies/ACT/condition-analyte-links.htm (accessed March 10, 2009).*

couples. A well-planned effort was made to educate members of the target population about risks, testing, and available options. As a result of screening, the number of Tay–Sachs disease births among Ashkenazi Jews in the United States and Canada declined by 90%, from 30 to 40 per year before 1970, to 3 to 5 per year in the 1990s, and to zero in 2003.

β-Thalassemia major, another serious autosomal recessive disorder, is especially common among many Mediterranean and South Asian populations (see Chapter 3). Effective carrier screening programs have produced a 75% decrease in the prevalence of newborns with this disorder in Greece, Cyprus, and Italy. Carrier screening is also possible for cystic fibrosis, another autosomal recessive disorder (Clinical Commentary 13-2). Table 13-4 presents a list of selected conditions for which heterozygote screening programs have been developed in industrial countries.

In addition to the criteria for establishing a population screening program for genetic disorders, guidelines have been developed regarding the ethical and legal aspects of heterozygote screening programs. These are summarized in Box 13-2.

> Heterozygote screening consists of testing (at the phenotype or genotype level) a target population to identify unaffected carriers of a disease gene. The carriers can then be given information about risks and reproductive options.

### Presymptomatic Diagnosis

With the development of genetic diagnosis through linkage analysis and direct mutation detection, presymptomatic diagnosis has become feasible for many genetic diseases. At-risk persons can be tested to determine whether they have inherited a disease-causing mutation before they develop clinical symptoms of the disorder. Presymptomatic diagnosis is available, for example, for Huntington disease, adult polycystic kidney disease, hemochromatosis, and autosomal dominant breast cancer. By informing persons that they do or do not carry a disease-causing mutation, presymptomatic diagnosis

## CLINICAL COMMENTARY   13-2
### Population Screening for Cystic Fibrosis

More than 1500 mutations have now been reported in the *CFTR* gene, and, although some are benign variants, most can cause cystic fibrosis. It would be technologically impractical to test for all reported mutations in a population screening program. However, among the mutations that can cause CF in whites, about 70% are the three-base deletion termed △F508 (see Chapter 4). In this population, carrier screening using PCR-based detection of △F508 would detect approximately 90% of couples in which one or both are heterozygous carriers of this mutation ($1 - 0.30^2$, where $0.30^2$ represents the frequency of carrier couples in whom neither carries the △F508 mutation). It is now recommended to test simultaneously for 25 of the most common *CFTR* mutations, which will detect approximately 85% of all *CFTR* mutations in persons of European descent (because mutation frequencies vary among populations, this figure is somewhat lower in other U.S. populations, such as blacks and Hispanics). Among whites, 98% of couples in which one or both members carry a CF mutation would be recognized (i.e., $1 - 0.15^2$), yielding a high level of sensitivity. The American College of Medical Genetics and the American College of Obstetricians and Gynecologists recommend that couples who are planning a pregnancy, or who are currently pregnant, should be offered a screen for CF carrier status. Couples in which both parents are heterozygotes would define a subset of the population in which prenatal diagnosis for CF might be offered. CF is now commonly screened in the newborn population, typically using an immunoreactive trypsinogen assay, followed, if indicated, by direct testing for *CFTR* mutations.

**TABLE 13-4**

## Selected Examples of Heterozygote Screening Programs in Specific Ethnic Groups

| Disease | Ethnic Group | Carrier Frequency | At-Risk Couple Frequency | Disease Incidence in Newborns |
|---------|-------------|-------------------|--------------------------|-------------------------------|
| Sickle cell disease | Blacks | 1/12 | 1/150 | 1/600 |
| Tay–Sachs disease | Ashkenazi Jews | 1/30 | 1/900 | 1/3600 |
| β-Thalassemia | Greeks, Italians | 1/30 | 1/900 | 1/3600 |
| α-Thalassemia | Southeast Asians, Chinese | 1/25 | 1/625 | 1/2500 |
| Cystic fibrosis | Northern Europeans | 1/25 | 1/625 | 1/2500 |
| Phenylketonuria | Northern Europeans | 1/50 | 1/2500 | 1/10,000 |

*Modified from McGinniss MJ, Kaback MM: Carrier screening. In Rimoin DL, Conner JM, Pyeritz RE, Korf BR (eds): Emery and Rimoin's Principles and Practice of Medical Genetics, 5th ed. New York: Churchill Livingstone, 2007, pp 752–762.*

## BOX 13-2 **Public Policy Guidelines for Heterozygote Screening**

Recommended guidelines:

Screening should be voluntary, and confidentiality must be ensured.

Screening requires informed consent.

Providers of screening services have an obligation to ensure that adequate education and counseling are included in the program.

Quality control of all aspects of the laboratory testing, including systematic proficiency testing, is required and should be implemented as soon as possible.

There should be equal access to testing.

*From Elias S, Annas GJ, Simpson JL: Carrier screening for cystic fibrosis: implications for obstetric and gynecologic practice. Am J Obstet Gynecol 1991;164:1077–1083.*

can aid in making reproductive decisions. It can provide reassurance to those who learn that they do not carry a disease-causing mutation. In some cases, early diagnosis can improve health supervision. For example, persons who inherit an autosomal dominant breast cancer mutation can undergo mammography at an earlier age to increase the chances of early tumor detection. Persons at risk for the inheritance of *RET* mutations (see Chapter 11), who are highly likely to develop multiple endocrine neoplasia type 2 (MEN2), can undergo a prophylactic thyroidectomy to reduce their chance of developing a malignancy. Those who inherit mutations that cause some forms of familial colon cancer (adenomatous polyposis coli [APC] and hereditary nonpolyposis colorectal cancer [HNPCC]; see Chapter 11) can also benefit from early diagnosis and treatment.

Because most genetic diseases are relatively uncommon in the general population, universal presymptomatic screening is currently impractical. It is usually recommended only for persons who are known to be at risk for the disease, generally because of a positive family history.

> Genetic testing can sometimes be performed to identify persons who have inherited a disease-causing gene before they develop symptoms. This is termed *presymptomatic diagnosis.*

### Psychosocial Implications of Genetic Screening and Diagnosis

Screening for genetic diseases has many social and psychological implications. The burden of anxiety, cost, and potential stigmatization surrounding a positive test result must be weighed against the need for detection. Often, screening tests are misperceived as diagnostically definitive. The concept that a positive screening test does not necessarily indicate disease presence must be emphasized to those who undergo screening (see Box 13-2).

The initial screening programs for sickle-cell carrier status in the 1970s were plagued by misunderstandings about the implications of carrier status. Occasionally, carrier detection led to cancellation of health insurance or employer discrimination. Such experiences underscore the need for effective genetic counseling and public education. Other issues include the right to choose *not* to be tested and the potential for invasion of privacy.

The social, psychological, and ethical aspects of genetic screening will become more complicated as newer techniques of DNA diagnosis become more accessible. For example, even though presymptomatic diagnosis of Huntington disease is available, several studies have shown that fewer than 20% of at-risk persons elected this option. Largely, this reflects the fact that no effective treatment is currently available for this disorder. Presymptomatic diagnosis of *BRCA1* and *BRCA2* carrier status in breast-and-ovarian cancer families has also met with mixed responses. In part, this is a reaction to the cost of the test: because of the large number of different mutations in *BRCA1* or *BRCA2* that can cause breast cancer, testing typically consists of sequencing all exons and promoters of both genes, as well as some intronic nucleotides near each exon. This is an expensive procedure. Preventive measures such as prophylactic mastectomy and oophorectomy (removal of the breasts and ovaries, respectively) are known to reduce cancer risk substantially, but they do not completely eliminate it.

For some genetic diseases, such as autosomal dominant colon cancer syndromes, early diagnosis can lead to improved survival because effective preventive treatments are readily available (colectomy or polypectomy for precancerous colon polyps). In addition, many at-risk persons will find that they do not carry the disease-causing gene, allowing them to avoid unpleasant (and possibly hazardous) diagnostic procedures such as colonoscopy or mammography. However, as screening for such diseases becomes more common, the issues of privacy and confidentiality and the need for accurate communication of risk information must also be addressed.

### MOLECULAR TOOLS FOR SCREENING AND DIAGNOSIS

Until recently, genetic screening usually relied on assays of the disease phenotype, such as a β-hexosaminidase assay for Tay–Sachs disease or a creatine kinase assay for Duchenne muscular dystrophy. Advances in DNA technology have led to diagnosis at the level of the genotype. In some cases, linkage analysis is used to determine whether a person has inherited a disease-causing gene, but in most cases direct assays of disease-causing mutations have been developed. Genetic diagnosis at the DNA level is now supplementing, and in many cases supplanting, tests based on phenotypic assays.

Linkage analysis and direct mutation diagnosis have been used for diagnostic testing within families, for prenatal diagnosis of genetic disorders, and, more recently, in population screening. Improved technology and an increased demand for testing have led to the establishment of clinical molecular laboratories in many medical centers throughout the world.

## Linkage Analysis

DNA polymorphisms (most commonly, short tandem repeat polymorphisms, STRPs), can be used as markers in linkage analysis, as described in Chapter 8. Once linkage phase is established in a family, the marker locus can be assayed to determine whether an at-risk person has inherited a chromosome segment that contains a disease-causing mutation or a homologous segment that contains a normal allele (Fig. 13-2). Because this approach uses linked markers but does not involve direct examination of the disease-causing mutations, it is a form of **indirect diagnosis**.

Linkage analysis has been employed successfully in diagnosing many of the genetic diseases discussed in this text. In principle, it can be used to diagnose any mapped genetic disease. It has the advantages that the disease gene and its product need not be known. The marker simply tells us whether or not the at-risk person has inherited the chromosome region that contains a disease-causing mutation.

The disadvantages of this approach are that several family members must be tested in order to establish linkage phase; not all markers are informative (sufficiently heterozygous) in all families (see Chapter 8 for a discussion of uninformative families); and recombination can occur between the marker and the disease-causing mutation, introducing a source of diagnostic error.

As discussed in Chapter 8, the highly polymorphic STRPs greatly increase the likelihood that a marker will be informative in a family. Informativeness can also be increased by using multiple marker polymorphisms, all of which are closely linked to the disease locus. The use of markers flanking both sides of the disease locus can alert the investigator to a recombination.

> Linkage analysis, a form of indirect genetic diagnosis, uses linked markers to determine whether a person has inherited a chromosome containing a disease gene from his or her parent. The need for typing of numerous family members and the possibilities of recombination and uninformative matings are disadvantages of this approach.

## Direct Mutation Analysis

Sometimes the disease-causing mutation happens to alter a recognition sequence for a restriction enzyme. In this case the mutation itself creates a restriction site polymorphism that can be detected after digestion with this enzyme. An example is given by the sickle-cell disease mutation, which alters an *Mst*II recognition site in the β-globin gene (see Fig. 3-18 in Chapter 3). Because the resulting RFLP reflects the disease-causing mutation directly, RFLP analysis in this context is an example of **direct diagnosis** of the disease. Direct diagnosis has the advantages that family information is not needed (the mutation is viewed directly in each individual), uninformative matings are not a problem, and there is no error resulting from recombination. (Table 13-5 summarizes the advantages and disadvantages of direct and indirect diagnosis.) The primary disadvantage of using RFLPs for direct diagnosis is that only about 5% of disease-causing mutations happen to affect known restriction sites.

> Direct genetic diagnosis is accomplished by typing the disease-causing mutation itself. It is potentially more accurate than indirect diagnosis and does not require family information. RFLP techniques can be used for direct diagnosis if the mutation affects a restriction site.

### Allele-Specific Oligonucleotides

If the DNA sequence surrounding a mutation is known, an oligonucleotide probe can be synthesized that will hybridize (undergo complementary base pairing) only to the mutated sequence (such probes are often termed **allele-specific oligonucleotides**, or ASOs). A second probe that will hybridize to the normal DNA sequence is also synthesized. Stringent hybridization conditions are used so that a one-base mismatch will prevent hybridization. DNA from persons who are homozygous for the mutation hybridizes only with the ASO containing the mutated sequence, whereas DNA from persons homozygous for the normal sequence hybridizes with the normal ASO. DNA from heterozygotes hybridizes with both probes (Fig. 13-3). The length of the ASO probes, usually about 18 to 20 nucleotides, is critical. Shorter probes

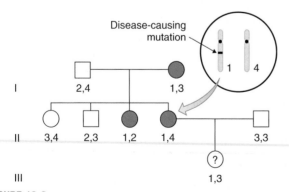

**FIGURE 13-2**

In this pedigree for autosomal dominant breast cancer, the analysis of a closely linked marker on chromosome 17 shows that the mutation is on the same chromosome as marker allele 1 in the affected mother in generation II. This indicates that the daughter in generation III has inherited the mutation-bearing chromosome from her mother and is highly likely to develop a breast tumor.

**TABLE 13-5**

## Summary of Attributes of Indirect and Direct Diagnosis

| Attribute | Indirect Diagnosis | Direct Diagnosis |
|---|---|---|
| Family information needed | Yes | No |
| Errors possible because of recombination | Yes | No |
| Markers may be uninformative | Yes | No |
| Single test can uncover multiple mutations | Yes | No |
| Disease-causing mutation must be known | No | Yes |

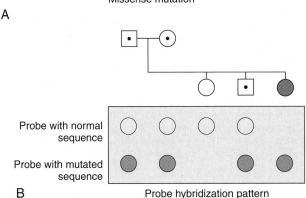

GTG CAC CTG ACT CCT GAG GAG  Normal probe
CAC GTG GAC TGA GGA CTC CTC  Normal β-globin sequence (antisense strand)

GTG CAC CTG ACT CCT GTG GAG  Probe containing mutation
CAC GTG GAC TGA GGA CAC CTC  Missense mutation in β-globin sequence

Missense mutation

A

Probe with normal sequence

Probe with mutated sequence

B                    Probe hybridization pattern

**FIGURE 13-3**
**A,** A 21-bp allele-specific oligonucleotide (ASO) probe *(yellow)* is constructed to undergo complementary base pairing only with the normal β-globin sequence, and another ASO probe *(green)* is constructed to undergo complementary base pairing only with a β-globin sequence that contains a missense mutation that produces a substitution of valine for glutamic acid at position 6 of the β-globin polypeptide (see Chapter 3), causing sickle cell disease in homozygotes. **B,** In this family, the parents are both heterozygous carriers of the missense mutation, so their DNA hybridizes to both ASO probes. The first female offspring has a homozygous normal genotype, the second offspring is heterozygous, and third offspring is an affected homozygote. A variety of methods, including microarrays (see Chapter 3), can be used to detect these ASO hybridization patterns.

would not be unique in the genome and would therefore hybridize to multiple regions. Longer probes are more difficult to synthesize correctly and could hybridize to both the normal and the mutated sequence.

The ASO method of direct diagnosis has the same advantages that were listed for direct diagnosis using RFLPs. It has the additional advantage that it is not limited to mutations that cause alterations in restriction sites. However, it does require that at least part of the disease gene has been cloned and sequenced. In addition, each disease-causing mutation requires a different oligonucleotide probe. For this reason, this approach, while powerful, can become difficult or impractical if the disease can be caused by a large number of different mutations that each have low frequency in the population.

> Direct diagnosis can be performed by hybridization of a person's DNA with allele-specific oligonucleotide probes. This approach is feasible if the DNA sequence causing a genetic disease is known and if the number of disease-causing mutations is limited.

Examples of diseases caused by a limited number of mutations include sickle cell disease and $\alpha_1$-antitrypsin deficiency (Clinical Commentary 13-3). Although more than 1500 cystic fibrosis mutations have been identified, 25 of the most common ones account for the great majority of mutations in many populations (see Clinical Commentary 13-2). Thus, direct diagnosis can be used to identify most cystic fibrosis homozygotes and heterozygous carriers. Prenatal diagnosis is also possible. Direct diagnosis, using polymerase chain reaction (PCR) or Southern blotting, can also be used to detect deletions or duplications (e.g., those of the

---

**CLINICAL COMMENTARY 13-3**
***The Genetic Diagnosis of $\alpha_1$-Antitrypsin Deficiency***

An inherited deficiency of $\alpha_1$-antitrypsin ($\alpha_1$-AT) is one of the most common autosomal recessive disorders among whites, affecting approximately 1 in 2000 to 1 in 5000. $\alpha_1$-AT, synthesized primarily in the liver, is a serine protease inhibitor. As its name suggests, it binds trypsin. However, $\alpha_1$-AT binds much more strongly to neutrophil elastase, a protease that is produced by neutrophils (a type of leukocyte) in response to infections and irritants. It carries out its binding and inhibitory role primarily in the lower respiratory tract, where it prevents neutrophil elastase from digesting the alveolar septa of the lung.

Persons with less than 10% to 15% of the normal level of $\alpha_1$-AT activity experience significant lung damage and typically develop emphysema during their 30s, 40s, or 50s. In addition, at least 10% develop liver cirrhosis as a result of the accumulation of variant $\alpha_1$-AT molecules in the liver. $\alpha_1$-AT deficiency accounts for almost 20% of all nonalcoholic liver cirrhosis in the United States. Cigarette smokers with $\alpha_1$-AT deficiency develop emphysema much earlier than do nonsmokers because cigarette smoke irritates lung tissue, increasing secretion of neutrophil elastase. At the same time, it inactivates $\alpha_1$-AT, so there is also less inhibition of elastase. One study showed that the median age of death of nonsmokers with

$\alpha_1$-AT deficiency was 62 years, whereas it was 40 years for smokers with this disease. The combination of cigarette smoking (an environmental factor) and an $\alpha_1$-AT mutation (a genetic factor) produces more-severe disease than either factor alone; thus, this is an example of a gene–environment interaction.

Testing for $\alpha_1$-AT deficiency usually begins with a form of protein electrophoresis, which is inexpensive and widely available. Direct DNA testing became feasible with the identification of *SERPINA1*, the gene that encodes $\alpha_1$-AT. More than 100 *SERPINA1* mutations have been identified, but only one of these, a missense mutation that produces the *Z* allele, is common and clinically significant. Ninety-five percent of cases of $\alpha_1$-AT deficiency are homozygous for this allele. Two large studies have indicated that the risk of developing emphysema among *ZZ* homozygotes is 70% for nonsmokers and 90% for smokers. Because the great majority of $\alpha_1$-AT cases are caused by a single mutation, this disease can be diagnosed efficiently by using ASO probes. A second mutation, termed *S*, is less common and severe, but it can also be detected by probe hybridization. ASO testing provides a rapid, sensitive (sensitivity >95%) method for detecting mutations that cause this important disease.

## BOX 13-3 Limitations of Genetic Testing

Although genetic testing offers many advantages, its limitations must also be borne in mind. These limitations can be summarized as follows:

- No genetic test is 100% accurate. Although most genetic tests do achieve a high level of accuracy, factors such as mosaicism can complicate cytogenetic diagnosis, and genotyping errors can occur in the diagnosis of single-gene disorders.
- Genetic tests reveal mutations, not the presence of disease, because many disease-causing mutations have incomplete penetrance. For example, approximately 50% to 80% of women with *BRCA1* or *BRCA2* mutations develop breast cancer, and 70% to 90% of persons with mutations in one of the HNPCC genes develop colorectal cancer. Even when penetrance approaches 100% (as in neurofibromatosis type 1 or Huntington disease), detection of the mutation often reveals little about the severity or age of onset of the disease.
- Genetic testing might not detect all of the mutations that can cause a disease. Even in the absence of genotyping or sequencing errors, many genetic tests lack sensitivity. For example, the panels commonly used to test for cystic fibrosis mutations have typically less than 90% sensitivity to detect homozygotes (see Clinical Commentary 13.2). When a large number of different mutations can produce a genetic disease (e.g., neurofibromatosis, autosomal dominant breast cancer, Marfan syndrome), it might not be practical to test for all possible mutations. In this case, the analysis of linked markers can provide additional diagnostic accuracy if multiple family members are affected. Other factors that can reduce accuracy include locus heterogeneity and the presence of phenocopies.
- Genetic testing can lead to complex ethical and social considerations. The results of a genetic test might lead to stigmatization or to discrimination by employers or insurance companies. Effective treatment is not available for some genetic diseases (e.g., Huntington disease, familial Alzheimer disease), decreasing the value of early diagnosis through genetic testing. Because genes are shared in families, the results of a genetic test might affect not only the tested person but also other members of the family (who might not wish to know about their risk for a genetic disease). These and other ethical and social issues are discussed further in Chapter 15.

dystrophin gene that cause most cases of Duchenne muscular dystrophy). Currently, clinical genetic testing is available for more than almost 1500 genetic diseases, including almost all of the single-gene conditions discussed in this textbook. Despite this wide availability, it should be kept in mind that genetic testing, like all testing procedures, has a number of limitations (Box 13-3).

### Other Methods of Direct Diagnosis

The ASO method (see Fig. 13-3) is commonly used in detecting direct mutations. Many other techniques can also be used to detect mutations, including several discussed in Chapter 3 (e.g., direct sequencing, DNA mismatch cleavage, and MLPA). Microarrays (DNA chips, also discussed in Chapter 3) are now widely used to detect large-scale mutations. Microarrays have many convenient properties, including miniaturization and automated computerized processing. They can be designed to assay large numbers of sequence variants (including disease-causing mutations) in a single rapid analysis (Box 13-4). For example, one microarray contains thousands of oligonucleotide probes that hybridize to large numbers of possible sequence variants in the *CYP2D6* and *CYP2C19* genes. The products of these genes influence the rate of metabolism of about 25% of all prescription drugs, and analysis of their variation might help to predict how individual patients will respond to these drugs.

Mass spectrometry, a method commonly used in chemistry, is also being explored as a rapid means of detecting mutations. This technique detects minute differences in the mass of PCR-amplified DNA molecules, which represent variations in DNA sequence. Mass spectrometry, which offers the advantages of high speed and great accuracy, has been used, for example, to detect mutations in the *CFTR* and *APOE* genes.

Another form of mass spectrometry, **tandem mass spectrometry**, is being used increasingly to screen newborns for protein variants that signal amino acid disorders (e.g., PKU, tyrosinemia, homocystinuria), organic acid disorders, and fatty acid oxidation disorders (e.g., MCAD and LCHAD deficiencies; see Chapter 7 and see earlier). This method begins with a sample of material from a dried blood spot, which is subjected to analysis by two mass spectrometers. The first spectrometer separates ionized molecules according to their mass, and the molecules are fragmented. The second spectrometer assesses the mass and charge of these fragments, allowing a computer to generate a molecular profile of the sample. Tandem mass spectrometry is highly accurate and very rapid: more than two dozen disorders can be screened in approximately 2 minutes.

> New methods of direct mutation detection, including the use of microarrays and mass spectrometry, have greatly increased the speed and accuracy of genetic diagnosis. Tandem mass spectrometry can be used to test for protein variants that are characteristic of a number of newborn disorders and is thus a useful screening tool.

## PRENATAL DIAGNOSIS OF GENETIC DISORDERS AND CONGENITAL DEFECTS

**Prenatal diagnosis** is a major focus of genetic testing, and several important areas of technology have evolved to provide this service. The principal aim of prenatal diagnosis is to supply at-risk families with information so that they can make informed choices during pregnancy. The potential benefits of prenatal testing include providing reassurance to at-risk families when the result is normal; providing risk

## BOX 13-4 **Direct-to-Consumer Genetic Testing**

Several private companies now offer microarray-based genetic testing on a direct-to-consumer basis. Typically, the customer collects and submits a cheek swab or mouthwash. DNA is extracted from the sample and hybridized to a microarray that can test simultaneously for a large number of DNA variants, including some of the variants associated, for example, with cystic fibrosis; hemochromatosis; age-related macular degeneration; type 1 diabetes; type 2 diabetes; psoriasis; and breast, prostate, and colorectal cancer. The customer is informed of the results and is given some explanatory information to aid in interpreting them. In some cases, genetic counseling is available. The cost of this procedure typically ranges from several hundred to several thousand dollars.

This type of testing, sometimes termed "recreational genomics," has understandable appeal. Many people want to know more about their genomes and how DNA variation might affect their health. Many will likely present the results of these tests to their primary care physicians, expecting explanations and even predictions. Several important considerations should be kept in mind.

For most disease conditions, these tests have relatively low sensitivity and low specificity. A positive result seldom predicts disease with precision (see Box 13.3), and a negative result should not induce a false sense of security. For common multifactorial diseases, most of the responsible genetic variants have not yet been identified, and, as discussed in Chapter 12, nongenetic factors typically play a large role in causing the disease. The relative increase in disease risk associated with most variants is quite small, on the order of a few percentage points. These risk estimates are usually based on specific populations, typically Europeans or Americans of European descent; they might not apply accurately to members of other populations. There is considerable potential for misunderstanding of these results, and many of the concerns discussed in Box 13-3 (stigmatization, potential for loss of privacy) apply to direct-to-consumer testing. For these reasons, this type of genetic testing should be regarded with considerable caution.

---

information to couples who, in the absence of such information, would not choose to begin a pregnancy; allowing a couple to prepare psychologically for the birth of an affected baby; helping the health care professional to plan delivery, management, and care of the infant when a disease is diagnosed in the fetus; and providing risk information to couples for whom pregnancy termination is an option.

Given the controversy surrounding the issue of pregnancy termination, it should be emphasized that the great majority of prenatal diagnoses yield a normal test result. Thus, most families receive reassurance, and only a small minority face the issue of considering pregnancy termination.

Both screening and diagnostic tests can be done prenatally. An example of a population screening test is the analysis of maternal serum at 15 weeks' gestation for increased or decreased levels of α-fetoprotein (AFP) and several other serum components that can indicate an abnormal pregnancy. A positive test result identifies a subgroup for further testing for aneuploidy syndromes or neural tube defects (NTDs). A subsequent **amniocentesis** (the withdrawal of amniotic fluid during pregnancy) would represent a more accurate, specific

diagnostic test. Prenatal diagnostic methods can be divided into two major types: analysis of fetal tissues (amniocentesis, chorionic villus sampling, cordocentesis, and preimplantation genetic diagnosis) and visualization of the fetus (ultrasonography, magnetic resonance imaging). In this section, each of these procedures is described, and their accuracy, safety, and feasibility are discussed.

### Amniocentesis

Amniocentesis is traditionally performed at 15 to 17 weeks after a pregnant woman's last menstrual period (LMP). After real-time ultrasound imaging localizes the placenta and determines the position of the fetus, a needle is inserted through the abdominal wall into the amniotic sac (Fig. 13-4). Between 20 and 30 mL of amniotic fluid is withdrawn; this fluid contains living cells (**amniocytes**) shed by the fetus. The amniocytes are cultured to increase their number (a procedure that requires up to 7 days), and standard cytogenetic studies are carried out on the cultured amniocytes. In addition, cells can be grown for biochemical assays or DNA-based diagnosis of any genetic disease for which mutation testing is available. The results of cytogenetic studies are typically available in 10 to 12 days. Because fluorescent in situ hybridization (FISH) can be carried out on a small number of uncultured amniocytes, it can provide an indication of fetal aneuploidy in just 1 to 2 days. If the FISH result is positive, subsequent, confirmatory diagnosis by routine cytogenetic methods is recommended. Indications for prenatal diagnosis by amniocentesis are listed in Box 13-5.

Amniocentesis is also used to measure AFP, a fetal protein that is produced initially by the yolk sac and subsequently by the fetal liver. The AFP level normally increases in amniotic fluid until about 10 to 14 weeks' gestation and then decreases steadily. Amniotic fluid AFP is significantly higher in pregnancies in which the fetus has an NTD. When an amniotic fluid AFP assay is used with ultrasonography (see later) in the second trimester, more than 98% of fetuses with an open spina bifida and virtually all of those with anencephaly can be recognized. Among women who undergo amniocentesis for cytogenetic analysis, it is routine to also measure their amniotic fluid AFP level.

In addition to a fetal NTD, there are several other causes of elevated (or apparently elevated) amniotic fluid AFP. These include underestimation of gestational age, fetal death, presence of twins, blood contamination, and several specific malformations (e.g., omphalocele or gastroschisis, which are abdominal wall defects). Usually, targeted ultrasonography can distinguish among these alternatives.

The safety and accuracy of amniocentesis have been established by several large collaborative studies. The risk of maternal complications is very low. Transient fluid leakage occurs in about 1% of mothers, and maternal infections are extremely rare. The risk of primary concern is fetal loss. Amniocentesis increases the risk of fetal loss by no more than 0.5% above the background risk at 15 to 17 weeks post-LMP (i.e., if the risk of pregnancy loss after 17 weeks were 3% in mothers who did not have an amniocentesis,

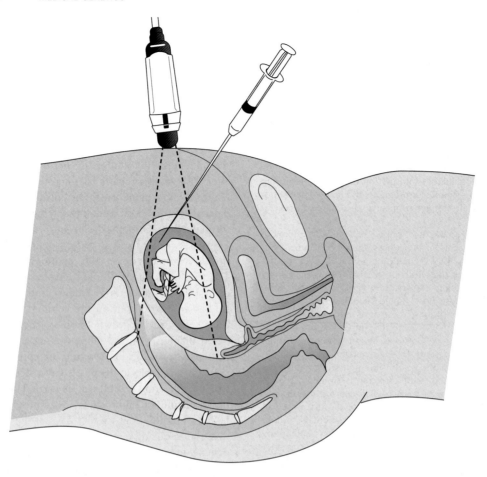

**FIGURE 13-4**
A schematic illustration of an amniocentesis, in which 20 to 30 mL of amniotic fluid is withdrawn transabdominally (with ultrasound guidance), usually at 15 to 17 weeks' gestation.

---

**BOX 13-5 Indications for Prenatal Diagnosis by Amniocentesis**

Maternal age >35 years
Previous child with chromosome abnormality
History of structural chromosome abnormality in one parent
Family history of genetic defect that is diagnosable by biochemical or DNA analysis
Increased risk of neural tube defect due to positive family history

---

the risk would increase to 3.5% in those who had the procedure). One must weigh the risk of fetal loss against the probability that the fetus is affected with a diagnosable condition (Clinical Commentary 13-4).

Although amniocentesis provides highly accurate results, chromosomal mosaicism can lead to misdiagnosis. Most apparent mosaicism is caused by the generation of an extra chromosome during in vitro cell culture and is labeled as **pseudomosaicism**. This can be distinguished easily from true mosaicism if techniques are used in which all cells in a colony are the descendants of a single fetal cell. If only some cells in the colony have the extra chromosome, it is assumed that pseudomosaicism exists. If, however, consistent aneuploidy is visualized in all cells of multiple colonies, then true fetal mosaicism is diagnosed. Further confirmation of fetal mosaicism (which is generally a rare condition) can be obtained by fetal blood sampling, as described later.

Some centers have evaluated amniocentesis performed earlier in pregnancy, at about 12 to 14 weeks post-LMP. Because less amniotic fluid is present at this time, the risk of fetal loss or injury may be higher. A number of large-scale evaluations have indicated significantly higher rates of fetal loss for **early amniocentesis**, and some studies have shown increased rates of specific congenital anomalies (club foot in particular).

Amniocentesis, the withdrawal of amniotic fluid during pregnancy, is performed at about 16 weeks post-LMP and is used to diagnose many genetic diseases. The amniotic α-fetoprotein level is elevated when the fetus has a neural tube defect and provides a reliable prenatal test for this condition. The rate of fetal loss attributable to this procedure is approximately 1/200 above the background risk level. Amniocentesis can also be performed earlier in the pregnancy; some studies indicate an elevated rate of fetal loss after early amniocentesis.

## CLINICAL COMMENTARY 13-4
### *The Amniocentesis Decision*

When a quadruple screen identifies a risk of fetal abnormality greater than 1 in 500, it is common for a pregnant woman to consider the possibility of amniocentesis. Several factors enter into the decision-making process. First is quantitative risk estimate, determined by the screening result, for Down syndrome and other chromosomal disorders. A second factor is the risk of fetal loss from the procedure (about 0.5% above the background risk). A third issue is the expense of an amniocentesis with ultrasound and cytogenetic analysis, which ordinarily costs about $2000. These factors must be weighed in terms of their relative costs and benefits for the woman and her family.

As this decision is explored in greater depth, other considerations often arise. If a woman has had previous miscarriages, the 0.5% risk of fetal loss may be weighed more heavily. In addition, the seriousness of bearing a child with disabilities may be perceived differently from family to family. Some couples are uncomfortable with the amount of time that elapses before test results are available (usually 10–12 days). This discomfort should be acknowledged and validated. The possibility of an ambiguous result (e.g., mosaicism) also deserves discussion. Finally, it is important for the clinician to specify that an amniocentesis typically detects only a specific class of disorders (i.e., chromosome abnormalities and neural tube defects) and not the entire range of birth defects and genetic disorders.

## Chorionic Villus Sampling

Chorionic villus sampling (CVS) is performed by aspirating fetal trophoblastic tissue (chorionic villi) by either a transcervical or transabdominal approach (Fig. 13-5). Because it is usually performed at 10 to 11 weeks post-LMP, CVS has the advantage of providing a diagnosis much earlier in pregnancy than second-trimester amniocentesis. This may be important for couples who consider pregnancy termination an option.

Cell culture (as in amniocentesis) and direct analysis from rapidly dividing trophoblasts can provide material for cytogenetic analysis. When chorionic villi are successfully obtained, CVS provides diagnostic results in more than 99% of cases. **Confined placental mosaicism** (mosaicism in the placenta but not in the fetus itself) is seen in about 1%

to 2% of cases in which direct analysis of villus material is performed. This can confuse the diagnosis, because the mosaicism observed in placental (villus) material is not actually present in the fetus. This problem can usually be resolved by a follow-up amniocentesis. A disadvantage of CVS is that amniotic fluid AFP cannot be measured. Women who undergo CVS may have their serum AFP level measured at 15 to 16 weeks after LMP as a screen for NTDs.

CVS, like amniocentesis, is generally a safe procedure. Several collaborative studies revealed a post-CVS fetal loss rate of approximately 1% to 1.5% above the background rate, compared with 0.5% above background for amniocentesis. Factors that increase the risk of fetal loss include a lack of experience with the procedure and an increase in the

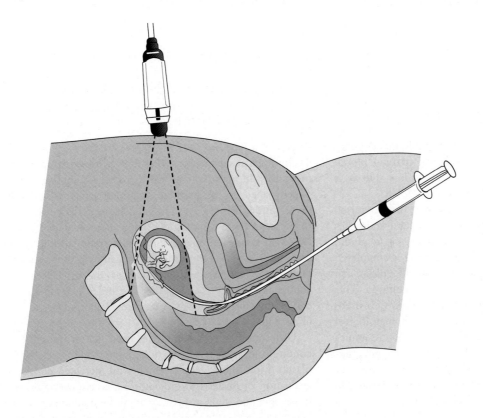

**FIGURE 13-5**
A schematic illustration of a transcervical chorionic villus sampling (CVS) procedure. With ultrasound guidance, a catheter is inserted, and several milligrams of villus tissue are aspirated.

number of transcervical passages used to obtain the villus sample. In experienced hands, transcervical and transabdominal procedures appear to entail similar risk levels.

Some studies have indicated that CVS can increase the risk of limb deficiencies. Although other investigations have not corroborated this result, the apparent association has been of concern because the proposed mechanism (vascular insult leading to hypoperfusion of the limb) is biologically plausible. The risk is greatest when CVS is performed earlier than 10 weeks post-LMP and decreases to no more than one in several thousand when the procedure is performed at 10 to 11 weeks post-LMP. Accordingly, many professionals now recommend against performing CVS before 10 weeks after LMP.

> CVS is performed earlier than amniocentesis (at 10 to 11 weeks post-LMP), using either a transcervical or a transabdominal approach. The risk of fetal loss attributable to CVS is approximately 1% to 1.5%. Confined placental mosaicism can confuse the diagnosis. There is some evidence that CVS can increase the risk of limb deficiencies; this risk is greatest when the procedure is performed before 10 weeks post-LMP.

Inborn errors of metabolism (see Chapter 7), which are usually autosomal or X-linked recessive diseases, can be diagnosed prenatally by amniocentesis or CVS if the specific metabolic defect is expressed in amniocytes or trophoblastic tissue. They can also be diagnosed prenatally by DNA-based methods if the disease-causing mutation can be identified. Table 13-6 lists selected inborn errors of metabolism and single-gene disorders for which amniocentesis or CVS is available. A comprehensive summary of conditions that may be prenatally diagnosed is provided by Weaver (1999).

### Other Methods of Fetal Tissue Sampling

**Cordocentesis**, or **percutaneous umbilical blood sampling** (PUBS), has become the preferred method to access fetal blood. PUBS is usually carried out after the 16th week of gestation and is accomplished by ultrasound-guided puncture of the umbilical cord and withdrawal of fetal blood. The fetal loss rate attributable to PUBS is low, but it is slightly higher than that of amniocentesis or CVS.

There are three primary applications of PUBS. It is used for cytogenetic analysis of fetuses with structural anomalies detected by ultrasound when rapid diagnosis is required. Cytogenetic analysis from fetal blood sampling is completed in 2 to 3 days, whereas diagnosis after amniocentesis can require 10 to 12 days if amniocytes must be cultured. This time difference can be critical in the later stages of a pregnancy. A second application is diagnosis of hematological diseases that are analyzed most effectively in blood samples or diagnosis of immunologic disorders such as chronic granulomatous disease (see Chapter 9). PUBS is also used to make a rapid distinction between true fetal mosaicism and false mosaicism caused by maternal contamination of an amniotic fluid sample.

**TABLE 13-6**

### Selected Inborn Errors of Metabolism that Are Diagnosable through Amniocentesis and/or Chorionic Villus Sampling

| Disease | Measurable Enzyme |
| --- | --- |
| **Disorders of Amino Acid or Organic Acid Metabolism** | |
| Maple syrup urine disease | Branched-chain ketoacid decarboxylase |
| Methylmalonic acidemia | Methylmalonic coenzyme A mutase |
| Multiple carboxylase deficiency | Biotin responsive carboxylase |
| **Disorders of Carbohydrate Metabolism** | |
| Glycogen storage disease, type 2 | α-Glucosidase |
| Galactosemia | Galactose-1-uridyl transferase |
| **Disorders of Lysosomal Enzymes** | |
| Gangliosidosis (all types) | β-Galactosidase |
| Mucopolysaccharidosis (all types) | Disease-specific enzyme (see Chapter 7) |
| Tay–Sachs disease | Hexosaminidase A |
| **Disorders of Purine and Pyrimidine Metabolism** | |
| Lesch–Nyhan syndrome | Hypoxanthine-guanine phosphoribosyl transferase |
| **Disorders of Peroxisomal Metabolism** | |
| Zellweger syndrome | Long-chain fatty acids |

> Percutaneous umbilical blood sampling (PUBS, or cordocentesis) is a method of direct sampling of fetal blood and is used to obtain a sample for rapid cytogenetic or hematological analysis or for confirmation of mosaicism.

### Ultrasonography

Technological advances in real-time **ultrasonography** have made this an important tool in prenatal diagnosis. A transducer placed on the mother's abdomen sends pulsed sound waves through the fetus. The fetal tissue reflects the waves in patterns corresponding to tissue density. The reflected waves are displayed on a monitor, allowing real-time visualization of the fetus. Ultrasonography can help to detect many fetal malformations, and it enhances the effectiveness of amniocentesis, CVS, and PUBS. Box 13-6 lists some of the congenital malformations that are diagnosable by fetal ultrasound.

Ultrasonography is sometimes used to test for a specific condition in an at-risk fetus (e.g., a short-limb skeletal dysplasia). More often, fetal anomalies are detected during the evaluation of obstetrical indicators such as uncertain gestational age, poor fetal growth, or amniotic fluid abnormalities. Second-trimester ultrasound screening has become routine in

## BOX 13-6 Selected Disorders Diagnosed by Ultrasound in the Second Trimester*

### SYMPTOM COMPLEX
Hydrops
Oligohydramnios
Polyhydramnios
Intrauterine growth retardation

### CENTRAL NERVOUS SYSTEM
Anencephaly
Encephalocele
Holoprosencephaly
Hydrocephalus

### CHEST
Congenital heart disease
Diaphragmatic hernia

### ABDOMEN, PELVIS
Gastrointestinal atresias
Gastroschisis
Omphalocele
Renal agenesis
Cystic kidneys
Hydronephrosis

### SKELETAL SYSTEM
Limb reduction defects
Many chondrodystrophies, including thanatophoric dysplasia and osteogenesis imperfecta

### CRANIOFACIAL
Cleft lip

*Detection rate varies by condition.

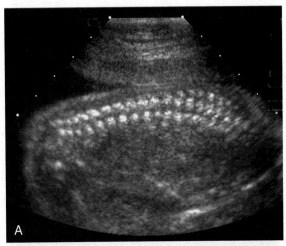

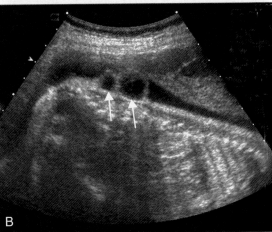

**FIGURE 13-6**
A, Photograph of an ultrasound result, revealing a fetus with a normal spinal column. B, Ultrasound result for a fetus with a meningomyelocele, visible as fluid-filled sacs *(arrow)* located toward the base of the spinal column.

developed countries. Studies of ultrasound screening suggest that sensitivity for the detection of most major congenital malformations ranges from 30% to 50%. Specificity, however, approaches 99%.

The sensitivity of ultrasonography is higher for some congenital malformations. In particular, ultrasound can detect virtually all fetuses with anencephaly and 85% to 90% of those with spina bifida (Fig. 13-6). It also sometimes identifies a fetus with a chromosome abnormality by detecting a congenital malformation, intrauterine growth retardation, hydrops (abnormal accumulation of fluid in the fetus), or an alteration of the amniotic fluid volume.

Ultrasonography is the technique used most commonly for fetal visualization, but other techniques are also used. Radiography is still used occasionally, for example, to evaluate a fetus for skeletal defects. Magnetic resonance imaging (MRI) offers much greater resolution than ultrasonography and is becoming more widely available for prenatal screening.

> Prenatal diagnosis includes invasive techniques designed to analyze fetal tissue (CVS, amniocentesis, PUBS) and noninvasive procedures that visualize the fetus (ultrasonography, MRI).

## Maternal Serum Screening in the First and Second Trimesters

Soon after the link between elevated amniotic fluid AFP and NTDs was recognized, an association between increased levels of **maternal serum AFP** (MSAFP) and NTDs was identified. AFP diffuses across the fetal membranes into the mother's serum, so MSAFP levels are correlated with amniotic fluid AFP levels. Thus, it is possible to measure amniotic fluid AFP noninvasively by obtaining a maternal blood sample at 15 to 17 weeks post-LMP.

Because 90% to 95% of NTD births occur in the absence of a family history of the condition, a safe, noninvasive population screening procedure for NTDs is highly desirable. However, there is considerable overlap of MSAFP levels in women carrying a fetus with an NTD and those carrying an unaffected fetus (Fig. 13-7). Thus, the issues of sensitivity and specificity must be considered. Typically, an MSAFP level is considered to be elevated if it is 2 to 2.5 times higher than the normal median level (adjustments for maternal weight, presence of diabetes mellitus, and ancestry are included in these calculations). Approximately 1% to 2% of pregnant women exhibit MSAFP levels above this cutoff level. After adjusting for advanced gestational age, fetal demise, and

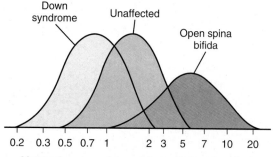

**FIGURE 13-7**

Maternal serum α-fetoprotein (MSAFP) levels in mothers carrying normal fetuses and in mothers carrying fetuses with Down syndrome and open spina bifida. MSAFP is somewhat lowered when the fetus has Down syndrome, and it is substantially elevated when the fetus has an open spina bifida.

*(From Milunsky A: Genetic Disorders and the Fetus: Diagnosis, Prevention, and Treatment, 4th ed. Baltimore: Johns Hopkins University Press, 1998.)*

presence of twins, about 1 in 15 of these women has an elevated amniotic fluid AFP. The positive predictive value of the MSAFP screening test is thus rather low, approximately 6% (1/15). However, the sensitivity of the test is fairly high: MSAFP screening identifies approximately 90% of anencephaly cases and about 80% of open spina bifida cases. Although this sensitivity level is lower than that of amniotic fluid AFP testing, MSAFP measurement poses no risk of fetal loss and serves as an effective screening measure. Women who have an elevated MSAFP may choose to undergo diagnostic amniocentesis to determine whether they are in fact carrying a fetus with an NTD.

An association was found in the 1990s between *low* MSAFP and the presence of a fetus with Down syndrome. Previously, population screening for Down syndrome consisted of amniocentesis for women older than 35 years. Although highly accurate, this screening strategy has a sensitivity of only 20%: because the great majority of births occur in women younger than 35 years, only about 20% of all babies with trisomy 21 are born to mothers older than age 35. MSAFP measurement has expanded the option of population screening for Down syndrome.

MSAFP levels overlap considerably in normal and Down syndrome pregnancies. The risk for Down syndrome in women younger than 35 years increases by a factor of 3 to 4 when the adjusted MSAFP value is lower than 0.5 multiples of the normal population median (see Fig. 13-7). In deriving a risk estimate, complex formulas take into account the mother's weight, age, and MSAFP level. A woman who is 25 years of age ordinarily has a risk of about 1/1250 for producing a fetus with Down syndrome, but if she has a weight-adjusted MSAFP of 0.35 multiples of the median, her risk increases to 1/171. This risk is higher than that of a 35-year-old woman in the general population. Most screening programs use a risk factor of 1/380 (equivalent to the average risk for a 35-year-old woman to produce a newborn with Down syndrome) as an indication for subsequent diagnostic evaluation by amniocentesis.

The accuracy of Down syndrome screening can be increased by measuring the serum levels of unconjugated estriol, human chorionic gonadotropin, and inhibin-A in addition to MSAFP (the **quadruple screen**). Although MSAFP alone identifies only about 40% of Down syndrome pregnancies, the four indicators together can identify approximately 80% (with a false-positive rate of 5%). The quadruple screen can also detect most cases of trisomy 18.

First-trimester maternal serum screening (at 10 to 13 weeks) for Down syndrome is being used increasingly in the United States and Europe. Three of the most useful measurements are the free β subunit of human chorionic gonadotropin (FβhCG), pregnancy-associated plasma protein A (PAPP-A), and an ultrasound assessment of nuchal translucency (NT, the abnormal accumulation of fluid behind the neck of a fetus). Measurement of these three quantities in the first trimester enables detection of 80% to 85% of Down syndrome cases (with a false-positive rate of 5%, or 95% specificity). Combining first- and second-trimester screening increases the sensitivity of Down syndrome detection to approximately 90%, with 95% specificity. Measurement of FβhCG and PAPP-A is also useful for detection of trisomy 13 and trisomy 18 in the first trimester. These screening results may be combined with CVS or amniocentesis to provide a more precise diagnostic test.

MSAFP provides a screening approach that increases the prenatal detection of fetuses with various abnormalities, including NTDs, trisomy 18, and Down syndrome. This noninvasive procedure entails virtually no risk, but its sensitivity and specificity for detecting NTDs are lower than those of amniotic AFP diagnosis. Use of additional markers (e.g., the quadruple screen) in the second trimester increases sensitivity for detecting Down syndrome. Screening of maternal serum for Down syndrome, trisomy 13, and trisomy 18 is now possible in the first trimester.

### Preimplantation Genetic Diagnosis

Several new approaches to prenatal diagnosis are now in the testing or early application stages. These include preimplantation genetic diagnosis (PGD) at three different stages: polar body, blastomere, and blastocyst. Research is also being done on genetic testing of fetal DNA obtained from the mother's circulation.

The most common type of PGD is carried out on a blastomere obtained in the course of **in vitro fertilization**. Diagnosis is begun 3 days after fertilization, when the embryo contains six or eight cells. One or two cells are removed from the embryo for diagnosis (this does not harm it). FISH analysis (see Chapter 6) can be used to diagnose aneuploidy. Also, DNA from the cell can be amplified using PCR, permitting the diagnosis of single-gene diseases. If the embryo is morphologically normal and neither the disease-causing mutation nor aneuploidy is detected, the embryo is implanted into the mother's uterus. Testing protocols have been developed for dozens of genetic diseases (e.g., cystic fibrosis, Tay–Sachs disease, β-thalassemia, myotonic dystrophy, Huntington disease, Duchenne muscular dystrophy), and more than 1000 normal babies have been born after blastomere diagnosis.

An occasional problem with blastomere PGD is that one of the two alleles of a locus may be undetectable, which could cause a heterozygote to appear as a homozygote. This phenomenon, termed "allelic dropout," occurs because of partial PCR amplification failure when using DNA from only a single cell. This has led to misdiagnosis in a small number of cases, and several methods are used to increase accuracy. For example, highly heterozygous STRPs closely linked to the disease-causing locus can be tested as part of the PCR analysis. If only one of the parents' STRP alleles can be observed in the blastomere's DNA, it is likely that allelic dropout has also occurred for the disease-causing locus. The testing of two cells, rather than one, helps to avoid allelic dropout.

PGD can also be carried out at the 100-cell blastocyst stage, using cells from the trophoectoderm of the blastocyst. This procedure has the advantage that a larger collection of cells is analyzed, helping to avoid allelic dropout. A disadvantage is that extraembryonic tissue (the trophoectoderm) is diagnosed, rather than the embryo itself.

**Polar body diagnosis** involves an examination of the first or second polar body produced along with the ovum (see Chapter 2). The polar body's DNA is tested to determine whether it contains a disease-causing mutation. If it does, it is assumed that the egg does not contain the mutation. This egg is then fertilized and implanted using the usual in vitro techniques. Because only the polar body is examined, paternal mutations cannot be evaluated. Polar body diagnosis is thus most useful when only the mother is at risk for transmitting a disease-causing mutation or when testing for aneuploidy (because most aneuploidies are contributed by the mother [see Chapter 6]).

PGD is most commonly used by couples who have resorted to in vitro fertilization and wish to test for diagnosable genetic conditions. It can also be useful for couples who want prenatal diagnosis but would not consider a pregnancy termination. PGD, however, is costly and technically challenging, and its availability is still limited.

> Preimplantation genetic diagnosis can be carried out on polar bodies, blastomeres, or blastocyst cells, on which PCR analysis and/or FISH is performed. Diagnosis of genetic conditions permits implantation of only unaffected embryos and avoids the issue of pregnancy termination.

### Analysis of Fetal DNA in Maternal Circulation

During pregnancy, a small number of fetal cells cross the placental barrier to enter the mother's circulation. Some of these fetal cells are nucleated red blood cells, which are otherwise rare in the adult circulation. These cells can be isolated as early as 6 to 8 weeks post-LMP and can be identified by cell-sorting techniques. Further specificity for fetal cells can be achieved by testing cells for surface proteins specific to the fetus. FISH analysis of these cells has been used to test for fetal conditions such as trisomies 13, 18, and 21. PCR has been used to test for a limited number of single-gene disorders, although this remains a challenge because of the difficulty of sorting pure populations of fetal cells. The major advantage of this approach is that it requires only a blood sample from the mother and thus poses no risk of fetal loss. Its accuracy and feasibility are being evaluated. Cell-free fetal DNA is also present in the mother's circulation and has been used to identify the sex of the fetus and its Rh blood type (especially important if the mother is Rh-negative and the fetus may be Rh-positive; see Chapter 9).

> Fetal cells or cell-free DNA that enter the maternal circulation can be isolated and evaluated for mutations using PCR or FISH. This experimental procedure entails no risk of fetal loss.

### FETAL TREATMENT

A potential goal of prenatal diagnosis is treatment of the affected fetus. Although this is not currently possible for most conditions, some examples can be cited. Many of these procedures are experimental.

Two of the best-established forms of in utero intervention are treatment for rare inborn errors of metabolism and treatment for hormone deficiencies. An important example of a treatable biochemical disorder is biotin-responsive multiple carboxylase deficiency, an autosomal recessive condition that can be diagnosed by amniocentesis. In one case report, oral administration of biotin to the mother was initiated at 23 weeks of pregnancy and resulted in the birth of a normal baby.

CAH is a second example of a condition for which in utero treatment has been successful after prenatal diagnosis. Because of excessive androgen secretion by the enlarged fetal adrenal glands, female fetuses with CAH become masculinized. Administration of dexamethasone to the mother beginning at 10 weeks post-LMP diminishes or prevents this masculinization.

Surgical treatment of fetuses, primarily for conditions involving urinary tract obstruction, has met with moderate success. Surgical correction of diaphragmatic hernia at 20 weeks' gestation has also been attempted, but results have been discouraging and this approach has been abandoned. Surgical closure of myelomeningocele (spina bifida) has been performed in more than 200 cases, and there is evidence that the procedure helps to restore the normal flow of cerebrospinal fluid. Clinical trials are under way to determine the effectiveness of this procedure. Some success has been achieved in transplantation of hematopoietic stem cells into fetuses with X-linked severe combined immune deficiency (see Chapter 9).

### GENE THERAPY

As we have seen, the identification of disease-causing genes provides opportunities for improved understanding and diagnosis of many diseases. The identification of these genes also leads to the possibility of genetic alteration of the cells of affected persons (**gene therapy**). Although gene therapy is still in its infancy and has only begun to affect the lives of patients, its potential for curing genetic diseases has excited a great deal of interest in both professional and lay circles. As of the end of 2008, nearly 1500 gene therapy protocols involving more than 100 different genes have been approved for experimental trials (see Table 13-7 for some examples). In this section we review gene therapy techniques and discuss their application in the treatment of disease.

## Somatic Cell Therapy

**Somatic cell gene therapy**, which has been the focus of gene therapy research in humans, consists of the alteration of genes in human somatic cells to treat a specific disorder. The patient's cells are extracted and manipulated outside the body (*ex vivo* therapy), or in some cases the cells are treated while they are in the body (*in vivo* therapy).

Some types of somatic cells are more amenable to gene therapy than others. Good candidates should be easily accessible and should have a long life span in the body. Proliferating cells are preferred for some gene-delivery systems, because the vector carrying the gene can then integrate into the cells' replicating DNA. The bone marrow stem cell meets all of these qualifications and thus has been a prime candidate for somatic therapy. Although these cells are difficult to manipulate and to isolate from bone marrow (the great majority of bone marrow cells are not stem cells), they have now been successfully isolated and genetically altered in several gene therapy treatments. Many other cell types have also been investigated as potential targets, including skin fibroblasts, muscle cells, vascular endothelial cells, hepatocytes, and lymphocytes. A disadvantage of using such cells is that their life span may be relatively short. Thus, therapy using them can require repeated treatment and administration of genetically altered cells.

## Gene Replacement Therapy

Most current gene therapy techniques involve replacing a missing gene product by inserting a normal gene into somatic cells. This approach is best suited to correcting loss-of-function mutations that result in a nonfunctional or missing gene product; inserting the normal gene supplies the missing product. Even a partially effective gene therapy strategy, producing perhaps 5% to 20% of the normal amount of the gene product, might provide significant health benefits.

There are many techniques for introducing genes into cells, but viruses, having naturally evolved clever strategies for inserting their genes into cells, are the most commonly used gene-therapy vectors. In the following paragraphs viral vectors are discussed first, followed by a discussion of some potentially effective nonviral delivery systems.

**TABLE 13-7**
## Partial List of Diseases for Which Somatic Cell Gene Therapy Protocols Are Being Tested

| Disease | Target Cell | Product of Inserted Gene |
| --- | --- | --- |
| Adenosine deaminase deficiency | Circulating lymphocytes, bone marrow stem cells | Adenosine deaminase |
| X-linked severe combined immunodeficiency (SCID) | Bone marrow stem cells | Gamma subunit of interleukin receptors |
| Hemophilia B | Hepatocytes, skin fibroblasts | Factor IX |
| Retinitis pigmentosa | Postmitotic retinal cells | Retinal pigment epithelium-specific protein |
| Epidermolysis bullosa | Skin stem cells | Type VII collagen |
| Familial hypercholesterolemia | Hepatocytes | Low-density lipoprotein receptor |
| Cystic fibrosis | Airway epithelial cells | Cystic fibrosis transmembrane conductance regulator (CFTR) |
| Malignant melanoma | Melanoma tumor cells | B7 costimulatory molecule |
| Duchenne muscular dystrophy | Myoblasts | Dystrophin; also antisense therapy to skip mutated exon |
| Gaucher disease | Macrophages | Glucocerebrosidase |
| Lung cancer | Lung cancer cells | Normal p53 |
| Brain tumors | Brain cells | Herpes thymidine kinase |
| Acquired immunodeficiency syndrome (AIDS) | Helper T lymphocytes | Dominant negative retrovirus mutations |
| Ischemic heart disease | Cardiomyocytes | Vascular endothelial growth factor, fibroblast growth factor |

## Retroviral Vectors

Retroviruses, a form of RNA virus, can insert copies of their genomes into the nuclei of host cells after reverse-transcribing their viral RNA into double-stranded DNA (see Chapter 11). The insertion of foreign DNA into a host cell via a viral vector is termed **transduction**. Retroviruses transduce host cells with a high degree of efficiency, and they seldom provoke immune responses, making them a reasonable choice as a gene-delivery vector (Fig. 13-8). Recombinant DNA techniques are used to create replication-defective retroviruses in which the three retroviral protein-coding genes are replaced with a normal copy of a human gene and a promoter element (the "insert," which can be as large as 8–12 kb in a retrovirus). The modified retroviruses are then incubated with the patient's somatic cells (e.g., bone marrow stem cells, lymphocytes) so that the retrovirus transduces the normal human gene into the DNA of the host cells. Ideally, the inserted gene will then encode a normal gene product in the patient's somatic cells. This type of protocol has been used experimentally with many diseases, including forms of severe combined immune deficiency (Clinical Commentary 13-5).

Although retroviruses offer the advantages of stable and efficient integration into the genome, they also present specific disadvantages. Because it integrates preferentially near promoter sequences, the retrovirus could locate near a proto-oncogene, activating it and thus causing tumor formation. Most types of retrovirus can enter the nucleus only when its membrane dissolves during cell division, so they can transduce only dividing cells and are ineffective in non-dividing or slowly dividing cells (e.g., neurons). Although this attribute is typically a disadvantage, it can be useful when the goal of therapy is to target only dividing cells and to avoid nondividing cells (e.g., in the treatment of a brain tumor, where tumor cells are dividing but nearby healthy neurons are not).

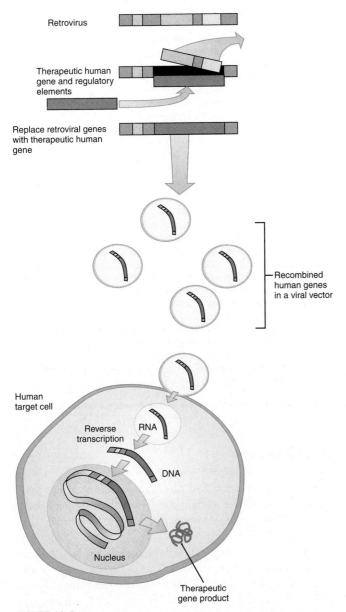

**FIGURE 13-8**

Gene therapy using a retroviral vector. The retrovirus is prevented from replicating by removal of most of its genome, and a normal human gene is inserted into the retrovirus. Incubated with human somatic cells allows the retrovirus to insert copies of the normal human gene into the cell. Once integrated into the cell's DNA, the inserted gene produces a normal gene product.

## Adenoviral Vectors

Because of the inability of most retroviruses to transduce nondividing cells, other delivery systems have been explored that are not limited in this way. An important example is the **adenovirus**, a double-stranded DNA virus that is often used in vaccine preparations. In addition to its ability to transduce nondividing cells, the adenoviral vector can now be designed to accept inserts of approximately 36 kb in size. Adenoviruses do not integrate into the host cell's DNA, which provides the advantage that they will not activate a proto-oncogene or otherwise disturb the genome. However, lack of integration is also a disadvantage because adenoviruses are eventually inactivated. This often results in transient gene expression (though long-term expression is sometimes achieved) and can require readministration of the vector. Because only part of the adenovirus genome is typically removed, the vector often provokes an immune response (e.g., inflammatory responses in the airways of cystic fibrosis patients in whom adenoviruses were used to introduce normal copies of the *CFTR* gene into airway epithelial cells). This problem increases with repeated introduction of the adenovirus, which stimulates further immune response to the foreign protein. Current research is focusing on "gutless" adenoviruses, in which nearly all of the viral genome is removed to reduce the immune response and increase the potential size of the insert.

## Adeno-associated Viral Vectors

Adeno-associated viruses (AAVs) are a type of parvovirus that requires the presence of adenoviruses for their normal

## CLINICAL COMMENTARY 13-5
### Gene Therapy and Severe Combined Immunodeficiency

Gene therapy has been attempted for several forms of severe combined immunodeficiency (SCID), including adenosine deaminase deficiency SCID (ADA-SCID) and X-linked SCID. ADA, which is produced primarily in lymphoid tissues, is an important component of the purine salvage pathway. ADA deficiency, an autosomal recessive disorder that accounts for about 15% of SCID cases, results in the abnormal accumulation of purine metabolites toxic to T lymphocytes. Subsequently, B lymphocytes are also reduced in function and number. The resulting SCID is usually fatal by the age of 2 years if untreated.

The preferred treatment for ADA-SCID is bone marrow transplantation. However, complications of bone marrow transplantation increase patient morbidity and are sometimes fatal. In addition, major histocompatibility complex (MHC)-compatible sibling donors are available for less than 30% of ADA-SCID patients. Patients may be treated with polyethylene glycol (PEG)-conjugated ADA (administered once or twice weekly by intramuscular injection), but the response to this treatment is variable, and some patients develop antibodies against PEG-ADA.

Because it is a systemic disorder caused by an enzyme deficiency, ADA-SCID represents a good candidate for gene replacement therapy. Ideally, proliferating bone marrow stem cells would be modified by retroviral vectors containing the normal ADA gene, resulting in a permanent cure for the disorder. Because of difficulties in dealing with bone marrow stem cells, gene therapy for ADA-SCID was initiated instead in 1990 by retroviral insertion of ADA genes into lymphocytes that were extracted from patients. After retroviral insertion, the lymphocytes were injected back into the patients' peripheral circulation. This was the first application of gene therapy to an inherited human disorder.

Lymphocyte gene therapy has been applied in more than a dozen patients with ADA deficiency. In some patients, ADA levels increased, T lymphocyte counts improved, and the number of infections decreased. Because of the limited life span of T lymphocytes, these patients were reinjected with modified T cells once every several months; however, the treated lymphocytes have displayed surprising longevity, surviving in the circulation for well over 1 year in some cases. These patients were also typically treated with PEG-ADA, so it has been somewhat difficult to establish the efficacy of gene therapy.

More recently, ADA-SCID has been treated by retroviral insertion of the ADA gene into bone marrow stem cells, rather than lymphocytes. This treatment has resulted in long-term increases in B cell and T cell counts (up to 9 years) and normal immune function in 11 treated patients.

X-linked SCID results from mutations in the gene, *SCIDX1*, that encodes subunits of the γ chain found in six different cytokine receptors (those of interleukins 2, 4, 7, 9, 15, and 21; see Chapter 9). Lacking these receptors, T cells and natural killer cells cannot receive the signals they need for normal maturation. The T cell deficiency in turn produces a deficiency of normal B cells, resulting in SCID. As with ADA deficiency, this disorder can be treated with bone marrow transplantation if an MHC-compatible donor is available. Without a bone marrow transplant, the disease is fatal early in childhood.

In 1999, retroviral therapy was initiated to introduce *SCIDX1* into patients' bone marrow stem cells. Less than 1% of bone marrow stem cells were effectively transduced with the therapeutic gene. However, the transduced cells enjoyed a selective growth advantage over other bone marrow stem cells because the inserted gene increased the cytokine signaling needed for normal cell function. In most treated patients, the number of natural killer cells, T cells, and B cells increased to near-normal levels, with sustained resistance to infections continuing for years after therapy.

These positive outcomes in most ADA and X-linked SCID patients have been widely heralded as the first successful uses of somatic cell gene therapy in the treatment of an inherited disease. However, five of the X-linked SCID patients developed leukemia-like disease (clonal T cell proliferation) as a result of random insertion of the retroviral vector in or near *LOM2*, a proto-oncogene that is activated in about half of all cases of acute lymphocytic leukemia. This was fatal in one patient, but the others were treated successfully by chemotherapy and continued to benefit from gene therapy. Although the cause of T cell proliferation in these patients remains somewhat unclear, there is evidence that a specific interaction occurs between the inserted γ-chain gene and *LOM2* to activate the proto-oncogene. This interaction might explain why, among the many different clinical trials involving retroviral transfer of genes to bone marrow stem cells, only this trial has resulted in cancer.

These examples illustrate some of the promise, as well as some of the perils, of somatic cell gene therapy. Clearly, gene therapy poses hazards that must be closely monitored. However, these protocols can lead to effective treatment of otherwise lethal diseases, and they provide invaluable information for the development of gene therapy protocols for other genetic diseases.

---

replication (hence the term *adeno-associated*). Like adenoviruses, AAVs are DNA viruses that can transduce nondividing cells. In addition, they elicit much less immune response than do adenoviruses and have little, if any, pathogenic effect. They are also capable of sustaining protracted therapeutic expression (months to years). These vectors, however, can accept a DNA insert of only about 4.5 kb. (In some cases, this problem can be circumvented by dividing the insert into two parts, placing each part in a vector, and designing the mRNA products to reassemble.) Because of their many useful properties, AAVs have become much more popular as a gene therapy vector during the past several years. They have been tested in clinical trials for the treatment of cystic fibrosis, hemophilia B, α₁-antitrypsin deficiency, Duchenne muscular dystrophy, Parkinson disease, Alzheimer disease, and many other disorders.

### Lentiviral Vectors

Lentiviral viruses are complex RNA retroviruses that, unlike simple retroviruses, can transduce nondividing cells through pores in the nuclear membrane (human immunodeficiency virus [HIV] is an example of a **lentivirus**). Like other retroviruses, lentiviruses can integrate stably into the genome, and they can accept reasonably large inserts (8 kb). Because they combine the desirable properties of stable integration and the ability to transduce nondividing cells, lentiviruses are currently the focus of much research and development.

### Challenges in Viral Gene Therapy

Although viral gene therapy holds considerable promise, there are several important challenges:

- *Transient and low-level expression.* The gene product may be expressed at subtherapeutic levels, often less than 1% of the normal amount. In part, this reflects the fact that only some of the target cells successfully incorporate the normal gene. In addition, random insertion of the virus into the host's genome can affect gene regulation (e.g., enhancer sequences required for normal expression levels are not present). Cells sometimes respond to foreign inserted DNA by methylating—and thus inactivating—it. For these reasons, transcription of the gene often ceases after a few weeks or months. It should be noted, however, that transient expression is sufficient, and even desirable, for some types of therapy, such as provoking an immune response against a tumor or generating new blood vessels (discussed later).
- *Difficulties in reaching or specifying target tissue.* Although some systemic disorders are relatively easy to target by modifying lymphocytes or bone marrow stem cells, others present formidable challenges. It may be difficult, for example, to target affected neurons responsible for central nervous system disorders. In addition, vectors must be modified so that they enter only the desired cell type.
- *Necessity for precise regulation of gene activity.* Accurate regulation of gene activity is not a concern for some diseases (e.g., a 50-fold overexpression of adenosine deaminase has no clinically significant effects). However, it is critical for diseases such as thalassemia, in which the number of α-globin and β-globin chains must be closely balanced (see Chapter 3). It is often difficult to achieve such precision using viral gene therapy.
- *Potential for insertional mutagenesis.* The unpredictable integration of a retroviral vector into the host's DNA can have undesired consequences, as discussed earlier. Although insertional mutagenesis appears to be a rare event, it has occurred in several patients (see Clinical Commentary 13-5).

Considerable research is being devoted to overcoming these and other problems. For example, the levels and permanence of gene expression are being increased by incorporating stronger promoter sequences in DNA inserts. Vectors are being modified to reduce immune responses and to increase target-cell specificity. Methods are being developed for targeted insertion of corrected DNA sequences. For example, proteins are engineered to bind to a specific mutated DNA sequence and to induce double-stranded DNA breaks followed by the insertion of a normal DNA sequence. With targeted insertion, the mutated DNA is corrected in situ, avoiding difficulties with random insertion of DNA and taking advantage of the native promoter and enhancer sequences in the host's genome.

> Viral vectors offer highly efficient transfer of therapeutic genes into somatic cells. However, they have several drawbacks, including low or transient expression of the gene product, limited insert size, generation of immune responses, difficulty in precise regulation, and, for some vectors, a lack of ability to transduce nondividing cells and the potential for oncogenesis.

## Nonviral Vectors

Although viral vectors provide the advantage of efficient gene transfer into cells, the disadvantages just listed have prompted researchers to investigate several types of nonviral vectors. One of the most extensively studied is the **liposome**, a fat body that can accept large DNA inserts. Liposomes sometimes fuse with cells, allowing the DNA insert to enter the cell. Because the liposome has no peptides, it does not elicit an immune response. Its primary disadvantage is that it lacks the transfer efficiency of viruses: most of the liposomes are degraded in the cytoplasm, and most of those that are not degraded are unable to enter the nucleus.

Surprisingly, it is possible to insert plasmids containing human DNA directly into cells without using any delivery vector at all. Although most "naked" DNA is repelled by the cell membrane, the DNA occasionally enters the cell, escapes degradation, and temporarily encodes proteins. Attempts are under way to use naked DNA as a vaccine that encodes a pathogenic protein against which the body mounts an immune response.

An intriguing development with potential for somatic cell therapy is the synthesis of **human artificial chromosomes**. Because these synthetically constructed chromosomes contain functional centromeres and telomeres, they should be able to integrate and replicate in human cell nuclei. Also, they are capable of accepting inserts as large as the entire 2.4 Mb Duchenne muscular dystrophy gene (*DMD*).

> Gene therapy using nonviral vectors, including liposomes and naked DNA, offers some advantages over viral vectors, but they currently lack the transfer efficiency of viral vectors.

### Gene-Blocking Therapies

Gene-replacement techniques are not effective in correcting gain-of-function or dominant negative mutations (e.g., Huntington disease, Marfan syndrome). To correct these conditions, the defective gene product must be blocked or disabled in some way. Although not as well developed as gene-replacement therapy methods, gene-blocking methods are being developed, and some show promise.

### Antisense Therapy

The principle behind **antisense therapy** is simple: an oligonucleotide is engineered whose DNA sequence is complementary to that of the messenger RNA (mRNA) sequence produced by a gain-of-function mutation. This antisense oligonucleotide binds to the abnormal mRNA, preventing its translation into a harmful protein (Fig. 13-9A). Antisense oligonucleotides can also be engineered to bind to double-stranded DNA containing the disease-causing mutation, creating a triple helix that cannot be transcribed into mRNA. A difficulty with this antisense therapy is that antisense oligonucleotides are often degraded before they can reach their target. Also, because of variation in the shape of the target DNA or RNA molecule, the antisense oligonucleotide might not be able to bind to its complementary sequence. Nevertheless, antisense therapy is being tested in a number

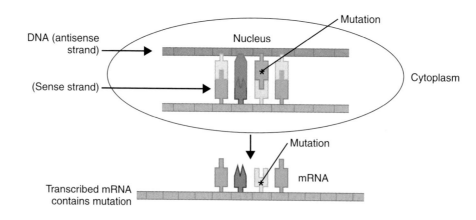

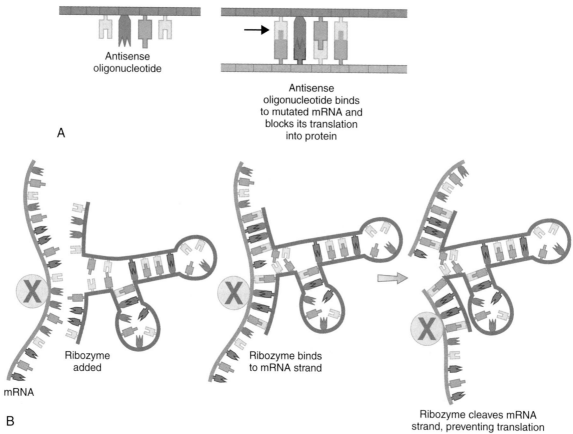

**FIGURE 13-9**
**A,** Gene therapy using an antisense technique. Binding of the abnormal mRNA by the antisense molecule prevents it from being translated into an abnormal protein. **B,** Gene therapy using a hammerhead ribozyme, which binds to a mutated mRNA, cleaving and eliminating it.

of experimental applications, including blocking *KRAS* onco-gene (see Chapter 9) expression in pancreatic and colorectal tumor cells.

## Ribozyme Therapy
**Ribozymes** are enzymatic RNA molecules, some of which can cleave mRNA. They can be engineered to disrupt specific mRNA sequences that contain a mutation, destroying them

before they can be translated into protein (see Fig. 13-9B). Ribozyme therapy is being tested, for example, as a method of countering the overexpression of epidermal growth factor receptor type 2, a feature of many breast tumors.

## RNA Interference
A third method of gene blocking involves **RNA interference** (RNAi; Fig. 13-10), a natural phenomenon that evolved to

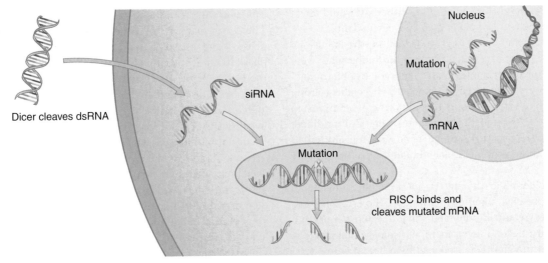

**FIGURE 13-10**
Gene-blocking therapy using RNA interference (RNAi). A dicer cleaves double-stranded RNA (dsRNA) into 20-bp single-stranded RNA fragments called *short interfering RNAs* (siRNAs). These fragments form a template that is recognized and bound by the RNA-induced silencing complex (RISC), which cleaves and destroys the complementary RNA strand. In RNAi, a dsRNA is engineered to produce siRNA strands that are complementary to a mutated mRNA, causing the RISC complex to destroy the mRNA.

defend cells against viral invasion. Because many viruses produce double-stranded RNA, cells of all multicellular organisms recognize this form of RNA and use an enzyme called *dicer* to digest it into small 20-bp pieces. These pieces are then used as a template to direct the destruction of any single-stranded RNA that has the same sequence as the double-stranded viral RNA (e.g., the single-stranded mRNA that the virus would use to encode viral proteins). By artificially synthesizing double-stranded RNA molecules that correspond to a disease-causing DNA sequence, RNAi can be induced to destroy the mRNA produced by the mutated sequence.

RNAi faces challenges similar to those of antisense and ribozyme therapy, such as degradation of the RNA molecule before it reaches the target. This difficulty is being overcome by inserting RNAi molecules into lentiviral and adeno-associated viral vectors. RNAi has shown some promise in reducing, for example, the number of transcripts produced by oncogenic *KRAS*, and it has also been shown to block transcripts of the *BCR-ABL* fusion gene, which causes chronic myelogenous leukemia (see Chapter 11). It is being tested for the treatment of age-related macular degeneration, asthma, hepatitis C, and Huntington disease.

> Gene-blocking techniques may be used to counter the effects of dominant-negative or gain-of-function mutations. They include the use of antisense molecules, RNA-cleaving ribozymes, and RNA interference.

### Gene Therapy for Noninherited Diseases

As indicated in Table 13-7, the application of gene therapy techniques is by no means limited to inherited diseases. Indeed, about two thirds of the gene therapy protocols now under way involve noninherited cancers, and approximately

10% involve acquired immunodeficiency syndrome (AIDS) therapy. For example, the *TP53* tumor suppressor gene, which is inactivated in about half of all cancers (see Chapter 11), has been inserted into lung tumors in an effort to halt tumor progression. As discussed in Chapter 9, some tumors evade immune system detection by discarding cell-surface molecules that are recognized by T cells. Liposomes containing DNA that encodes the B7 costimulatory molecule (see Chapter 9) have been introduced into malignant melanoma cells, resulting in B7 expression on the cell surface and subsequent cell destruction by cytotoxic T cells. In some cases, this has led to regression of the melanoma.

A variety of gene therapy approaches are being formulated to combat HIV. Most of these efforts are aimed at halting replication of the virus or preventing its spread to healthy cells. For example, a dominant-negative mutation introduced into HIV-infected T cells produces a protein that interferes with proteins produced by HIV, blocking their normal action. Trials are also in progress to reduce expression of CCR5, a chemokine coreceptor used by HIV to enter cells of the immune system (see Chapter 9).

Another example of gene therapy for a noninherited disease involves the treatment of coronary artery disease. Copies of the genes that encode members of the vascular endothelial growth factor (VEGF) and fibroblast growth factor (FGF) families have been injected into ischemic myocardium (using viral vectors or as naked DNA) with the hope of producing new coronary vessels.

### Germline Therapy

Somatic cell therapy consists of altering only specific somatic cells and thus differs little in principle from many other types of medical intervention (e.g., bone marrow transplantation). In contrast, **germline gene therapy** involves altering all cells of the body, including those that give rise to the gametes. Thus, this type of gene therapy would affect not only the patient but also his or her descendants.

Germline therapy was first achieved in the mouse in 1983, when copies of a human growth hormone gene were successfully introduced into mouse embryos by microinjection (the gene was inserted directly into the embryo using a very small needle). Among the minority of embryos in which the gene integrated, the gametes were also modified, and the human growth hormone gene was transmitted to future generations (the mice, incidentally, were abnormally large).

Although germline therapy is, in principle, possible in humans, it presents significant problems (Box 13-7). First, injected embryos usually die, and some develop tumors and malformations. Second, even in an autosomal dominant disorder, half of the embryos produced by a heterozygous parent are genetically normal. If it were possible to distinguish the genetically normal embryos (e.g., through preimplantation genetic diagnosis), it would be simpler to implant the normal embryos than to alter the abnormal ones. Finally, numerous ethical questions are associated with the permanent alteration of a human's genetic legacy. For these reasons, it appears unlikely that human germline therapy would be useful or desirable.

## Gene Therapy: A Perspective

The great majority of gene therapy protocols are still in phase I and phase II trials, but more than 30 are now in phase III clinical trials. The past several years have witnessed the first arguable successes of gene therapy, some of which have been discussed in this chapter (therapy for X-linked SCID and ADA deficiency; evidence of therapeutic effects in various cancers). However, success has been achieved thus far in only a relatively small number of persons.

Gene therapy is not without risk. In addition to the insertional mutagenesis potential already discussed, a young man with ornithine transcarbamylase deficiency (see Chapter 7) died as a result of an adverse immune reaction to an adenovirus vector. In addition, retroviral therapy resulted in leukemia-like disease in several X-linked SCID patients (see Clinical Commentary 13-5). It thus remains unclear whether gene therapy will provide a safe treatment or cure for a reasonable cost.

Despite these reservations, gene therapy research is providing many new insights of fundamental biological significance. As with many avenues of biomedical research, the potential of gene therapy research is considerable, and current progress suggests strongly that it may provide efficacious treatment of some important human diseases.

---

**BOX 13-7** **Germline Therapy, Genetic Enhancement, Human Cloning, and Embryonic Stem Cells: Controversial Issues in Medical Genetics**

For reasons outlined in the text, germline gene therapy is not being undertaken in humans. Nevertheless, germline gene therapy is in many ways technically easier to perform than is somatic cell therapy. Germline therapy also offers (in theory) the possibility of "genetic enhancement," the introduction of favorable genes into the embryo. However, a gene that is favorable in one environment may be unfavorable in another (e.g., the sickle-cell mutation, which is advantageous only for heterozygotes in a malarial environment). And, because of pleiotropy, the introduction of advantageous genes can have completely unintended consequences (e.g., a gene thought to enhance one characteristic could negatively affect another). For these reasons, and because germline therapy usually destroys the targeted embryo, neither germline therapy nor genetic enhancement is advocated by the scientific community.

Controversy also surrounds the prospect of cloning humans. Many mammalian species (e.g., sheep, pigs, cattle, goats, mice, cats, dogs) have been successfully cloned by introducing a diploid nucleus from an adult cell into an egg cell from which the original haploid nucleus was removed (a technique termed **somatic cell nuclear transfer**, or SCNT; see the figure that follows). The cell is manipulated so that all of its genes can be expressed (recall that most genes in a typical differentiated adult cell are transcriptionally silent). This procedure, when allowed to proceed through a full-term pregnancy, could likely be used to produce a human being (reproductive cloning). Some argue that human cloning offers childless couples the opportunity to produce children to whom they are biologically related or even to replace a child who has died. It is important to keep in mind, however, that a clone is only a *genetic* copy. The environment of the individual, which also plays a large role in development, cannot be replicated. Furthermore, the great majority of cloning attempts in mammals fail: in most cases, the embryo either dies or has gross malformations. Because the

consequences of human reproductive cloning would almost certainly be similar, reproductive cloning of humans is condemned almost universally by scientists.

It is important to distinguish reproductive cloning from the cloning and cultivation of cells for therapeutic purposes. **Embryonic stem cells** (ESCs), which are derived from the inner-cell mass of blastocyst-stage embryos, can be cloned and have the unique potential to differentiate into any type of cell in the human body (pluripotency). For example, they can potentially form neurons for the treatment of Parkinson disease or cardiac myocytes for the treatment of ischemic heart disease. However, with current technology, the embryo is destroyed to obtain ESCs, and this is controversial in many circles. Ongoing research efforts are aimed at inducing pluripotency in differentiated adult cells. Research is also under way to extract usable single cells from 3-day blastomere embryos (as in preimplantation genetic diagnosis) so that embryos are not destroyed. It remains to be seen whether these technologies can produce cells that have the same flexibility and utility as ESCs.

One difficulty in using cells derived from ESCs is that they might induce an immune response in the recipient. This problem could be largely overcome if ESC clones were available from many persons with different MHC types. The recipient would then be immunologically matched to the appropriate ESC. However, only a limited number of ESC lines are currently available to most researchers. Another suggestion is that SCNT could be used with a patient's own cells to create ESCs that would be identical in DNA sequence to the patient.

Although these technologies offer the hope of effective treatment for some recalcitrant diseases, they also present thorny ethical issues. Clearly, decisions regarding their use must be guided by constructive input from scientists, legal scholars, philosophers, and others.

BOX 13-7 **Germline Therapy, Genetic Enhancement, Human Cloning, and Embryonic Stem Cells: Controversial Issues in Medical Genetics—cont'd**

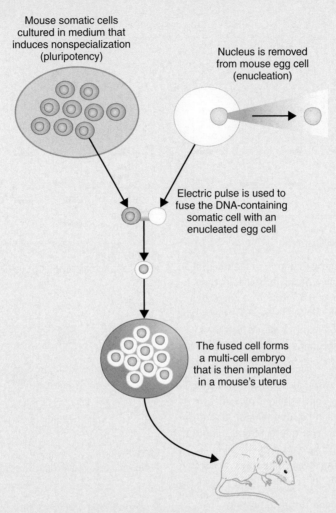

Mouse somatic cells cultured in medium that induces nonspecialization (pluripotency)

Nucleus is removed from mouse egg cell (enucleation)

Electric pulse is used to fuse the DNA-containing somatic cell with an enucleated egg cell

The fused cell forms a multi-cell embryo that is then implanted in a mouse's uterus

Somatic cell nuclear transfer (SCNT) to create a mouse clone. A mouse diploid somatic cell (e.g., a fibroblast) is cultured and grown in media that cause it to become pluripotent. It is fused with an enucleated egg cell, creating a diploid one-cell embryo. This embryo is allowed to develop to a multicell stage and implanted in a mouse's uterus. The resulting mouse is genetically identical (a clone) to the mouse that provided the somatic cell.

## Study Questions

**1.** A newborn-screening program for a metabolic disease has just been initiated. Of 100,000 newborns, 100 were shown by a definitive test to have been affected with the disease. The screening test identified 93 of these neonates as affected and 7 as unaffected. It also identified 1000 neonates as affected who were later shown to be unaffected. Calculate the sensitivity, specificity, and positive predictive value of the screening test, and specify the rate of false positives and false negatives.

**2.** Study the family shown in the pedigree in Figure 13-11. Individual 3 has PKU, an autosomal recessive disease. A two-allele RFLP closely linked to the PKU locus has been assayed for each family member, and the figure shows the genotypes of each individual. The marker alleles are 5 kb and 3 kb in size. Based on the genotypes of the linked marker, is individual 6 affected, a heterozygous carrier, or a normal homozygote?

*Continued*

## Study Questions—cont'd

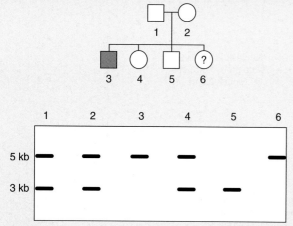

**FIGURE 13-11**
Pedigree to accompany Study Question 2.

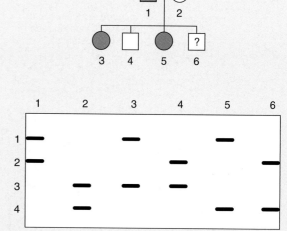

**FIGURE 13-12**
Pedigree to accompany Study Question 3.

**3.** Study the family shown in the pedigree in Figure 13-12. The affected individuals have neurofibromatosis type 1 (NF1), an autosomal dominant condition. A four-allele microsatellite system closely linked to the NF1 locus has been typed for each family member. Based on the genotypes shown in the accompanying figure, will individual 6 develop NF1?

**4.** In the pedigree for an autosomal dominant disorder shown in Figure 13-13, a tightly linked two-allele RFLP has been typed in each family member. Based on this information, what can you tell the family about the risk that the offspring in generation III will develop the disorder? How might diagnostic accuracy be improved in this case?

**5.** Compare the advantages and disadvantages of amniocentesis and chorionic villus sampling (CVS).

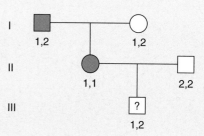

**FIGURE 13-13**
Pedigree to accompany Study Question 4.

**6.** What type of gene therapy would be most appropriate for Huntington disease? Why?

## Suggested Readings

Aitken DA, Crossley JA, Spencer K. Prenatal screening for neural tube defects and aneuploidy. In: Rimoin DL, Connor JM, Pyeritz RE, Korf BR, editors. Emery and Rimoin's Principles and Practice of Medical Genetics. 5th ed., vol. 1. New York: Churchill Livingstone; 2007. p. 636–78.

Alexander BL, Ali RR, Alton EW, et al. Progress and prospects: gene therapy clinical trials (part 1). Gene Ther 2007;14(20):1439–47.

Cavazzana-Calvo M, Fischer A. Gene therapy for severe combined immunodeficiency: are we there yet? J Clin Invest 2007;117(6):1456–65.

Edelstein ML, Abedi MR, Wixon J. Gene therapy clinical trials worldwide to 2007—an update. J Gene Med 2007; 9(10):833–42.

Farrell PM, Rosenstein BJ, White TB, et al. Guidelines for diagnosis of cystic fibrosis in newborns through older adults: Cystic Fibrosis Foundation Consensus Report. J Pediatr 2008;153(2):S4–14.

Fischer A, Cavazzana-Calvo M. Gene therapy of inherited diseases. Lancet 2008;371(9629):2044–7.

Fragouli E. Preimplantation genetic diagnosis: Present and future. J Assist Reprod Genet 2007;24(6):201–7.

Gaffney MM, Hynes SO, Barry F, O'Brien T. Cardiovascular gene therapy: Current status and therapeutic potential. Br J Pharmacol 2007;152(2):175–88.

Gross S, Cuckle H. Prenatal screening and diagnosis—an introduction. Amer J Med Genet Part C 2007;145C:1–4.

Heshka JT, Palleschi C, Howley H, et al. A systematic review of perceived risks, psychological and behavioral impacts of genetic testing. Genet Med 2008;10(1):19–32.

Hochedlinger K, Jaenisch R. Nuclear reprogramming and pluripotency. Nature 2006;441(7097):1061–7.

Hunter DJ, Khoury MJ, Drazen JM. Letting the genome out of the bottle—will we get our wish? N Engl J Med 2008;358(2):105–7.

Jaenisch R. Human cloning—the science and ethics of nuclear transplantation. N Engl J Med 2004;351: 2878–91.

Kim DH, Rossi JJ. Strategies for silencing human disease using RNA interference. Nat Rev Genet 2007; 8(3): 173–84.

Lau TK, Leung TN. Genetic screening and diagnosis. Curr Opin Obstet Gynecol 2005;17(2):163–9.

Malone FD, Canick JA, Ball RH, et al. First-trimester or second-trimester screening, or both, for Down syndrome. N Engl J Med 2005;353(19):2001–11.

McCabe LL, McCabe ER. Expanded newborn screening: implications for genomic medicine. Annu Rev Med 2008; 59:163–75.

O'Connor TP, Crystal RG. Genetic medicines: treatment strategies for hereditary disorders. Nat Rev Genet 2006;7 (4):261–76.

Pagon RA, Tarczy-Hornoch P, Baskin PK, et al. Gene tests—gene clinics: genetic testing information for a growing audience. Hum Mutat 2002;19:501–9.

Riordan JR. CFTR function and prospects for therapy. Annu Rev Biochem 2008;77:701–26.

Scheuner MT, Sieverding P, Shekelle PG. Delivery of genomic medicine for common chronic adult diseases: a systematic review. JAMA 2008;299(11):1320–34.

Sekizawa A, Purwosunu Y, Matsuoka R, et al. Recent advances in non-invasive prenatal DNA diagnosis through analysis of maternal blood. J Obstet Gynaecol Res 2007; 33(6):747–64.

Sermon K, Van Steirteghem A, Liebaers I. Preimplantation genetic diagnosis. Lancet 2004;363:1633–41.

Shulman LP, Simpson JL. Techniques for prenatal diagnosis. In: Rimoin DL, Connor JM, Pyeritz RE, Korf BR, editors. Emery and Rimoin's Principles and Practice of Medical Genetics. 5th ed., vol. 1. New York: Churchill Livingstone; 2007. p. 679–702.

South ST, Chen Z, Brothman AR. Genomic medicine in prenatal diagnosis. Clin Obstet Gynecol 2008; 51(1):62–73.

Spencer K. Aneuploidy screening in the first trimester. Amer J Med Genet Part C 2007;145C:18–32.

Stoller JK. Aboussouan LS: α1-Antitrypsin deficiency. Lancet 2005;365(9478):2225–36.

Van Voorhis BJ. Clinical practice. In vitro fertilization. N Engl J Med 2007;356(4):379–86.

Verma IM, Weitzman MD. Gene therapy: twenty-first century medicine. Annu Rev Biochem 2005; 74: 711–38.

Waisbren SE. Expanded newborn screening: information and resources for the family physician. Am Fam Physician 2008;77:987–94.

Warrington Jr KH, Herzog RW. Treatment of human disease by adeno-associated viral gene transfer. Hum Genet 2006;119(6):571–603.

Weaver D. Catalog of Prenatally Diagnosed Conditions. 3rd ed. Baltimore: Johns Hopkins University Press; 1999.

Wilcken B. Recent advances in newborn screening. J Inherit Metab Dis 2007;30(2):129–33.

Wolfberg AJ. Genes on the web—direct-to-consumer marketing. N Engl J Med 2006;355:543–5.

## Internet Resources

Gene Therapy Clinical Trials Worldwide (updated list of all gene therapy protocols) *http://www.wiley.co.uk/genetherapy/clinical/*

Gene Clinics/Gene Tests (reviews of genetic diseases and lists of laboratories that perform diagnostic tests) *http://www.geneclinics.org/*

Human Genome Project Information (includes information on genetic testing and gene therapy, with relevant links)*http://www.ornl.gov/sci/techresources/Human_Genome/medicine/genetest.shtml*
*http://www.ornl.gov/sci/techresources/Human_Genome/medicine/genetherapy.shtml*

National Newborn Screening and Genetics Resource Center *http://genes-r-us.uthscsa.edu/*

National Organization for Rare Diseases (database of rare disorders that includes brief reviews and information on diagnostic tests and treatment for families and professionals) *http://www.rarediseases.org*

# Chapter 14

# GENETICS AND PERSONALIZED MEDICINE

Scientific, technological, and medical advances have made it possible to detect, diagnose, and treat most common diseases (e.g., asthma, diabetes, hypertension) early in their course and more effectively than ever before. Such advances are, however, highly dependent on the skills and knowledge of clinicians, access to health care services, and the availability and affordability of diagnostic technologies. Most health care providers follow a conventional model in which a patient presents with a set of symptoms and signs, which the provider uses to make a "most likely" diagnosis. The provider then prescribes a treatment that he or she thinks will be most effective. If this treatment fails, the process is repeated until a correct diagnosis or more effective treatment is found. In such a model, preventive health maintenance is encouraged. Compliance, however, is challenging because information about risk factors, as well as the patient's perception of risk, is at best approximate.

**Personalized medicine** is a model of practice in which each patient's personal risk for common diseases and the effectiveness of various treatments are estimated directly from the patient's unique combination of genetic and environmental risk factors. Accordingly, a health care provider can predict a person's risk for common diseases, select diagnostic tests to confirm the presence of disease, and prescribe the best therapeutic regimen to treat it. Ideally, knowledge of disease risk promotes interventions (e.g., modification of diet, choice of drug therapy) that not only can treat disease early in its course but also can delay its onset or prevent it altogether.

The effectiveness of personalized medicine depends on a number of factors. These include identifying genetic and environmental risk factors (and their interactions) that enable accurate prediction of clinically significant risk; demonstrating that individual risk assessment improves diagnostic accuracy and treatment outcome; developing technologies for cost-efficient assessment of a person's genome; building an infrastructure for clinicians to access risk data, interpret risk information, and explain risk estimates to patients; and developing guidelines and policies for how risk assessment information should be used in clinical and research applications. Not all of these aims will be achieved for every common disease. Indeed, for many complex diseases, it is likely that there will be no alternative to

the conventional model of practice in the near future because so little is known about their etiology and pathophysiology. Nevertheless, for some common diseases and drug responses, genetic testing and, in several cases personalized medicine, are already being adapted to the clinical setting.

In this chapter, we discuss how new technologies are making the assessment of individual human genomes widely accessible, how such genome-wide information is being used currently to make personal decisions about health, and the implications of personalized care.

> Personalized medicine is the use of each person's unique combination of genetic and environmental risk factors to make predictions about the individual person's disease risk and response to various treatments.

## A TECHNOLOGY-DRIVEN TRANSFORMATION

Traditionally, the search for genetic variants that influence common complex diseases has been a daunting task and has been one of the major obstacles to developing personalized medicine. The most common approach to finding such variants involved testing whether polymorphisms in candidate genes were associated with disease risk in a small group of unrelated patients with the same phenotype (e.g., diabetes, obesity). This was problematic, in part because choosing the most appropriate candidate genes was difficult, small cohorts provided limited statistical power, and the process of genotyping or sequencing was labor intensive and expensive. This situation has changed dramatically over the past decade with the development of technologies to interrogate millions of polymorphisms per person cheaply and efficiently (Box 14-1). These technologies, coupled with advances in statistics and computing, enabled the application of new approaches such as genome-wide association studies (see Chapter 8), as well as the study of much larger cohorts of thousands or tens of thousands of persons. In addition, these new technologies for genotyping and DNA sequencing make it possible to develop cost-effective clinical tests that take advantage of newly discovered risk variants.

## Assessing Your Genome

Knowledge of a person's genetic makeup will clearly be an important tool for making better decisions about health, medical care, and perhaps lifestyle as well. Until recently, assessing the genome as a whole was fairly expensive and done only in research laboratories. However, new technologies have dramatically lowered the cost of whole-genome analysis and have spurred the development of consumer services that offer whole-genome studies directly to the public (see Chapter 13). These services have rapidly made headlines, as much for their novelty as for their potential to inform persons about their genetic composition.

Most whole-genome consumer services offer to genotype hundreds of thousands to millions of common single nucleotide polymorphisms (SNPs). The SNPs typed for consumers are the same ones commonly being used by researchers to identify disease–gene associations for common multifactorial disorders such as hypertension, diabetes, and obesity. As gene–disease associations are reported, consumers who have access to their genetic information can evaluate their risk for genetic diseases. Also, because each person's genetic data are permanent, evaluation of risk can be reassessed with each new discovery. However, many of the SNP–disease associations reported to consumers are relatively weak and may be misunderstood or misinterpreted by the lay consumer (see Chapter 13).

More recently, whole-genome sequencing has been made available to the public. This service is still expensive and therefore very limited in application. Also, it is debatable whether one's understanding of health-related risks will be increased by knowing about the 99% of the genome that does not encode proteins. An alternative strategy is to sequence only the protein-coding exons. In any case, the same caveats raised in the previous paragraph for whole-genome SNP typing apply equally to whole-genome sequencing.

## THE IMPACT OF GENOMICS

### Pharmacogenetics

Many of the drinks and foods that we ingest each day (e.g., coffee, tea) contain thousands of complex compounds that each of us must process. Some of these compounds never leave the gastrointestinal tract, but most are absorbed, distributed, metabolized, and eliminated (i.e., biotransformed) to a variety of products that are used immediately, stored, or excreted. Exogenously synthesized compounds that are administered to achieve a specific effect on the human body (e.g., pharmaceuticals) also undergo biotransformation, and humans vary in the efficiency and speed with which they do this. Moreover, the response of a drug's target (e.g., enzymes, receptors) can also vary among individuals. The study of the individual genetic variants that modify human responses to pharmacological agents is called **pharmacogenetics**; the assessment of the action of many genes simultaneously is called **pharmacogenomics.**

### Genetic Prediction of Serious Adverse Drug Responses

Over the last decade, ambitious efforts have been undertaken to advance the knowledge of pharmacogenetics. This has been driven, in part, by the expectation that through the use of pharmacogenetics, we will be able to profile DNA differences among individuals and thereby predict responses to different medicines. For example, a genetic profile (i.e., a summary of a person's risk alleles) might predict who is more or less likely to respond to a drug or to suffer a **serious adverse drug reaction** (SADR).

Many drugs have a response rate between 25% and 75%. For example, ACE inhibitors and beta blockers have been found to be ineffective or only partially effective in up to 70% of hypertensive patients. The use of such drugs in persons who are unlikely to respond increases the incidence of SADRs and adds to the burden of health care costs. Yet, for most drugs, no tests are available to determine who will or will not respond, so these drugs are administered largely on a trial-and-error basis.

Many drugs have adverse effects that are of clinical importance, and of the approximately 1200 drugs approved for use in the United States, about 15% are associated with a significant incidence of SADRs. A widely cited analysis conducted in the mid-1990s suggested that nearly 2 million people are hospitalized each year as a result of adverse drug effects, and approximately 100,000 people die from them, even when the drugs are appropriately prescribed and administered. Studies in Europe and Australia have yielded similar results. Thus, identification of genetic profiles that predict a person's response to drugs is likely to increase the overall efficacy and safety of pharmaceuticals.

Testing is currently available for a handful of alleles that predict SADRs. For example, thiopurine methyl transferase (TPMT) is an enzyme that inactivates thiopurine drugs (e.g., 6-mercaptopurine, azathioprine), which are frequently used to treat acute lymphatic leukemia and to prevent rejection of organ transplants. A mutation of the *TPMT* gene reduces enzyme activity. About one in 300 persons of European ancestry is homozygous for this mutation, and these patients can experience life-threatening bone marrow suppression upon exposure to thiopurine drugs. The presence of such variants can be assessed by genotyping or by enzyme assays, which are now commonly done before administering thiopurines.

> Each person's response to natural and synthetic chemicals is determined in part by polymorphisms in genes that control pathways of biotransformation and the chemical's target.

### Personalized Drug Therapy

One of the major challenges of pharmacogenetics is the selection of appropriate targets (e.g., a specific enzyme, cytokine, or cell-surface receptor) that might be amenable to manipulation by a drug. The results of genetic studies are used to identify polymorphisms associated with varying susceptibility to disease (i.e., a potential target for a drug) or polymorphisms that modify the human response to a drug. For example, long QT syndrome (LQT syndrome; see Chapter 12) can be caused by mutations in one of at least

10 different genes whose protein products affect ion channel function in heart cells (e.g., sodium and calcium channels). Because sodium channels and calcium channels are blocked by different drugs, a person's genetic profile can be used to choose the best drug for treatment of LQT syndrome. In this case, the relationship between disease and target is well characterized.

Polymorphisms in genes that encode angiotensinogen, angiotensin-converting enzyme (ACE), and the angiotensin II type 1 receptor have been associated with differential responses to antihypertensive agents. For example, the ACE gene contains a 190-bp sequence that can be either present (the I allele) or deleted (the D allele). Persons who are homozygous for the D allele are more responsive to ACE inhibitors. Response to antihypertensive beta blockers has been associated with polymorphisms in genes that encode subunits of the β-adrenergic receptor (Table 14-1). None of these variants are commonly tested prior to initiating antihypertensive therapy, but studies are under way to determine when such information, in conjunction with environmental risk factors such as smoking and diet, might facilitate the development of personalized treatment.

Many of the physiological effects of variation in drug response have been known for decades. A deficiency of glucose-6-phosphate dehydrogenase (G6PD), which is estimated to affect more than 200 million people worldwide, causes increased sensitivity to the antimalarial drug, primaquine, producing an acute hemolytic anemia. The metabolism of isoniazid (a drug commonly used to treat tuberculosis) is strongly influenced by an allele of the gene that encodes N-acetyltransacetylase 2 (NAT2), the enzyme that is used to acetylate, and thereby inactivate, isoniazid. Persons who are homozygous for this allele are known as slow inactivators and are at higher risk for developing side effects than persons who metabolize isoniazid more quickly. About half of persons of European or African ancestry are slow inactivators, but this figure is lower among East Asians. Succinylcholine is a drug widely used in anesthesia to induce short-term muscle paralysis. Typically, the effects of succinylcholine last only a few minutes before it is rapidly degraded in the plasma by circulating butyrylcholinesterase. Several alleles of the gene that encodes butyrylcholinesterase cause reduced enzyme activity. Persons who are homozygotes or compound heterozgotes for such alleles have a diminished ability to inactivate succinylcholine. This can result in prolonged paralysis and respiratory failure that requires mechanical ventilation for up to several hours.

In each of these examples, a person who has a relatively common allele might, upon exposure to a specific chemical, experience an unanticipated pharmacological effect. Variants have been discovered in enzymes that produce a much broader effect on the body's response to multiple drugs. An example is debrisoquine hydroxylase, an enzyme encoded by the gene CYP2D6. This gene is a member of the cytochrome P450 superfamily, which encodes many different enzymes responsible for the biotransformation of compounds with widely divergent chemical structures. Polymorphisms of CYP2D6 affect the metabolism of more than 25% of all pharmaceuticals, including β-adrenergic receptor antagonists, neuroleptics, and tricyclic antidepressants (Fig. 14-1). All of

**TABLE 14-1**

## Examples of Effects of Gene Polymorphisms on Drug Response

| Gene | Enzyme/Target | Drug | Clinical Response |
|------|---------------|------|-------------------|
| CYP2D6 | Cytochrome P4502D6 | Codeine | Persons homozygous for an inactivating mutation do not metabolize codeine to morphine and thus experience no analgesic effect |
| CYP2C9 | Cytochrome P4502C9 | Warfarin | Persons heterozygous for a polymorphism need a lower dose of warfarin to maintain anticoagulation |
| VKORC1 | Vitamin K epoxide reductase | Warfarin | Persons heterozygous for a polymorphism need a lower dose of warfarin complex, subunit 1, to maintain anticoagulation |
| NAT2 | N-Acetyl transferase 2 | Isoniazid | Persons homozygous for slow-acetylation polymorphisms are more susceptible to isoniazid toxicity |
| TPMT | Thiopurine S-methyltransferase | Azathioprine | Persons homozygous for an inactivating mutation develop severe toxicity if treated with standard doses of azathioprine |
| ADRB2 | β-Adrenergic receptor | Albuterol | Persons homozygous for a polymorphism get worse with regular use of albuterol |
| KCNE2 | Potassium channel, voltage-gated | Clarithromycin | Persons heterozygous for a polymorphism are more susceptible to life-threatening arrhythmias |
| SUR1 | Sulfonylurea receptor 1 | Sulfonylureas | Persons heterozygous for polymorphisms exhibit diminished sensitivity to sulfonylurea-stimulated insulin secretion |
| F5 | Coagulation factor V (Leiden) | Oral contraceptives | Persons heterozygous for a polymorphism are at increased risk for venous thrombosis |

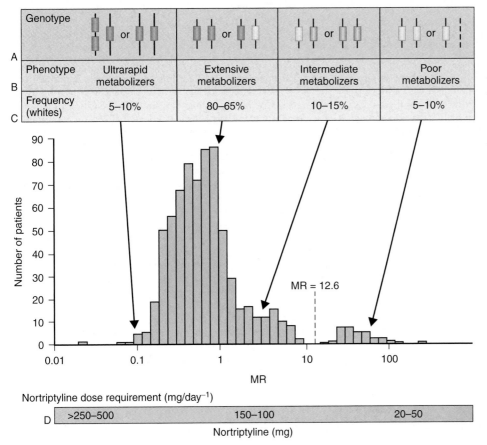

| | | | | |
|---|---|---|---|---|
| **A** Genotype | Ultrarapid | Extensive | Intermediate | Poor |
| **B** Phenotype | Ultrarapid metabolizers | Extensive metabolizers | Intermediate metabolizers | Poor metabolizers |
| **C** Frequency (whites) | 5–10% | 80–65% | 10–15% | 5–10% |

MR = 12.6

Nortriptyline dose requirement (mg/day⁻¹)

**D** | >250–500 | 150–100 | 20–50 |

Nortriptyline (mg)

**FIGURE 14-1**
Genotype–phenotype relationships between *CYP2D6* polymorphisms and drug metabolism. **A,** Possible genotypes at the *CYP2D6* locus. Fully functional alleles of the *CYP2D6* gene are indicated by *red* boxes, alleles with reduced function in *orange,* and null (i.e., inactive) *CYP2D6* alleles are shown in *yellow.* **B,** The ability to metabolize many pharmaceuticals drugs varies depending on an individual's *CYP2D6* genotype. **C,** Distribution of phenotype frequencies assessed in a population of European Americans as determined by the urinary metabolic ratio of debrisoquine to 4-hydroxy-debrisoquine. **D,** Poor metabolizers require a smaller dose of the antidepressant drug nortriptyline, and ultrarapid metabolizers require a higher dose to achieve the same plasma concentration.
*(Adapted from Meyers U: Pharmacogenetics—five decades of therapeutic lessons from genetic diversity. Nat Rev Genet 2004;5:669-676.)*

these are examples of relatively simply genetic profiles (i.e., single polymorphisms) that affect drug response. Many drug responses are likely to be determined by much more complex profiles that are composed of multiple polymorphisms at multiple loci.

Two common variants of *CYP2C9* (*CYP2C9*2* and *CYP2C9*3*), another cytochrome P450 gene, influence the metabolism of warfarin, an anticoagulant drug. The frequencies of these alleles vary between 6% and 12% in populations of European origin, but each is found at a substantially lower frequency in sub-Saharan Africans and East Asians. Warfarin is widely used to prevent thrombosis, but because of variation in dose requirements, hemorrhagic complications from warfarin therapy are common. Therefore, a person's level of anticoagulation needs to be checked regularly so that warfarin is given at a dose that prevents thrombosis but avoids excessive bleeding. Persons with at least one copy of either *CYP2C9*2* or *CYP2C9*3* require less warfarin for effective anticoagulation than the general population. Consistent with this observation, hemorrhagic complications are, on standard dosing, more common in persons who carry the *CYP2C9*2* or the *CYP2C9*3* alleles. Thus, *CYP2C9* variants influence both warfarin metabolism and adverse outcomes associated with warfarin. Genetic variation in one of warfarin's pharmacologic targets, vitamin K epoxide reductase (VKORC1; see Table 14-1), also helps to predict a person's response to this drug. Genetic testing

can be done on both CYP2C9 and VKORC1 to help to calibrate warfarin dosage.

Pharmacogenetics and pharmacogenomics are slowly beginning to change the way that medicine is practiced, although the pace of change is likely to accelerate over the next few decades (Box 14-2). A primary issue for all alleles that are associated with drug response is whether testing these alleles will affect the clinical management of patients and, if so, to what extent. The genetic profile of a drug response may be important if the drug is widely used in clinical practice and the response to the drug is medically important, if the drug's therapeutic and toxic effects are difficult to assess and titrate clinically, if adverse effects are difficult to predict with existing information, and if a profile provides easily interpretable results with high sensitivity and specificity. To date, there is no estimate of how many drug–and–genetic profile combinations are likely to meet these criteria. However, it is probable that these pharmacogenetic profiles will be useful in at least some clinical circumstances.

> Genetic testing for polymorphisms associated with variation in drug metabolism or effectiveness can lead to better predictions of a person's response to drugs and can reduce the incidence of drug-related side effects.

---

**BOX 14-2**
## Personal Genomics

It is 2025. Jonathan is a 1 hour-old baby sleeping comfortably in his mother's arms in the room in which he was just born. A nurse enters and swabs the inside of Jonathan's mouth with a brush to collect buccal epithelial cells. DNA is extracted from these cells, and a week later an electronic summary of Jonathan's complete genome sequence is deposited in a national health information database (NHID). A subset of genotypes that represent a unique genetic profile are put into a national forensics database. Mutation data for conditions covered in the newborn screening program, including PKU, galactosemia, cystic fibrosis, and sickle-cell disease, are forwarded to the state health department. Jonathan's parents are notified that he is a carrier for sickle-cell disease.

Jonathan's parents control access to genetic risk data banked in the NHID for disorders that commonly manifest in childhood, and they elect to provide Jonathan's pediatric care providers with these data. On Jonathan's well-child visit at 1 month of age, a genetic counselor explains that Jonathan has a higher-than-average genetic risk for autism, peanut allergy, chronic otitis media, and adverse responses to penicillin. It is recommended to his parents that he avoid both penicillin and products containing peanuts until he can undergo direct testing. Jonathan is also found to have a lower-than-average genetic risk for asthma.

At 1 year of age, it is apparent that Jonathan's speech and language development is delayed. His genetic profile confirms that he has no risk variants known to be associated with hearing loss, suggesting that his delay might be an early indication of autism. An optimal therapy for autism is selected based on his genetic profile. Jonathan responds well to this intervention and, in conjunction with speech training, his development by the age of 5 years is appropriate.

Jonathan remains healthy throughout childhood, and, when he turns 18 years old, control over his banked genetic risk data is switched from his parents to him. At the same time, Jonathan's medical care is transferred to a family physician. At his first appointment, Jonathan's physician explains his risk for heart disease, hypertension, obesity, type 2 diabetes, and colon cancer. Jonathan is alerted to his high risk of developing both diabetes and obesity, and a program of exercise and diet that has been shown to delay disease onset is recommended to him.

Ten years later, Jonathan informs his physician that he and his wife are planning to start a family. His wife is also a carrier for sickle-cell disease and has several risk variants for asthma, so they are referred for further counseling about prenatal genetic testing options. When Jonathan is 45, he develops hypertension, and, based on his profile of drug response variants, therapy is initiated with a specific antihypertensive agent to which he is likely to respond.

---

### Diagnosing and Monitoring Common Disease

In the previous sections, we have explained how genomic information can be used to personalize risk assessments for common diseases and drug responses. Genomic information can also be used to facilitate disease diagnosis and to monitor therapeutic responses. For example, a microarray (Chapter 3) can be used to estimate the expression level of each gene (i.e., the amount of mRNA that is transcribed) in a specific tissue. These gene-expression profiles can be used to identify patterns of gene expression that are associated with specific diseases (e.g., increased transcription of an oncogene or reduced transcription of a tumor suppressor gene in tumor tissue). Such information can help to distinguish different types of cancers, different types of infections, or other phenotypes associated with disease.

### Cancer Genomics

Every cancer cell harbors numerous alterations in DNA sequence and copy number that affect genes or regulatory sequences, often accompanied by reversible, epigenetic, modifications. These changes perturb the expression and/or function of hundreds to thousands of genes. Collectively, these changes result in the activation or inhibition of various cellular pathways that control the characteristics of cancers such as growth or metastasis, and they determine, in part, prognosis and response to treatment. Cancer genomics is the study of the DNA-associated changes that accompany cancer with the overall goal of better preventing, detecting, diagnosing, and treating common cancers.

A particularly powerful application of genomics to cancer has been the use of genome-wide gene expression analyses

to provide a snapshot of gene activity within a tumor at a given point in time. This has facilitated the development of classification schemes based on expression profiles for many types of cancer, including leukemia, lymphoma, and cancers of the breast, lung, colon, and brain. This information can be used, for example, in refining prognosis, directing the application of conventional and targeted biological therapies, and identifying targets for new drug development (Fig. 14-2).

Currently, it is often difficult to predict the prognosis of cancer patients based on traditional phenotypic information such as the type of tumor (T), whether the cancer is found in nearby lymph nodes (N), and evidence of metastasis (M). Staging using this TMN system is currently the standard for most solid tumors, yet these stages are often not predictive of prognosis or treatment response. Gene-expression profiling can help to distinguish between cancers that are easily confused (e.g., Burkitt lymphoma vs. diffuse large B-cell lymphoma). It can also facilitate the identification of subsets of tumors of the same TMN stage that might have quite different outcomes. Several gene-expression profiles are currently available for assessing breast cancer prognosis, and gene-expression profiles that predict recurrence of several other types of cancer have been established. Prospective trials will determine the extent to which the use of expression profiling is of clinical benefit, but it is anticipated that its use will lead to a substantial improvement in cancer management.

The conventional approach to cancer therapy has been to provide treatment based on the tissue or organ in which the cancer originated. However, persons with the same type of cancer often have different genetic abnormalities in their tumors, resulting in differential responses to treatment. For example, among young women whose breast cancer has not spread to their lymph nodes and who are treated by resection of the tumor and local radiation, only 20% to 30% will experience a recurrence. This subgroup of women might benefit the most from receiving adjuvant chemotherapy, and those at lower risk of recurrence (the majority) might benefit less from chemotherapy. Yet, because the high- and low-risk groups cannot be distinguished reliably, 85% to 95% of all women with this type of breast cancer receive adjuvant chemotherapy. This means that many women might undergo such treatment unnecessarily, putting them at risk for drug-related complications and increasing the overall cost of health care. Expression profiling has the potential to help delineate subsets of cancers that are likely to be more responsive to various therapeutic regimens and to guide the optimal selection of agents for each individual.

▶ Gene-expression profiling of cancers is helping to improve the classification of different types of tumors and may help to guide therapy.

## Common Disease

Gene-expression profiling is being used to study the pathogenesis of common diseases and to monitor tissue-specific gene activity in order to facilitate diagnosis and monitor disease progression. For example, expression profiling of circulating white blood cells in patients with type 1 diabetes has revealed increased expression of a large number of proinflammatory genes. The expression of some of these genes is also increased in persons with rheumatoid arthritis, suggesting that some autoimmune disorders might share expression profiles. A screening test based on these profiles might enable earlier diagnosis and/or identify high-risk persons who could benefit from preventive care. Studies are also under way to identify whether gene-expression profiles can predict outcome in persons infected with pathogens such as malaria, HIV-1, and tuberculosis.

## Race and Genetic Assessment of Individual Ancestry

An important and controversial issue in personalized medicine is whether a person's race—using its historical meaning as a descriptor of Africans, Asians, Europeans, Native Americans, and Pacific Islanders—and/or ancestry is useful for making predictions about health-related risks. Traditionally, it has been commonplace to use race to predict the likelihood that a person carries a particular genetic variant that influences susceptibility to disease or drug response. This practice is based partly on the observation that disparities in health are common among racial groups. For example, the incidence of prostate cancer is two-fold higher in African American men than in European American men. Other disorders that vary in prevalence or outcome among racial groups include hypertension, end-stage renal disease, preterm birth, and type 2 diabetes. It remains unclear, however, whether genetic risk factors explain, even partly, these disparities. Many health-related disparities probably are influenced more strongly instead by

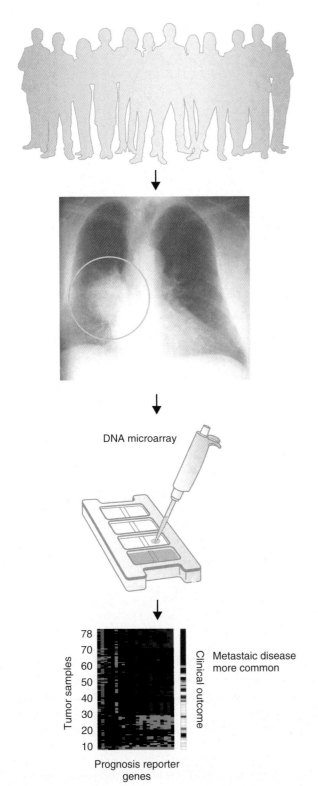

**FIGURE 14-2**
Prediction of disease outcome by gene-expression profiling. The clinical outcome of individuals with lung cancer *(circled tumor on radiograph)* is predicted by testing the expression of a set of genes known to be abnormally regulated in lung cancer cells. For each individual tumor, RNA is extracted and placed onto a microarray, and the expression of each gene is measured. *Bottom,* Each column represents the expression profile of a different tumor. Diminished expression of a gene in one lung tumor compared to other lung tumors is indicated by *green*, and increased expression is indicated by *red*. The outcome of the disease is shown at the *right*, where *white* indicates persons with metastatic disease (poor outcome) and *black* indicates no metastasis (good outcome).

environmental factors such as dietary differences and inequities in the provision of health care services. Accordingly, the use of race to make predictions about whether a person has such risk factors is still the subject of considerable debate.

It is important to distinguish between **race** and **ancestry**. Race has traditionally been used to categorize large groups of persons and can reflect geographic origin, language, and various cultural attributes that describe a group (e.g., Native Americans or Asians). Ancestry refers to the geographic, historical, or biological origins of one's ancestors and, for any person, can be complex. For example, a person might have ancestors from Africa, Europe, and North America (i.e., a complex ancestry), but he or she might still self-identify as an African American. Therefore, race captures some biological information about ancestry, but the two concepts are not equivalent. Knowledge of a person's ancestry can provide information about his or her genetic makeup and thus can be useful for identifying genetic and environmental factors that underlie common diseases. Accordingly, over the past several years, it has become increasingly common to use several hundred single nucleotide polymorphisms (SNPs) to directly estimate the genetic ancestry of a person (Fig. 14-3). The extent to which race helps us to predict genetic differences that influence health depends partly on how well traditional classifications of race correspond with such genetic inferences of individual ancestry.

On average, persons chosen at random from different populations, such as sub-Saharan Africans, Europeans, and East Asians, will be only slightly more different from one another than persons from the same population, reflecting the fact that all humans are quite similar in their DNA sequence (see Chapter 3). Common disease-associated polymorphisms, such as those associated with response to antihypertensive agents (see earlier), differ in these populations only in their frequency. Few, if any, genetic variants are found in all members of one major population and in no members of another major population. For this reason, population affiliation or race is not a reliable predictor of individual genotypes.

It is possible, however, to allocate individuals to groups that correspond to different geographic regions by simultaneously analyzing several hundred or more variants, such as SNPs (see Fig. 14-3). These variants differ in frequency among geographic regions because our ancestors were more likely to mate with close neighbors than with distant ones. Thus, proxies for geographic ancestry such as race can sometimes be used to make reasonably accurate predictions of a person's genetic ancestry. Indeed, several studies done in the United States have reported that classification of persons by self-identified population group is highly correlated with inferences based on genetic data.

In many circumstances, however, race is not a good predictor of ancestry. For example, populations from neighboring geographical regions typically share more recent common ancestors, and therefore their allele frequencies can be very similar. Consequently, persons sampled at regular intervals across some intercontinental regions (e.g., the

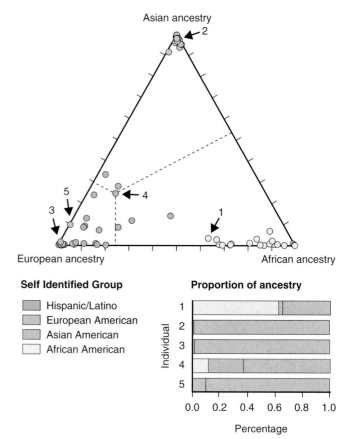

**FIGURE 14-3**
Genetically inferred ancestry fractions for persons (*colored circles*) sampled from the United States and genotyped for 6000 SNPs. Each circle represents one person, color-coded to correspond to one of four self-identified groups. The distance of a circle to the edge of the triangle is proportional to the amount of the person's ancestry contributed by each of the three ancestral populations in the corners of the triangle (African, Asian, and European). For example, the Hispanic/Latino American labeled number 4 received about 60% of his genetic ancestry from Europe, 30% from Asia (due to Native American ancestry), and 10% from Africa. The circles representing Hispanic/Latino and African Americans are less tightly clustered because the proportion of ancestry among persons is more varied than in Asian Americans and Americans of European descent. A bar graph indicates the estimated ancestry proportions for each of the subjects labeled 1-5.

Middle East or Central Asia) are difficult to allocate into genetic groups that are concordant with common notions of race. Correspondence with geography is also less apparent for populations (e.g., Latin Americans, South Asians) that have been influenced by recent historical mixtures of multiple ancestral populations.

In the United States, race is only a crude predictor of a person's genetic ancestry. For example, the average portion of African ancestry among self-identified African Americans is about 80%, but it ranges from 100% to 20% or even less in some persons. The genetic composition of self-identified European Americans also varies, with about 30% of European Americans estimated to have less than 90% European ancestry. Similarly, Hispanics from different regions of the United States have highly variable ancestries (e.g., more African ancestry in Hispanics living in the Southeast and more Native American ancestry in the Southwest). Accordingly, membership in a group does not mean that all members of the group necessarily have similar genetic ancestries.

Although it is clear that explicit genetic information, rather than race, can be used to make more accurate inferences of ancestry, it is not yet known whether personal ancestry information can make useful predictions about one's risk of common disease. The consequences of using detailed ancestry information in a clinical setting are also largely unknown. It is possible that personal ancestry information could have adverse effects on a person's perception of risk and cultural identity. Similarly, such information could reinforce unfair stereotypes about specific populations. Further research is needed to examine the potential benefits and risks of using ancestry information in clinical practice.

▶ The relationship between ancestry and traditional concepts of race is complex. Genetic information, rather than race, is a better predictor of ancestry.

## THE FUTURE OF PERSONALIZED MEDICINE

Genetic variants that increase the risk of common disease are now being found with increasing speed and efficiency (see Chapter 12), but only a small fraction of the genetic basis of disease risk has thus far been defined. In addition, the interactions of multiple disease-predisposing gene products, and their interactions with nongenetic factors, remain almost completely unknown. Thus, the promise of personalized medicine, in which a detailed genetic profile can provide clinically useful risk information for common diseases such as diabetes, cancer, or heart disease, remains largely unfulfilled. It is hoped that with increasing knowledge of alleles that predispose persons to disease, genetic testing will begin to contribute more substantially to the diagnosis and treatment of common disease. It must also be borne in mind that nongenetic factors, such as diet and exercise, are also part of each person's risk profile. These factors can and should be assessed and modified to maximize each person's potential for a healthy life.

## Study Questions

1. Explain how genetic information can be used to enhance the practice of preventive medicine compared to a conventional model of medical service. Give at least one example.

2. Different individuals with the same type of cancer often respond differently to therapy. Provide at least two possible explanations of this observation.

3. Define race and ancestry; explain the differences between them.

4. Consider how explicit genetic information about your ancestry might change your perception of your biological and cultural identities.

5. Give an example of a polymorphism that affects drug metabolism and/or the response to a drug.

6. Explain some of the possible obstacles to the use of genetic information in practicing personalized medicine.

7. Distinguish between genetic medicine and genomic medicine.

8. Give examples of how the availability of whole-genome data from individuals might change the ways medicine is currently practiced.

## Suggested Readings

Bamshad MJ. Genetic influences on health: Does race matter? JAMA 2006;294:937–46.

Belle DJ, Singh H. Genetic factors in drug metabolism. Am Fam Physician 2008;77:1553–60.

Chin L, Gray JW. Translating insights from the cancer genome to clinical practice. Nature 2008;452: 553–63.

Feero GW, Guttmacher AE, Collins FS. The genome gets personal—almost. JAMA 2008;299:1351–2.

Hunter DJ, Khoury MJ, Drazen JM. Letting the genome out of the bottle—will we get our wish? N Engl J Med 2008;358:105–7.

Khoury MJ, Gwinn M, Yoon PW, et al. The continuum of translation research in genomic medicine: how can we accelerate the appropriate integration of human genome discoveries into health care and disease prevention? Genet Med 2007;9:665–74.

Olson MV. Dr. Watson's base pairs. Nature 2008; 452:819–20.

Rothstein MA. Keeping your genes private. Sci Amer 2008;299(3):64–9.

Swen JJ, Huizinga TW, Gelderblom H, et al. Translating pharmacogenomics: challenges on the road to the clinic. PLOSMedicine 2008;4:1317–24.

Wheeler DA, Srinivasan M, Egholm M, et al. The complete genome of an individual by massively parallel DNA sequencing. Nature 2008;452:872–6.

Wilke RA, Lin DW, Roden DM, et al. Identifying genetic risk factors for serious adverse drug reactions: current progress and challenges. Nat Rev Drug Discovery 2007;6:904–16.

## Internet Resources

National Institutes of Health–sponsoredpharmacogenetics research network *http://www.nigms.nih.gov/Initiatives/PGRN*

National Cancer Institute–sponsored tutorial on cancer genomics *http://www.cancer.gov/cancertopics/understandingcancer/cancergenomics*

# Chapter 15

## CLINICAL GENETICS AND GENETIC COUNSELING

Medical genetics has recently emerged as a true specialty in mainstream medicine. In the 1960s, the fields of biochemical genetics, clinical cytogenetics, and **dysmorphology** (the study of abnormal physical development) developed. The 1970s witnessed the establishment of the techniques necessary for prenatal diagnosis of genetic disorders. By the end of the 1970s, discussions about forming the American Board of Medical Genetics had occurred, and in 1981 the first certification examination was administered. The American Board of Genetic Counseling was established in the early 1990s, and now various types of geneticists, including genetic counselors, medical geneticists, and basic human geneticists, can be certified. In 1991 the American Board of Medical Specialties recognized this new field, and medical genetics has now become an integral part of medicine.

Whereas medical genetics involves the study of the genetics of human disease, **clinical genetics** deals with the direct clinical care of persons with genetic diseases. The diagnostic, counseling, and management issues surrounding genetic disease are the principal foci of clinical genetics.

In this chapter, we summarize the principles of clinical genetics and the process of genetic counseling. In addition, we provide an overview of the field of dysmorphology, because the growth of this area has influenced and paralleled the emergence of clinical genetics.

### THE PRINCIPLES AND PRACTICE OF CLINICAL GENETICS

As mentioned in Chapter 1, genetic conditions as a group are common and are a significant cause of human mortality and morbidity. Typically, genetic disorders are complex, multiorgan, systemic conditions, and the care of persons with these disorders can also involve multiple medical specialties. Thus, genetic disorders are among the differential diagnosis of most symptoms and clinical presentations. For example, when evaluating an infant with a blistering skin disease, the ability to distinguish between one of the many forms of epidermolysis bullosa (an inherited disorder of keratinocytes in which skin blisters develop after mild trauma) and staphylococcal skin disease must be part of the clinician's repertoire.

Because of the complexity and number of human genetic diseases, their clinical diagnosis and treatment can seem overwhelming. To help manage this information, we provide an overview of the most important concepts, including the importance of accurate diagnosis, the application of the tenets of medical genetics to medical practice, and the role of genetic counseling in the care of persons with genetic disease.

### Accurate Diagnosis

The significance of the basic medical principle of accurate diagnosis cannot be overemphasized. The process of genetic counseling, one of the principal services of medical genetics, begins with correct diagnosis. All discussions of natural history, prognosis, management, risk determination, options for prenatal diagnosis, and referral to genetic advocacy groups (also termed *genetic support groups*) depend on an accurate diagnosis of the patient's condition. For example, genetic counseling for a family who has a son with mental retardation usually involves questions of risk for this condition in future offspring. An accurate answer requires the clinician to identify a condition of known etiology. If a specific diagnosis (e.g., fragile X syndrome) is made, then the rest of the genetic counseling process starts: Current information can be shared and management can be initiated (Clinical Commentary 15-1).

> In clinical genetics, as in all of medicine, accurate diagnosis is the most important first step in patient care.

The process of diagnosing a genetic disorder is a complex sequence of events. It depends upon diagnostic decision making, recognition of important phenotypic signs, application of principles of dysmorphology and medical genetics, and laboratory diagnosis. For diseases in which the diagnostic criteria are well established, the practitioner has guidelines for making a diagnosis. An example of such criteria is the criteria recommended by the National Institutes of Health Consensus Development Conference for the diagnosis of neurofibromatosis type 1 (NF1; see Chapter 4). For conditions that are defined by a specific laboratory marker, such as an abnormal karyotype or biochemical assay, the diagnostic procedure is generally straightforward. For many genetic diseases, however, there are no well-established

## CLINICAL COMMENTARY 15-1
### *Reasons for Making a Diagnosis of a Syndrome*

The long list of syndromes associated with congenital malformations is overwhelming to the clinician. More than 400 conditions are listed in *Smith's Recognizable Patterns of Human Malformations*, and more than 1000 are accessible through the POSSUM or London Dysmorphology computerized databases (see Internet Resources at the end of this chapter). This number imparts a sense that the diagnosis of a malformation syndrome lies in the arena of academic trivia. However, this is not the case.

Consider, for example, the child who is large for gestational age and has a number of physical abnormalities: omphalocele (intestinal protrusion at the umbilicus), large tongue, facial hemangioma, flank mass, and asymmetrical limb length. His family has questions such as, "What does he have?," "How will he do?," "Will he look different?," "Will he have mental retardation?," "What is the chance of his condition occurring again in another child?"

By putting these features together and making the pattern recognition diagnosis of the Beckwith–Wiedemann syndrome, the clinician is able to answer all of the parents' questions fairly precisely. Most cases of Beckwith–Wiedemann syndrome occur sporadically, but some are inherited. In addition, the genes that cause the disease exhibit imprinting effects (see Chapter 5). If there is no family history, however, the sibling recurrence risk is less than 5%. If there is a family history, the recurrence risk is higher, and linkage or mutation analysis can provide a more precise risk estimate. In future pregnancies, prenatal diagnosis using ultrasound can test for an omphalocele in the second trimester and for large size for gestational age, excessive amniotic fluid (polyhydramnios), and large tongue. If a fetus is thought to have Beckwith–Wiedemann syndrome, then the delivery plan would change and the baby should be born in a tertiary care center

Children with the Beckwith–Wiedemann syndrome do not usually have mental retardation. Although the large tongue can cause orthodontic problems, speech difficulties, and occasionally upper airway problems, these conditions usually improve as the child gets older. The facial appearance is not strikingly abnormal in later childhood.

Chromosome analysis should be considered, although most Beckwith–Wiedemann patients do not have the chromosome 11 duplication that has been reported in a small number of cases. Otherwise, the main emphases of the medical care plan include regular abdominal sonogram to look for intra-abdominal malignancies, especially Wilms' tumor and hepatoblastoma. Children with Beckwith–Wiedemann syndrome have a 5% to 10%

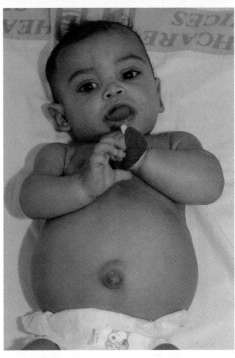

A child with Beckwith–Wiedemann syndrome. Note the prominent eyes and large, protruding tongue.

risk of developing these tumors, and both types are treatable if detected early.

In this example, it was important to diagnose the Beckwith–Wiedemann syndrome. The correct label led to precise information for genetic counseling, prediction of natural history (including reassurance), organization of appropriate laboratory studies, a health supervision plan, and referral to a lay advocacy group. Diagnosis was helpful to the parents, the family physician, and the child.

---

criteria, the definition and delineation of the disorder are not clear-cut, and diagnosis can be challenging.

Dysmorphic syndromes require knowledge and skills in the recognition of mild malformations, minor anomalies, and phenotypic variations. The diagnosis of other genetic diseases, including cancer syndromes and inborn errors of metabolism, can require expertise from a variety of disciplines. For instance, the diagnosis of any of the forms of retinitis pigmentosa (see Chapter 8) requires input from an ophthalmologist who is familiar with this group of retinal degenerative conditions. The diagnostic process is further complicated by the variable expression, incomplete penetrance, and heterogeneity of many genetic diseases. These concepts are discussed in Chapter 4.

### Application of the Principles of Medical Genetics
Developing a genetic approach to human disease in the clinical setting requires the application of all of the basic principles of medical genetics discussed in this book. For example, making or excluding the diagnosis of NF1 requires knowledge of the clinical variability and age of onset of certain features of the condition (Clinical Commentary 15-2). Recognition of the various forms of neurofibromatosis (i.e., heterogeneity) is also important.

**CLINICAL COMMENTARY   15-2**
*The Negative Family History*

One of the common discussions on ward rounds is the notation that a person's family history is negative or noncontributory. This is often thought to rule out a genetic disorder. However, the majority of persons who have a genetic disease do not have a positive family history. A quick review of the mechanisms of mendelian, chromosomal, and multifactorial disease inheritance shows that a lack of other affected persons in the family is common and does not by any means exclude the presence of a genetic disease. For example, the sibling recurrence risk is 25% for diseases with autosomal recessive inheritance. Thus, a significant number of families with multiple offspring have only one affected child and no family history. Even some well-established autosomal dominant disorders often present a negative family history because of high proportions of new mutations (examples include Marfan syndrome, neurofibromatosis type 1 [NF1], and achondroplasia, in which the percentages of cases caused by new mutations are 30%, 50%, and 80%, respectively). Chromosomal syndromes usually have a low recurrence risk. Even when a parent carries a balanced chromosome rearrangement, the recurrence risk among the offspring is usually less than 15%. The sibling recurrence risks for multifactorial conditions are usually 5% or less.

### CASE

A family comes in with a 6-year-old boy who has 10 *café-au-lait* spots exceeding 0.5 cm in diameter and an optic glioma. The family has questions about the diagnosis and the recurrence risk in future pregnancies. On initial telephone contact it is learned that there is no history of a family member with similar features.

There are several possible explanations for this finding. Exploring them underscores the implications of a negative family history:

- *New mutation of the NF1 gene*. Because of the relatively high percentage of new mutations for this disorder, this is the most likely explanation.
- *Variable expression*. It is also possible that one of the parents carries the gene but has mild expression of the phenotype. Occasionally a parent has multiple *café-au-lait* spots and a few neurofibromas, but a diagnosis of NF1 has never been made. Thus, it is important to evaluate the parents for mild expression of NF1.
- *Incomplete penetrance*. This is a possibility; however, it is unlikely for NF1, in which penetrance is close to 100%. If a family has two children with NF1 and neither parent has the gene, germline mosaicism would be the more likely explanation.
- *Incorrect diagnosis*. One of the assumptions and basic principles of medical genetics is accurate diagnosis. This patient meets the National Institutes of Health established criteria for NF1 (see Chapter 4). However, if this patient had only *café-au-lait* spots, then the diagnosis would have

been an issue. One would need to know the differential diagnosis of multiple *café-au-lait* spots.
- *False paternity*. Although it is relatively unlikely, this possibility must be kept in mind.

We began with an individual who had a classic autosomal dominant disorder with no family history. This can be explained in a number of ways. The statement that there is "a negative family history" should not be considered conclusive evidence against the presence of a heritable condition.

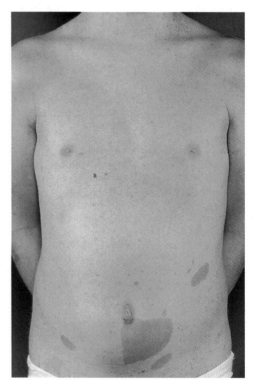

A 6-year-old boy with multiple *café-au-lait* spot.
*(From Burger P, Scheithauer B, Vogel FS: Surgical Pathology of the Nervous System and its Coverings, 4th ed. Philadelphia: Churchill Livingstone, 2002.)*

Knowledge of the other formal principles of medical genetics is also necessary in the care of persons with genetic conditions. The accumulation of family history data and the interpretation of pedigree information are important in answering a family's questions regarding risk of recurrence. An understanding of the various modes of inheritance is necessary in any explanation of recurrence risk. Discussion of the concepts of new mutation and pleiotropy are commonplace in reviewing the cause and pathogenesis of a genetic disease with a family. Even an understanding of meiosis is a requirement for discussions of etiology with the family of a newborn with Down syndrome (Clinical Commentary 15-3).

### Genetic Counseling: Definition and Principles

Genetic counseling represents one of the central foci of medical genetics. At first glance, use of the term "counseling" implies that this service lies in the domain of mental health, social work, or psychotherapy. In fact, genetic counseling is centered in the conventional medical model because it depends significantly on accurate diagnosis and knowledge of medical genetics. As a tradition, genetic counseling grew out of the field of human genetics rather than from behavioral science, unlike other counseling disciplines.

## CLINICAL COMMENTARY 15-3
### *Talking to the Parents of a Newborn with Down Syndrome*

The birth of a newborn with Down syndrome presents many challenges. Typically, the infant is not acutely ill and the parents are not aware of the diagnosis before the birth. Thus, the practitioner must approach the parents, often strangers, with unexpected and sometimes disappointing news. The family can experience a series of emotions that are somewhat similar to the reactions after a loss: anger, denial, sadness, and then usually reorganization and adaptation. Families face these situations with markedly different backgrounds: varying attitudes toward crisis, varied demographic and socioeconomic circumstances, and even a wide range of differences in the cultural meaning of a disability or defect. All of these variables, plus the fact that physicians are often not trained in being the bearers of difficult news, can make this a challenging situation. Parents remember in detail the way the news is presented. The practitioner has both the opportunity and the challenge to help the family through these events.

A number of practical suggestions have come from studies investigating the recommendations of parents who have experienced this event:

- *Prepare yourself.* Set up the interview scenario, and think about how you will begin the discussion.
- *Talk to both parents together whenever possible.* This is sometimes not practical, but when it can be accomplished, it is critical.
- *Communicate the diagnosis as soon as possible.* All studies of parental interviews show that they prefer early communication of the diagnosis.
- *Choose a place that is private and quiet where both the parents and the professionals can sit down.* Avoid standing up with the parents seated. Always be sure to introduce yourself. Structure the interview from the beginning.
- *Humanize the situation as much as possible.* Learn the baby's first name if it has been decided on, and always know the baby's gender. Refer to the infant by name or as a son or daughter, and be aware of the use of all language. Phrases such as "mental retardation" have great impact. Terms such as "mongolism" are not appropriate because they are stigmatizing, pejorative, and incorrect.
- *Develop a sense of realistic positivism.* It is important to discuss the developmental limitations in a patient with Down syndrome, but it is also important to have an optimistic and positive attitude. This suggestion comes from the advocacy and parents' organizations that have developed in the last three decades.
- *Answer the parents' questions, but avoid technical overload.* It is important to be accurate and current on the biological and medical aspects of the condition under discussion. When an answer is not known, mention that the question can be reviewed or referred to a consultant.
- *Listen actively.* Assume that almost all feelings are natural and that parents will be wrestling with their own guilt and shame. Validate all feelings that arise. Most parents can meet this challenge quite effectively and do not require psychiatric consultation.
- *Refer the family to the appropriate resources early.* This would include parents' advocacy groups or even individual parents who have a child with Down syndrome. Share available written material or web pages, but make certain that it is accurate and current.

Above all, be aware of the unique plight of families in such a situation, and make an effort to spend time with them. Although it is difficult to present in written form how one can develop attributes such as kindness and empathy, it is important for physicians in training to learn from their mentors and use their own individual communication style as a strength. Clearly, the recommendations provided here apply not only to genetic counseling but also to any situation in which difficult information is presented to patients or families.

---

In 1975 the American Society of Human Genetics adopted a definition of genetic counseling. Newer language has been proposed recently to modernize and simplify this definition, but the original language stands the test of time:

> "Genetic counseling is a communication process that deals with the human problems associated with the occurrence or risk of occurrence of a genetic disorder in a family. This process involves an attempt by one or more appropriately trained persons to help the individual or family to (1) comprehend the medical facts, including the diagnosis, probable course of the disorder, and the available management; (2) appreciate the way heredity contributes to the disorder and the risk of recurrence in specified relatives; (3) understand the alternatives for dealing with the risk of recurrence; (4) choose a course of action that seems to them appropriate in their view of their risk, their family goals, and their ethical and religious standards, and act in accordance with that decision; and (5) make the best possible adjustment to the disorder in an affected family member and/or to the risk of recurrence of that disorder."

This definition illustrates the complex tasks that face the practitioner. The first task involves establishing the diagnosis and discussing the natural history and management of the condition. In this regard, the medical care of a patient with a genetic disease does not differ from that of a patient with any other type of disease.

The second task requires an understanding of the basic tenets of medical genetics, especially the principles of human genetics and risk determination. For chromosomal and multifactorial disorders, empirical figures are used to estimate recurrence risk. Inheritance patterns are used to predict the recurrence risk of mendelian disorders. However, the clinical issues are often complicated by incomplete penetrance, variable expression, delayed age of onset, and allelic and locus heterogeneity. In some cases, incorporation of additional information using the Bayesian probability approach can significantly alter estimates (Box 15-1).

The third and fourth objectives of the genetic counseling process underlie the primary differences between the genetic model and the traditional biomedical approach. These tasks involve discussing reproductive options and facilitating decision making. Implicit in the fourth part of the definition is the notion of respect for the family's autonomy and their perceptions of risk and of the disorder itself. This approach has been called **nondirectiveness**: the counselor leaves all decisions about future reproduction up to the family. This differs somewhat from the more traditional medical approach, in which *recommendations* for treatment or intervention are often made in a more directed fashion. This is an important issue, because nondirectiveness sometimes conflicts with the broader view of preventive medicine, which

## BOX 15-1
## Recurrence Risks and Bayes' Theorem

The estimation of recurrence risks was treated at some length in Chapters 4 and 5. A typical example of recurrence risk estimation is a case in which a man with hemophilia A, an X-linked recessive disorder, produces a daughter (individual II-1 in the following diagram). Because the man can transmit only the X chromosome carrying the hemophilia A mutation to his daughter, she must be a carrier. The carrier's daughter, individual III-6, has a 50% chance of receiving the X chromosome carrying the mutation and being herself a carrier. Even though the daughter in generation III has five normal brothers, her risk remains 50% because we know that the mother in generation II is a carrier.

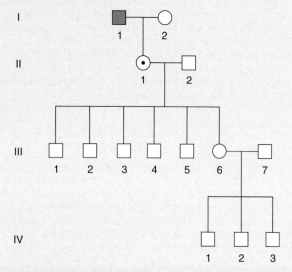

Suppose now that the woman in generation III produces three sons (generation IV), none of whom has hemophilia A. Intuitively, we might begin to suspect that she is not a carrier after all. How can we incorporate this new information into our recurrence risk estimate?

A statistical principle that allows us to make use of such information is called **Bayes' theorem** (the application of Bayes' theorem is often termed Bayesian analysis or Bayesian inference). The table below summarizes the basic steps involved in Bayesian analysis. We begin with the **prior probability** that the woman in generation III is a carrier. As its name suggests, the prior probability denotes the probability that she is a carrier before we account for the fact that she has produced three normal sons. Because we know her mother is a carrier, this woman's prior probability must be 1/2. Then the prior probability that she is not a carrier is $1 - 1/2$, or 1/2.

|  | She is a carrier | She is not a carrier |
|---|---|---|
| Prior probability | 1/2 | 1/2 |
| Conditional probability | 1/8 | 1 |
| Joint probability | 1/16 | 1/2 |
| Posterior probability | 1/9 | 8/9 |

Next we take into account the woman's three normal sons by estimating the probability that all three of them would be normal given that she is a carrier. Because this probability is *conditioned* on

her carrier status, it is termed a **conditional probability**. If she is a carrier, the conditional probability that all three of her sons are normal would be $(1/2)^3$, or 1/8. We also estimate the probability that all of her sons would be normal given that she is *not* a carrier. This conditional probability is, of course, very close to 1.

Next we want to find the probability that the woman is a carrier *and* that she is a carrier with three normal sons. To obtain the probability of the co-occurrence of these two events, we multiply the prior probability times the conditional probability to derive a **joint probability** (i.e., the probability of both events occurring together, a concept discussed in Chapter 4). The joint probability that she is a carrier is then $1/2 \times 1/8 = 1/16$. Similarly, the joint probability that she is not a carrier is $1/2 \times 1 = 1/2$. These joint probabilities indicate the woman is 8 times more likely not to be a carrier than to be a carrier.

The final step is to standardize the joint probabilities so that the two probabilities under consideration (i.e., being a carrier versus not being a carrier) sum to 1. To do this, we simply divide the joint probability that the woman is a carrier (1/16) by the sum of the two joint probabilities $(1/16 + 1/2)$. This yields a **posterior probability** of 1/9 that she is a carrier and 8/9 that she is not a carrier. Notice that this standardization process allows us to provide a risk estimate (1/9, or 11%), while preserving the odds of noncarrier versus carrier status indicated by the joint probabilities.

Having worked through the Bayesian analysis, we see that our intuition was confirmed: The fact that the woman in question produced three normal sons reduced her risk of being a carrier substantially, from an initial estimate of 50% to a final probability of only 11%.

Another common application of Bayesian analysis is illustrated in part A of the diagram that follows. The male in generation II is affected with Duchenne muscular dystrophy (DMD), a lethal X-linked recessive disease (see Chapter 5). Either his unaffected mother is a carrier of the mutation, or he received a new mutation on the X chromosome transmitted by his mother. It is important to determine whether the mother is a carrier or not, because this fact will influence recurrence risks for DMD in her subsequent offspring. If the mother has only one affected offspring, the probability that she is a carrier can be evaluated directly, because one third of all cases of X-linked lethal recessive disorders arise as a result of new mutations. (To understand this, consider the fact that because females have two X chromosomes and males have only one, 2/3 of all X-linked disease-causing mutations in a population must be found in females. For a lethal X-linked recessive, all of the male X chromosomes are eliminated from the population in each generation. Yet the frequency of the mutation remains the same, generation after generation. This is because new disease-causing mutations arise at the same rate as the loss of mutation-containing X chromosomes. Because one third of the mutation-containing X chromosomes are lost each generation, it follows that one third of the mutations in the population must occur as the result of new mutation.) If the probability that the affected son received a new mutation is 1/3, then the probability that the mother is a carrier—the alternative possibility—must be $1 - 1/3$, or 2/3.

In the table below, we use Bayesian analysis to evaluate the probability that the mother is a carrier. As in the previous example, we derive a prior probability that she is a carrier, assuming no knowledge that she has produced an affected son. This probability is given by $4\mu$, where $\mu$ is the mutation rate for the *DMD* locus (i.e., the probability, per generation, that a disease-causing mutation arises at this locus in an individual). The derivation of the probability, $4\mu$, is beyond the scope of this text, but it can be found elsewhere (Hodge, 1998). Because the prior probability that the mother is a carrier is 4μ, the prior probability that she is not a carrier is 1 − 4μ, which is approximately equal to 1 because μ is very small. The conditional probability that the woman transmits the mutation given that she is a carrier is 1/2 (there is also a very small probability that she transmits her normal allele, which is then mutated, but this can be ignored). The conditional probability that she transmits a mutation given that she is not a carrier (i.e., the probability that a new mutation arises in the gamete she transmits) is μ. We then multiply the prior probability that she is a carrier, 4μ, by the corresponding conditional probability, 1/2, to obtain a joint probability of 2μ. The same procedure produces a joint probability of μ that she is not a carrier. Finally, we standardize the joint probabilities to get the posterior probabilities. The posterior probability that she is a carrier is 2μ ÷ (2μ + μ) = 2/3, and the posterior probability that she is not a carrier is μ ÷ (2μ + μ) = 1/3. As expected, these probabilities correspond to the ones we obtained by simple direct observation.

|  | She is a carrier | She is not a carrier |
|---|---|---|
| Prior probability | 4μ | 1-4μ ≈ 1 |
| Conditional probability | 1/2 | μ |
| Joint probability | 2μ | μ |
| Posterior probability | 2/3 | 1/3 |

Suppose, however, that the woman has had an affected son *and* an unaffected son (see B in the figure below). This gives us additional information, and intuitively it increases the possibility that she is not a carrier (i.e., that the one affected offspring is the result of a new mutation). In the table below, we incorporate this new information. The prior probabilities remain the same as before (i.e., we assume no knowledge of either of her offspring). But the conditional probability of transmission, given that she is a carrier, changes to account for the fact that she now has two offspring: 1/2 × 1/2 = 1/4 (i.e., the probability that she did not transmit the mutation to one offspring times the probability that she did transmit the mutation to the other offspring). The conditional probability that she transmitted a new mutation to the affected offspring is μ, and the probability that she did not transmit a mutation to the unaffected offspring is 1 − μ. Thus, the probability of both events, given that she is not a carrier, is μ × (1 − μ) ≈ μ. The joint and posterior probabilities are obtained as before, and we see that the woman's chance of being a carrier is now reduced from 2/3 to 1/2. Again, this confirms (and quantifies) our expectation.

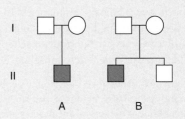

|  | She is a carrier | She is not a carrier |
|---|---|---|
| Prior probability | 4μ | 1-4μ ≈ 1 |
| Conditional probability | 1/4 | μ × (1-μ) ≈ μ |
| Joint probability | μ | μ |
| Posterior probability | 1/2 | 1/2 |

Before the advent of disease diagnosis through linked markers or mutation detection, Bayesian analysis was often the only way to derive a risk estimate in situations such as these. Now, of course, an attempt would be made to identify the factor VIII or *DMD* mutation that causes hemophilia A or DMD in these families directly, or, failing that, linked markers would be used. This is a much more direct and accurate approach for determining carrier status. However, as discussed in Chapter 13, it is not always possible to identify the responsible mutation, particularly when a large number of mutations can cause the disorder (as is the case for hemophilia A, DMD, or cystic fibrosis). Bayesian inference can be used in such cases to incorporate the sensitivity of the genetic test (e.g., if a standard mutation analysis of the *CFTR* gene reveals 85% of the mutations [see Chapter 13], there is a 15% probability that the person in question has the mutation even though the test did not reveal it). In addition, linkage analysis is not always informative. Thus, Bayesian analysis is still sometimes a useful tool for refining risk estimates.

The additional information incorporated in Bayesian analysis is not confined to the assessment of health status in relatives, as was shown in these examples. Another type of information is a biochemical assay, such as factor VIII activity level, that could help to indicate carrier status. Because there is usually overlap between carriers and normal homozygotes for such tests, the assay cannot determine carrier status with certainty, but it does provide a *probability* estimate for incorporation into Bayesian analysis. In diseases with delayed age of onset, such as adult polycystic kidney disease, the probability of being affected at a certain age can be used in a Bayesian analysis. Here, one considers the fact that the at-risk person is less and less likely to possess the disease gene if he or she remains unaffected beyond a certain age.

might suggest that the principal goal of genetic counseling should be the reduction of the incidence of genetic diseases.

Historically, the principle of nondirectiveness developed in the arena of reproductive counseling and in the context of decisions surrounding prenatal diagnosis. If prevention or reduction of disease is the primary goal, then one's approach would logically be more directive. However, the main goal of genetic counseling is to help an individual family understand and cope with genetic disease, not to reduce the incidence of genetic disease.

Although most geneticists subscribe to the principles of autonomy and nondirectiveness, it may be challenging to

the clinician to be entirely nondirective, simply because of the limitations of a time-restricted session. For example, an explanation of nutritional management of an infant with a disease detected by newborn screening (see Chapter 13) would require a more directive approach than the discussion of disease risks in future pregnancies. Nondirectiveness may be challenging when the consequences of disease are severe, as in high-risk cancer counseling (see the discussion of the newly proposed definition later). In addition, information may be presented quite differently in different contexts. Information about Down syndrome, for example, may be communicated differently depending on whether the diagnosis was made prenatally or after the birth of an affected newborn (see Clinical Commentary 15-3).

In 2006, leaders in the field of genetic counseling and the National Society of Genetic Counselors reached a consensus on a modern definition of genetic counseling:

"Genetic counseling is the process of helping people understand and adapt to the medical, psychological, and familial implications of genetic contributions of disease. The process integrates the following: (1) interpretation of family and medical histories to assess the chance of disease occurrence or recurrence; (2) education about inheritance, testing, management, prevention, resources, and research; and (3) counseling to promote informed choices and adaptation to the risk of the condition."

The majority of geneticists subscribe to the principle of nondirectiveness in reproductive counseling: Information about risks, natural history, treatment, and outcome is presented in a balanced and neutral manner, and decisions about reproduction are left to the family.

The facilitation of discussion about reproductive decision making is central to the task of genetic counseling. Several factors are involved in a family's decision about future pregnancies when there is an increased risk. The obvious ones are the magnitude of the risk figure and the burden or impact of the disorder. However, these are not the only significant issues. The individual family's perception of the impact of the condition is probably more important in their decision making than the professional's perception of the burden. The meaning of children to the individual family, according to their own cultural, religious, or personal preferences, is weighed heavily in the reproductive decision-making process. In addition, families frequently play out the scenario of coping with a recurrence of the condition in another child. Identification of these issues for a family often helps to stimulate their own discussions. Some families perceive a recurrence risk qualitatively rather than quantitatively: They consider themselves to be either "at risk" or not, with the actual risk estimate being a secondary consideration. The fact that there is so much variation in the importance people assign to each of these factors (perception of risk, perception of impact, meaning of children, and possibility of recurrence) underscores the point that the professional should be a facilitator and not the decision maker.

The final task of genetic counseling is to help the family cope with the presence of the disorder or its recurrence risk, or both. This task is similar to the physician's support of a family dealing with any chronic disease or disability. What is unique, perhaps, is the family's perception of the meaning of a genetic disorder (Boxes 15-2 and 15-3). In many

---

## BOX 15-2
## Birth of a Child with Trisomy 18

Our daughter Juliett arrived on a beautiful summer afternoon in 1984. Late in my pregnancy, an ultrasound had showed an enlarged heart, dilated left kidney, and possible malformation of the cerebellum. During labor, our daughter's heart rate decelerated significantly, and we were given the option of an emergency C-section. Without hesitation, we opted for the C-section. A drape was hung so that I could not see, and I only knew that the baby was born when the pediatrician ran out of the room with something in his arms. My husband quickly followed, and then I waited for what seemed an eternity.

Juliett weighed 4 pounds, 6 ounces and was 18 inches long. I had graduated as an RN just before Juliett was conceived and had worked in a pediatric ICU. I had just enough experience to pick up a few of her obvious problems, but many escaped me. She was clearly much too thin, and her rib cage looked too short and prominent. But compared to the mental images I had formed after the ultrasound, I was relieved to see how beautiful she was. Her most striking feature was her incredible blue eyes, which were wide open and very alert. Her nose and mouth were beautifully formed and very petite. As my husband and I sat in awe over her, a neonatologist entered the room. He pointed out several characteristics, the only one of which I clearly remember was the clenched fist with the index finger lying over the middle finger. He concluded that she probably had trisomy 18. Of the grim things he rattled off, the only thing I remembered was that he said she would be a vegetable and that she would most likely die within the next couple of days. He then walked away, and we sat there, stunned. In this state of grief and turmoil I tried to understand how this clenched fist could lead to death, and how these bright, alert eyes could belong to a vegetable.

In the days that followed, I often opened up her fist and laid her fingers straight, hoping that the blood tests would not confirm the doctor's suspicions. Our bonding with Juliett had been instant, and our great desire was to be able to take her home before she died. As she began to tolerate feedings and was weaned away from oxygen, our pleas to take her home were granted.

We left with no follow-up or care plan. Each time she went to sleep, we prayed she would wake up again and that we could complete another feeding. At 3 months of age, she started to smile at us, and our hopes brightened. We have been fortunate to see her outlive the grim statistics, and we have learned that there is no clear explanation why some children with this condition live longer than others. Juliett's heart was enlarged because of a defect similar to a tetralogy of Fallot. Mild scoliosis at birth has now progressed to a 100-degree lumbar curve and a 90-degree thoracic curve. Despite her many physical challenges, Juliett has continued at her slower pace to learn and develop new skills. Her personality is delightful, and people are often surprised to see how responsive and interactive she is.

We have often been asked if we were afraid to have more children. Perhaps we were crazy, but we felt that another like Juliett would be great. We have had four more girls. To everyone's surprise, our fifth child, Camille, was born with Down syndrome. With Juliett, the grieving process had been covered up with the gratitude we felt that she was even alive. With Camille, we experienced the more typical grieving process.

On the day of Juliett's birth, a pediatrician came forward, put his arms around us, and told us he thought she was beautiful and to love her for as long as she could be with us. He turned her into a human being with a life to be highly valued. In the 13 years that have followed, Juliett has seen many doctors. Most of them, although they could not cure her problems, gave us the most important thing we needed: to know that our daughter's life was of great worth and that, if they could, they would do anything to help her.

## BOX 15-3
## Raising a Child with Bloom Syndrome

Tommy was born via an emergency cesarean section, because 1 week before his delivery date his fetal movements markedly decreased. At birth, he weighed only 4 pounds, and the first time that I saw him, he was in an incubator connected to all sorts of tubes. He spent his first month of life in the neonatal intensive care unit so that his weight gain could be closely monitored. Because he was so small, he was fed through a feeding tube for many months, and as a consequence, he refused to drink from a bottle. Eventually, he overcame his aversion to using his mouth to eat but only after substantial training. Nevertheless, despite our care, Tommy remained small for his age.

The following summer, Tommy developed dark red marks on his cheeks and under his eyes. Our pediatrician referred us to a dermatologist, who suspected that the marks on Tommy's face were related to his growth failure. We were very surprised. How could these two findings be related? That is when we were told that Tommy might have a genetic disorder called Bloom syndrome. We hoped that the doctor was wrong, but soon thereafter Tommy had a genetic test that measured the number of sister chromatid exchanges per cell (see Chapter 2). This test confirmed that Tommy had Bloom syndrome. Although I insisted that it was a false-positive result, I learned to accept that our son had a very rare cancer syndrome.

We were barraged by questions from family, friends, and doctors. As a result, we became very protective of our son and his privacy. Nevertheless, there was only so much we could do to protect him because he is such a social little boy, loving to play with family and friends. This also made choosing an appropriate elementary school a very difficult decision for us. We expected that children would pick on him because of his small size. However, to our surprise, he easily developed friendships and adjusted well to his classmates. In fact, the problems that he did develop were largely because of his misbehavior. Thus, we struggled to find a balance between protecting Tommy while not permitting him special privileges because of his small stature.

In our home, we try to treat Tommy like any of our other children. One challenge for us is that because of Tommy's small size, people wrongly perceive that he is much younger than his chronological age. This is very frustrating for Tommy, yet we occasionally reinforce this image because of our concern for his safety. For example, although Tommy is 6 years old, he weighs only 21 pounds. Thus, he must sit in an infant seat when he travels in a car and we explain to Tommy's friends that it helps him see through the windows. Another safety problem is that many of the sensors for the automatic doors at supermarkets cannot detect his presence and easily slam into him.

Overall, Tommy has adapted well. He climbs or jumps in order to reach things. To keep up with his peers, he often runs, hops, or jumps instead of walking. We constantly worry about his safety, but we cannot control all that happens to him. To date, he has been healthy, and although it seems like we have been riding an emotional roller-coaster, we wouldn't trade our experiences for anything.

---

acquired conditions, such as infections or accidents, the ultimate meaning of the condition is externalized. In genetic disorders, the condition is more intrinsic to the individual and the family; it thus often presents a complex personal dilemma. Validation of the plight of families is vital and is probably more effective than simplistic attempts to wipe away guilt. Feelings of guilt and shame are natural to the situation and also need acknowledgment.

The primary care practitioner plays a vital role in the ongoing support of families in which a member has a genetic disease. Additional support strategies include referral of the family to a genetic advocacy group, distribution of current printed and Internet information on the disorder, referral to mental health professionals for ongoing counseling, and frequent follow-up visits that include time for discussions of feelings and thoughts.

> Genetic counseling includes many themes: medical diagnosis and management, determining risk of recurrence, options for addressing the risk, reproductive decision making, and support services.

Numerous studies in the past 3 decades have attempted to evaluate the effectiveness of genetic counseling. The methodology of these studies is complicated, and the evaluation of the results depends on one's interpretation of the goal of genetic counseling. A few general points, however, can be made. Families tend to recall recurrence risks relatively well. A letter sent to them after the visit improves this recall.

Families who perceive their offspring's condition as being serious and one of "burden" recall risk figures better. Most studies suggest that genetic counseling is relatively effective in providing information about the medical aspects and genetic risks of the condition. Issues surrounding decision making and psychosocial support require additional investigation.

### Genetic Counselors and the Delivery of Genetic Counseling

As the discipline of medical genetics (including genetic counseling) evolved in the 1970s, it became clear that the delivery of this service is complex and time-consuming. Not only did the geneticist need to have skills in most specialties of medicine, but facilitation of decision making and provision of psychological support were also necessary. As a need for genetics professionals other than physicians became apparent, a number of genetic counseling training programs emerged in the United States and Canada. Currently, more than 30 accredited programs in North America provide master's level training in genetic counseling. Genetic counselors have become integral partners with physicians and other professionals in the delivery of medical genetic services. From this growth evolved a professional society, The National Society of Genetic Counselors, and, the certifying and accrediting body, the American Board of Genetic Counseling. Although the range of skills is wide and job descriptions vary in different medical centers, genetic counselors have established themselves as experts in the determination of recurrence chances, reproductive decision making, and psychosocial support (Box 15-4). In the prenatal and cancer genetics settings, genetic counselors function relatively independently as practitioners. More recently, genetic counselors have become important professionals on research teams and in genetics laboratory services.

## Genetic Advocacy Groups

Genetic advocacy groups can provide critical support in assisting families who have a member with a genetic disorder (see, for example, Genetic Alliance under Internet Resources at the end of this chapter). These support organizations provide the family with the sense of a fellow traveler in a way that the professional is not able to do. The sense of isolation that often accompanies genetic disorders (and rare conditions in general) is often alleviated by meeting someone else in the same situation. Immediate bonds are formed

that often assist in the coping process. In the last few decades, a clear partnership of professionals and persons with genetic disorders and disabilities has developed. Not only have these groups provided a needed service, but they also have promoted the establishment of databases and research studies. Referral to a genetic support group and distribution of their written information are now a routine part of the care and management of all genetic disorders.

> The delivery of genetic services including genetic counseling involves a partnership of physicians, genetic counselors, and genetic advocacy groups.

## Clinical Genetics Evaluation and Services

With the development of medical genetics as a medical specialty, clinical genetics services have become part of the

---

BOX 15-4
## An Insider's View of Genetic Counseling

### What Is a Genetic Counselor?

In its most general usage, the term *genetic counselor* refers to any medical professional who is professionally qualified to provide genetic counseling. Typically, a genetic counselor is a genetics professional with a master's degree or Ph.D. in genetic counseling. Degree programs in genetic counseling provide education and clinical training in medical genetics and counseling. A certified genetic counselor has also passed a certification examination administered by the American Board of Genetic Counseling or the American Board of Medical Genetics.

### What Do Genetic Counselors Do?

Some of the primary responsibilities of a genetic counselor are to interview individuals and families with genetic disorders and to answer questions about the possibilities of a genetic disorder. A genetic counselor often works as part of a team that may include medical geneticists, other physicians (e.g., obstetricians, oncologists, neurologists), social workers, psychologists, nutritionists, or nurses. Genetic counselors help to collect and assess medical information leading to a diagnosis, provide patient education, provide psychosocial support and counseling, provide genetic counseling and risk assessment for genetic testing, and help physicians with the management of genetic conditions. They often triage inquiries and referrals to the genetics service in which they practice. They might manage or coordinate clinics and personnel. They are active in genetic education programs for medical professionals and the lay public.

### In What Settings Do Genetic Counselors Work?

Genetic counselors often work in general genetics settings in pediatrics and adult medicine. They also work in obstetrics settings, providing counseling for prenatal diagnosis and screening, genetic testing for couples with multiple pregnancy loss, diagnosis and management of pregnancies affected with abnormalities detected by radiological imaging, and, most recently, alternative reproductive technologies. They work in multidisciplinary specialty clinics for groups of diseases (e.g., metabolic, craniofacial, bone dysplasia, neurogenetic) or for single diseases (e.g., Down syndrome, neurofibromatosis, hemophilia). More recently, genetic counselors are becoming increasingly integrated into cancer genetics clinics.

Many counselors participate in research related to clinical genetics and genetic counseling. For example, predictive testing

for disorders such as Huntington disease and hereditary cancers has occurred mainly in research studies aimed at assessing the medical, ethical, legal, and social consequences. Within this research setting, genetic counselors provide counseling and help to design, implement, and evaluate research protocols.

Some genetic counselors work in laboratories to provide an interface between the laboratory and its clients and to help develop laboratory protocols. A small percentage of genetic counselors are in private practice, and some work in administrative positions for the state or federal government. Many genetic counselors are active at regional and national levels in professional organizations, and some counselors help to start, maintain, or advise lay advocacy groups for genetic disorders.

### What Skills and Personal Qualities Make a Good Genetic Counselor?

A good genetic counselor needs both a strong background in the biological sciences and genetics and training in the theory and practice of psychosocial techniques (e.g., family systems, crisis counseling, interviewing skills). Because most genetic counselors provide direct patient service, it is essential to work well with people. Genetic counselors must work well both independently and on a team. There is a high level of responsibility involved in patient care aspects, and counselors must successfully learn to handle the stress of the difficult situations of the families with whom they work.

### What Is the Future of Genetic Counseling?

It is difficult to predict the extent to which medical genetics will continue to move into mainstream medicine. Will geneticists and genetic counselors increase in number, or will genetics professionals remain small in number and limit their role to advising generalists and seeing only the most complicated cases? In either case, there is clearly a need for an increase in the genetics education of medical professionals and the public. Many observers think that medical genetics and genetic counseling have a high potential for expansion. What is undisputed is the striking emergence of medical genetics from an obscure medical subspecialty to an area of knowledge that is fast becoming integrated into every field of medicine.

_____

*(Courtesy of Bonnie J. Baty, M.S.)*

health care delivery system. Most university medical centers in North America include a genetics clinic whose major objective is to provide genetic diagnosis, management, and counseling services.

As in all medical visits, evaluation of a person or family for a potential genetic condition requires a thorough history and physical examination. The history includes information about the family's concerns, the prenatal period, labor, delivery, and documentation of family relationships (the pedigree). The physical examination should focus on the physical variations or minor anomalies that provide clues to a diagnosis. Additional family members might need evaluation for the presence or absence of a genetic disorder. Photographs and recording of certain physical measurements are a standard component of the genetic evaluation. Ancillary tests may be required to document specific physical features (e.g., an echocardiogram or MRI for aortic dilatation in Marfan syndrome or radiographs to diagnose achondroplasia).

An important type of clinical data gathered in this process is the family history (Box 15-5). The data obtained in a family history are often useful in obtaining an accurate diagnosis of a condition. For example, a strong family history of early-onset coronary disease might indicate the presence of a low-density lipoprotein–receptor defect causing familial hypercholesterolemia. A family history of early-onset colon cancer could indicate that a gene for familial adenomatous polyposis or hereditary nonpolyposis colorectal cancer is present in the family. Family history information can also guide the estimation of recurrence risks by helping to determine whether a genetic disease has been transmitted by one's parents or has occurred as a new mutation (this is especially important for diseases with reduced penetrance). The knowledge and skills required to take an accurate and thorough family history are important for all clinicians, not only clinical geneticists.

Routinely, the clinician sends the family a letter summarizing the diagnosis, natural history, and risk information regarding the condition. This letter is a valuable resource for the family, because it helps to document the risk information for later review. Information regarding lay advocacy groups, including pamphlets, booklets, and brochures, is often provided. Follow-up visits are recommended depending on the individual situation. Box 15-6 provides a list of clinical genetics services.

> Clinical genetic evaluations include physical examination, detailed family history, ancillary tests as needed, and communication of information to the family through letters and the distribution of published literature.

In recent years, the care of persons with genetic disease has included the development of guidelines for follow-up and routine care. Knowledge of the natural history of a condition, coupled with a critical review of screening tests and interventions, can provide a framework for health supervision and anticipatory guidance. The management plan can subsequently be used by the primary care provider. It is primarily for this purpose that many of the specialized clinics, such as those for NF1 or hemophilia, have been established. An example of this approach is the management checklist for the health maintenance of infants and children with Down syndrome (see Chapter 6).

As treatment options for mendelian disorders become more numerous (e.g., the treatment of aortic dilatation in Marfan syndrome, see Chapter 4), the role of clinical geneticists will likely change. Since the turn of the 21st century, geneticists have become increasingly involved in the design and implementation of clinical trials, and this trend will certainly continue and change the nature of the practice.

Traditionally, genetic counseling involves the family who comes in with questions about the diagnosis, management, and recurrence risk of the condition in question. Thus, in the majority of situations, genetic counseling is carried out retrospectively. With the increased availability of prenatal, carrier, and presymptomatic testing, prospective genetic counseling will become more common. Box 15-7 lists common reasons for referral for genetic evaluation.

---

## BOX 15-5
## The Family History

A thorough, accurate family history is an indispensable part of a medical evaluation, and a pedigree should be part of the patient's chart. At a minimum, the following items should be included:

- The sex of each individual and his or her relationship to other family members. This information should be indicated using standard pedigree symbols (see Chapter 4).
- A three-generation family history should be obtained. For example, male relatives on the mother's side of the family will be especially important when considering an X-linked recessive disorder.
- The age of each individual. A record must be kept of whether each individual is affected with the disease in question, and inquiries should be made about diseases that may be related

to the disease in question (e.g., ovarian cancer in a family being seen for familial breast cancer).
- All known miscarriages and stillbirths.
- The ethnic origin of the family. This is important because many diseases vary considerably in prevalence among different ethnic groups.
- Information about consanguinity. Although it is relatively rare in most Western populations, consanguinity is common in many of the world's populations, and immigrant populations often maintain relatively high rates of consanguinity (see Chapter 4).
- Changes in family histories. Family members develop newly diagnosed diseases, and additional children are born. These changes can affect diagnosis and risk estimation, so the family history and pedigree should be updated periodically.

## BOX 15-6
## Types of Clinical Genetic Services and Programs

Center-based genetics clinics
Outreach clinics
Inpatient consultations
Specialty clinics
- Metabolic clinics
- Cancer genetics clinics
- Spina bifida clinics
- Hemophilia clinics
- Craniofacial clinics
- Other single-disorder clinics (e.g., NF1 clinics)

Prenatal diagnosis programs: Perinatal and reproductive genetics clinics
- Amniocentesis and chorionic villus sampling clinics
- Ultrasound programs
- Maternal serum triple screen programs
- Preimplantation diagnosis programs
- Presymptomatic diagnosis in families (e.g., familial breast cancer diagnosis)

Genetic screening
- Newborn screening program/follow-up clinics
- Other population screening programs (e.g., Tay–Sachs disease)

Education and training
- Health care professionals, including clinical geneticists and genetic counselors
- General public
- School system
- Teratology information services

## BOX 15-7
## Common Indications for Genetics Referral

Evaluation of a person with mental retardation or developmental delay
Evaluation of a person with single or multiple malformations; question of a dysmorphic syndrome
Evaluation of a person with a possible inherited metabolic disease
Presence of a possible single-gene disorder
Presence of a chromosomal disorder, including balanced rearrangements
Person at risk for a genetic condition, including questions of presymptomatic diagnosis or cancer risk
Person or family with questions about the genetic aspects of any medical condition
Couples with a history of recurrent miscarriages
Consanguinity in a couple, usually first cousin or closer relationship
Teratogen counseling
Preconceptional counseling and risk-factor counseling, including advanced maternal age and other potential indications for prenatal diagnosis

## DYSMORPHOLOGY AND CLINICAL TERATOLOGY

Dysmorphology was defined at the beginning of this chapter as the study of abnormal physical development (**morphogenesis**). Congenital defects are caused by altered morphogenesis. Although the term dysmorphology may seem synonymous with **teratology**, the latter term usually implies the study of the environmental causes of congenital anomalies, even though its literal meaning does not refer to etiology. (The term

*teratology* is derived from *teras,* the Greek word for "monster." The term dysmorphology was proposed by Dr. David Smith as a reaction to the pejorative connotation of teratology.)

Congenital defects represent an important cause of infant mortality and morbidity. Current studies indicate that the frequency of medically significant malformations diagnosed in the newborn period is 2% to 3%. Investigations that have observed children for a longer period demonstrate that this frequency increases to 3% to 4% by the age of 1 year. In the United States, congenital malformations represent the most common cause of mortality during the first year of life. Table 15-1 lists some of the most common and important malformation syndromes.

## TABLE 15-1
## Examples of Common Multiple Congenital Anomaly and Dysplasia Syndromes Referred to a Medical Genetics Clinic

| Syndromes | Etiology |
|---|---|
| Down syndrome | Chromosomal |
| Neurofibromatosis type 1 | Single gene (AD) |
| Angelman syndrome | Microdeletion of chromosome 15q; uniparental disomy |
| Amnion disruption sequence | Unknown |
| Osteogenesis imperfecta | Single gene; heterogeneous, AD, AR; type I collagen, related genes |
| Trisomy 18 | Chromosomal |
| VATER association | Unknown |
| Marfan syndrome | Single gene (AD) |
| Prader–Willi syndrome | Microdeletion of chromosome 15q; uniparental disomy |
| Noonan syndrome | Single gene (AD) |
| Williams syndrome | Microdeletion of chromosome 7 |
| Achondroplasia | Single gene (AD) |
| Trisomy 13 | Chromosomal |
| Turner syndrome | Chromosomal (45,X) |
| Rett syndrome | X-linked gene |
| Rubinstein–Taybi syndrome | Single gene (AD) |
| Klippel–Trenaunay | Heterogeneous; one susceptibility syndrome gene identified |
| Fetal alcohol syndrome | Excessive alcohol |
| Cornelia de Lange syndrome | Single gene (AD) |

*AD, Autosomal dominant, AR autosomal recessive; VATER, vertebral anomalies, anal atresia, tracheoesophageal fistula, esophageal atresia, renal anomalies.*

**TABLE 15-2**
## Causes of Malformations among Affected Infants

| Genetic Cause | Number* | Percentage |
|---|---|---|
| Chromosome abnormalities | 157 (45) | 10.1 |
| Single mutant genes | 48 | 3.1 |
| Familial | 225 (3) | 14.5 |
| Multifactorial inheritance | 356 (23) | 23.0 |
| Teratogens | 49 | 3.2 |
| Uterine factors | 39 (5) | 2.5 |
| Twinning | 6 (2) | 0.4 |
| Unknown cause | 669 (24) | 43.2 |
| Total | 1549 (102) | |

*Values in parentheses denote therapeutic abortions; of the 69,277 infants studied, 1549 had malformations, for an incidence of 2.24%.
Data from Nelson K, Holmes LB: Malformations due to spontaneous mutations in newborn infants. N Engl J Med 1989;320:19-23.

There are several ways to classify congenital abnormalities. The most common classification approach is by organ system or body region (e.g., craniofacial, limb, heart). More clinically useful classification schemes include (1) single defect versus multiple congenital anomaly syndrome, (2) major (medically or surgically significant defects) versus minor anomalies, (3) categorization by pathogenic process, and (4) an etiological classification.

Table 15-2 lists the causes of main defects in a major study conducted in Boston. Three key messages emerged from these data: the etiology of two thirds of congenital defects is unknown or multifactorial (see Chapter 12), well-established environmental causes of congenital malformations are infrequent, and a known genetic component is identified in approximately 30% of cases.

### Principles of Dysmorphology

In discussing the basic principles of dysmorphology, it is important to define certain key terms. The following definitions, based on pathogenic processes, are used in clinical practice:

- **Malformation** is a primary morphologic defect of an organ or body part resulting from an intrinsically abnormal developmental process (e.g., cleft lip, polydactyly).
- **Dysplasia** is a primary defect involving abnormal organization of cells into tissue (e.g., vascular malformation).
- **Sequence** is a primary defect with its secondary structural changes (e.g., Pierre Robin sequence, a disorder in which a primary defect in mandibular development produces a small jaw, secondary glossoptosis, and a cleft palate)
- **Syndrome** is a pattern of multiple primary malformations with a single etiology (e.g., trisomy 13 syndrome).
- **Deformation** is alteration of the form, shape, or position of a normally formed body part by mechanical forces. It

usually occurs in the fetal period, not in embryogenesis. It is a secondary alteration. It can be extrinsic, as in oligohydramnios (reduced amniotic fluid), or intrinsic, as in congenital myotonic dystrophy.
- **Disruption** is a morphological defect of an organ, part of an organ, or a larger region of the body resulting from the extrinsic breakdown of, or interference with, an originally normal developmental process. It is a secondary malformation (e.g., secondary limb defect resulting from a vascular event).

Note that malformations and dysplasias are primary events in embryogenesis and histogenesis, whereas disruptions and deformations are secondary.

When evaluating a child with a congenital malformation, the most important question is whether the abnormality represents a single, isolated anomaly or is instead one component of a broader, organized pattern of malformation (i.e., a syndrome). An example is given by the evaluation of a baby with a cleft lip. If a baby has an isolated, nonsyndromic cleft lip with no other malformations, the discussion of natural history, genetics, prognosis, and management is markedly different than if the baby's cleft lip is one feature of the trisomy 13 syndrome (see Chapter 6). The former condition can be repaired surgically and has a relatively low recurrence risk (see Chapter 12) and few associated medical problems. Trisomy 13 is a serious chromosomal disorder. In addition to oral–facial clefts, these infants usually have a congenital heart defect and central nervous system malformations. More importantly, 50% of children with trisomy 13 die in the newborn period, and 90% die before 1 year of age. Thus, prediction of the natural history and medical management in these two examples are quite different.

Another example is a child with cleft lip who also has pits or fistulas of the lower lip. The combination of orofacial clefts and lip pits signifies an autosomal dominant condition called the van der Woude syndrome. Although the natural history of this condition differs little from that of nonsyndromic cleft lip, the discussion of genetic recurrence risks is much different. In the evaluation of a child with van der Woude syndrome, it is very important to determine whether one of the parents carries the gene. If so, the sibling recurrence risk is 50%. This is much greater than the 4% sibling recurrence risk usually given for nonsyndromic cleft lip. Because van der Woude syndrome has highly variable expression and often is manifested only by lip fistulas, it is quite commonly overlooked. Thus, a careful physical examination, combined with a knowledge of the genetics of isolated malformations and syndromes, is necessary to determine accurate recurrence risks.

> The most important question to ask when evaluating a child with a congenital malformation is whether the defect is isolated or part of a syndrome pattern.

Increasing knowledge of the pathogenesis of human congenital defects has led to a better understanding of the developmental relationship of the defects in multiple congenital anomaly patterns. Some well-established conditions that appear to be true syndromes at first glance are really a constellation of defects consisting of a primary malformation with its secondary, localized effects (i.e., a sequence). In a sequence, the pattern is a developmental unit in which the cascade of secondary pathogenic events is well understood. In contrast, the pathogenic relationship of the primary malformations in a syndrome is not as well understood, although pathogenesis may be clarified when the syndrome is the result of the pleiotropic effects of a single gene (e.g., Marfan syndrome; see Chapter 4).

One of the best examples of a sequence is the Potter phenotype or oligohydramnios sequence. It is currently believed that any significant and persistent condition leading to oligohydramnios can produce this sequence, whether it be intrauterine renal failure due to kidney malformations (such as missing kidneys, renal agenesis) or chronic leakage of amniotic fluid. The fetus will develop a pattern of secondary growth deficiency, joint contractures (deformations), characteristic facial features, and pulmonary hypoplasia (Fig. 15-1).

Before the cause of these features was understood, the phenotype was termed the *Potter syndrome*. Now, with the understanding that all of the features are secondary to oligohydramnios, the disorder is more properly called the *oligohydramnios sequence*. As with any malformation, the renal defect can occur by itself, or it can be a part of any number of syndromes in which renal malformations are component features (such as autosomal recessive Meckel–Gruber syndrome or the more common nonsyndromic disorder, bilateral renal agenesis). Distinguishing between syndromes and sequences can often improve understanding of the underlying cause of a disorder and aid in prediction of prognosis.

> It is important to distinguish between a sequence, which is a primary defect with secondary structural changes, and a syndrome, which is a collection of malformations whose relationship to one another tends to be less well understood.

## Clinical Teratology

A **teratogen** is an agent external to the fetus's genome that induces structural malformations, growth deficiency, and/or functional alterations during prenatal development. Although teratogens cause only a small percentage of all birth defects, the preventive potential alone makes them worthy of study. Table 15-3 lists the well-established human teratogens.

It is important to understand the reasoning process that leads to the designation of a substance as a teratogen. This process is based on an evaluation of epidemiological, clinical, biochemical, and physiological evidence. Animal studies can also help to establish whether an agent is teratogenic.

Some of the issues involved in determining whether an agent is teratogenic are summarized in Clinical Commentary 15-4. A key clinical point is that it is common for families to ask their doctors questions about the risks of certain

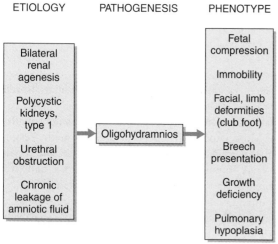

**FIGURE 15-1**
The oligohydramnios sequence. Oligohydramnios can arise from a number of distinct causes. It produces a constellation of secondary phenotypic features.

**TABLE 15-3**
## Well-Established Human Teratogens*

| Drug | Potential Defect | Critical Exposure Period | Percentage of Those Exposed Who Are Affected |
|---|---|---|---|
| Angiotensin-converting enzyme (ACE) inhibitors | Renal dysgenesis; oligohydramnios; skull ossification defects | Second to third trimester | NE |
| Alcohol, chronic | Craniofacial and central nervous system anomalies; heart defects | <12 wk | 10-15 |
| | Low birth weight; developmental delay | >24 wk | NE |
| Aminopterin | Spontaneous abortion | <14 wk | NE |
| | Craniofacial anomalies; limb defects; craniosynostosis; neural tube defects | First trimester | NE |
| | Low birth weight | >20 wk | NE |

**TABLE 15-3**
## Well-Established Human Teratogens—cont'd

| Drug | Potential Defect | Critical Exposure Period | Percentage of Those Exposed Who Are Affected |
|------|------------------|--------------------------|-----------------------------------------------|
| High-dose androgens or norprogesterones | Masculinization of external female genitalia | >10 wk | 0.3 |
| Carbamazepine | Spina bifida | <30 days after conception | ≈1 |
| Carbimazole/ methimazole | Hypothyroidism; goiter | NE | NE |
| Cocaine | Abruptio placentae | Second to third trimester | NE |
| | Intracranial hemorrhage; premature labor and delivery | Third trimester | NE |
| Diethylstilbestrol | Uterine abnormalities; vaginal adenosis; vaginal adenocarcinoma; cervical ridges; male infertility | <12 wk | NE |
| Fluconazole (high-dose) | Limb and craniofacial defects | First trimester | NE |
| Isotretinoin | Fetal death; hydrocephalus; central nervous system defects; microtia or anotia; small or missing thymus; conotruncal heart defects; micrognathia | >15 days after conception | 45-50 |
| Methotrexate | Craniosynostosis; underossified skull; craniofacial dysmorphology; limb defects | 6-9 wk after conception | NE |
| Penicillamine | Cutis laxa, joint contractures | NE | NE |
| Phenytoin | Craniofacial anomalies; hypoplastic phalanges and nails | First trimester | 10-30 |
| Solvents, abuse (entire pregnancy) | Small size for gestational age; developmental delay | | NE |
| Streptomycin | Hearing loss | Third trimester | NE |
| Tetracycline | Stained teeth and bone | >20 wk | NE |
| Thalidomide | Limb deficiencies; ear anomalies | 38-50 days post-LMP | 15-25 |
| Thiouracil | Spontaneous abortion | First trimester | NE |
| | Stillbirth | >20 wk | NE |
| | Goiter | | NE |
| Trimethadione | Developmental delay; V-shaped eyebrows; low-set ears; irregular teeth | First trimester | NE |
| Valproic acid | Spina bifida | <30 days after conception | <1 |
| | Craniofacial anomalies; preaxial defects | First trimester | NE |
| Warfarin | Nasal hypoplasia; stippled epiphyses | 6-9 wk | NE |
| | Central nervous system defects secondary to cerebral hemorrhage | >12 wk | NE |

*Other established teratogens include maternal infections (rubella, cytomegalovirus, toxoplasmosis, varicella, Venezuela equine encephalitis, syphilis, parvovirus), maternal disease states (diabetes mellitus, phenylketonuria, systemic lupus erythematosus, Graves' disease), and ionizing radiation.*

*NE, Not established; post-LMP, after last menstrual period.*

*Data from Martinez LP, Robertson J, Leen-Mitchell M: Environmental causes of birth defects. In Rudolph CD, Rudolph AM, Hostetter MK, Lister G, Siegel NM [eds]: Rudolph's Pediatrics, 21st ed. New York, McGraw-Hill, 2002, p 9.*

## CLINICAL COMMENTARY  15-4
### The Bendectin Saga

Bendectin, or doxyclamine, was an agent introduced in the 1960s for "morning sickness" (nausea and vomiting during pregnancy). The agent was particularly efficacious, and during the 1970s about one third of American women took Bendectin at some time during their first trimester of pregnancy.

Bendectin has probably been studied more than any other individual pregnancy medication. Several epidemiological studies have found no conclusive evidence of an increased risk of congenital malformations with use of Bendectin during pregnancy. The few studies that have demonstrated weak associations between Bendectin and birth defects did not reveal any consistent patterns. Animal studies also indicate no association. Despite these data, a number of lawsuits were filed in the 1980s against the company that marketed Bendectin. As a result, the company removed the drug from the market.

The reasoning process used to determine causation in such cases is complex. Before judgments about etiology are made, a critical review of the literature is required. The available epidemiological studies must be evaluated in terms of their methodology, design, and biases. Up-to-date knowledge of the etiology and pathogenesis of congenital malformations must be included in the design of the study. The basic principles of teratology must be applied. This includes an evaluation of the *critical period* (i.e., did exposure occur during the period of pregnancy in which the malformed fetal structures were developing?). Clinical evidence includes the search for a pattern of defects (i.e., a specific syndrome), because all well-established teratogens produce consistent patterns (see Table 15-3). Animal models never prove causation in humans but can provide supportive evidence and information about pathogenesis. In addition, the proposed effect of the agent should be biologically plausible.

A review of the evidence regarding Bendectin shows that it meets none of these criteria. Indeed, because of its extensive testing, Bendectin satisfies standard criteria for safety as well as any known medication. What, then, could have produced this litigation? An important part of the answer to this question involves the *coincidental* occurrence of congenital malformations and Bendectin exposure. Considering that a major congenital malformation will be diagnosed in 3% of infants by 1 year of age *and* that about one third of women were taking Bendectin in the first trimester of pregnancy, then about 1% (1/3 × 3%) of all pregnancies in the 1970s would have experienced the co-occurrence of these two events by chance alone. Because about two thirds of congenital malformations have no known cause, it is not surprising that many families of children with these disorders (and their lawyers) would attribute them to Bendectin use.

Bendectin was removed from the market in 1983. Since then, the percentage of babies born with congenital malformations has remained the same, and the number of women admitted to hospitals for complaints of morning sickness has doubled.

It is very difficult to prove epidemiologically that any exposure is "safe." The statistical power of such studies does not permit an absolute statement that there is no effect. All that can be demonstrated is that there is no definitive evidence that a particular agent (in this case, Bendectin) causes an adverse outcome. Blanket reassurance or absolute statements of complete safety are not appropriate. On the other hand, when the evidence is relatively conclusive, as in the case of Bendectin, it is clinically appropriate to have a reassuring tone in discussing exposure in the pregnancy setting.

---

exposures during pregnancy. When faced with such a question, the physician has a number of options. One is to review the literature on particular exposures in humans, make a judgment regarding the degree of risk, and then provide counseling. Various online and printed sources available to the clinician are listed at the end of the chapter. An alternative is to refer the patient to a teratology information service or to a clinical genetics unit. Because of the complexity of these issues, teratology information services have sprung up throughout the United States, Canada, and Europe. The Organization of Teratology Information Services provides up-to-date and comprehensive information on exposures in pregnancy and lists the available teratology information services in North America.

> Teratogens are external agents that cause a small but important percentage of congenital malformations. It is often difficult to prove conclusively that a substance is teratogenic.

## Prevention of Congenital Malformations

Because the majority of structural defects have no obvious cause, their prevention presents a challenge. An example is fetal alcohol syndrome (FAS), one of the more common preventable causes of human malformation (Clinical Commentary 15-5). The institution of rubella immunization programs and the preconceptional administration of folic acid exemplify successful prevention (Clinical Commentary 15-6).

Preconceptional counseling is a model for primary prevention. Women who have diabetes mellitus, phenylketonuria, or systemic lupus erythematosus (an autoimmune disorder involving production of autoantibodies that affect multiple organs) can decrease their risk of producing an infant with a structural defect through appropriate preconceptional management, a primary prevention strategy. Examples of secondary and tertiary levels of prevention, respectively, include newborn screening for hearing loss and the provision of high-quality medical care for infants and children with congenital malformations. Instituting appropriate guidelines for health supervision and anticipatory guidance can decrease some of the complications of these disorders. Also important is public education regarding the limitations of scientific knowledge and the emotional difficulties of families in which a child has a congenital defect. Such information has the potential to diminish anxiety, improve the family's coping process, and reduce the stigma that surrounds congenital malformations and genetic disorders.

## CLINICAL COMMENTARY 15-5
### *Fetal Alcohol Syndrome*

Among the human teratogens, one of the most common and potentially preventable exposures is excessive alcohol consumption. Women who are chronic alcoholics are at significant risk of bearing a child with the fetal alcohol syndrome (FAS). This condition consists of prenatal and postnatal growth deficiency, microcephaly (small head), a wide range of developmental disabilities, and a constellation of facial alterations. The more distinctive and consistent facial features include short palpebral fissures, low nasal root, upturned nose, simple or flat philtral folds, and a thin upper lip. Although most of these signs are not specific for FAS, their mutual occurrence in the context of maternal alcohol abuse allows the clinician to make the diagnosis.

In addition to these findings, infants and children with FAS are at risk for a number of structural defects, including congenital heart defects, neural tube defects, and renal malformations. Most children with FAS have a mild degree of developmental delay, ranging from mild mental retardation to learning disabilities.

There are still many unanswered questions regarding alcohol use in pregnancy. These include the extent of genetic predisposition to FAS, the risk of binge drinking, the role of moderate and social drinking, and the safe level of alcohol in pregnancy. Although there is no conclusive evidence that light to moderate drinking in early pregnancy is harmful, avoidance of alcohol during pregnancy is the most prudent approach.

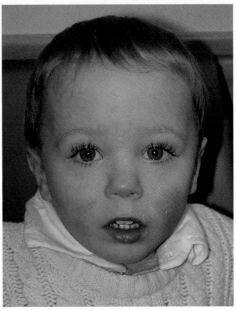

A 2-year-old child with fetal alcohol syndrome. Note the low nasal root, short palpebral fissures, smooth philtrum, and thin upper lip.

## CLINICAL COMMENTARY 15-6
### *Folate and the Prevention of Neural Tube Defects*

The primary prevention of congenital malformations is an important goal of clinical genetics. Because the ultimate cause of most congenital malformations is currently unknown, there are relatively few opportunities for primary prevention. A recent approach to the prevention of congenital malformations is the periconceptional use of folate and multivitamin regimens to prevent the occurrence and recurrence of neural tube defects (NTDs).

NTDs consist of malformations of the developing neural tube and express themselves as anencephaly, encephalocele, and spina bifida (see Chapter 12). Their impact is serious: Anencephaly is invariably fatal, and the medical complications of spina bifida (lower limb paralysis, hydrocephalus, urinary obstruction) are significant. Because of the potential influence of nutritional elements on embryogenesis, a series of epidemiological studies were performed in the 1970s and 1980s. With one exception, they demonstrated that the use of vitamins and folate in the periconception period reduced the recurrence risk of spina bifida and anencephaly in families that had previously had a child with one of these conditions. In 1991 the Medical Research Council of the United Kingdom published a double-blind study* in which 4 mg of folate with or without vitamins was administered to women who had had a child with an NTD. The group receiving folate alone experienced a 70% reduction in the recurrence of these malformations in their offspring. In 1992, a Hungarian group demonstrated the usefulness of vitamins and folic acid in preventing the initial occurrence of NTDs. In this study, two groups of women, one receiving vitamins and folic acid and the other not, were followed for the duration of their pregnancies. The vitamin-and-folic acid regimen significantly decreased the occurrence of NTDs. Numerous additional studies have confirmed these results.

Although it is still not clear whether the purported protective effect is due to folic acid or a combination of folic acid and other vitamins, these data indicate that periconceptional vitamin use is an effective prevention strategy. The mechanism for the apparent effect remains unknown. Nonetheless, the encouraging results of these studies have prompted the Centers for Disease Control and Prevention to publish two recommendations regarding the use of folate. The first is that all women who have previously had a child with an NTD should take 4 mg/day of folic acid if they are planning to become pregnant. The second recommendation is that all women of reproductive age should take 0.4 mg/day of folic acid (the amount available in a typical multivitamin tablet) throughout their reproductive years. The latter recommendation is prudent in light of the fact that approximately half of pregnancies in the United States are unplanned. These recommendations have led to the folic acid fortification of wheat and other grain products in the United States and other nations. In the past decade, studies in a variety of countries throughout the world have demonstrated a decrease in the occurrence of NTDs after the initiation of a food fortification program

---

*A **double-blind** study is one in which, during the treatment phase of the study, neither the subjects nor the investigator knows which subjects are receiving an active ingredient and which are receiving placebo.

## BIOETHICS AND MEDICAL GENETICS

With new discoveries and advances in medical technology come new choices for patients, families, and society. As medical genetics came to be defined as a medical specialty during the past several decades, a number of new issues in bioethics also emerged. Because of the significance and complexity of these issues, a significant portion of the budget of the Human Genome Project has been devoted to the ethical, legal, and social implications of human genetics. Some of these implications have been touched upon previously in this textbook (e.g., genetic testing, gene therapy, embryonic stem cell research). Our goal here is to provide a sampling of the major ethical questions currently confronting the medical and genetics communities.

The combination of advances in prenatal diagnosis (e.g., ultrasound, amniocentesis), the ability to determine the human karyotype, and the option of pregnancy termination set the stage for the rise of prenatal diagnosis as a clinical service in the 1970s. By that time, most tertiary medical centers in developed nations offered amniocentesis for various indications, most commonly advanced maternal age (see Chapter 13). In the early discussions of the ethics of prenatal diagnosis, the central controversy involved a woman's (or couple's) right to terminate a pregnancy. In the 1990s, the issue took on a new dimension with the concern that persons with disabilities may be devalued by society when prenatal diagnosis can lead to the selective termination of fetuses with a disability. Similar concerns revolve around the issue of withdrawal of support for newborns with severe birth defects (e.g., trisomy 13, some neural tube defects). The main principles guiding decisions about these issues are to consider the interests of the child and to provide genetic counseling so that the parents can make an informed decision.

Ethical concerns have arisen about other types of genetic testing, including carrier testing and presymptomatic testing (see Chapter 13). Genetic tests differ from other types of medical testing because genes (including mutations that predispose to disease) can be shared in families. Thus, a genetic test performed in one person might reveal risk information about a relative who might not wish to know about his or her risk (e.g., testing of a young adult for an autosomal dominant disease might indicate that one of the parents must have transmitted the disease-causing mutation). In addition, many people perceive that their genetic inheritance is very much an intrinsic part of themselves (and their families). Genetic risk is often misperceived as "unchangeable." All of these factors can lead to unfair stigmatization of individuals, families, and even whole populations. To help counter this, care providers must be sensitive to the needs and concerns of individuals and their families. They should avoid making value judgments that could lead to or reinforce stigmatization. They should dispel notions of genetic determinism, making it clear to families that genes are but one part of the cause of a disease. Nongenetic factors, which often can be altered, can also play an important role. As with all medical information, privacy and confidentiality must be respected.

Genetic testing also raises the specter of discrimination by insurance companies or employers. Insurance companies have long collected information about family history as a means of assessing risk. In some cases (e.g., an individual at risk for inheriting a *BRCA1* mutation), a genetic test can provide a much more accurate measure of disease risk. At-risk persons are understandably concerned about the possibility of losing their insurance benefits or their employment because of the outcome of a genetic test. A paradoxical result is that some choose not to be tested, even when it could lead to a potentially life-saving intervention. Insurance providers and employers reason that denying coverage (or increasing premiums) to at-risk individuals serves the general interest by minimizing costs. Others respond that, unlike lifestyle choices such as cigarette smoking or exercise, one does not choose one's genes, so it is unfair to discriminate on the basis of genetic tests.

Because of concerns in the United States regarding discrimination in employment and health insurance, the genetics community and supportive legislators worked together during the early 2000s to enact legislation to ensure the confidentiality of genetic test results. The Genetic Information Nondiscrimination Act (GINA) was formulated to prevent the discriminatory use of genetic test results by employers or health insurers. GINA became a federal law in 2008 and took effect in 2009.

Preimplantation genetic testing (see Chapter 13) has also been the focus of ethical debates. For example, this form of testing can be used to determine the sex of an embryo. Indeed, an early application was to avoid implanting male embryos that had an increased risk of carrying an X-linked recessive mutation. Many scientists and ethicists believe that preimplantation diagnosis solely for sex selection is inappropriate, and the practice is currently banned in the United Kingdom. Preimplantation diagnosis could also be used to select preferentially for embryos that *do* carry disease-causing mutations. For example, an embryo that is homozygous for mutations causing autosomal recessive deafness could be selected to match the phenotypes of deaf parents (a case was reported in which deaf parents deliberately conceived a deaf child by artificial insemination). Such applications might put the interests of the parents and the interests of the child in conflict. Another controversial application of preimplantation diagnosis was in the selection of a human leukocyte antigen (HLA)-matched embryo that could later provide bone marrow cells for an older sibling who had Fanconi anemia. It is argued that such persons could feel that their life was devalued because they had been selected partly on the basis of their suitability as a bone marrow donor.

Questions have arisen about the genetic testing of children. When such testing can lead to useful diagnostic measures or interventions, it may be warranted. An example would be the genetic testing of children at risk for inheriting a mutation in the adenomatous polyposis coli (*APC*) gene. As discussed in Chapter 11, colonoscopy should begin in gene carriers by age 12 years and is potentially life-saving. In contrast, childhood diagnosis of Huntington disease currently provides no preventive or therapeutic benefits and increases the potential for anxiety and stigmatization. A consensus has emerged that childhood genetic testing should not be pursued unless it provides an avenue to clinically beneficial intervention.

Intense controversy has surrounded the issues of cloning and embryonic stem cell research (see Chapter 13). It is important to repeat the distinction between *reproductive* cloning and *therapeutic* cloning. Because of the high failure rate of reproductive cloning in other mammals, and because the benefits of reproductive cloning are unclear, the scientific community is nearly unanimous in its opposition to the creation of human beings through reproductive cloning. The use of cloning to create embryonic stem cells for therapeutic purposes (e.g., pancreatic islet cells for patients with diabetes or neurons for those with Alzheimer disease) is more controversial and, like pregnancy termination, involves highly charged questions about the definition of human life and the limits of medical intervention. As noted

in Chapter 13, these issues require informed and thoughtful input from the scientific community and from patient advocacy groups, ethicists, philosophers, legal scholars, the clergy, and others.

The science of genetics is no stranger to controversy and even to abuse. The **eugenics** movement (Greek, "good birth"), popular in the United States and some European countries in the early part of the 20th century, advocated both "positive eugenics" (preferential reproduction of those deemed to be genetically more fit) and "negative eugenics" (the prevention of reproduction of those thought to be genetically less fit). Eugenics, in concert with the political thinking of the times, led to a series of abuses that culminated in the atrocities of Nazi Germany. These events are a sobering reminder of the potential for misuse of genetic information. Geneticists must help to ensure that their science is used for maximum benefit while adhering to the time-honored maxim *primum non nocere* ("first, do no harm").

## Study Questions

1. Allen, a 40-year-old man, comes to your office because he is concerned about his family history of heart disease. His father had a fatal myocardial infarction (MI) at 45 years of age, and his paternal grandfather had a fatal MI at age 47 years. Allen's father had two brothers and two sisters. One of the brothers had an MI at age 44 years, and one of the sisters had an MI at age 49 years. Allen's mother had a brother and sister, both of whom are still alive. Allen's mother's parents both survived into their 80s and died of "natural causes." Draw a pedigree summarizing the information you have taken about Allen's family, and make a recommendation for further study and/or treatment.

2. Mary's two brothers and her mother's brother all had Duchenne muscular dystrophy (DMD) and are now dead. Based on only this information, what is the probability that Mary is a heterozygous carrier for this disorder? What is the probability that she will produce affected offspring? Suppose Mary has a serum creatine kinase (CK) test and is told that her level is above the 95th percentile for homozygous normal individuals. Approximately two thirds of DMD carriers have CK levels above the 95th percentile. Given this information, use Bayes' theorem to calculate the probability that Mary is a carrier and the probability that she will produce affected offspring.

3. Bob's father had Huntington disease and is now deceased. Bob is now 51 years old and has no symptoms of Huntington disease. Age-of-onset curves show that approximately 85% of individuals with an affected father show symptoms by this age (the percentage is slightly lower, about 80%, if the mother is affected). Based on this information, use Bayes' theorem to estimate the probability that Bob inherited the Huntington disease mutation from his father.

## Suggested Readings

Aase J. Dysmorphologic diagnosis for the pediatric practitioner. Pediatr Clin North Am 1992;39:135–56.

Baker DL, Schuette JL, Uhlmann WR. A Guide to Genetic Counseling. New York: John Wiley; 1998.

Baty B Biesecker B. Evidence-based genetic counseling. (Special issue on genetic counseling). Am J Med Genet C 2007;142C:220.

Biesecker BB. Goals of genetic counseling. Clin Genet 2001;60:323–30.

Biesecker BB, Peters KF. Process studies in genetic counseling: peering into the black box. (Special issue

on genetic counseling). Am J Med Genet 2001; 106: 191–8.

Boulet SL, Yang Q, Mai CL, et al. Trends in the postfortification prevalence of spina bifida and anencephaly in the United States. Birth Defects Res A Clin Mol Teratol 2008;82:527–32.

Brent RL. Environmental causes of human congenital malformations: the pediatrician's role in dealing with these complex clinical problems caused by a multiplicity of environmental and genetic factors. Pediatrics 2004; 113:957–68.

Brent RL. The Bendectin saga: an American tragedy. Teratology 1983;27:283–6.

Carey JC, Viskochil DH. Status of the human malformation map: 2007. Am J Med Genet A 2007;143A(24):2868–85.

Cassidy SB, Allanson JE. Management of Genetic Syndromes. 3rd ed. Hoboken, NJ: John Wiley; 2009.

Clarke A. Ethical and social issues in clinical genetics. In: Rimoin DL, Connor JM, Pyeritz RE, Korf BR, editors. Emery and Rimoin's Principles and Practice of Medical Genetics. 4th ed., vol. 1. Philadelphia: Churchill Livingstone; 2002. p. 897–928.

Clayton EW. Ethical, legal, and social implications of genomic medicine. N Engl J Med 2003;349:562–9.

Cohen MM. The Child with Multiple Birth Defects. New York: Oxford; 1997.

Dent K, Carey JC. Breaking difficult news in the newborn setting: Down Syndrome as a paradigm. Am J of Med Genet C 2006;142C:173–9.

Donnai D. Genetic services. Clin Genet 2002;61:1–6.

Friedman JM, Polifka JE. Teratogenic Effects of Drugs: A Resource for Clinicians (TERIS). Baltimore: Johns Hopkins University Press; 2000.

Hennekam RCM, Allanson J, Krantz I. Gorlin's Syndromes of the Head and Neck. 5th ed. New York: Oxford University Press; 2009.

Harper PS. Practical Genetic Counseling. 5th ed. Oxford: Butterworth Heineman; 1999.

Hodge SE. A simple, unified approach to Bayesian risk calculations. J Genet Couns 1998;7:235–61.

Hunter AG. Medical genetics: 2. The diagnostic approach to the child with dysmorphic signs. CMAJ 2002;167:367–72.

Jones KL. Smith's Recognizable Patterns of Human Malformation. 6th ed. Philadelphia: WB Saunders; 2006.

Koren G. Medication Safety in Pregnancy and Breast Feeding. New York: McGraw-Hill; 2007.

Mahowald MB, Verp MS, Anderson RR. Genetic counseling: clinical and ethical challenges. Annu Rev Genet 1998;32:547–59.

Martinez LP, Robertson J, Leen-Mitchell M. Environmental causes of birth defects. In:Rudolph CD, Rudolph AM, Hostetter MK, Lister G, editors. Rudolph's Pediatrics. 22nd ed. New York: McGraw-Hill; 2009, p. 774–9.

Nelson K, Holmes LB. Malformations due to presumed spontaneous mutations in newborn infants. N Engl J Med 1989;320:19–23.

Nowlan W. Human genetics: a rational view of insurance and genetic discrimination. Science 2002; 297:195–6.

Polifka JE, Friedman JM. Medical genetics: 1. Clinical teratology in the age of genomics. CMAJ 2002; 167:265–73.

Resta R. Defining and redefining the scope and goals of genetic counseling. Am J Med Genet Part C 2007; 142C:269–75.

Rothenberg KH, Terry SF. Human genetics: before it's too late—addressing fear of genetic information. Science 2002;297:196–7.

Schneider KA. Counseling About Cancer: Strategies for Genetic Counselors. 2nd ed. New York: John Wiley; 2001.

Walker AP. Genetic counseling. In: Rimoin DL, Connor JM, Pyeritz RE, Korf BR, editors. Emery and Rimoin's Principles and Practice of Medical Genetics. 4th ed., vol 1. New York: Churchill Livingstone; 2002, p. 842–74.

Weil J. Psychosocial Genetic Counseling. New York: Oxford University Press; 2000.

Weiss JO, Mackta JS. Starting and Sustaining Genetic Support Groups. Baltimore: Johns Hopkins University Press; 1996.

## Internet Resources

Gene Clinics (reviews of tests for genetic diseases and laboratories that perform them) *http://www.geneclinics.org/*

Genetic Alliance (descriptions of genetic conditions, information on health insurance, and links to lay advocacy groups) *http://www.geneticalliance.org/*

Genetic and Rare Conditions Site (links to lay advocacy groups for a large number of genetic disorders) *http://www.kumc.edu/gec/support/*

Organization of Teratology Information Specialists (reviews of various exposures in pregnancy): *http://www.otispregnancy.org*

Genetic Interest Group (alliance of organizations, with a membership of more than 120 charities that support children, families, and individuals affected by genetic disorders) *http://www.gig.org.uk/*

National Institutes of Health Office of Rare Diseases (information on more than 6000 rare diseases) *http://rarediseases.info.nih.gov/*

POSSUM (computer-aided diagnosis of genetic disorders and other syndromes; paid subscription required) *http://www.possum.net.au/*

NOTE: When boldface type is used to highlight a word or phrase within a definition, it signifies that the highlighted word or phrase is defined elsewhere in this glossary.

**5′ cap**   A chemically modified guanine nucleotide added to the 5′ end of a growing mRNA molecule.

**acceptor site**   AG sequence that defines the splice site at the 3′ end of an intron.

**acetylation**   The addition of an acetyl group to a molecule (as in histone acetylation).

**acrocentric**   A chromosome whose centromere is close to the end of one arm.

**activator**   A **specific transcription factor** that binds to **coactivators** and to **enhancers** to help regulate the transcriptional activity of certain genes.

**adaptive immune system**   The portion of the immune system that is capable of changing its DNA sequence to bind foreign particles more effectively. It includes the humoral and cellular components. Compare with **innate immune system**.

**addition rule**   A law of probability that states that the probability of one event or another event occurring is derived by adding the probabilities of the two events together, assuming the events occur independently of one another.

**adenine**   One of the four DNA bases (abbrev.: A).

**adeno-associated virus**   A type of parvovirus sometimes used as a vector for somatic cell gene therapy.

**adenovirus**   A double-stranded RNA virus that is sometimes used in gene therapy.

**adjacent segregation**   A meiotic segregation pattern in which the pairing of translocated chromosomes leads to unbalanced gametes. Compare with **alternate segregation**.

**affected sib-pair method**   A linkage analysis method in which sib pairs who are both affected with a disease are assessed for the extent to which they share alleles at various marker loci. If alleles are shared significantly more often than the expected 50%, linkage of the disease to the marker is indicated.

**affinity**   The binding power of an antibody with an antigen (low affinity indicates poor binding; high affinity indicates precise binding).

**affinity maturation**   A phase of B cell development in which the cell undergoes **somatic hypermutation** so that some cells are formed that can achieve high-affinity binding to peptides from a pathogen.

**allele**   Conventional abbreviation for "allelomorph." Refers to the different forms, or DNA sequences, that a gene can have in a population.

**allele-specific oligonucleotide**   A short DNA sequence, usually 18 to 20 nucleotides, that can hybridize with either disease-causing or normal DNA sequences. Used in **direct diagnosis** of mutations.

**alpha-fetoprotein**   Albumin-like protein produced by the fetus. The level of alpha-fetoprotein is elevated in pregnancies with neural tube defects and may be decreased in pregnancies with Down syndrome.

**alpha-satellite DNA**   A type of repetitive DNA sequence found near centromeres.

**alternate segregation**   A meiotic segregation pattern in which the pairing of translocated chromosomes leads to balanced gametes. Compare with **adjacent segregation**.

**alternative splice site**   A variation in the location of intron–exon splice sites in some genes that allow one gene to produce multiple different protein products.

***Alu* family**   A major group of dispersed repetitive DNA sequences.

**amino acids**   The major building blocks of polypeptides. Each of the 20 amino acids is encoded by one or more mRNA codons.

**amniocentesis**   A prenatal diagnostic technique in which a small amount of amniotic fluid is withdrawn transabdominally at about 16 weeks after the last menstrual period. Fetal cells can then be tested for some genetic diseases.

**amniocentesis, early**   Amniocentesis carried out at approximately 12 to 14 weeks after the last menstrual period.

**amniocyte**   Fetal cell found in the amniotic fluid.

**anaphase**   One of the stages of cell division, in which sister chromatids separate and move toward opposite sides of the cell.

**ancestry**   The origins of an individual's ancestors; most commonly used to refer to the geographic origins of one's ancestors.

**aneuploid**   The condition in which the number of chromosomes is not a multiple of 23, as in trisomy and monosomy. Compare with **euploid**. (n.: aneuploidy).

**antibody**   Molecule produced by plasma cells; antibodies bind to invading antigens.

**anticipation**   A feature of pedigrees in which a disease is seen at earlier ages or with increased severity in more recent generations.

**anticodon**   A three-nucleotide DNA sequence in a tRNA molecule that undergoes complementary base pairing with an mRNA codon.

**antigen**   A molecule that provokes antibody formation (from "*antibody generator*").

**antigen-presenting cell**   A cell that engulfs foreign bodies, digests them, and then displays the foreign antigens on its cell surface for recognition by T lymphocytes.

**antisense strand**   In a double-stranded DNA molecule, the strand from which mRNA is transcribed. See **sense strand**.

**antisense therapy**   A type of somatic cell gene therapy in which an **oligonucleotide** is synthesized that can hybridize with a mutant mRNA sequence, blocking its translation into protein.

**apoptosis**   Programmed cell death.

**association**   The co-occurrence of two traits or events more often than expected by chance.

**autoimmunity**   The condition in which one's immune system attacks one's own cells.

**autoradiogram**   The image produced by exposing a radioactively labeled substance, such as a probe, to x-ray film (used, for example, in detecting RFLPs and in performing in situ hybridization).

**autosomes**   The 22 pairs of chromosomes excluding the sex chromosomes (X and Y).

**axis specification**   Definition, during embryonic development, of the major axes of the embryo: ventral/dorsal and anterior/posterior.

**B lymphocyte** (also, **B cell**)   A component of the adaptive immune system that produces antibodies.

**bacterial artificial chromosome (BAC)**   A recombinant plasmid inserted in bacteria that serves as a cloning **vector** capable of accepting DNA inserts of 50 to 200 kb.

**bacteriophage**   A virus that infects bacteria. In recombinant DNA technology, bacteriophages are used as vectors to carry inserted DNA sequences.

**bacteriophage P1 artificial chromosome (PAC)**   A cloning vector that consists of bacteriophage P1 and is inserted into a plasmid; it accepts DNA inserts up to 100 kb.

**bands**   (1) Visibly darkened areas on **autoradiograms** that represent the location of alleles on a gel. (2) Alternating dark and light areas visible on chromosomes after certain types of stains are used.

**Barr body**   The inactive X chromosome, visible as a densely staining chromatin mass in the somatic cells of normal females. Also known as **sex chromatin**.

**base**   One of the four nitrogenous substances (adenine, cytosine, guanine, or thymine) that make up part of the DNA molecule. Combinations of bases specify amino acid sequences.

**base analog**   A substance that can mimic the chemical behavior of one of the four DNA bases. Base analogs are a type of **mutagen**.

**base pair (bp)**   A unit of complementary DNA bases in a double-stranded DNA molecule (A-T or C-G).

**base pair substitution**   The replacement of one base pair by another. A type of mutation.

**Bayes' theorem**   A statistical procedure in which prior and conditional probabilities are used to derive an improved estimate of probability or risk.

**benign**   Describes a neoplasm (tumor) that does not invade surrounding tissue or **metastasize** to other parts of the body. Compare with **malignant**.

**bivalent**   A pair of intertwined homologous chromosomes seen in prophase I of meiosis. Synonymous with **tetrad**.

**blood group**   Molecules found on the surfaces of erythrocytes, some of which (ABO and Rh) determine blood transfusion compatibility.

**body plan**   The pattern and arrangement of body segments during embryonic development.

**bp**   Abbreviation for **base pair**.

**breakpoint**   The location on a chromosome at which a translocation has occurred.

**C-banding**   A type of chromosome staining that highlights the constitutive heterochromatin that lies at and near centromeres.

**candidate gene**   A gene that, on the basis of known properties or protein product, is thought to be the gene causing a specific genetic disease.

**capillary sequencing**   A method of DNA sequencing in which DNA migrates through a thin capillary tube in order to separate DNA fragments of different lengths.

**carcinogen**   A substance that can produce cancer (adj.: carcinogenic).

**carcinogenesis**   The process of cancer development.

**carrier**   A person who has a copy of a disease-causing gene but does not express the disease. The term is usually used to denote heterozygotes for a recessive disease gene.

**catalyst**   A substance that increases the rate of a chemical reaction. Enzymes are an example of a catalyst.

**cDNA**   Complementary DNA, formed by reverse transcription of mRNA purified from a collection of cells. This type of DNA corresponds only to coding sequence (**exons**).

**cDNA library**   A collection of segments of complementary DNA (cDNA) cloned into vehicles such as phages or plasmids. Compare with **genomic library**.

**cell adhesion molecule**   Cell-surface molecule that participates in the interaction of T cells and their targets.

**cell cycle**   The alternating sequence of mitosis and interphase.

**cell fate**   The location and function of cells, programmed during embryonic development.

**cellular immune system**   The T-cell component of the adaptive immune system.

**centimorgan (cM)**   A unit of measure of the frequency of recombination between two loci, also known as a **map unit**. One cM corresponds to a recombination frequency of 1%.

**centriole**   Structure in cells that helps to pull chromosomes apart during meiosis and mitosis.

**centromere**   The region of a chromosome that separates the two arms; centromeres are the sites of attachment of spindle fibers during cell division.

**CG islands**   Unmethylated CG sequences that are found near the 5′ ends of many genes.

**chiasma (**pl. **chiasmata)**   The location of a **crossover** between two homologous chromosomes during meiosis.

**chorionic villus sampling (CVS)**   A prenatal diagnostic technique in which a small sample of chorionic villi is aspirated. Usually performed at 10 to 12 weeks' gestation.

**chromatin**   The combination of proteins (e.g., **histones**) and nucleic acids that makes up chromosomes.

**chromatin loop**   A unit of DNA coiling consisting of a group of solenoids. Each loop is approximately 100 kb.

**chromatin opener**   Regulatory element that is capable of decondensing or "opening" regions of chromatin.

**chromosome**   Threadlike structure (literally "colored body") consisting of **chromatin**. Genes are arranged along chromosomes.

**chromosome abnormalities**   A major group of genetic diseases, consisting of microscopically observable alterations of chromosome number or structure.

**chromosome banding**   The process of applying specific stains to chromosomes in order to produce characteristic patterns of **bands** (example: G-banding).

**chromosome breakage**   The fracture of chromosomes; breakage is increased in the presence of **clastogens**.

**chromosome instability syndrome**   Diseases characterized by the presence of large numbers of chromosome breaks or exchanges, such as **sister chromatid exchange** (example: Bloom syndrome).

**chromosome-specific library**   A collection of DNA fragments from a single chromosome.

**class switching**   The process in which the heavy chains of B lymphocytes change from one class, or **isotype**, to another (e.g., IgM to IgG).

**clastogen**   A substance that can induce **chromosome breakage** (example: radiation).

**clone**   (1) A series of identical DNA fragments created by recombinant DNA techniques. (2) Identical cells that are descended from a single common ancestor.

**cloning, functional**   A method of isolating genes in which a gene whose protein product's function is already known is evaluated as a **candidate gene** responsible for a trait or disease.

**cloning, positional**   The isolation and cloning of a disease gene after determining its approximate physical location; the gene product is subsequently determined. Formerly termed "reverse genetics."

**coactivator**   A type of **specific transcription factor** that binds to **activators** and to the **general transcription factor** complex to regulate the transcription of specific genes.

**codominant**   Alleles that are both expressed when they occur together in the heterozygous state (example: *A* and *B* alleles of the ABO blood group system).

**codon**   A group of three mRNA bases, each of which specifies an amino acid when translated.

**coefficient of relationship**   A statistic that measures the proportion of genes shared by two individuals as a result of descent from a common ancestor.

**cofactors**   Substances that interact with enzymes to produce chemical reactions, such as various metabolic processes (examples: dietary trace elements, vitamins).

**colcemid** or **colchicine**   A spindle poison that arrests cells in metaphase, making them easily discernible microscopically.

**comparative genomic hybridization (CGH)**   Technique in which a mixture of DNA from a test source (e.g., a tumor) and a normal control are differentially labeled, mixed, and hybridized with normal metaphase chromosomes or a microarray (**array CGH**). Differences in color reveal chromosome losses or duplications in the test DNA.

**complement system**   A component of the immune system, encoded by genes in the class III MHC region, that can destroy invading organisms. The complement system also interacts with other components of the immune system, such as antibodies and phagocytes.

**complementary base pairing**   A fundamental process in which adenosine pairs only with thymine and guanine pairs only with cytosine. Also sometimes known as *Watson–Crick* pairing.

**compound heterozygote**   An individual who is heterozygous for two different disease-causing mutations at a locus.

Compare with **homozygote**. Compound heterozygotes for recessive disease mutations are usually affected with the disorder.

**concordant** Refers to two individuals who have the same trait (e.g., monozygotic twins may be concordant for a disease such as diabetes). Compare with **discordant**. (n.: concordance).

**conditional probability** The probability that an event will occur, given that another event has already occurred. Conditional probabilities are used, for example, in **Bayes' theorem**.

**confined placental mosaicism** A form of **mosaicism** that is observed in the placenta but not in the fetus.

**congenital** Present at birth.

**consanguinity** The mating of related individuals (adj.: consanguineous).

**consensus sequence** A sequence that denotes the DNA bases most often seen in a region of interest. The sequences that are found near **donor** and **acceptor sites** are a type of consensus sequence.

**conservation** The preservation of highly similar DNA sequences among different organisms; conserved sequences are usually found in functional genes.

**conserved** See **conservation**.

**constitutional** or **constitutive** Pertaining to DNA in normal cells of the body, usually used in contrast to tumor DNA.

**constitutive heterochromatin** See **heterochromatin, constitutive**.

**contig map** A physical map of a chromosome region constructed by isolating overlapping (contiguous) DNA segments.

**contiguous gene syndrome** A disease caused by the deletion or duplication of multiple consecutive genes. See also **microdeletion**.

**copy number variant (CNV)** DNA sequences of 1000 or more base pairs that are present in variable numbers in different individuals.

**cordocentesis** See **percutaneous umbilical blood sampling (PUBS)**.

**cosmid** A phage–plasmid hybrid capable of accepting larger DNA inserts (up to 40-50 kb) than either phage or plasmids.

**costimulatory molecule** Cell-surface molecule that participates in the binding of T-cell receptors to MHC–antigen complexes.

**cross** Mating between organisms in genetic studies.

**cross-reaction** The binding of an antibody to an antigen other than the one that originally stimulated antibody formation. Such antigens are usually very similar to the original antibody-generating antigen.

**crossing over** or **crossover** The exchange of genetic material between homologous chromosomes during meiosis (also occurs rarely during mitosis); produces **recombination**.

**cryptic splice site** Site at which intron–exon splicing can occur when the usual splice site is altered.

**cyclin-dependent kinases** Enzymes that form complexes with specific **cyclins** to phosphorylate regulatory proteins (such as pRb) at specific stages of the cell cycle.

**cyclin-dependent kinase inhibitors** Proteins that inactivate **cyclin-dependent kinases**. Many of these are tumor suppressors (examples: p16, p21).

**cyclins** Proteins that interact with specific **cyclin-dependent kinases** to regulate the cell cycle at specific stages.

**cytogenetics** The study of chromosomes and their abnormalities. Combines cytology, the study of cells, and genetics.

**cytokine** A growth factor that causes cells to proliferate (example: interleukins).

**cytokinesis** Cytoplasmic division that occurs during mitosis and meiosis.

**cytosine** One of the four DNA bases (abbrev.: C).

**cytotoxic T lymphocyte** A type of T lymphocyte that destroys a cell when the cell presents a complex of MHC class I molecule and foreign peptide. Part of the cellular immune system.

**daughter cells** Cells that result from the division of a parent cell.

**deformation** Alteration of the form, shape, or position of a normally formed body part by mechanical forces (example: oligohydramnios sequence).

**deletion** The loss of chromosome material. May be terminal or interstitial. Compare with **duplication**.

**deletion, interstitial** Deletion that removes part of the interior of the chromosome.

**deletion, terminal** Deletion that removes part of a chromosome, including one telomere.

**denaturing gradient gel electrophoresis (DGGE)** See **electrophoresis, denaturing gradient gel**.

**deoxyribonucleic acid** See **DNA**.

**derivative chromosome** A chromosome that has been altered as a result of a translocation [example: derivative 9, or der(9)].

**dideoxy method** A technique for sequencing DNA in which dideoxynucleotides, which terminate replication, are incorporated into replicating DNA strands.

**diploid** Having two copies of each chromosome. In humans, the diploid number is 46. Compare with **haploid, polyploid**.

**direct cDNA selection** A method of exon detection in which a genomic DNA fragment inserted into a vector such as a YAC is hybridized with cDNA; hybridization and selection identify the regions of the genomic DNA fragment that correspond to cDNA (i.e., exon-containing DNA).

**direct diagnosis**   A form of DNA-based disease diagnosis in which the mutation itself is examined directly. Compare with **indirect diagnosis**.

**discordant**   Refers to two individuals who do not share the same trait. Compare with **concordant**. (n.: discordance).

**dispermy**   Fertilization of a single ovum by two sperm cells.

**dispersed repetitive DNA**   A class of repeated DNA sequences in which single repeats are scattered throughout the genome. Compare with **tandem repeat**.

**disruption**   Morphological defect that results from the breakdown of an otherwise normal developmental process (example: limb reduction defect resulting from poor vascularization).

**dizygotic**   A type of twinning in which each twin is produced by the fertilization of a different ovum. Synonymous with fraternal twin. Compare with **monozygotic**.

**DNA (deoxyribonucleic acid)**   A double-helix molecule that consists of a sugar-phosphate backbone and four nitrogenous bases (A, C, G, and T). DNA bases encode **messenger RNA (mRNA)**, which in turn encodes amino acid sequences.

**DNA-binding motifs**   Portions of transcription factors that allow them to interact with specific DNA sequences (examples: helix-loop-helix motif, zinc finger motif).

**DNA chips**   See **microarrays**.

**DNA looping**   The formation of looped structures in DNA; sometimes permits the interaction of various regulatory elements.

**DNA mismatch repair**   A type of **DNA repair** in which nucleotide mismatches (i.e., violations of the G-C, A-T complementary base-pairing rule) are corrected by specialized enzymes.

**DNA polymerase**   An enzyme involved in DNA replication and repair.

**DNA profile**   A series of DNA polymorphisms (usually VNTRs or microsatellites) typed in an individual. Because these polymorphisms are highly variable, the combined genotypes are useful in identifying individuals for forensic purposes.

**DNA repair**   A process in which mistakes in the DNA sequence are altered to represent the original sequence.

**DNA sequence**   The order of DNA bases along a chromosome.

**DNA sequencing, automated**   DNA sequencing techniques in which automated procedures such as computer-guided laser scanning are used to provide more rapid, accurate DNA sequence results.

**dominant**   An allele that is expressed in the same way in single copy (heterozygotes) as in double copy (homozygotes). Compare with **recessive**.

**dominant negative**   A type of mutation in which the altered protein product in a heterozygote forms a complex with the normal protein product produced by the homologous normal gene, thus disabling it.

**donor site**   The GT sequence that defines the splice site at the 5′ end of an intron.

**dosage compensation**   The situation in which, as a consequence of X inactivation, the amount of X chromosome–encoded gene product in females is roughly equal to that in males.

**dosage mapping**   A technique for mapping genes in which excess or deficient gene product is correlated with a chromosomal duplication or deletion.

**dosage sensitivity**   A condition in which alteration of the level of a gene product (e.g., a **deletion** resulting in 50% expression or a **duplication** resulting in 150% expression of the gene product) causes a substantial alteration of the phenotype (including disease).

**double helix**   The "twisted-ladder" shape of the double-stranded DNA molecule.

**double-stranded DNA breaks**   A type of DNA breakage in which both strands are broken at a specific location.

**duplication**   The presence of an extra copy of chromosome material. Compare with **deletion**.

**dysmorphology**   The study of abnormal physical development.

**dysplasia**   A defect in which cells are abnormally organized into tissue (example: bone dysplasia).

**ectopic expression**   Expression of a gene product in an abnormal location or tissue type.

**electrophoresis**   A technique in which charged molecules are placed in a medium and exposed to an electrical field, causing them to migrate through the medium at different rates, according to charge, length, or other attributes.

**electrophoresis, denaturing gradient gel (DGGE)**   A method of mutation detection in which DNA fragments are electrophoresed through a gel in which there is a changing denaturing factor, such as temperature.

**embryonic stem cell**   Cells, found in early embryos, that have the potential to become any cell type (**pluripotency**).

**empirical risk**   A risk estimate based on direct observation of data.

**endocytosis**   A process in which molecules are transported into the interior of cells.

**enhancer**   A regulatory DNA sequence that interacts with **specific transcription factors** to increase the transcription of genes. Compare with **silencer**.

**equational division**   The second major cycle of meiosis: meiosis II. Compare with **reduction division**.

**equatorial plane**   The middle of the cell's spindle, along which homologous chromosomes are arranged during metaphase.

**erythroblast** A nucleated red blood cell; the precursor of an **erythrocyte**.

**erythrocyte** A red blood cell.

**euchromatin** Chromatin that is light-staining during interphase and tends to be transcriptionally active. Compare with **heterochromatin**.

**eugenics** The use of controlled breeding to increase the prevalence of "desirable" genetic traits (positive eugenics) and to decrease the prevalence of "undesirable" traits (negative eugenics).

**eukaryotes** Organisms whose cells have true nuclei.

**euploid** Refers to cells whose chromosome number is a multiple of 23 (in humans). (n.: euploidy).

**exons** Portions of genes that encode amino acids and are retained after the primary mRNA transcript is spliced. Compare with **intron**.

**exon trapping** A method for isolating exons in a fragment of genomic DNA by using an in vitro cell system to artificially splice out the introns.

**expanded newborn screening** See **newborn screening**.

**expanded repeat** A type of mutation in which a tandem trinucleotide repeat increases in number (example: Huntington disease).

**expressed sequence tag (EST)** Several hundred bp of known **cDNA** sequence, flanked by **PCR** primers. Because they are derived from **cDNA** libraries, these sequences represent portions of expressed genes.

**false negative** A test result in which an affected individual is incorrectly identified as being unaffected with the disease in question. Compare with **false positive**.

**false positive** A test result in which an unaffected individual is incorrectly identified as being affected with the disease in question. Compare with **false negative**.

**fetoscopy** A fetal visualization technique in which an endoscope is inserted through the abdominal wall. Sometimes used in prenatal diagnosis.

**fiber FISH** See **fluorescence in situ hybridization, fiber**.

**fibrillin** A component of connective tissue; mutations in the fibrillin gene can cause Marfan syndrome.

**flow cytometry** A technique whereby chromosomes can be individually sorted.

**fluorescence in situ hybridization (FISH)** A molecular cytogenetic technique in which labeled probes are hybridized with chromosomes and then visualized under a fluorescence microscope.

**fluorescence in situ hybridization, fiber (fiber FISH)** A version of the **FISH** technique in which fluorescent probes are hybridized to interphase chromosomes that have been manipulated so that the chromatin fibers are stretched, allowing high-resolution visualization and mapping.

**founder effect** A large shift in gene frequencies that results when a small "founder" population, which contains limited genetic variation, is derived from a larger population. Founder effect can be considered a special case of **genetic drift**.

**frameshift mutation** An alteration of DNA in which a duplication or deletion occurs that is not a multiple of three base pairs.

**fusion gene** A gene that results from a combination of two genes or parts of two genes.

**gain of function** A class of mutations that results in a protein product that either is increased in quantity or has a novel function. Compare with **loss of function**.

**gamete** The haploid germ cell (sperm and ovum).

**gametogenesis** The process of gamete formation.

**gastrulation** The embryonic stage in which cells of the blastula are arranged to become a three-layer structure consisting of endoderm, mesoderm, and ectoderm.

**gene** The fundamental unit of heredity.

**gene–environment interaction** A mutual phenotypic effect of a gene and an environmental factor that is greater than the single effect of either factor alone (example: the effect of $\alpha_1$-antitrypsin deficiency and cigarette smoking on pulmonary emphysema).

**gene family** A group of genes that are similar in DNA sequence and have evolved from a single common ancestral gene; they may or may not be located in the same chromosome region.

**gene flow** The exchange of genes between different populations.

**gene frequency** In a population, the fraction of chromosomes that contain a specific gene.

**gene therapy** The insertion or alteration of genes to correct a disease.

**gene therapy, germline** Gene therapy that alters all cells of the body, including the germline. Compare with **gene therapy, somatic cell**.

**gene therapy, somatic cell** Gene therapy that alters somatic cells but not cells of the germline. Compare with **gene therapy, germline**.

**genetic code** The combinations of mRNA **codons** that specify individual amino acids.

**genetic counseling** The delivery of information about genetic diseases (risks, natural history, and management) to patients and their families.

**genetic drift** An evolutionary process in which gene frequencies change as a result of random fluctuations in the transmission of genes from one generation to the next. Drift is greater in smaller populations.

**genetic engineering** The alteration of genes; it typically involves **recombinant DNA** techniques.

**genetic mapping**   The ordering of genes on chromosomes according to recombination frequency. Compare with **physical mapping**.

**genetic screening**   The large-scale testing of defined populations to identify those who are at increased risk of having a disease-causing gene.

**genetic testing**   The analysis of DNA, RNA, or proteins to test for the presence of differences that might cause a genetic disease.

**genome**   The totality of an organism's DNA.

**genome scan**   A gene-mapping approach in which markers from the entire human genome are tested for linkage with a disease phenotype.

**genome-wide association studies (GWAS)**   A study design in which allele frequencies at large numbers of loci (typically **single nucleotide polymorphisms, SNPs**) are compared in disease cases and unaffected controls. SNPs that show large frequency differences between cases and controls are likely to be located in or near genes that are responsible for the disease.

**genomic library**   A collection of DNA fragments from an organism's entire genome. It includes cDNA as well as noncoding DNA. Compare with **cDNA library**.

**genomic instability**   An abnormal condition in which there is a substantial increase in mutations throughout the genome. It can occur, for example, when a DNA repair system is disabled.

**genotype**   An individual's allelic constitution at a locus.

**genotype frequency**   The fraction of individuals in a population that carry a specific genotype.

**genotype–phenotype correlation**   The relationship between different possible genotypes (i.e., different alleles) at a locus and the individual's phenotype. Because of **allelic heterogeneity**, different alleles at a locus can produce more or less severe expression of a disease phenotype (e.g., missense vs. nonsense mutations).

**germline**   Cells responsible for the production of gametes.

**Giemsa**   A type of staining that produces G-bands in chromosomes.

**globin**   A major component of the hemoglobin molecule. Globin is also found in the vertebrate myoglobin molecule.

**growth factor**   A substance capable of stimulating cell proliferation.

**growth factor receptor**   A structure on cell surfaces to which growth factors can bind.

**guanine**   One of the four DNA bases (abbrev.: G).

**guanosine diphosphate (GDP)**   A partially dephosphorylated form of **guanosine triphosphate**.

**guanosine triphosphate (GTP)**   A molecule required for the synthesis of peptide bonds during translation.

**haploid**   Refers to cells that have one copy of each chromosome, the typical state for gametes. In humans, the haploid number is 23.

**haploinsufficiency**   Describes the situation in which 50% of the normal level of gene expression (i.e., in a heterozygote) is not sufficient for normal function.

**haplotype**   The allelic constitution of multiple loci on a single chromosome. Derived from "haploid genotype."

**Hardy–Weinberg principle**   Specifies an equilibrium relationship between gene frequencies and genotype frequencies in populations.

**heavy chain**   A major structural component of an antibody molecule, having a higher molecular weight than the other major component, the light chain. There are five major types of heavy chains in the human: $\gamma$, $\mu$, $\alpha$, $\delta$, and $\varepsilon$.

**helper T lymphocyte**   A type of T lymphocyte whose receptors bind to a complex of MHC class II molecule and foreign peptide on the surfaces of antigen-presenting cells. Part of the cellular immune system.

**heme**   The iron-containing component of the hemoglobin molecule; binds with oxygen.

**hemizygous**   Refers to a gene that is present in only a single copy (*hemi* = "half"). Most commonly refers to genes on the single male X chromosome but can refer to other genes in the haploid state, such as the genes homologous to a deleted region of a chromosome.

**heritability**   The portion of population **variance** in a trait that can be ascribed to genetic factors.

**heterochromatin**   Dark-staining chromatin that is usually transcriptionally inactive and consists mostly of repetitive DNA. Compare with **euchromatin**.

**heterochromatin, constitutive**   Heterochromatin that consists of **satellite DNA**; located near centromeres and on the short arms of acrocentric chromosomes.

**heterodisomy**   The presence in a cell of two chromosomes derived from a single parent and none from the other parent (disomy). In heterodisomy, the two chromosomes are the nonidentical homologous chromosomes. Compare with **isodisomy**.

**heterogeneity, allelic**   Describes conditions in which different alleles at a locus can produce variable expression of a disease. Depending on phenotype definition, allelic heterogeneity can cause two distinct diseases, as in Duchenne and Becker muscular dystrophy.

**heterogeneity, locus**   Describes diseases in which mutations at distinct loci can produce the same disease phenotype (examples: osteogenesis imperfecta; retinitis pigmentosa).

**heteromorphism**   Variation in the microscopic appearance of a chromosome.

**heteroplasmy**   The existence of differing DNA sequences at a locus within a single cell. Often seen in mitochondrial genes.

**heterotetramer**   A molecule consisting of four (*tetra*) subunits, at least one of which differs from the others. Compare with **homotetramer**.

**heterozygote**   An individual who has two different alleles at a locus. Compare with **homozygote**.

**high-resolution banding**   Chromosome banding using prophase or prometaphase chromosomes, which are more extended than metaphase chromosomes and thus yield more bands and greater resolution.

**histone**   The protein core around which DNA is wound in a chromosome.

**holandric**   Refers to Y-linked inheritance; a transmission exclusively from father to son.

**homeodomain**   A DNA-binding portion of transcription factor proteins that are involved in embryonic development.

**homologous**   (1) Refers to DNA or amino acid sequences that are highly similar to one another. (2) Describes chromosomes that pair during meiosis, one derived from the individual's father and the other from the mother.

**homologs**   Chromosomes that are **homologous**.

**homotetramer**   Molecule consisting of four (*tetra*) identical subunits. Compare with **heterotetramer**.

**homozygote**   An individual in whom the two alleles at a locus are the same. Compare with **heterozygote**.

**housekeeping genes**   Genes whose protein products are required for cellular maintenance or metabolism. Because of their central role in the life of the cell, housekeeping genes are transcriptionally active in all cells.

**human artificial chromosome**   A synthetic chromosome consisting of an artificial centromere and telomeres and an insert of human DNA that can be 5 to 10 Mb.

**human leukocyte antigen (HLA)**   Older term for **major histocompatibility complex (MHC)**.

**humoral immune system**   The B-cell component of the adaptive immune system, termed humoral because antibodies are secreted into the circulation.

**immunodeficiency disease**   A class of diseases characterized by insufficiencies in the immune response (example: severe combined immune deficiency).

**immunodeficiency disease, primary**   Disorder of the immune system that is caused directly by defects (usually genetic) in components or cells of the immune system.

**immunodeficiency disease, secondary**   Disorder of the immune system that is a consequence of an agent or defect that originates outside the immune system (e.g., infection, radiation, drugs).

**immunogenetics**   The study of the genetic basis of the immune system.

**immunoglobulin**   Receptor found on the surfaces of B cells. When secreted into the circulation by B cells that have matured into plasma cells, immunoglobulins are known as **antibodies**.

**imprinting, genomic**   Describes process in which genetic material is expressed differently when inherited from the mother than when inherited from the father.

**in situ hybridization**   Molecular gene mapping technique in which labeled probes are hybridized to stained metaphase chromosomes and then exposed to x-ray film to reveal the position of the probe.

**in vitro fertilization (IVF)**   A procedure in which the fertilization of an egg cell by a sperm cell is carried out in the laboratory. The embryo is then implanted in the mother's uterus.

**incest**   The mating of closely related individuals, usually describing the union of first-degree relatives.

**independence**   A principle, often invoked in statistical analysis, indicating that the occurrence of one event has no effect on the probability of occurrence of another (adj.: independent).

**independent assortment**   One of Mendel's fundamental principles; dictates that alleles at different loci are transmitted independently of one another.

**index case**   See **proband**.

**indirect diagnosis**   A form of genetic diagnosis in which the disease-causing mutation is not directly observed; usually refers to diagnosis using linked markers. Compare with **direct diagnosis**.

**induction**   The influence or determination of development of one group of cells by a second group of cells.

**innate immune system**   The portion of the immune system that does not change its characteristics to respond to infections. Major part of the initial immune response. Compare with **adaptive immune system**.

**insert**   A DNA sequence that is placed into a vector, such as a plasmid or cosmid, using recombinant DNA techniques.

**interphase**   The portion of the cell cycle that alternates with **meiosis** or **mitosis** (cell division). DNA is replicated and repaired during this phase.

**intraclass correlation coefficient**   A statistical measure that varies between $-1$ and 1 and specifies the degree of similarity of two quantities in a sample or population.

**intron**   DNA sequence found between two **exons**. It is transcribed into primary mRNA but is spliced out in the formation of the mature mRNA transcript.

**inversion**   A structural rearrangement of a chromosome in which two breaks occur, followed by reinsertion of the chromosome segment, but in reversed order. It may be either paracentric or pericentric. See also **inversion, paracentric** and **inversion, pericentric**.

**inversion, paracentric**   An inversion that does not include the centromere.

**inversion, pericentric**   An inversion that includes the centromere.

**isochromosome** A structural chromosome rearrangement caused by the division of a chromosome along an axis perpendicular to the usual axis of division; results in chromosomes with either two short arms or two long arms.

**isodisomy** The presence in a cell of two identical chromosomes derived from one parent and none from the other parent. Compare with **heterodisomy**.

**isotype** Classes of immunoglobulin molecules (e.g., IgA, IgE, IgG), determined by the type of heavy chain present in the molecule.

**joint probability** The probability of two events both occurring.

**karyotype** A display of chromosomes, ordered according to length.

**kilobase (kb)** One thousand DNA base pairs.

**knockout** An animal model in which a specific gene has been disabled.

**lentivirus** A type of **retrovirus** that can enter nondividing cells.

**light chain** A major structural component of the antibody molecule, consisting of either a κ or a λ chain. The light chain has a lower molecular weight than the other major component, the heavy chain.

**likelihood** A statistic that measures the probability of an event or a series of events.

**LINEs (long interspersed elements)** A class of dispersed repetitive DNA in which each repeat is relatively long, up to 7 kb. Compare with **SINEs**.

**linkage** Describes two loci that are located close enough on the same chromosome that their recombination frequency is less than 50%.

**linkage disequilibrium** A nonrandom association of alleles at linked loci in populations. Compare with **linkage equilibrium**.

**linkage equilibrium** Lack of preferential association of alleles at linked loci. Compare with **linkage disequilibrium**.

**linkage phase** The arrangement of alleles of linked loci on chromosomes.

**liposome** A fat body sometimes used as a vector for somatic cell gene therapy.

**locus** The chromosome location of a specific gene (pl.: loci).

**locus control region** A DNA sequence in the 5′ region of the globin gene clusters that is involved in transcriptional regulation.

**LOD score** The common logarithm of the ratio of the likelihood of linkage at a specific recombination fraction to the likelihood of no linkage.

**loss of function** A class of mutation in which the alteration results in a nonfunctional protein product. Compare with **gain of function**.

**loss of heterozygosity** Describes a locus or loci at which a deletion or other process has converted the locus from heterozygosity to homozygosity or hemizygosity.

**Lyon hypothesis** A proposal (now verified) that one X chromosome is randomly inactivated in each somatic cell of the normal female embryo (lyonization).

**macrophage** A type of **phagocyte** that ingests foreign microbes and displays them on its surface for recognition by T-cell receptors.

**major gene** A single locus responsible for a trait (sometimes contrasted with a polygenic component).

**major histocompatibility complex (MHC), class I** A membrane-spanning glycoprotein, found on the surfaces of nearly all cells, that presents antigen for recognition by cytotoxic T lymphocytes. Compare with **MHC class II**.

**major histocompatibility complex (MHC), class II** A membrane-spanning glycoprotein, found on the surfaces of **antigen-presenting cells**, that presents antigen for recognition by helper T cells.

**malformation** A primary morphologic defect resulting from an intrinsically abnormal developmental process (example: polydactyly).

**malignant** Describes a tumor that is capable of invading surrounding tissue and metastasizing to other sites in the body. Compare with **benign**.

**manifesting heterozygote** An individual who is heterozygous for a recessive trait but displays the trait. Most often used to describe females who are heterozygous for an X-linked trait and display the trait.

**markers** Polymorphisms, such as RFLPs, VNTRs, microsatellite repeats, and blood groups, that are linked to a disease locus.

**mass spectrometry** Analysis of the mass-to-charge ratio of molecules; can be used to sequence DNA and detect mutations.

**mass spectrometry, tandem** A form of mass spectrometry in which two mass spectrometry machines are used, the first of which separates molecules according to mass, and the second of which assesses the mass and charge of the molecules after they have been fragmented.

**maternal serum alpha-fetoprotein (MSAFP)** Alpha-fetoprotein present in the serum of pregnant women; used in prenatal screening for fetal disorders such as neural tube defects and Down syndrome.

**mature transcript** Describes mRNA after the introns have been spliced out. Before splicing, the mRNA is referred to as a **primary transcript**.

**maximum likelihood estimate** A statistical procedure in which the likelihoods of a variety of parameter values are estimated and then compared to determine which likelihood is the largest. Used, for example, in evaluating LOD scores to determine which recombination frequency is the most likely.

**megabase (Mb)**   One million base pairs.

**meiosis**   Cell division process in which haploid gametes are formed from diploid germ cells.

**meiotic failure**   Aberrant meiosis in which a diploid gamete is produced rather than the normal haploid gamete.

**memory cells**   A class of high-affinity-binding B cells that remain in the body after an immune response has ended; they provide a relatively rapid, high-affinity response should the same disease antigen be encountered a second time.

**mendelian**   Referring to Gregor Mendel; describes a trait that is attributable to a single gene.

**mesenchyme**   Tissue that forms connective tissues and lymphatic and blood vessels during embryonic development.

**messenger RNA (mRNA)**   An RNA molecule that is formed from the **transcription** of DNA. Before intron splicing, mRNA is termed a **primary transcript**; after splicing, the **mature transcript** (or mature mRNA) proceeds to the cytoplasm, where it is translated into an amino acid sequence.

**metacentric**   A chromosome in which the centromere is located approximately in the middle of the chromosome arm.

**metaphase**   A stage of mitosis and meiosis in which homologous chromosomes are arranged along the equatorial plane, or metaphase plate, of the cell. This is the mitotic stage at which chromosomes are maximally condensed and most easily visualized.

**metastasis**   The spread of malignant cells from one site in the body to another (v.: metastasize).

**methylation**   The attachment of methyl groups; in genetics, refers especially to the addition of methyl groups to cytosine bases, forming 5-methylcytosine. Methylation is correlated with reduced transcription of genes.

**MHC restriction**   The limiting of immune response functions to MHC-mediated interactions (e.g., the binding of T-cell receptors is MHC restricted because it requires the presentation of antigen by MHC class I or MHC class II molecules).

**microarrays**   Arrangement of large numbers of DNA sequences, such as oligonucleotides consisting of normal and mutated sequences, on glass slides or silicone chips (DNA chips). These oligonucleotides can be hybridized to labeled DNA from subjects to test for sequence variants, to sequence DNA, or to analyze gene expression patterns.

**microdeletion**   A chromosome deletion that is too small to be visible under a microscope (examples: DiGeorge syndrome; Prader–Willi syndrome). See also **contiguous gene syndrome**.

**microsatellite**   A type of **satellite DNA** that consists of small repeat units (usually 2, 3, 4, or 5 bp) that occur in tandem.

**microsatellite repeat polymorphism**   A type of genetic variation in populations consisting of differing numbers of **microsatellite** repeat units at a locus. Also known as short tandem repeat polymorphisms (STRPs).

**minisatellite**   A type of **satellite DNA** that consists of tandem repeat units that are each about 20 to 70 bp long. Variation in the number of minisatellite repeats is the basis of VNTR polymorphisms.

**mismatch**   The presence in one chain of double-stranded DNA of a base that is not complementary to the corresponding base in the other chain. Also known as mispairing.

**mismatch repair**   A DNA repair process in which mismatched nucleotides are altered so that they become complementary.

**missense**   A type of mutation that results in a single amino acid change in the translated gene product. Compare with **nonsense mutation**.

**mitochondria**   Cytoplasmic organelles that are important in cellular respiration. The mitochondria have their own unique DNA.

**mitosis**   The cell division process in which two identical progeny cells are produced from a single parent cell. Compare with **meiosis**.

**mobile elements**   DNA sequences that are capable of inserting themselves (or copies of themselves) into other locations in the genome.

**modifier gene**   A gene that alters the expression of a gene at another locus.

**molecular genetics**   Study of the structure and function of genes at the molecular level.

**monoclonal**   Refers to a group of cells that consist of a single clone (i.e., all cells are derived from the same single ancestral cell).

**monogenic**   Describing a single-gene, or mendelian, trait.

**monosomy**   An aneuploid condition in which a specific chromosome is present in only a single copy, giving the individual a total of 45 chromosomes.

**monozygotic**   Describes a twin pair in which both members are derived from a single zygote. Synonymous with identical twin. Compare with **dizygotic**.

**morphogenesis**   The process of development of a cell, organ, or organism.

**mosaic**   The existence of two or more genetically different cell lines in an individual.

**mosaic, germline**   A type of **mosaic** in which the germ line of an individual contains an allele not present in the somatic cells.

**multifactorial**   Describes traits or diseases that are the product of the interactions of multiple genetic and environmental factors (example: neural tube defects).

**multi-hit concept of carcinogenesis**   The principle that most tumors arise from a series of errors, or "hits," occurring in a cell.

**multiplication rule**   A law of probability that states that the probability of co-occurrence of two or more independent events can be obtained by multiplying the individual probabilities of each event.

**multipoint mapping**  A type of genetic mapping in which the recombination frequencies among three or more loci are estimated simultaneously.

**mutagen**  A substance that causes a **mutation**.

**mutation**  An alteration in DNA sequence.

**mutation, induced**  A mutation that is caused by an exogenous factor, such as radiation. Compare with **mutation, spontaneous**.

**mutation, spontaneous**  A mutation that is not known to be the result of an exogenous factor. Compare with **mutation, induced**.

**mutation–selection balance**  A state in which the rate of elimination of an allele from a population (due to natural selection) is equal to the rate of introduction of the allele in a population (due to mutation). Mutation–selection balance can predict the **gene frequency** of an allele in a population.

**natural killer cell**  A type of lymphocyte that is involved in the early phase of defense against foreign microbes and tumors and is not MHC restricted.

**natural selection**  An evolutionary process in which individuals with favorable genotypes produce relatively greater numbers of surviving offspring.

**negative predictive value**  In disease screening, the percentage of subjects with a negative test result who truly do not have the disease. Compare with **positive predictive value**.

**neoplasm or tumor**  A group of cells characterized by unregulated proliferation (may be **benign** or **malignant**).

**neurofibromin**  The protein product of the neurofibromatosis type 1 gene.

**neurulation**  Formation of the neural tube during embryonic development.

**new mutation**  An alteration in DNA sequence that appears for the first time in a family as the result of a mutation in one of the parents' germ cells.

**newborn screening**  Testing the population of newborn infants for a condition, such as PKU, that is detectable shortly after birth. **Expanded newborn screening** refers to the use of techniques such as **tandem mass spectrometry** that allow a larger number of conditions to be screened in the newborn population.

**nondirectiveness**  Describes the genetic counseling approach in which information is provided to a family, while decisions about reproduction are left to the family.

**nondisjunction**  Failure of homologous chromosomes (in mitosis or meiosis I) or sister chromatids (in meiosis II) to separate properly into different progeny cells. Can produce **aneuploidy**.

**nonsense mutation**  A type of mutation in which an mRNA stop codon is produced, resulting in premature termination of translation, or removed, resulting in an elongated protein product. Compare with **missense mutation**.

**Northern blotting**  A gene expression assay in which mRNA on a blot is hybridized with a labeled probe.

**nucleosome**  A structural unit of chromatin in which 140 to 150 bp of DNA are wrapped around a core unit of eight histone molecules.

**nucleotide**  A basic unit of DNA or RNA, consisting of one deoxyribose (or ribose in the case of RNA), one phosphate group, and one nitrogenous base.

**nucleotide excision repair**  A type of **DNA repair** in which altered groups of nucleotides, such as pyrimidine dimers, are removed and replaced with properly functioning nucleotides.

**obligate carrier**  An individual who is known to possess a disease-causing gene (usually on the basis of pedigree examination) but may or may not be affected with the disease phenotype.

**oligonucleotide**  DNA sequence consisting of a small number of nucleotide bases.

**oncogene**  A gene that can transform cells into a highly proliferative state, causing cancer.

**oogenesis**  The process by which ova are produced.

**oogonium**  The diploid germline stem cell from which ova are ultimately derived.

**organogenesis**  The formation of organs during embryonic development.

**packaging cells** or **helper cells**  Cells in which replication-deficient viruses are placed so that the replication machinery of the packaging cell can produce viral copies.

**palindrome**  A DNA sequence whose complementary sequence is the same if read backward (e.g., 5′-AATGCG-CATT-3′).

**panmixia**  Describes a population in which individuals mate at random with respect to a specific genotype.

**paralog**  Within a species, a member of a set of homologous genes (example: *HOXA13* and *HOXD13*).

**partial trisomy**  A chromosome abnormality in which a portion of a chromosome is present in three copies; may be produced by reciprocal translocation or unequal crossover. See also **translocation, reciprocal**; and **crossover**.

**pattern formation**  The spatial arrangement of differentiated cells to form tissues and organs during embryonic development.

**pedigree**  A diagram that describes family relationships, sex, disease status, and other attributes.

**penetrance**  The probability of expressing a phenotype, given that an individual has inherited a predisposing genotype. If this probability is less than 1.0, the disease genotype is said to have reduced or incomplete penetrance.

**penetrance, age-dependent**  Describes disease phenotypes that have a higher probability of occurrence as the age of the individual with the at-risk genotype increases (examples: Huntington disease, familial breast cancer).

**percutaneous umbilical blood sampling (PUBS)**  Prenatal diagnostic technique in which fetal blood is obtained by puncture of the umbilical cord. Also called cordocentesis.

**personalized medicine**  An approach to medical care in which treatment is designed specifically for the individual patient. In genetics, the goal is to incorporate the genetic profile of the patient into decisions about diagnosis and treatment.

**phagocyte**  A cell that engulfs foreign particles.

**pharmacogenetics**  Study of genetic variation in drug response.

**pharmacogenomics**  Study of genetic variation in drug response, using data from many genes assayed across the genome (compare with **pharmacogenetics**).

**phenocopy**  A phenotype that resembles the phenotype produced by a specific gene but is caused instead by a different, typically nongenetic, factor.

**phenotype**  The observed characteristics of an individual, produced by the interaction of genes and environment.

**Philadelphia chromosome**  A reciprocal translocation between the long arms of chromosomes 9 and 22 in somatic cells; produces chronic myelogenous leukemia.

**phosphorylation**  The addition of a phosphate group to a molecule.

**physical mapping**  The determination of physical distances between genes using cytogenetic and molecular techniques. Compare with **genetic mapping**, in which recombination frequencies are estimated.

**plasma cell**  A mature B lymphocyte capable of secreting antibodies.

**plasmid**  A circular double-stranded DNA molecule found in bacteria; capable of independent replication. Plasmids are often used as cloning vectors in recombinant DNA techniques.

**pleiotropy**  Describes genes that have multiple phenotypic effects (examples: Marfan syndrome, cystic fibrosis). (adj.: pleiotropic)

**pluripotency**  The ability of a cell to develop into more than one type of mature, differentiated cell type.

**point mutation**  (1) In molecular genetics, the alteration of a single nucleotide to a different nucleotide. (2) In classical genetics, an alteration of DNA sequence too small to be detected under a light microscope.

**polar body**  A cell produced during oogenesis that has a nucleus but very little cytoplasm.

**polar body diagnosis**  A prenatal diagnostic technique in which DNA from a polar body is subjected to PCR amplification and assayed using molecular methods.

**polarity**  Direction (e.g., definition of anterior versus posterior in **axis specification**).

**poly-A tail**  The addition of several adenine nucleotides to the 3' end of a primary mRNA transcript.

**polygenic**  Describes a trait caused by the combined additive effects of multiple genes.

**polymerase chain reaction (PCR)**  A technique for amplifying a large number of copies of a specific DNA sequence flanked by two oligonucleotide primers. The DNA is alternately heated and cooled in the presence of DNA polymerase and free nucleotides so that the specified DNA segment is denatured, hybridized with primers, and extended by DNA polymerase.

**polymorphism**  A locus in which two or more alleles have gene frequencies greater than 0.01 in a population. When this criterion is not fulfilled, the locus is monomorphic.

**polypeptide**  A series of amino acids linked together by peptide bonds.

**polyploidy**  A chromosome abnormality in which the number of chromosomes in a cell is a multiple of 23 but is greater than the diploid number (examples are **triploidy** [69 chromosomes in the human] and **tetraploidy** [92 chromosomes in the human]).

**population genetics**  The branch of genetics dealing with genetic variation and genetic evolution of populations.

**population screening**  The large-scale testing of populations for a disease.

**positional candidate**  A gene-mapping approach in which linkage analysis is used to define a gene's approximate location. Known **candidate genes** in the region are then evaluated for their possible role in causing the trait or disease being analyzed.

**positional cloning**  See **cloning, positional**.

**positive predictive value**  Among the individuals identified by a test as having a disease, the percentage that actually have the disease. Compare with **negative predictive value**.

**posterior probability**  In Bayesian analysis, the final probability of an event after taking into account prior, conditional, and joint probabilities.

**posttranslational modification**  Various types of additions and alterations of a polypeptide that take place after the mature mRNA transcript is translated into a polypeptide (e.g., hydroxylation, glycosylation, cleavage of portions of the polypeptide).

**preimplantation genetic diagnosis (PGD)**  A form of genetic testing in which one or two cells obtained from an early embryo (created by **IVF**) are checked for chromosome abnormalities (typically by **FISH**) or single-gene mutations (by **PCR** amplification of DNA).

**prenatal diagnosis**  The identification of a disease in a fetus or embryo.

**presymptomatic diagnosis**   The identification of a disease before the phenotype is clinically observable.

**primary oocyte**   The diploid product of an oogonium. All primary oocytes are produced in the female during prenatal development; they undergo meiosis I to produce secondary oocytes when ovulation begins.

**primary spermatocyte**   The diploid progeny cell of a spermatogonium, which then undergoes meiosis I to produce secondary spermatocytes.

**primary transcript**   The mRNA molecule directly after transcription from DNA. A mature mRNA transcript is formed from the primary transcript when the introns are spliced out.

**primer**   An oligonucleotide sequence that flanks either side of the DNA to be amplified by **polymerase chain reaction**.

**primer extension**   Part of the **polymerase chain reaction** process, in which DNA polymerase extends the DNA sequence beginning at an oligonucleotide primer.

**primitive streak**   A structure formed during mammalian **gastrulation**, consisting of thickened epiblast tissue along the anterior/posterior axis.

**prior probability**   In Bayesian analysis, the probability that an event will occur, estimated before any additional information, such as a biochemical carrier test, is incorporated.

**probability**   The proportion of times that a specific event occurs in a series of trials.

**proband**   The first person in a pedigree to be identified clinically as having the disease in question. Synonymous with **propositus** and **index case**.

**probe**   In molecular genetics, a labeled substance, such as a DNA segment, that is used to identify a gene, mRNA transcript, or gene product (usually through hybridization of the probe with the target).

**promoter**   A DNA sequence located 5′ of a gene to which RNA polymerase binds in order to begin transcription of the DNA into mRNA.

**proofreading**   The correction of errors that occur during replication, transcription, or translation.

**prophase**   The first stage of mitosis and meiosis.

**propositus**   See **proband**.

**protein electrophoresis**   A technique in which amino acid variations are identified on the basis of charge differences that cause differential mobility of polypeptides through an electrically charged medium.

**protein kinase**   An enzyme that phosphorylates serine, threonine, or tyrosine residues in proteins.

**protein truncation test**   A mutation detection test in which the encoded protein product is artificially translated to reveal the presence of truncation-causing mutations (e.g., nonsense or frameshift mutations).

**proto-oncogene**   A gene whose protein product is involved in the regulation of cell growth. When altered, a proto-oncogene can become a cancer-causing **oncogene**.

**pseudoautosomal region**   The distal tip of the Y chromosome short arm, which undergoes crossover with the distal tip of the X chromosome short arm during meiosis in the male.

**pseudogene**   A gene that is highly similar in sequence to another gene or genes but has been rendered transcriptionally or translationally inactive by mutations.

**pseudomosaicism**   A false indication of fetal mosaicism, caused by an artifact of cell culture.

**pulsed-field gel electrophoresis**   A type of electrophoresis suitable for relatively large DNA fragments; the fragment is moved through a gel by alternating pulses of electricity across fields that are 90 degrees in orientation from one another.

**Punnett square**   A table specifying the genotypes that can arise from the gametes contributed by a mating pair of individuals.

**purine**   The two DNA (also RNA) bases, adenine and guanine, that consist of double carbon–nitrogen rings. Compare with **pyrimidine**.

**pyrimidine**   The bases (cytosine and thymine in DNA; cytosine and uracil in RNA) that consist of single carbon–nitrogen rings. Compare with **purine**.

**quadruple screen**   A screening test for fetal Down syndrome, and several other conditions, that can be carried out on maternal serum during pregnancy. The quadruple screen tests maternal serum levels of unconjugated estriol, human chorionic gonadotropin, inhibin-A, and maternal serum alpha-fetoprotein.

**quantitative trait**   A characteristic that can be measured on a continuous scale (e.g., height, weight).

**quantitative trait locus methods**   Methods for finding genes (quantitative trait loci [QTLs]) that underlie complex multifactorial traits.

**quasidominant**   A pattern of inheritance that appears to be autosomal dominant but is actually autosomal recessive. Usually the result of a mating between an affected homozygote and a heterozygote.

**quinacrine banding (Q-banding)**   Chromosome staining technique in which a fluorochrome dye (quinacrine compound) is added to chromosomes, which are then viewed under a fluorescence microscope.

**race**   A grouping of human populations that may be based on geographic origin, language, or other attributes. Most commonly, human races have corresponded roughly to continental origins.

**radiation hybrid mapping**   A technique in which ionizing radiation is used to break human chromosomes into small pieces that are then hybridized with rodent chromosomes. The physical distance between loci is estimated by assessing how frequently two loci are found together on the same human chromosome fragment.

**radiation, ionizing**   A type of energy emission that is capable of removing electrons from atoms, thus causing ions to form (example: x-rays).

**radiation, nonionizing**   A type of energy emission that does not remove electrons from atoms but can change their orbits (example: ultraviolet radiation).

**random mating**   See **panmixia**.

**receptor**   A cell-surface structure that binds to extracellular particles.

**recessive**   An allele that is phenotypically expressed only in the homozygous or hemizygous state. The recessive allele is masked by a dominant allele when the two occur together in a heterozygote. Compare with **dominant**.

**recognition site**   See **restriction site**.

**recombinant DNA**   A DNA molecule that consists of components from more than one parent molecule (e.g., a human DNA insert placed in a plasmid vector).

**recombinase**   An enzyme that helps to bring about somatic recombination (especially important in B and T lymphocytes).

**recombination**   The occurrence among offspring of new combinations of alleles, resulting from crossovers that occur during parental meiosis.

**recombination frequency**   The percentage of meioses in which recombinants between two loci are observed. Used to estimate genetic distances between loci. See also **centimorgan**.

**recombination hot spot**   A region of a chromosome in which the recombination frequency is elevated.

**recurrence risk**   The probability that another affected offspring will be produced in families in which one or more affected offspring have already been produced.

**reduction division**   The first stage of meiosis (meiosis I), in which the chromosome number is reduced from diploid to haploid.

**redundancy, genetic**   The existence of alternate genetic mechanisms or pathways that can compensate when a mechanism or pathway is disabled.

**repetitive DNA**   DNA sequences that are found in multiple copies in the genome. They may be dispersed or repeated in **tandem**.

**replication**   The process in which the double-stranded DNA molecule is duplicated.

**replication bubble**   A replication structure that occurs in multiple locations on a chromosome, allowing replication to proceed more rapidly.

**replication origin**   The point at which replication begins on a DNA strand; in eukaryotes, each chromosome has numerous replication origins.

**replicative segregation**   Refers to changes in the proportions of different mitochondrial DNA alleles as the mitochondria reproduce.

**restriction digest**   Process in which DNA is exposed to a restriction enzyme, causing it to be cleaved into **restriction fragments**.

**restriction endonuclease**   A bacterial enzyme that cleaves DNA at a specific DNA sequence (restriction site).

**restriction fragment**   A piece of DNA that has been cleaved by a restriction endonuclease.

**restriction fragment length polymorphism (RFLP)**   Variations in DNA sequence in populations, detected by digesting DNA with a restriction endonuclease, electrophoresing the resulting restriction fragments, transferring the fragments to a solid medium (blot), and hybridizing the DNA on the blot with a labeled probe.

**restriction site**   A DNA sequence that is cleaved by a specific restriction endonuclease.

**restriction site polymorphism (RSP)**   A variation in DNA sequence that is caused by the presence or absence of a restriction site. This type of polymorphism is the basis for most traditional RFLPs.

**retrovirus**   A type of RNA virus that can reverse-transcribe its RNA into DNA for insertion into the genome of a host cell; useful as a vector for gene therapy.

**reverse banding (R-banding)**   A chromosome banding technique in which chromosomes are heated in a phosphate buffer; produces dark and light bands in patterns that are the reverse of those produced by G-banding.

**reverse transcriptase**   An enzyme that transcribes RNA into DNA (hence "reverse").

**ribonucleic acid (RNA)**   A single-stranded molecule that consists of a sugar (ribose), a phosphate group, and a series of bases (adenine, cytosine, guanine, and uracil). There are three basic types of RNA: **messenger RNA (mRNA)**, **ribosomal RNA (rRNA)**, and **transfer RNA (tRNA)**.

**ribosomal RNA (rRNA)**   In conjunction with protein molecules, composes the **ribosome**.

**ribosome**   The site of translation of mature messenger RNA into amino acid sequences.

**ribozyme**   An mRNA molecule that has catalytic activity. Some ribozymes can be used to cleave mRNA in somatic cell gene therapy.

**ring chromosome**   A structural chromosome that is abnormally formed when both ends of a chromosome are lost and the new ends fuse together.

**RNA interference (RNAi)**   A method in which a specific mRNA sequence is recognized and destroyed by a complex of proteins that exist naturally in eukaryotic cells. RNAi can be used to block the expression of specific genes or, in the context of gene therapy, gain-of-function mutations.

**RNA polymerase**  An enzyme that binds to a promoter site and synthesizes messenger RNA from a DNA template.

**satellite DNA**  A portion of the DNA that differs enough in base composition so that it forms a distinct band on a cesium chloride gradient centrifugation; usually contains highly repetitive DNA sequences.

**secondary oocyte**  A cell containing 23 double-stranded chromosomes, produced from a primary oocyte after meiosis I in the female.

**secondary spermatocyte**  A cell containing 23 double-stranded chromosomes, produced from a primary spermatocyte after meiosis I in the male.

**segregation**  The distribution of genes from homologous chromosomes to different gametes during **meiosis**.

**selection coefficient**  A numerical measure of the degree of natural selection against a specific genotype, typically measured as the number of offspring produced by individuals who have the genotype, relative to other genotypes at the locus. A coefficient of zero indicates that there is no selection against a genotype, and a coefficient of one indicates lethality of the genotype.

**senescent**  Aged (e.g., an aged, or senescent, cell).

**sense strand**  In a double-stranded DNA molecule, this is the strand from which messenger RNA is not transcribed. Because of complementary base pairing, the sense strand is identical in sequence to the transcribed mRNA (with the exception that mRNA has uracil instead of thymine). See antisense strand.

**sensitivity**  The percentage of affected individuals who are correctly identified by a test (true positives). Compare with **specificity**.

**sequence**  (formerly "anomalad") A primary defect with secondary structural changes in development (examples: oligohydramnios sequence, Pierre–Robin sequence).

**sequence tagged sites (STSs)**  DNA sequences of several hundred base pairs that are flanked by PCR primers. Their chromosome location has been established, making them useful as indicators of physical positions on the genome.

**sex chromatin**  See Barr body.

**sex chromosomes**  The X and Y chromosomes in humans. Compare with **autosomes**.

**sex-influenced**  A trait whose expression is modified by the sex of the individual possessing the trait.

**sex-limited**  A trait that is expressed in only one sex.

**sex-linked**  A trait that is caused by genes on the sex chromosomes (X or Y).

**short tandem repeat polymorphism (STRP)**  A DNA sequence that contains multiple repeated short sequences, one after another. These sequences are polymorphic because the number of repeats varies among individuals.

**signal transduction**  A process in which biochemical messages are transmitted from the cell surface to the nucleus.

**silencer**  A DNA sequence that binds to **specific transcription factors** to decrease or repress the activity of certain genes. Compare with **enhancer**.

**silent substitution**  A DNA sequence change that does not change the amino acid sequence because of the degeneracy of the genetic code.

**SINEs (short interspersed elements)**  A class of dispersed repetitive DNA in which each repeat is relatively short. Compare with **LINEs**.

**single-strand conformation polymorphism (SSCP)**  A technique for detecting variation in DNA sequence by running single-stranded DNA fragments through a nondenaturing gel; fragments with differing secondary structure (conformation) caused by sequence variation migrate at different rates.

**single-copy DNA**  DNA sequences that occur only once in the genome. Compare with **repetitive DNA**.

**single-gene disorder or trait**  A feature or disease that is caused by a single gene. Compare with **polygenic** and **multifactorial**.

**single nucleotide polymorphisms (SNPs)**  Polymorphisms that result from variation at a single nucleotide. Compare with **microsatellites** and **VNTRs**.

**sister chromatid exchange**  Crossover between **sister chromatids**; can occur either in the sister chromatids of a tetrad during **meiosis** or between sister chromatids of a duplicated somatic chromosome.

**sister chromatids**  The two identical strands of a duplicated chromosome, joined by a single centromere.

**solenoid**  A structure of coiled DNA, consisting of approximately six **nucleosomes**.

**somatic cell**  A cell other than those of the gamete-forming germline. In humans, most somatic cells are diploid.

**somatic cell hybridization**  A physical gene-mapping technique in which somatic cells from two different species are fused and allowed to undergo cell division. Chromosomes from one species are selectively lost, resulting in clones with only one or a few chromosomes from one of the species.

**somatic hypermutation**  An extreme increase in the mutation rate of somatic cells; observed in B lymphocytes as they achieve increased binding affinity for a foreign antigen.

**somatic recombination**  The exchange of genetic material between homologous chromosomes during mitosis in somatic cells; much rarer than meiotic recombination.

**Southern transfer** (also **Southern blot**)  Laboratory procedure in which DNA fragments that have been electrophoresed through a gel are transferred to a solid membrane, such as nitrocellulose. The DNA can then be hybridized with a labeled probe and exposed to x-ray film (an **autoradiogram**).

**specificity**  The percentage of unaffected individuals who are correctly identified by a test (**true negatives**). Compare with **sensitivity**.

**spectral karyotype** A chromosome display (**karyotype**) in which combinations of fluorescent probes are used with special cameras and image-processing software so that each chromosome has a unique color.

**spermatid** One of the four haploid cells formed from a primary spermatocyte during spermatogenesis. Spermatids mature into spermatozoa.

**spermatogenesis** The process of male gamete formation.

**spermatogonia** Diploid germline stem cells from which sperm cells (spermatozoa) are ultimately derived.

**spindle fiber** One of the microtubular threads that form the spindle in a cell.

**splice site mutation** DNA sequence alterations in **donor** or **acceptor sites** or in the consensus sites near them. Produces altered intron splicing, such that portions of exons are deleted or portions of introns are included in the mature mRNA transcript.

**sporadic** Refers to the occurrence of a disease in a family with no apparent genetic transmission pattern (often the result of a new mutation).

**stop codon** mRNA base triplets that specify the point at which translation of the mRNA ceases.

**structural genes** Genes that encode protein products.

**submetacentric** A chromosome in which the centromere is located closer to one end of the chromosome arm than the other. Compare with **metacentric** and **acrocentric**.

**subtelomeric rearrangements** Alterations of chromosomes, principally **deletions** and **duplications**, that occur close to the **telomeres** and can cause genetic disease.

**synapsis** The pairing of homologous chromosomes during prophase I of meiosis.

**syndrome** A pattern of multiple primary malformations or defects all resulting from a single underlying cause (examples: Down syndrome, Marfan syndrome).

**syntenic** Describes two loci located on the same chromosome; they may or may not be linked.

**T lymphocyte or T cell** A component of the adaptive immune system whose receptors bind to a complex of MHC molecule and foreign antigen. There are two major classes of T lymphocytes, **helper T lymphocytes** and **cytotoxic T lymphocytes**.

**tandem repeat** DNA sequences that occur in multiple copies located directly next to one another. Compare with **dispersed repetitive DNA**.

**targeted disruption** The disabling of a specific gene so that it is not expressed.

**telomerase** A transferase enzyme that replaces the DNA sequences in **telomeres** during cell division.

**telomere** The tip of a **chromosome**.

**telophase** The final major stage of mitosis and meiosis, in which daughter chromosomes are located on opposite edges of the cell and new nuclear envelopes form.

**template** A strand of DNA that serves as the model for replication of a new strand. Also denotes the DNA strand from which mRNA is transcribed.

**teratogen** A substance in the environment that can cause a birth defect.

**teratology** The study of environmental factors that cause birth defects or congenital malformations.

**termination sequence** A DNA sequence that signals the cessation of transcription.

**tetrad** The set of four homologous chromatids (two sister chromatids from each homologous chromosome) observed during meiotic prophase I and metaphase I. Synonymous with **bivalent**.

**tetraploidy** A polyploid condition in which the individual has four copies of each chromosome in each cell, for a total of 92 chromosomes.

**thymine** One of the four DNA bases (abbrev.: T).

**tissue-specific mosaic** A mosaic in whom the mosaicism is confined to only specific tissues of the body.

**transcription** The process in which an mRNA sequence is synthesized from a DNA template.

**transcription factor** Protein that binds to DNA to influence and regulate transcription.

**transcription factor, general** A class of transcription factors that are required for the transcription of all structural genes.

**transcription factor, specific** A class of transcription factors that activate only specific genes at specific points in time.

**transduction** The transfer of DNA from one cell to another by a vector such as a **plasmid** or **bacteriophage**.

**transfection** The transfer of a DNA sequence into a cell.

**transfer RNA (tRNA)** A class of RNA that helps to assemble a polypeptide chain during translation. The anticodon portion of the tRNA binds to a complementary mRNA codon, and the 3′ end of the tRNA molecule attaches to a specific amino acid.

**transformation** The oncogenic conversion of a normal cell to a state of unregulated growth.

**transgenic** Refers to an organism into which a gene has been introduced from an organism of another species (e.g., a transgenic mouse could contain an inserted human gene).

**translation** The process in which an amino acid sequence is assembled according to the pattern specified by the mature mRNA transcript.

**translocation**   The exchange of genetic material between nonhomologous chromosomes.

**translocation, reciprocal**   A translocation resulting from breaks on two different chromosomes and a subsequent exchange of material. Carriers of reciprocal translocations maintain the normal number of chromosomes and the normal amount of chromosome material.

**translocation, Robertsonian**   A translocation in which the long arms of two acrocentric chromosomes are fused at the centromere; the short arms of each chromosome are lost. The translocation carrier has 45 chromosomes instead of 46 but is phenotypically normal because the short arms contain no essential genetic material.

**transposon**   See mobile element.

**triploidy**   A polyploid condition in which the individual has three copies of each chromosome in each cell, for a total of 69 chromosomes.

**trisomy**   An aneuploid condition in which the individual has an extra copy of one chromosome, for a total of 47 chromosomes in each cell.

**true negative**   Individual who is correctly identified by a test as not having a disease. See also **specificity**.

**true positive**   Individual who is correctly identified by a test as having a disease. See also **sensitivity**.

**tumor**   See neoplasm.

**tumor suppressor**   A gene whose product helps to control cell growth and proliferation; mutations in tumor suppressors can lead to cancer (example: retinoblastoma gene, *RB1*).

**two-hit model**   A model of carcinogenesis in which both copies of a gene must be altered before a neoplasm can form.

**ultrasonography**   A technique for fetal visualization in which sound waves are transmitted through the fetus and their reflection patterns are displayed on a monitor.

**unequal crossover**   Crossing over between improperly aligned DNA sequences; produces **deletions** or **duplications** of genetic material.

**uninformative mating**   A mating in which **linkage phase** cannot be established.

**uniparental disomy**   A condition in which two copies of one chromosome are derived from a single parent, and no copies are derived from the other parent. May be either **heterodisomy** or **isodisomy**.

**variable expression**   A trait in which the same genotype can produce phenotypes of varying severity or expression (example: neurofibromatosis type 1).

**variable number of tandem repeats (VNTRs)**   A type of polymorphism created by variations in the number of minisatellite repeats in a defined region.

**variance**   A statistical measure of variation in a quantity; estimated as the sum of the squared differences from the average value.

**vector**   The vehicle used to carry a DNA insert (e.g., phage, plasmid, cosmid, BAC, or YAC).

**X inactivation**   Process by which genes from one X chromosome in each cell of the female embryo are rendered transcriptionally inactive.

**X inactivation center**   The location on the X chromosome from which the X inactivation signal is transmitted (includes the *XIST* gene).

**X-linked**   Refers to genes that are located on the X chromosome.

**yeast artificial chromosome (YAC)**   A synthesized yeast chromosome capable of carrying a large DNA insert (up to 1000 kb).

**zygote**   The diploid fertilized ovum.

# ANSWERS TO STUDY QUESTIONS

## CHAPTER 2

1. The mRNA sequence is: 5'-CAG AAG AAA AUU AAC AUG UAA-3' (remember that transcription moves along the 3'–5' DNA strand, allowing the mRNA to be synthesized in the 5' to 3' direction). This mRNA sequence is translated in the 5' to 3' direction to yield the following amino acid sequence: Gln-Lys-Lys-Ile-Asn-Met-STOP.

2. The *genome* is the sum total of our genetic material. It is composed of 23 pairs of nuclear *chromosomes* and the mitochondrial chromosome. Each chromosome contains a number of *genes*, the basic unit of heredity. Genes are composed of one or more exons; *exons* alternate with introns. Exons encode mRNA *codons*, which consist of three *nucleotides* each. It is important to remember that DNA coiling patterns also produce a hierarchy: Chromosomes are composed of 100-kb chromatin loops, which are in turn composed of solenoids. Each solenoid contains approximately six nucleosomes. Each nucleosome contains about 150 DNA base pairs and may or may not include coding material.

3. Approximately 55% of human DNA consists of repetitive sequences whose function is largely unknown. Single-copy DNA includes protein-coding genes, but it consists mostly of extragenic sequences and introns that do not encode proteins. Because individual cells have specialized functions, most make only a limited number of protein products. Thus only a small percentage of the cell's coding DNA is transcriptionally active at any given time. This activation is controlled by elements such as transcription factors, enhancers, and promoters.

4. Mitosis is the cell division process whereby one diploid cell produces two diploid daughter cells. In meiosis, a diploid cell produces haploid cells (gametes). Meiosis produces haploid cells because the centromeres are not duplicated in meiosis I and because there is no replication of DNA in the interphase stage between meiosis I and meiosis II. Another difference between mitosis and meiosis is that the homologous chromosomes form pairs and exchange material (crossing over) during meiosis I. Homologs do not pair during mitosis, and mitotic crossing over is very rare.

5. Each mitotic division doubles the number of cells in the developing embryo. Thus the embryo proceeds from 1 to 2 to 4 to 8 cells, and so on. After $n$ cell divisions, there are $2^n$ cells. For example, after 10 divisions, there are $2^{10}$, or 1024, cells. We want a value of $n$ that satisfies the simple relationship $2^n = 10^{14}$. One way to find our answer is simply to plug in values of $n$ until we get $10^{14}$. A more elegant approach is to take the common logarithms of both sides of this equation, yielding $n\log(2) = 14\log(10)$. Because the common logarithm of 10 is 1, we obtain the relationship $n = 14/\log(2)$. Thus, $n = 46.5$. This result, approximately 46 to 47 cell divisions, is only an *average* value. Some cell lineages divide more times than others, and many cells are replaced as they die.

6. A total of 400 mature sperm and 100 mature egg cells will be produced. Each primary spermatocyte produces four mature sperm cells, and each primary oocyte produces only one mature egg cell (the other products of meiosis are polar bodies, which degenerate).

## CHAPTER 3

1. Mutation 1 is a nonsense mutation in the fourth codon, which produces premature termination of translation. Mutation 2 is a frameshift mutation in the third codon, and mutation 3 is a missense mutation in the second codon.

2. Transcription mutations generally lower the production of a gene product, but often they do not eliminate it completely. Transcription mutations in the β-globin gene usually produce $β^+$-thalassemia, a condition in which there is some production of β-globin chains. $β^+$-thalassemia tends to be less severe than $β^0$-thalassemia. Missense mutations alter only a single amino acid in a polypeptide chain, and when they occur in the β-globin chain they can produce $β^+$-thalassemia. (However, keep in mind that sickle cell disease, which is relatively severe, is also caused by a missense mutation.) In contrast, frameshift mutations alter many or all codons downstream from the site of the mutation, so a large number of amino acids may be changed. Frameshifts can also produce a stop codon. Nonsense mutations produce truncated polypeptides, which are often useless (especially if the nonsense mutation

occurs near the 5′ end of the gene, eliminating most of the polypeptide chain). Donor and acceptor mutations can delete whole exons or large portions of them. This deletion can substantially alter the amino acid composition of the polypeptide. Nonsense, frameshift, and donor or acceptor mutations all tend to produce the more-severe $\beta^0$-thalassemia, in which no $\beta$-globin chains are present.

3. In thalassemia conditions, one of the chains, $\alpha$- or $\beta$-globin, is reduced in quantity. Most of the harmful consequences are caused by the relative excess of the chain that is produced in normal quantity. If both chains are reduced in quantity, there may be a rough balance between the two, resulting in less accumulation of excess chains.

4. Restriction site polymorphisms (RSPs) reflect the presence or absence of a restriction site. They thus can have only two alleles. They are detected with the use of RFLP technology. VNTRs are also a type of RFLP, but here the polymorphism is in the number of tandem repeats that lie between two restriction sites, rather than in the presence or absence of a restriction site. Because the number of tandem repeats can vary considerably, VNTRs can have many different alleles in populations. VNTRs are found in minisatellite regions. STRPs consist of variations in the number of shorter microsatellite repeats (usually dinucleotides, trinucleotides, and tetranucleotides). They are detected with the use of the polymerase chain reaction (PCR). RSPs and VNTRs can also be detected using PCR (instead of Southern blotting), provided that the DNA sequences flanking the polymorphism are known. The autoradiogram represents an STRP. This is indicated by the fact that there are multiple alleles (distinguishing it from an RSP) and by the fact that the various alleles differ in size by only 4 bp (recall from Chapter 2 that the tandem repeat units in minisatellite regions are generally 20 to 70 bp long).

5. Because the disease mutation destroys a recognition site, those who have the disease allele have a longer restriction fragment. This fragment migrates more slowly on the gel and is seen higher on the autoradiogram. Individual A has only the longer fragment and thus has two copies of the disease mutation. This individual has $\alpha_1$-antitrypsin deficiency. Individual B has only the short fragment and is genetically and physically unaffected. Individual C has both fragments and is thus a clinically unaffected heterozygote.

6. In this sample of 100 individuals, there are $88 \times 2$ *HbA* alleles in the *HbA* homozygotes and 10 *HbA* alleles in the heterozygotes. There are thus 186 *HbA* alleles in the population. The frequency of *HbA*, $p$, is $186/200 = 0.93$, and the frequency of *HbS*, $q$, is $1 - 0.93 = 0.07$. The genotype frequencies in the population are $88/100 = 0.88$, $10/100 = 0.10$, and $2/100 = 0.02$ for the *HbA/HbA*, *HbA/HbS*, and *HbS/HbS* genotypes, respectively.

Assuming Hardy–Weinberg proportions, the expected genotype frequencies are given by $p^2$, $2pq$, and $q^2$, respectively. This yields expected genotype frequencies of $(0.93)^2 = 0.865$, $2 \times 0.93 \times 0.07 = 0.130$, and $(0.07)^2 = 0.005$, respectively. In this population, the observed and expected genotype frequencies are fairly similar to each other.

7. For an autosomal recessive disease, the prevalence $(1/10,000)$ equals the recessive genotype frequency, $q^2$. Thus the PKU gene frequency, $q$, is given by $\sqrt{q^2} = \sqrt{1/10,000} = 1/100 = 0.01$. The carrier frequency is given by $2pq$, which is approximately $2q$, or 0.02 (i.e., 1/50).

## CHAPTER 4

1. Because this is an autosomal dominant disorder, and because affected homozygotes die early in life, the man is a heterozygote and has a 50% chance of passing the disease-causing allele to each of his offspring. The probability that all four will be affected is given by the product of each probability: $(1/2)^4 = 1/16$. The probability that none will be affected (1/16) is obtained in exactly the same way.

2. The probability that the offspring will inherit the retinoblastoma susceptibility allele is 0.50, because familial retinoblastoma is an autosomal dominant disease. However, we must also consider the penetrance of the disorder. The probability of both inheriting the disease-causing allele (0.50) *and* expressing the disease phenotype (0.90) is given by multiplying the two probabilities together: $0.90 \times 0.50 = 0.45$.

3. Because the woman's sister had Tay–Sachs disease, both parents must be heterozygous carriers. This means that, at birth, one fourth of their offspring will be affected, one half will be carriers, and one fourth will be genetically normal. Note, however, that the woman in question is 30 years old. She cannot possibly be an affected homozygote because affected individuals die by age 6 years. There are thus three equally likely possibilities: (1) the disease allele was inherited from the mother and a normal allele was inherited from the father; (2) the disease allele was inherited from the father and a normal allele was inherited from the mother; (3) normal alleles were inherited from both parents. Because two of these three possibilities lead to the carrier state, the woman's probability of being a heterozygous carrier is 2/3.

4. Because the mother has neurofibromatosis and can be assumed to be a heterozygote (homozygotes for this condition have not been reported), the probability that her daughter (the man's sister) inherits the disease-causing allele is 1/2. The probability that the sister transmits the disease-causing allele to her daughter is again 1/2, so the probability that both events occur is $1/2 \times 1/2 = 1/4$. If we knew that the man's sister was affected, then the probability that his sister's daughter is affected is simply 1/2.

5. The probability that the woman's disease-causing allele is transmitted to her offspring is 1/2, and the probability that this offspring in turn transmits the disease-causing allele to his or her offspring (i.e., the grandchild) is again 1/2. Thus, the probability that one grandchild has inherited the allele is $1/2 \times 1/2$, or 1/4. Similarly, the probability that the other grandchild has inherited the allele is 1/4. The probability that both grandchildren have inherited the allele is $1/4 \times 1/4 = 1/16$. If the grandmother has PKU, she must be homozygous for the disease-causing allele. Thus, both of her children must be heterozygous carriers (probability = 1). The probability that one of these individuals transmits the disease-causing allele to their offspring is 1/2. The probability that they both transmit the disease-causing allele to their offspring (i.e., that both grandchildren are heterozygous carriers) is $1/2 \times 1/2 = 1/4$.

6. The coefficient of relationship is $(1/2)^6$, or 1/64. This gives the probability that the second member of the couple carries the PKU allele. The probability that two carriers will produce an affected offspring is 1/4. The overall probability that this couple will produce a baby affected with PKU is given by multiplying the probability that the mate also carries the allele (1/64) by the probability that the couple both transmit the allele to their offspring (1/4): $1/64 \times 1/4 = 1/256$. This demonstrates that the probability of producing an affected offspring in this consanguineous mating is actually quite small.

7. The frequency of the heterozygote genotype in the general population is given by $2pq$, according to the Hardy–Weinberg law. Thus, the frequency of the first genotype in the general population is predicted to be $2 \times 0.05 \times 0.10 = 0.01$. Similarly, for the second system, the frequency of heterozygotes in the general population is $2 \times 0.07 \times 0.02 = 0.0028$. The perpetrator was a homozygote for the third system; the frequency of the homozygote genotype in the general population is given by $p^2$, or $0.08^2 = 0.0064$. If we can assume independence of the three STR loci in the general population, we can then multiply the frequencies of the three genotypes together to get the probability that a randomly chosen individual in the general population would have the same genotypes as the perpetrator. We thus multiply $0.01 \times 0.0028 \times 0.0064$ to obtain a probability of 0.000000179, or 1/5,580,357. This probability can be made smaller by typing additional loci.

8. As in question 7, we assume independence of the four STR loci. In this case, we multiply the four allele frequencies together to get the probability: $0.05 \times 0.01 \times 0.01 \times 0.02 = 0.0000001 = 1/10,000,000$. Notice a key difference between question 8 and question 7: In the paternity case, the father has contributed only half of the baby's genotype at each locus, the other half being contributed by the mother. Thus, we examine only a single allele for each locus. In contrast, the rapist has contributed both alleles of each genotype to the evidentiary sample, so we need to know the frequency of each genotype in the general population. We use the Hardy–Weinberg law to estimate the population frequency of each genotype, based on the known allele frequencies.

## CHAPTER 5

1. There are four Barr bodies, always one less than the number of X chromosomes.

2. This is most likely a result of X inactivation. The heterozygotes with muscle weakness are the ones with relatively large proportions of active X chromosomes containing the mutant allele.

3. The disease frequency in males is $q$, and in females it is $q^2$. The male-to-female ratio is thus $q/q^2$ or $1/q$. Thus, as $q$ decreases, the male-to-female ratio increases.

4. Because the male's grandfather is affected with the disorder, his mother must be a carrier. His father is phenotypically normal and therefore does not have the disease gene. Thus, the male in question has a 50% risk of developing hemophilia A. His sister's risk of being a heterozygous carrier is also 50%. Her risk of being affected with the disorder is close to zero (barring a new mutation on the X chromosome transmitted by her father).

5. Male-to-male transmission can be observed in autosomal dominant inheritance, but it is not observed in X-linked dominant inheritance. Thus, males affected with X-linked dominant disorders must always have affected mothers, unless a new mutation has occurred. Males and females are affected in approximately equal proportions in autosomal dominant inheritance, but there are twice as many affected females as males in X-linked dominant inheritance (unless the disorder is lethal prenatally in males, in which case only affected females are seen). In X-linked dominant inheritance, *all* of the sons of an affected male are normal, and *all* of the daughters are affected. In X-linked dominant diseases, heterozygous females tend to be more mildly affected than are hemizygous males. In autosomal dominant inheritance, there is usually no difference in severity of expression between male and female heterozygotes.

6. In mitochondrial inheritance, the disease can be inherited only from an affected mother. In contrast to all other types of inheritance, no descendants of affected fathers can be affected. Note that males with an X-linked recessive disease who mate with normal females cannot transmit the disease to their offspring, but their grandsons can be affected with the disease.

7. Because Becker muscular dystrophy is relatively rare, it is reasonable to assume that the woman is a normal homozygote. Thus, she can transmit only normal X chromosomes to her offspring. Her male offspring, having received a Y chromosome from the father and a

normal X from the mother, will all be unaffected. All female offspring of this mating will receive a mutated X chromosome from the father and all will be heterozygous carriers of the disorder.

8. If the man is phenotypically normal, his X chromosome cannot carry a Duchenne muscular dystrophy mutation. The female will transmit her mutation-carrying chromosome to half of her offspring, on average. Thus, half of the sons will be affected with the disorder, and half of the daughters will be heterozygous carriers.

9. The boys' mother must be a heterozygous carrier for a factor VIII mutation. Thus, one of the mother's parents must also have carried the mutation. Consequently, the probability that the mother's sister is a carrier is 1/2.

10. This is explained by genomic imprinting. For normal development, differential expression of genes inherited from the father and mother is necessary. If the expression pattern from only one parent is inherited, the embryo cannot develop normally and dies. The experiment works in amphibians because genomic imprinting does not occur in these animals.

## CHAPTER 6

1. Euploid cells have a multiple of 23 chromosomes. Haploid ($n = 23$), diploid ($n = 46$), and polyploid (triploid and tetraploid) cells are all euploid. Aneuploid cells do not have a multiple of 23 chromosomes and include trisomies (47 chromosomes in a somatic cell) and monosomies (45 chromosomes in a somatic cell).

2. Conventional FISH analysis involves the hybridization of a fluorescently labeled probe to denatured metaphase chromosomes and is most useful in detecting aneuploidy, deletions, and chromosome rearrangements. FISH can also be extended to use multiple differently colored probes, allowing the detection of several different aneuploidies simultaneously. Spectral karyotyping is an extension of FISH in which each chromosome can be visualized as a different color. This enables accurate and easy characterization of aneuploidy, loss or duplication of chromosome material, and (especially) chromosome rearrangements such as translocations. CGH is particularly useful in detecting the gain or loss of chromosome material, but it cannot detect balanced rearrangements of chromosomes, such as reciprocal translocations. This is because, despite the rearrangement, the test sample and the reference sample each have the same amount of DNA from each chromosome region.

3. A normal egg can be fertilized by two sperm cells (dispermy, the most common cause of triploidy). An egg and polar body can fuse, creating a diploid egg, which is then fertilized by a normal sperm cell. Diploid sperm or egg cells can be created by meiotic failure; subsequent union with a haploid gamete would produce a triploid zygote.

4. The difference in incidence of various chromosome abnormalities reflects the fact that embryos and fetuses with chromosome abnormalities are spontaneously lost during pregnancy. The rate and timing of loss vary among different types of chromosome abnormalities.

5. A karyotype establishes whether the condition is the result of a true trisomy or a translocation. If the trisomy is a translocation, the recurrence risk in future pregnancies is greatly elevated. A karyotype also helps to establish whether the patient is a mosaic. This can help to predict and explain the severity of expression of the disorder.

6. The risk, from lowest risk to highest, is: (1) 25-year-old woman with one previous Down syndrome child (approximately 1%); (2) 25-year-old male carrier of a 21/14 translocation (1%-2%); (3) 45-year-old woman with no family history (approximately 3%); (4) 25-year-old female carrier of a 21/14 translocation (10%-15%).

7. Nondisjunction of the X chromosome can occur in both meiosis I and meiosis II. If these two nondisjunctions occur in the same cell, an ovum with four X chromosomes can be produced. If this is fertilized by an X-bearing sperm cell, the zygote will have the 49, XXXXX karyotype.

8. The meiotic error must have occurred in the father, because his X chromosome carries the gene for hemophilia A. Because the daughter has normal factor VIII activity, she must have inherited her single X chromosome from her mother.

9. A loss of genetic material usually produces more severe consequences than does a gain of material, so one would expect the patient with the deletion—46,XY,del(8p)—to be more severely affected than the patient with the duplication.

10. A translocation can place a proto-oncogene near a sequence that activates it, producing a cancer-causing oncogene. It can also interrupt a tumor-suppressor gene (see Chapter 11), inactivating it. Because these genes encode tumor-suppressing factors, their inactivation can also lead to cancer.

## CHAPTER 7

1. In consanguineous matings, a higher fraction of genes in the offspring are identical by descent. The identical fraction is measured by the coefficient of relationship. The carrier frequency of autosomal recessive inborn errors of metabolism such as alkaptonuria decreases as the prevalence of the disorder diminishes. Thus, the carrier frequency of very rare disorders is very low. When alkaptonuria is diagnosed in a child, it is reasonable to suspect that parents who might be related are more likely to share an individual's alkaptonuria gene than two individuals chosen at random from the population.

2. Many metabolic reactions can proceed in the complete absence of an enzyme. For example, a hydroxide ion can

combine with carbon dioxide to form bicarbonate. Yet this reaction occurs much more efficiently in the presence of a catalyst, in this case the enzyme carbonic anhydrase. Although many reactions in the human body would continue in the absence of an enzymatic catalyst, they would not do so at a rate high enough to support normal metabolism and physiology.

3. Although most inborn errors of metabolism are rare, in aggregate they contribute substantially to the morbidity and mortality of children and adults. Additionally, understanding the pathogenetic basis of rare metabolic disorders has the potential to help physicians and scientists understand similar processes that contribute to common diseases. For example, by understanding how mutations in glucokinase cause hyperglycemia, we might better understand the pathogenesis of diabetes mellitus.

4. Metabolic disorders have been classified in many different ways. In Chapter 7, we categorized them by the types of metabolic processes that are affected. Some examples include carbohydrate metabolism (e.g., galactosemia, hereditary fructose intolerance, glycogen storage disorders); amino acid metabolism (e.g., PKU, MSUD, tyrosinemia); lipid metabolism (e.g., MCAD, LCAD); degradative pathways (e.g., Hurler syndrome, OTC deficiency, Gaucher disease); energy production (e.g., OXPHOS defects); and transport systems (e.g., cystinosis, cystinuria).

5. Mutations in *GAL-1-P uridyl transferase* are the most common cause of galactosemia. However, mutations in genes encoding other enzymes necessary for the metabolism of galactose, such as galactokinase and uridine diphosphate galactose-4-epimerase, can also result in galactosemia. This is an example of genetic heterogeneity. That is, indistinguishable phenotypes may be produced by mutations in different genes. Other examples of metabolic disorders that are genetically heterogeneous include hyperphenylalaninemia, MSUD, and cystinuria.

6. The prevalence rates for many inborn errors of metabolism vary widely among ethnic groups. In many instances this is due to founder effect and genetic drift (see Chapter 3). For example, in Finland more than 30 autosomal recessive disorders are found at an unusually high prevalence compared with closely related populations. A similar scenario partly explains the finding of only a single or few disease-causing mutations in Ashkenazi Jews (lysosomal storage disorders), Mennonites (MSUD), and French Canadians (tyrosinemia type 1). In other cases, it appears that there may be a selective advantage for carriers of a recessive allele (remember that heterozygotes for recessive disorders are unaffected). This might explain the varying frequency distribution of *LAC*P*, which confers persistent lactase activity and may have been advantageous in populations in which dairy products were a valuable nutritional source.

7. Although the mitochondrial genome is inherited only from the mother, most of the proteins in the mitochondria are encoded by genes in the nuclear genome. Thus, disorders of mitochondrial fatty acid oxidation are inherited in an autosomal recessive pattern. Given the age of this young woman, it is unlikely that she is affected with a mitochondrial fatty acid oxidation disorder, so we can assume that she either is a heterozygous carrier or is homozygous for the normal allele. Consequently, the probability that she is a carrier is 0.67. The probability that her mate is a carrier is given by the carrier rate for MCAD in the general population: 1 in 70 ($\sim 0.014$). In a carrier $\times$ carrier mating, there is a 25% chance of having an affected child (probability of two gametes bearing mutant alleles uniting); then the woman's risk of having an affected child is $0.67 \times 0.014 \times 0.25 = 0.002$ or 1/500. Note that her risk of being a carrier (0.67/0.014) is 48 times higher than in the general population.

8. Five different enzymes control the urea cycle pathway. Deficiencies of four of these enzymes (AS, ASA, arginase, and CPS) are inherited in an autosomal recessive pattern. OTC deficiency is an X-linked recessive condition. Most women who are carriers for an X-linked recessive disorder are not symptomatic. However, there are at least two explanations for a symptomatic female. First, one X chromosome is normally inactivated (lyonized) at random in each somatic cell. Sometimes, inactivation is skewed, with more normal than abnormal X chromosomes being inactivated, and the woman becomes symptomatic. Testing for skewed inactivation is not commonly performed in diagnostic laboratories. A second possible explanation is that only a single X chromosome is present in each somatic cell (Turner syndrome; see Chapter 6). Thus, X inactivation does not take place because each cell must contain one functional copy of the X chromosome. If this X chromosome contains a mutated gene (e.g., *OTC*), a woman will be symptomatic. The simplest test to confirm this diagnosis is a karyotype.

9. A functioning OXPHOS system is necessary to metabolize pyruvate aerobically. Defects of the OXPHOS system cause pyruvate to be metabolized anaerobically into lactate, increasing the level of circulating lactate. Thus, an elevated concentration of circulating lactate might signal the presence of a disorder of energy production. Increased levels of lactate are also produced by decreased tissue oxygenation due to decreased circulation (e.g., strenuous exercise, shock). Often there are other biochemical abnormalities that suggest a defect of OXPHOS; however, the diagnosis can be challenging.

10. Polymorphisms in the genes controlling the metabolism of drugs (e.g., *CYP2D6*) can affect a patient's therapeutic response as well as the profile of side effects.

Natural foods contain thousands of chemical compounds, many of which have pharmacological properties similar to, but less potent than, contemporary drugs used by health care providers. Thus, polymorphisms in the genes might have enabled some groups of hunter-gatherers to use resources that other groups found inedible. Over time this might have been enough of a selective advantage that the size of one group grew more quickly than the size of others.

## CHAPTER 8

**1.** The affected male in generation II inherited the disease allele and marker allele *1* from his affected father, and he inherited a normal allele and marker allele *2* from his mother. Therefore, the disease allele must be on the chromosome that contains marker allele *1* in this male (linkage phase). Because he married a female who is heterozygous for marker alleles *3* and *4*, we expect to observe allele *1* in the affected offspring under the hypothesis of linkage. Individual III-5 has the *2,4* marker genotype but is affected, and individual III-7 has the *1,3* genotype but is normal. They both represent recombinants. Thus, there are two recombinations observed in eight meioses, giving a recombination frequency of 2/8, or 25%.

**2.** For marker *A*, the affected mother in generation II must carry allele *2* on the same chromosome as the Huntington disease allele. Under the hypothesis that $\theta = 0.0$, all of her children must also inherit allele *2* if they are affected. Marker *A* shows a recombinant with the disease allele in individual III-5 and thus produces a likelihood of zero for a recombination frequency of 0.0. The LOD score is $-\infty$ (the logarithm of 0). For marker *B*, the disease allele is on the chromosome bearing marker allele *1* in the affected mother in generation II. All offspring inheriting allele *1* also inherit the disease allele, so there are no recombinants. Under the hypothesis that $\theta = 0.0$, the affected mother can transmit only two possible haplotypes: the disease allele with marker allele *1* and the normal allele with marker allele *2*. The probability of each of these events is 1/2. Thus, the probability of observing six offspring with the marker genotypes shown is $(1/2)^6 = 1/64$. This is the numerator of the likelihood ratio. Under the hypothesis that $\theta = 0.5$ (no linkage), four possible haplotypes can be transmitted, each with probability 1/4. The probability of observing six children with these haplotypes under the hypothesis of no linkage is then $(1/4)^6 = 1/4096$. This is the denominator of the likelihood ratio. The ratio is then $(1/64)/(1/4096) = 64$. The LOD score is given by the common logarithm of 64, 1.8.

**3.** The table shows a maximum LOD score, 3.5, at a recombination frequency of 10%. Thus, the two loci are most likely to be linked at a distance of approximately 10 cM. The odds in favor of linkage at this *q* value, versus nonlinkage, are 3162 (or $10^{3.5}$) to 1.

**4.** Linkage phase can be established in both families, and no recombinants are observed in either family. Thus, the estimated recombination frequency is 0.0. The LOD score for $\theta = 0.0$ in the first family is given by $\log_{10}(1/2)^5/(1/4)^5 = \log_{10}(32) = 1.5$. In the second family the LOD score for $\theta = 0.0$ is $\log_{10}(1/2)^6/(1/4)^6 = \log_{10}(64) = 1.8$. These two LOD scores can then be added to obtain an overall LOD score of 3.3.

**5.** These matings allow us to establish linkage phase in individuals II-1 and II-2, the parents of the individuals in generation III. The disease allele is on the same chromosome as marker *4* in individual II-1 and on the same chromosome as marker *5* in individual II-2. Under the hypothesis of linkage, we would predict that offspring who inherit markers *4* and *5* will be homozygous for the disease allele and, thus, affected, offspring who inherit either *4* or *5* will be heterozygous carriers, and offspring who inherit neither *4* nor *5* will be homozygous normal. Note a key difference between autosomal recessive and autosomal dominant pedigrees in estimating recombination frequencies: Here, both parents in generation II contribute informative meioses because both parents carry a disease-causing allele. Thus, we can evaluate all 10 meioses contributing to generation III for recombination between the disease and marker loci. No recombinations are seen in the first four offspring; however, individual III-5 is homozygous normal but has inherited allele *5* from his mother. Thus, one recombination in five meioses has occurred in the mother, and no recombinations in five meioses have occurred in the father. One recombination in 10 meioses yields a recombination frequency of 1/10 = 10%.

**6.** These matings allow us to establish linkage phase in individual II-1. Under the hypothesis of linkage, she carries the disease-causing allele (labeled *D*) on the same chromosome as marker allele *1*. Her haplotypes are thus *D1/d2*. Based on her haplotypes and on those of her unaffected mate, we can predict that the offspring who inherit the *1,1* genotype will be affected, and those who inherit the *1,2* genotype will not be affected. We see that this is the case for all but one of the offspring (III-5). This means that the likelihood that $\theta = 0.0$ is zero, so the LOD score for this recombination frequency is $-\infty$. To assess the LOD score for $\theta = 0.1$, we consider that the probability that the father passes each recombinant haplotype (*D2* or *d1*) is $\theta/2 = 0.05$. The probability that he transmits each nonrecombinant haplotype (*D1* or *d2*) is $(1 - \theta)/2 = 0.45$ (see Box 8-1 for the details of this reasoning process). There is one recombinant and seven nonrecombinants among the offspring. The probability of observing these eight events is $0.05 \times (0.45)^7 = 0.00019$. This is the numerator of the likelihood ratio. For eight offspring, the probability that $\theta = 0.5$ is $(1/4)^8 = 0.000015$. This is the denominator of the ratio. The logarithm of the odds ratio is given by $\log_{10}(12.2) = 1.09$. This is the LOD score for $\theta = 0.1$.

7. The mating in generation II is uninformative because the father is a homozygote for the marker allele. It will be necessary to type the family for another closely linked marker (preferably one with more alleles) before any risk information can be given. At this point, the only risk information that can be given is that each child has a 50% risk of inheriting the disease gene from the affected father.

8. *Synteny* refers to loci that are on the same chromosome. *Linkage* refers to loci that are less than 50 cM apart on a chromosome; alleles at such loci tend to be transmitted together within families. Linked loci are thus syntenic, but syntenic loci are not necessarily linked. *Linkage disequilibrium* is the nonrandom association of alleles at linked loci, observable when chromosome haplotypes are examined in populations. *Association* indicates that two traits are observed together in a population more often than expected by chance; the traits might or might not be genetic. Association thus does not necessarily have anything to do with linkage, unless we are referring to linkage disequilibrium.

9. Linkage disequilibrium can arise when a single disease-causing mutation first occurs on a chromosome copy that contains specific nearby marker alleles. At first, the mutation will be seen only on copies of the chromosome that contain the specific marker alleles. On the other hand, if there are multiple disease-causing mutations at a locus like *NF1*, they are likely to occur on chromosome copies with different marker alleles, and little association will be observed between the disease genotype (which actually consists of a collection of different mutations at the disease locus) and a specific marker allele.

## CHAPTER 9

1. The class I molecules present peptides at the surfaces of nearly all of the body's cells. The peptide–class I molecule complex is recognized by cytotoxic T cells, which kill the cell if the class I MHC molecule presents foreign peptide. The class II MHC molecules also present peptides at the cell surface, but only in antigen-presenting cells of the immune system (e.g., dendritic cells, macrophages, B cells). If the class II molecule presents foreign peptides derived from an invading microbe, they are bound by the receptors of helper T-cells; this in turn stimulates appropriate B cells to proliferate and to produce antibodies that will help to kill the microbes.

2. Immunoglobulins differ among B cells within individuals, so that a large variety of infections can be combated. MHC molecules are identical on each cell surface within an individual, but they vary a great deal among individuals. This interindividual variability may have evolved to prevent infectious agents from spreading easily through a population.

3. T-cell receptors and immunoglobulins are similar in that they are both cell-surface receptors that bind to foreign peptides as part of the immune response. Diversity in both types of molecules is generated by multiple germline genes, VDJ recombination, and junctional diversity. They differ in that immunoglobulins are secreted into the circulation (as antibodies) and can bind directly to foreign peptide, whereas T-cell receptors are not secreted and must "see" foreign peptides in conjunction with MHC molecules to recognize them. Also, somatic hypermutation generates diversity in immunoglobulins but not in T-cell receptors.

4. Somatic recombination alone could produce $30 \times 6 \times 80 = 14,400$ different heavy chains of this class.

5. The probability that a sibling will be HLA-identical is 0.25. The probability that two siblings share one gene or haplotype is 0.50 (the coefficient of relationship for siblings; see Chapter 4). Then, the probability that they share two haplotypes and are HLA-identical is $0.50 \times 0.50 = 0.25$.

6. If a homozygous Rh-positive (*DD*) man mates with an Rh-negative woman, all of the offspring will be Rh-positive heterozygotes (*Dd*) and incompatible with the mother. If the man is an Rh-positive heterozygote (*Dd*), half of the children, on average, will be incompatible Rh-positive heterozygotes. However, if the couple is ABO-incompatible, this will largely protect against Rh incompatibility.

## CHAPTER 10

1. Animals such as roundworm, fruit fly, frog, zebra fish, chick, and mouse are commonly used to model human development. Each of these organisms has a relatively short generation time and can be selectively bred, and large populations can be maintained in captivity. Furthermore, embryos from each of these animals can be manipulated either in vitro or in vivo using a variety of techniques (e.g., surgical ablation or transplantation of cells or tissues, ectopic expression of native genes or transgenes, knockouts). Of course, each organism has advantages and disadvantages that must be considered when choosing an appropriate model. In some animals, naturally occurring mutant strains have proved to be valuable models of human genetic disease. For example, mutations in murine *Pax6* produce an abnormally small eye. Mutations in the human homolog, *PAX6*, cause hypoplasia or absence of the iris. Most of what we understand about axis determination and pattern formation has been learned from studies of nonhuman animal models.

2. Although identical mutations in a developmental gene can produce different birth defects, the defects are usually the same within each family. One possible explanation is that other genes may be modifying the phenotype differently in each family. In other words, the same mutation is occurring in different genetic

backgrounds, resulting in different phenotypes. To date, only a few examples of different human conditions caused by identical mutations are known. However, strain-dependent phenotypic effects are well described in other organisms such as mice and fruit flies.

3. Transcription factors commonly activate or repress more than a single gene. Often they affect the transcription of many genes that may be members of different developmental pathways. This maintains developmental flexibility and genomic economy. These developmental pathways are used to build many different tissues and organs. Thus, a mutation in a transcription factor can affect the growth and development of many body parts. For example, mutations in *TBX5* cause defects of the limbs and the heart in a pleiotropic disorder called Holt–Oram syndrome.

4. Pattern formation is the process by which ordered spatial arrangements of differentiated cells create tissues and organs. Initially, the general pattern of the animal body plan is established. Then, semiautonomous regions are formed that will pattern specific organs and appendages. Thus, pattern formation requires that complex temporal–spatial information be available to different populations of cells at different periods during development. This information is communicated between cells by signaling molecules. Mutations in the genes encoding these signaling molecules can disrupt signaling pathways, leading to abnormal pattern formation. For example, Sonic hedgehog (SHH) is widely used in many different patterning processes, including that of the brain. Mutations in *SHH* can cause failure of the forebrain to divide (holoprosencephaly). Patterning defects are also produced by mutations in genes encoding transcription factors (e.g., Hox, T-box) that are activated in response to signals from other cells.

5. Because of the tight constraints on developmental programs, the role of some elements in developmental pathways can be played by more than one molecule. This is known as *functional redundancy*. Hox paralogs appear to act in some developmental pathways in just this manner. For example, mice with *Hoxa11* or *Hoxd11* mutations have only minor abnormalities, whereas *Hoxa11/Hoxd11* double mutants exhibit a marked reduction in the size of the radius and ulna. Thus, *Hoxa11* and *Hoxd11* may be able to partially compensate for each other in some developmental programs.

6. There are many reasons why it may be impractical to use an organism to study loss-of-function mutations by creating knockouts. For example, baboons have small nuclear families and long generation times and are expensive to maintain in captivity. A short generation time is critical, because complete loss of function is achieved by backcrossing chimeras and subsequently mating heterozygotes with a genetic modification to produce animals homozygous for a genetic modification (i.e., a knockout). One way to circumvent the problem

of long generation times would be to produce sperm from the model animal (e.g., baboon) in an organism with a short generation time (e.g., mouse). In this technique, immature sperm-making cells from an infant baboon would be placed in the testicles of mice to make mature sperm. These sperm could be used for in vitro fertilization of eggs harvested from female baboons with a genetic modification (i.e., chimeras or heterozygotes). In this manner, the generation time would be substantially reduced. This type of experiment has not yet been successfully accomplished, although the technology is forthcoming.

7. Cells communicate with each other through many different signaling pathways. Signaling requires that a ligand bind to a receptor molecule. This generates a response that might activate or inhibit a variety of processes within a cell. Mutations in fibroblast growth factor receptor (FGFR) genes cause a variety of craniosynostosis syndromes, and mutations in fibroblast growth factor genes have been associated with cleft lip and/or cleft palate. Mutations in the genes encoding endothelin-3 and its receptor, endothelin-B, cause defects of the enteric cells of the gastrointestinal tract, leading to severe, chronic constipation (Hirschsprung disease). This demonstrates that a mutation in a ligand or its receptor can produce the same phenotype.

8. There are many obstacles to treating birth defects with gene therapy. Many developmental genes are expressed early in development. Thus, the phenotype would have to be identifiable at a very early age. This is certainly feasible (e.g., via preimplantation testing) in the gametes or zygotes of a known carrier. Because many of these genes encode axis, pattern, or organ-specific information, gene therapy would have to be initiated very early in the developmental process. In addition, the therapy might have to be targeted to specific areas, at critical times during development, and at appropriate levels required for interaction with other developmental genes. Consequently, it will be challenging to develop strategies that may be useful for treating birth defects with gene therapy.

## CHAPTER 11

1. *G6PD* is on a portion of the X chromosome that is inactivated in one copy in normal females. Thus, any single cell will express only one *G6PD* allele. If *all* tumor cells express the same *G6PD* allele, this indicates that they all arose from a single ancestral cell. This evidence was used to support the theory that most tumors are monoclonal.

2. The probability that a single cell will experience two mutations is given by the square of the mutation rate per cell (i.e., the probability of one mutation and a second mutation in the same cell): $(3 \times 10^{-6})^2 \approx 10^{-11}$. We then multiply this probability by the number of retinoblasts to obtain the probability that an individual

will develop sporadic retinoblastoma: $10^{-11} \times 2 \times 10^6 = 2 \times 10^{-5}$. We would thus expect 2 in 100,000 individuals to develop sporadic retinoblastoma, which is consistent with observed prevalence data (i.e., retinoblastoma is seen in about 1/20,000 children, and about half of these cases are sporadic). If an individual has inherited one copy of a mutant retinoblastoma gene, then the number of tumors is given by the rate of somatic mutation (second hit) per cell times the number of target cells: $3 \times 10^{-6} \times 2 \times 10^6 = 6$ tumors per individual.

3. Oncogenes are produced when proto-oncogenes, which encode substances that affect cell growth, are altered. Oncogenes usually act as dominant genes at the level of the cell and help to produce transformation of a normal cell into one that can give rise to a tumor. Tumor suppressor genes are also involved in growth regulation, but they usually act as recessive genes at the level of the cell (i.e., both copies of the gene must be altered before progression to a tumor can proceed). Because only a single oncogene need be altered to initiate the transformation process, oncogenes have been detected by using transfection and retroviral assays and by observing the effects of chromosome translocations. Such methods are less effective for uncovering tumor suppressors, because two altered copies of these genes must be present before their effects can be observed in a cell. As a consequence, most tumor suppressor genes have been detected by studying relatively rare cancer syndromes in which one mutant copy of the tumor suppressor is inherited and the second alteration occurs during somatic development.

4. Li–Fraumeni syndrome is caused by the inheritance of a mutation in the *TP53* gene. The inherited mutation is present in all cells and greatly increases predisposition to tumor formation. However, because of the multistep nature of carcinogenesis, this inherited event is not sufficient to produce a tumor. Other events, occurring in somatic cells, must also take place. These somatic events are rare, so the probability of their occurrence in any given cell is small, explaining the low frequency of any specific tumor type. The involvement of many different tumor types in Li–Fraumeni syndrome is explained by the fact that normal p53 activity is required for growth regulation in many different tissues. Thus, the likelihood that an individual will develop at least one primary tumor, in one of many tissues, is very high.

## CHAPTER 12

1. Because the trait is more common in females than in males, we infer that the threshold is lower in females than in males. Thus, an affected father is at greater risk for producing affected offspring than is an affected mother. The recurrence risk is higher in daughters than in sons.

2. For a multifactorial trait, the recurrence risk decreases rapidly in more remote relatives of a proband (as shown

in Table 12-2). In contrast, for an autosomal dominant gene, the recurrence risk is 50% smaller with each degree of relationship, reflecting the coefficient of relationship (see Chapter 4). Recall that first-degree relatives (parents, offspring, and siblings) share 50% of their DNA because of descent from common ancestors, second-degree relatives (uncles and nieces, grandparents and grandchildren) share 25% of their DNA, and so on. Thus, for a disease-causing genotype with 10% penetrance, the recurrence risk is obtained by multiplying the penetrance times the percentage of shared DNA between relatives: 5% (10% × 50%) for first-degree relatives, 2.5% (10% × 25%) for second-degree relatives, 1.25% (10% × 12.5%) for third-degree relatives, and so on. In addition, if the disease is multifactorial, the recurrence risk should increase in populations in which the disease is more common. There is typically no relationship between disease frequency and recurrence risk for an autosomal dominant disease.

3. (1) The disease-causing genotype may have reduced penetrance, due, for example, to environmental factors. (2) A somatic mutation might have occurred after cleavage of the embryo, such that one twin is affected by the disorder and the other is not.

4. These results imply that shared environmental factors are increasing the correlations among siblings, because siblings tend to share a more common environment than do parents and offspring. The spouse correlation reinforces the interpretation of a shared environment effect, although it is also possible that persons with similar body fat levels marry preferentially.

## CHAPTER 13

1. The sensitivity of the test is 93% (93 of the 100 true disease cases were detected). The specificity is 99% (98,900 of 99,900 unaffected neonates were correctly identified). The positive predictive value is 8.5% (93 of the 1093 neonates with positive tests actually had the disease). The false-positive rate is 1% (1 − specificity, or 1000 of 99,900), and the false-negative rate is 7% (1 − sensitivity, or 7 of 100).

2. Because individual 3 is homozygous for the 5-kb allele, we infer that the disease gene is on the same chromosome as the 5-kb allele in both parents. Thus, individual 6, who inherited both copies of the 5-kb allele, also inherited both copies of the PKU disease gene and is affected.

3. The *NF1* disease gene is on the same chromosome as allele *1* in the affected father. Thus, individual 6, who inherited allele *2* from his father, should be unaffected. Note that our degree of confidence in the answers to both question 2 and question 3 depends on how closely linked the marker and disease loci are.

4. The mating in generation I is uninformative, so we cannot establish linkage phase in the female in

generation II. Thus, the risk estimate for her daughter cannot be improved from the usual 50% figure used for autosomal dominant disease genes. Diagnostic accuracy could be improved by assaying another, more polymorphic marker (e.g., an STRP, which would more likely permit the accurate definition of linkage phase).

5. The primary advantages of amniocentesis are a lower rate of fetal loss (approximately 0.5% versus 1% to 1.3% for CVS) and the ability to do an AFP assay to detect neural tube defects. CVS offers the advantages of diagnosis earlier in the pregnancy and a more rapid laboratory diagnosis. The CVS diagnosis may be complicated by confined placental mosaicism, and there is limited evidence for an association between early (before 10 weeks post-LMP) CVS and limb reduction defects.

6. Huntington disease (HD) is a poor candidate for gene replacement therapy because it is caused at least in part by a gain-of-function mutation. Thus, simple gene insertion is unlikely to correct the disorder. It would be a better candidate for antisense, ribozyme, or RNAi treatment, in which the defective gene product would be disabled. An additional consideration is that this disorder affects primarily neurons, which are relatively difficult to manipulate or target. The fact that HD has a delayed age of onset, however, is encouraging because identification of the action of the HD gene might lead to drug therapy to block the gene product's effects before neuronal damage occurs.

## CHAPTER 14

1. Genetic information can be used to assess a person's risk of disease and response to therapy. Persons at high risk for a disease could therefore be encouraged to alter their health management to reduce their risk. Similarly, genetic information can be used to predict a person's response to therapy and whether the person is at increased risk for a serious adverse drug response. Drugs predicted to be less effective or likely to cause an adverse event could be avoided. For example, a person predicted to be at high risk for diabetes might be tested for hyperglycemia more frequently, placed on a medical diet earlier, and treated more aggressively for glucose intolerance.

2. Persons with the same type of cancer might respond differently to therapy because the genetic abnormalities in their tumors are different or because of differences, for example, in the way that chemotherapeutic drugs are metabolized.

3. Race has traditionally been used to categorize large groups of individuals and reflects geographic origin, language, and various cultural attributes that describe a group (e.g., Native Americans or South Asians). Ancestry refers to the geographic, historical, or biological origins of one's ancestors and, for any individual, can be complex.

4. Genetic information about a person's ancestry can influence the person's perception of his or her biological and/or cultural identity. For example, some people who self-identify as African American have sought to find the specific geographic regions of sub-Saharan Africa where some of their ancestors once lived. In some cases, this information suggests specific populations, and therefore cultural links, with which an individual might identify. On the other hand, genetic information sometimes suggests that a person has little, if any, biological ancestry from the populations with which he or she self-identifies. Ultimately, every person is a member of many different populations and has multiple identities—social, economic, religious—and genetic ancestry information provides little insight about who they are but provides, instead, some information about where they came from.

5. Examples of polymorphisms that influence drug metabolism include variants of *CYP2C9* and *VKORC1* that influence the metabolism of warfarin, an anticoagulant drug; *CYP2D6* variants that affect the biotransformation of β-adrenergic receptor antagonists, neuroleptics, and tricyclic antidepressants; *NAT2* variants that affect the inactivation of isoniazid, a drug commonly used to treat tuberculosis; and *G6PD* that influences sensitivity to the antimalarial drug, primaquine. Response to antihypertensive β-blockers has been associated with variants in genes that encode subunits of the β-adrenergic receptor.

6. Potential obstacles to the use of genetic information in personalized medicine include an inability to identify genetic and environmental risk factors (and their interactions) that enable accurate prediction of clinically significant risk; lack of evidence demonstrating that individual risk assessment improves diagnostic accuracy and treatment outcome; lack of technologies for cost-efficient assessment of an individual's genome; building an infrastructure for clinicians to access risk data, interpret risk information, and explain risk estimates to patients; and the need for guidelines and policies for how risk-assessment information should be used in clinical and research applications.

7. The use of individual genetic variants to predict risk of disease and/or response to pharmacologic agents can be considered the practice of genetic medicine, whereas the assessment of the action of many genes simultaneously to predict disease risk or drug response distinguishes genomic medicine.

8. Potential uses of whole-genome data from an individual include screening for inborn errors, metabolism in newborns (i.e., newborn screening), testing for carriers of genetic disorders (e.g., sickle-cell disease, cystic fibrosis), assessing risk for common diseases, predicting drugs that might influence risk for a serious adverse drug, and forensic identification.

## CHAPTER 15

**1.** Allen's family includes several members who have had myocardial infarctions at a relatively young age. The pedigree suggests that an autosomal dominant gene predisposing family members to heart disease *may* be segregating in this family. This could be caused by autosomal dominant familial hypercholesterolemia, or possibly by another disorder of lipid metabolism. Allen should be encouraged to have his serum lipid levels tested (total cholesterol, LDL, HDL, and triglycerides). If his LDL level is abnormally high, intervention may be necessary (e.g., dietary modification, cholesterol-lowering drugs).

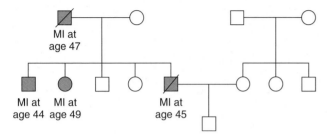

**2.** Based on the fact that Mary's two brothers and one uncle were all affected, we can be quite certain that her mother is a carrier of the DMD gene. (If only one of Mary's brothers had been affected, we would have to consider the possibility that he was the product of a new mutation.) If Mary's mother is a carrier, then there is a probability of 1/2 that Mary is a carrier. As we saw in Chapter 5, a female carrier will transmit the disease gene to half of her sons, on average. Thus, the probability that one of her sons will be affected with DMD is 1/2 × 1/2 = 1/4. The creatine kinase test contributes additional information. We can set up the Bayesian calculation as follows:

| | Mary is a carrier | Mary is not a carrier |
|---|---|---|
| Prior probability | 1/2 | 1/2 |
| Conditional probability that her CK is in the 95th percentile | 2/3 | 0.05 |
| Joint probability | 1/3 | 0.025 |
| Posterior probability | 0.93 | 0.07 |

It should be clear that the conditional probability of being a carrier, given a CK level above the 95th percentile, is 2/3. The conditional probability of *not* being a carrier with this CK level must be 0.05, because only 5% of CK values in normal persons fall above the 95th percentile. Thus, the information derived from the CK test has increased Mary's probability of being a carrier from 1/2 to 0.93. Because the probability that she would transmit the DMD gene to her male offspring is 1/2, the probability of producing an affected male increases from 0.25 to 0.47. Currently, it is likely that additional tests, such as screening for DMD mutations and performing a dystrophin assay, would yield even more precise information.

**3.** The prior probability that Bob has inherited the gene from his father is 1/2. Because 85% of gene carriers manifest symptoms by age 51 if they inherited the gene from an affected father, the conditional probability that Bob is 51 years old and unaffected, given that he inherited the gene, is 0.15. The probabilities can be set up as in the table below. In this example, the incorporation of age-of-onset information decreased Bob's chance of inheriting the disease gene from 50% to only 13%. With the cloning of the Huntington disease gene, Bob would likely be given a DNA diagnostic test to determine with certainty whether he had inherited an expanded repeat mutation from his father.

| | Bob carries the HD gene | Bob does not carry the HD gene |
|---|---|---|
| Prior probability | 1/2 | 1/2 |
| Conditional probability that Bob is normal at age 51 years | 0.15 | 1 |
| Joint probability | 0.075 | 1/2 |
| Posterior probability | 0.13 | 0.87 |

# INDEX